ABSTRACT ALGEBRA

ABSTRACT ALGEBRA

Paul B. Garrett

University of Minnesota
Minneapolis, U.S.A.

CRC Press
Taylor & Francis Group
Boca Raton London New York

CRC Press is an imprint of the
Taylor & Francis Group, an **informa** business

A CHAPMAN & HALL BOOK

CRC Press
Taylor & Francis Group
6000 Broken Sound Parkway NW, Suite 300
Boca Raton, FL 33487-2742

First issued in paperback 2019

© 2008 by Taylor & Francis Group, LLC
CRC Press is an imprint of Taylor & Francis Group, an Informa business

No claim to original U.S. Government works

ISBN-13: 978-1-58488-689-1 (hbk)
ISBN-13: 978-0-367-38858-4 (pbk)

Library of Congress Cataloging-in-Publication Data

Garrett, Paul B.
   Abstract algebra / Paul B. Garrett.
      p. cm.
   Includes index.
   ISBN 978-1-58488-689-1 (hardback : alk. paper)
   1. Algebra, Abstract--Textbooks. I. Title.

QA162.G375 2007
512'.02--dc22
                                                      2007028460

Visit the Taylor & Francis Web site at
http://www.taylorandfrancis.com

and the CRC Press Web site at
http://www.crcpress.com

I covered this material in a two-semester graduate course in abstract algebra in 2004-05. My goal was to rethink the standard course from scratch, ignoring traditional prejudices.

The level of symbolism is as low as possible, transferring some of the load into the ambient natural language. I tried to minimize occurrence of needless common degeneracies in the natural language in which mathematics is imbedded.

I tried to write proofs which would be natural outcomes of the viewpoint I promoted, with the idea that from *some* viewpoints there is less to remember. *Robustness*, as opposed to *fragility*, is a desirable feature of an argument. It is burdensome to remember arguments that require persistent cleverness, as opposed to a calm *naturality*. Of course, it is non-trivial to arrive at a viewpoint that allows proofs to seem natural. Thus, giving such proofs is revisionism. However, there are much worse revisionisms which are popular, most notably the misguided impulse to ascetic *logical* perfection of the development of subjects. Logical streamlining is not reliably the same as optimizing for performance. Further, logical flawlessness of an argument is not necessarily *persuasive*, especially if inscrutable.

The worked examples are meant to be model solutions for many of the standard problems traditionally posed in such a course. I no longer believe that everyone is obliged to redo everything themselves. Hopefully it is possible for us to learn from others' efforts.

I learned abstract algebra from an early edition of S. Lang's *Algebra*, from Lang's *Algebraic Number Theory*, and from B.L. van der Waerden's *Algebra*. These books of Lang were strongly influenced by Emil Artin and Bourbaki, and van der Waerden's book originated in lectures of Emmy Noether.

Paul Garrett
July 2007, Minneapolis

Abstract Algebra is not a conceptually well-defined body of material, but a conventional name that refers roughly to one of the several lists of things that mathematicians need to know to be competent, effective, and insightful. This material fits a two-semester beginning graduate course in abstract algebra. It is a *how-to* manual, not a monument to traditional icons. Rather than an encyclopedic reference, it tells a story, with plot-lines and character development propelling it forward.

The main novelty is that most of the standard exercises in abstract algebra are given here as *worked* examples. The exercises are variations on the worked examples. The reader might contemplate the examples before reading the solutions, but this is not mandatory. The examples are given to *assist*, not necessarily *challenge*, the reader. The point is *not* whether or not the reader can do the problems on their own, since all of these are at least fifty years old, but, rather, whether the *viewpoint* is assimilated. In particular, it can happen that a logically correct solution is conceptually regressive and should not be considered satisfactory.

The approach here promotes taking an efficient, abstract viewpoint whenever it is purposeful to do so, especially when it happens that letting go of appealingly tangible but insidiously burdensome details brings advantage. Some things often not mentioned in an algebra course are included, to address typical weaknesses in contemporary curriculums. Some naive set theory, developing ideas about ordinals, is occasionally necessary, and the abstraction of this setting makes the set theory seem less farfetched or baffling than it might in a more elementary context. Equivalents of the Axiom of Choice are described, with sample proofs exemplifying the techniques. Quadratic reciprocity is useful in understanding quadratic and cyclotomic extensions of the rational numbers, and we give the proof by Gauss' sums. An economical proof of Dirichlet's theorem on primes in arithmetic progressions is included, with associated discussion of related complex analysis, since existence of primes satisfying linear congruence conditions comes up in practice. Other small enrichment topics are treated briefly at opportune moments in examples and exercises. Again, algebra is not a unified or linearly ordered body of knowledge, but only a rough naming convention for an ill-defined landscape of ideas. Further, as with all parts of the basic graduate mathematics curriculum, many important things are inevitably left out. For algebraic geometry or algebraic number theory, much more commutative algebra is useful than is presented here. Only vague hints of representation theory are detectable here.

Far more systematic emphasis is given to finite fields, cyclotomic polynomials (divisors of $x^n - 1$), and cyclotomic fields than is usual, and less emphasis is given to *abstract* Galois theory. Ironically, there are many more explicit Galois theory *examples* here than in sources that do pretend to emphasize Galois theory. After proving Lagrange's theorem and the Sylow theorem, the pure theory of finite groups is not especially emphasized. After all, the Sylow theorem is not interesting because it allows classification of groups of small order, but because its *proof* illustrates the critical notion of *a group acting on a set*, a ubiquitous mechanism in mathematics. A strong and recurring theme is the characterization of objects by *(universal) mapping properties*, rather than by curious constructions whose salient feature becomes clear only after a gauntlet of lemmas. Nevertheless, formal category theory does not appear. A greater emphasis is put on linear and multilinear algebra, while doing little with general commutative algebra apart from Gauss' lemma and Eisenstein's criterion, which are immediately useful. It is difficult to achieve *closure* in an introduction to Noetherian rings, despite their central role in modern mathematics. The *Nullstellensatz* is hard to appreciate at first approach. The systematic beginnings of commutative algebra for algebraic geometry and algebraic number theory deserve their own many chapters, which would not fit into this book.

Introduction

Students need good role models for writing mathematics. This is another reason for the complete write-ups of solutions to many examples, since most traditional situations do not provide students with good models for solutions to the standard problems. This is bad. Even worse, lacking full solutions written by a more practiced hand, inferior and retrogressive solutions propagate. I did not insist that my students give solutions in the style I wished, but I thought it would be very desirable to provide them with good examples.

The reader is assumed to have *some* prior acquaintance with introductory abstract algebra and linear algebra, not to mention other standard courses that are considered preparatory for graduate school. This is not so much for specific information as for maturity.

Contents

Contents

Contents

Contents

Contents

Contents

1. The integers

1.1 *Unique factorization*

Let $\mathbb{Z}$ denote the integers. Say d **divides** m, equivalently, that m is a **multiple** of d, if there exists an integer q such that $m = qd$. Write $d|m$ if d divides m.

It is easy to prove, from the definition, that if $d|x$ and $d|y$ then $d|(ax + by)$ for any integers x, y, a, b: let $x = rd$ and $y = sd$, and

$$ax + by = a(rd) + b(sd) = d \cdot (ar + bs)$$

1.1.1 Theorem: Given an integer N and a non-zero integer m there are unique integers q and r, with $0 \le r < |m|$ such that

$$N = q \cdot m + r$$

The integer r is the **reduction modulo** m of N.

Proof: Let S be the set of all non-negative integers expressible in the form $N - sm$ for some integer s. The set S is non-empty, so by well-ordering has a least element $r = N - qm$. Claim that $r < |m|$. If not, then still $r - |m| \ge 0$, and also

$$r - |m| = (N - qm) - |m| = N - (q \pm 1)m$$

(with the sign depending on the sign of m) is still in the set S, contradiction. For uniqueness, suppose that both $N = qm + r$ and $N = q'm + r'$. Subtract to find

$$r - r' = m \cdot (q' - q)$$

Thus, $r - r'$ is a multiple of m. But since $-|m| < r - r' < |m|$ we have $r = r'$. And then $q = q'$.   ///

1.1.2 Remark: The conclusion of the theorem is that in $\mathbb{Z}$ one can divide and obtain a remainder *smaller* than the divisor. That is, $\mathbb{Z}$ is **Euclidean**.

As an example of nearly trivial things that can be proven about divisibility, we have:

A divisor d of n is **proper** if it is neither $\pm n$ nor ± 1. A positive integer p is **prime** if it has no proper divisors and if $p > 1$.

1.1.3 Proposition: A positive integer n is prime if and only if it is not divisible by any of the integers d with $1 < d \le \sqrt{n}$.

Proof: Suppose that n has a proper factorization $n = d \cdot e$, where $d \le e$. Then

$$d = \frac{n}{e} \le \frac{n}{d}$$

gives $d^2 \le n$, so $d \le \sqrt{n}$.   ///

1.1.4 Remark: The previous proposition suggests that to test an integer n for primality we attempt to divide n by all integers $d = 2, 3, \ldots$ in the range $d \le \sqrt{n}$. If no such d divides n, then n is prime. This procedure is **trial division**.

Two integers are **relatively prime** or **coprime** or **mutually prime** if for every integer d if $d|m$ and $d|n$ then $d = \pm 1$.

An integer d is a **common divisor** of integers $n_1, \ldots, n_m$ if d divides each n_i. An integer N is a **common multiple** of integers $n_1, \ldots, n_m$ if N is a multiple of each. The following peculiar characterization of the greatest common divisor of two integers is fundamental.

1.1.5 Theorem: Let m, n be integers, not both zero. Among all *common* divisors of m, n there is a unique $d > 0$ such that for *every* other common divisor e of m, n we have $e|d$. This d is the *greatest common divisor* of m, n, denoted $\gcd(m, n)$. And

$$\gcd(mn) = \text{least positive integer of the form } xm + yn \text{ with } x, y \in \mathbb{Z}$$

Proof: Let $D = x_o m + y_o n$ be the least positive integer expressible in the form $xm + yn$. First, we show that any divisor d of both m and n divides D. Let $m = m'd$ and $n = n'd$ with $m', n' \in \mathbb{Z}$. Then

$$D = x_o m + y_o n = x_o(m'd) + y_o(n'd) = (x_o m' + y_o n') \cdot d$$

which presents D as a multiple of d.

On the other hand, let $m = qD + r$ with $0 \le r < D$. Then

$$0 \le r = m - qD = m - q(x_o m + y_o n) = (1 - qx_o) \cdot m + (-y_o) \cdot n$$

That is, r is expressible as $x'm + y'n$. Since $r < D$, and since D is the smallest positive integer so expressible, $r = 0$. Therefore, $D|m$, and similarly $D|n$.   ///

Similarly:

1.1.6 Corollary: Let m, n be integers, not both zero. Among all *common* multiples of m, n there is a unique positive one N such that for *every* other common multiple M we have $N|M$. This multiple N is the *least common multiple* of m, n, denoted $\operatorname{lcm}(m, n)$. In particular,

$$\operatorname{lcm}(m, n) = \frac{mn}{\gcd(m, n)}$$

Proof: Let

$$L = \frac{mn}{\gcd(m, n)}$$

First we show that L is a multiple of m and n. Indeed, let

$$m = m' \cdot \gcd(m, n) \qquad n = n' \cdot \gcd(m, n)$$

Then

$$L = m \cdot n' = m' \cdot n$$

expresses L as an integer multiple of m and of n. On the other hand, let M be a multiple of both m and n. Let $\gcd(m, n) = am + bn$. Then

$$1 = a \cdot m' + b \cdot n'$$

Let $N = rm$ and $N = sn$ be expressions of N as integer multiples of m and n. Then

$$N = 1 \cdot N = (a \cdot m' + b \cdot n') \cdot N = a \cdot m' \cdot sn + b \cdot n' \cdot rm = (as + br) \cdot L$$

as claimed.   ///

The innocent assertion and perhaps odd-seeming argument of the following are essential for what follows. Note that the key point is the peculiar characterization of the *gcd*, which itself comes from the Euclidean property of $\mathbb{Z}$.

1.1.7 Theorem: A prime p divides a product ab if and only if $p|a$ or $p|b$.

Proof: If $p|a$ we are done, so suppose p does not divide a. Since p is prime, and since $\gcd(p, a) \neq p$, it must be that $\gcd(p, a) = 1$. Let r, s be integers such that $1 = rp + sa$, and let $ab = kp$. Then

$$b = b \cdot 1 = b(rp + sa) = p \cdot (rb + sk)$$

so b is a multiple of p.   ///

Granting the theorem, the proof of unique factorization is nearly an afterthought:

1.1.8 Corollary: *(Unique Factorization)* Every integer n can be written in an *essentially unique* way (up to reordering the factors) as $\pm$ a product of primes:

$$n = \pm \, p_1^{e_1} \, p_2^{e_2} \ldots p_m^{e_m}$$

with positive integer exponents and primes $p_1 < \ldots < p_m$.

Proof: For *existence,* suppose $n > 1$ is the least integer *not* having a factorization. Then n cannot be prime itself, or just '$n = n$' is a factorization. Therefore n has a proper factorization $n = xy$ with $x, y > 1$. Since the factorization is *proper,* both x and y are strictly smaller than n. Thus, x and y both can be factored. Putting together the two factorizations gives the factorization of n, contradicting the assumption that there exist integers lacking prime factorizations.

Now *uniqueness.* Suppose

$$q_1^{e_1} \ldots q_m^{e_m} = N = p_1^{f_1} \ldots p_n^{f_n}$$

where $q_1 < \ldots < q_m$ are primes, and $p_1 < \ldots < p_n$ are primes, and the exponents e_i and f_i are positive integers. Since q_1 divides the left-hand side of the equality, it divides the right-hand side. Therefore, q_1 must divide one of the factors on the right-hand side. So q_1 must divide some p_i. Since p_i is prime, it must be that $q_1 = p_i$.

If $i > 1$ then $p_1 < p_i$. And p_1 divides the left-hand side, so divides one of the q_j, so is some q_j, but then

$$p_1 = q_j \geq q_1 = p_i > p_1$$

which is impossible. Therefore, $q_1 = p_1$.

Without loss of generality, $e_1 \leq f_1$. Thus, by dividing through by $q_1^{e_1} = p_1^{e_1}$, we see that the corresponding exponents e_1 and f_1 must also be equal. Then do induction.   ///

1.1.9 Example: The simplest meaningful (and standard) example of the failure of unique factorization into primes is in the collection of numbers

$$\mathbb{Z}[\sqrt{-5}] = \{a + b\sqrt{-5} : a, b \in \mathbb{Z}\}$$

The relation

$$6 = 2 \cdot 3 = (1 + \sqrt{-5})(1 - \sqrt{5})$$

gives two different-looking factorizations of 6. We must verify that 2, 3, $1 + \sqrt{-5}$, and $1 - \sqrt{-5}$ are *primes* in R, in the sense that they cannot be further factored.

To prove this, we use *complex conjugation,* denoted by a bar over the quantity to be conjugated: for real numbers a and b,

$$\overline{a + b\sqrt{-5}} = a - b\sqrt{-5}$$

For α, β in R,

$$\overline{\alpha \cdot \beta} = \overline{\alpha} \cdot \overline{\beta}$$

by direct computation. Introduce the **norm**

$$N(\alpha) = \alpha \cdot \overline{\alpha}$$

The multiplicative property

$$N(\alpha \cdot \beta) = N(\alpha) \cdot N(\beta)$$

follows from the corresponding property of conjugation:

$$N(\alpha) \cdot N(\beta) = \alpha\overline{\alpha}\beta\overline{\beta} = (\alpha\beta) \cdot (\overline{\alpha}\,\overline{\beta})$$

$$= (\alpha\beta) \cdot (\overline{\alpha\beta}) = N(\alpha\beta)$$

Note that $0 \le N(\alpha) \in \mathbb{Z}$ for α in R.

Now suppose $2 = \alpha\beta$ with α, β in R. Then

$$4 = N(2) = N(\alpha\beta) = N(\alpha) \cdot N(\beta)$$

By unique factorization in $\mathbb{Z}$, $N(\alpha)$ and $N(\beta)$ must be $1, 4$, or $2, 2$, or $4, 1$. The middle case is impossible, since no norm can be 2. In the other two cases, one of α or β is ± 1, and the factorization is not *proper*. That is, 2 cannot be factored further in $\mathbb{Z}[\sqrt{-5}]$. Similarly, 3 cannot be factored further.

If $1 + \sqrt{-5} = \alpha\beta$ with α, β in R, then again

$$6 = N\left(1 + \sqrt{-5}\right) = N(\alpha\beta) = N(\alpha) \cdot N(\beta)$$

Again, the integers $N(\alpha)$ and $N(\beta)$ must either be $1, 6$, $2, 3$, $3, 2$, or $6, 1$. Since the norm cannot be 2 or 3, the middle two cases are impossible. In the remaining two cases, one of α or β is ± 1, and the factorization is not *proper*. That is, $1 + \sqrt{-5}$ cannot be factored further in R. Neither can $1 - \sqrt{-5}$. Thus,

$$6 = 2 \cdot 3 = \left(1 + \sqrt{-5}\right)\left(1 - \sqrt{5}\right)$$

is a factorization of 6 in two different ways *into primes* in $\mathbb{Z}[\sqrt{-5}]$.

1.1.10 Example: The **Gaussian integers**

$$\mathbb{Z}[i] = \{a + bi : a, b \in \mathbb{Z}\}$$

where $i^2 = -1$ do have a Euclidean property, and thus have unique factorization. Use the *integer-valued* norm

$$N(a + bi) = a^2 + b^2 = (a + bi) \cdot \overline{(a + bi)}$$

It is important that the notion of size be integer-valued and respect multiplication. We claim that, given $\alpha, \delta \in \mathbb{Z}[i]$ there is $q \in \mathbb{Z}[i]$ such that

$$N(\alpha - q \cdot \delta) < N(\delta)$$

Since N is *multiplicative* (see above), we can divide through by δ inside

$$\mathbb{Q}(i) = \{a + bi : a, b, \in \mathbb{Q}\}$$

(where $\mathbb{Q}$ is the rationals) to see that we are asking for $q \in \mathbb{Z}[i]$ such that

$$N(\frac{\alpha}{\delta} - q) < N(1) = 1$$

That is, given $\beta = \alpha/\delta$ in $\mathbb{Q}(i)$, we must be able to find $q \in \mathbb{Z}[i]$ such that

$$N(\beta - q) < 1$$

With $\beta = a + bi$ with $a, b \in \mathbb{Q}$, let

$$a = r + f_1 \qquad b = s + f_2$$

with $r, s \in \mathbb{Z}$ and f_1, f_2 rational numbers with

$$|f_i| \leq \frac{1}{2}$$

That this is possible is a special case of the fact that any *real* number is at distance at most $1/2$ from some integer. Then take

$$q = r + si$$

Then

$$\beta - q = (a + bi) - (r + si) = f_1 + if_2$$

and

$$N(\beta - q) = N(f_1 + if_2) = f_1^2 + f_2^2 \leq \left(\frac{1}{2}\right)^2 + \left(\frac{1}{2}\right)^2 = \frac{1}{2} < 1$$

Thus, indeed $\mathbb{Z}[i]$ has the Euclidean property, and, by the same proof as above, has unique factorization.

1.2 *Irrationalities*

The usual proof that there is no square root of 2 in the rationals $\mathbb{Q}$ uses a little bit of unique factorization, in the notion that it is possible to put a fraction into lowest terms, that is, having relatively prime numerator and denominator.

That is, given a fraction a/b (with $b \neq 0$), letting $a' = a/\gcd(a, b)$ and $b' = b/\gcd(a, b)$, one can and should show that $\gcd(a', b') = 1$. That is, $a'b/b'$ is **in lowest terms**. And

$$\frac{a'}{b'} = \frac{a}{b}$$

1.2.1 Example: Let p be a prime number. We claim that there is no $\sqrt{p}$ in the rationals $\mathbb{Q}$. Suppose, to the contrary, that $a/b = \sqrt{p}$. Without loss of generality, we can assume that $\gcd(a, b) = 1$. Then, squaring and multiplying out,

$$a^2 = pb^2$$

Thus, $p | a^2$. Since $p | cd$ implies $p | c$ or $p | d$, necessarily $p | a$. Let $a = pa'$. Then

$$(pa')^2 = pb^2$$

or

$$pa'^2 = b^2$$

Thus, $p | b$, contradicting the fact that $\gcd(a, b) = 1$.   ///

The following example illustrates a possibility that will be subsumed later by *Eisenstein's criterion*, which is itself an application of *Newton polygons* attached to polynomials.

1.2.2 Example: Let p be a prime number. We claim that there is no rational solution to

$$x^5 + px + p = 0$$

Indeed, suppose that a/b were a rational solution, in lowest terms. Then substitute and multiply through by b^5 to obtain

$$a^5 + pab^4 + pb^5 = 0$$

From this, $p|a^5$, so, since p is prime, $p|a$. Let $a = pa'$. Then

$$(pa')^5 + p(pa')b^4 + pb^5 = 0$$

or

$$p^4 a'^5 + p^2 a' b^4 + b^5 = 0$$

From this, $p|b^5$, so $p|b$ since p is prime. This contradicts the lowest-terms hypothesis.

1.3 $\mathbb{Z}/m$, *the integers mod* m

Recall that a *relation* R on a set S is a subset of the cartesian product $S \times S$. Write

$$x \, R \, y$$

if the ordered pair (x, y) lies in the subset R of $S \times S$. An **equivalence relation** R on a set S is a relation satisfying
- **Reflexivity**: $x \, R \, x$ for all $x \in S$
- **Symmetry**: If $x \, R \, y$ then $y \, R \, x$
- **Transitivity**: If $x \, R \, y$ and $y \, R \, z$ then $x \, R \, z$

A common notation for an equivalence relation is

$$x \sim y$$

that is, with a tilde rather than R.

Let $\sim$ be an equivalence relation on a set S. For $x \in S$, the $\sim$ - **equivalence class** $\bar{x}$ containing x is the subset

$$\bar{x} = \{x' \in S : x' \sim x\}$$

The **set of equivalence classes** of $\sim$ on S is denoted by

$$S/\sim$$

(as a quotient). Every element $z \in S$ is contained in an equivalence class, namely the equivalence class $\bar{z}$ of all $s \in S$ so that $s \sim z$. Given an equivalence class A inside S, an x in the set S such that $\bar{x} = A$ is a **representative** for the equivalence class. That is, any element of the subset A is a representative.

A set $\mathcal{S}$ of non-empty subsets of a set S whose union is the whole S, and which are mutually disjoint, is a **partition** of S. One can readily check that the equivalence classes of an equivalence relation on a set S form a partition of S, and, conversely, any partition of S defines an equivalence relation by positing that $x \sim y$ if and only if they lie in the same set of the partition. ///

If two integers x, y differ by a multiple of a non-zero integer m, that is, if $m|(x - y)$, then x is **congruent to** y **modulo** m, written

$$x \equiv y \bmod m$$

Such a relation is a **congruence** modulo m, and m is the **modulus**. When Gauss first used this notion 200 years ago, it was sufficiently novel that it deserved a special notation, but, now that the novelty has worn off, we will simply write

$$x = y \bmod m$$

and (unless we want special emphasis) simply say that x is **equal to** y modulo m.

1.3.1 Proposition: (For fixed modulus m) equality modulo m is an equivalence relation.   ///

Compatibly with the general usage for equivalence relations, the **congruence class** (or **residue class** or **equivalence class**) of an integer x modulo m, denoted $\bar{x}$ (with only implicit reference to m) is the set of all integers equal to x mod m:

$$\bar{x} = \{y \in \mathbb{Z} : y = x \bmod m\}$$

The **integers mod** m, denoted $\mathbb{Z}/m$, is the collection of *congruence classes* of integers modulo m. For some $X \in \mathbb{Z}/m$, a choice of ordinary integer x so that $\bar{x} = X$ is a **representative** for the congruence class X.

1.3.2 Remark: A popular but unfortunate notation for $\mathbb{Z}/m$ is $\mathbb{Z}_m$. We will not use this notation. It is unfortunate because for primes p the notation $\mathbb{Z}_p$ is the *only* notation for the *p-adic integers*.

1.3.3 Remark: On many occasions, the bar is dropped, so that x-mod-m may be written simply as 'x'.

1.3.4 Remark: The traditionally popular collection of representatives for the equivalence classes modulo m, namely

$$\{\bar{0}, \bar{1}, \bar{2}, \ldots \overline{m-2}, \overline{m-1}\}$$

is not the only possibility.

The benefit Gauss derived from the explicit notion of congruence was that congruences behave much like equalities, thus allowing us to benefit from our prior experience with equalities. Further, but not surprisingly with sufficient hindsight, congruences behave nicely with respect to the basic operations of addition, subtraction, and multiplication:

1.3.5 Proposition: Fix the modulus m. If $x = x' \bmod m$ and $y = y' \bmod m$, then

$$x + y = x' + y' \bmod m$$

$$xy = x'y' \bmod m$$

Proof: Since $m|(x'-x)$ there is an integer k such that $mk = x' - x$. Similarly, $y' = y + \ell m$ for some integer ℓ. Then

$$x' + y' = (x + mk) + (y + m\ell) = x + y + m \cdot (k + \ell)$$

Thus, $x' + y' = x + y \bmod m$. And

$$x' \cdot y' = (x + mk) \cdot (y + m\ell) = x \cdot y + xm\ell + mky + mk \cdot m\ell = x \cdot y + m \cdot (k + \ell + mk\ell)$$

Thus, $x'y' = xy \bmod m$. ///

As a corollary, congruences *inherit* many basic properties from ordinary arithmetic, simply because $x = y$ implies $x = y \bmod m$:
- *Distributivity:* $x(y + z) = xy + xz \bmod m$
- *Associativity of addition:* $(x + y) + z = x + (y + z) \bmod m$
- *Associativity of multiplication:* $(xy)z = x(yz) \bmod m$
- *Property of 1:* $1 \cdot x = x \cdot 1 = x \bmod m$
- *Property of 0:* $0 + x = x + 0 = x \bmod m$

In this context, a **multiplicative inverse mod** m to an integer a is an integer b (if it exists) such that

$$a \cdot b = 1 \bmod m$$

1.3.6 Proposition: An integer a has a multiplicative inverse modulo m if and only if $\gcd(a, m) = 1$.

Proof: If $\gcd(a, m) = 1$ then there are r, s such that $ra + sm = 1$, and

$$ra = 1 - sm = 1 \bmod m$$

The other implication is easy. ///

In particular, note that if a is invertible mod m then any a' in the residue class of a mod m is likewise invertible mod m, and any other element b' of the residue class of an inverse b is also an inverse. Thus, it makes sense to refer to elements of $\mathbb{Z}/m$ as being invertible or not. Notation:

$$(\mathbb{Z}/m)^{\times} = \{\bar{x} \in \mathbb{Z}/m : \gcd(x, m) = 1\}$$

This set $(\mathbb{Z}/m)^{\times}$ is the **multiplicative group** or **group of units** of $\mathbb{Z}/m$.

1.3.7 Remark: It is easy to verify that the set $(\mathbb{Z}/m)^{\times}$ is **closed under multiplication** in the sense that $a, b \in (\mathbb{Z}/m)^{\times}$ implies $ab \in (\mathbb{Z}/m)^{\times}$, and is **closed under inverses** in the sense that $a \in (\mathbb{Z}/m)^{\times}$ implies $a^{-1} \in (\mathbb{Z}/m)^{\times}$.

1.3.8 Remark: The superscript is not an 'x' but is a 'times', making a reference to multiplication and multiplicative inverses mod m. Some sources write $\mathbb{Z}/m^{*}$, but the latter notation is inferior, as it is too readily confused with other standard notation (for *duals*).

1.4 *Fermat's little theorem*

1.4.1 Theorem: Let p be a prime number. Then for any integer x

$$x^{p} = x \bmod p$$

Proof: First, by the binomial theorem

$$(x + y)^{p} = \sum_{0 \leq i \leq p} \binom{p}{i} x^{i} y^{p-i}$$

In particular, the binomial coefficients are *integers*. Now we can show that the prime p divides the binomial coefficients

$$\binom{p}{i} = \frac{p!}{i!\,(p-i)!}$$

with $1 \le i \le p-1$. We have

$$\binom{p}{i} \cdot i! \cdot (p-i)! = p!$$

(Since we know that the binomial coefficient is an integer, the following argument makes sense.) The prime p divides the right-hand side, so divides the left-hand side, but does not divide $i!$ nor $(p-i)!$ (for $0 < i < p$) since these two numbers are products of integers smaller than p and (hence) not divisible by p. Again using the fact that $p|ab$ implies $p|a$ or $p|b$, p does not divide $i! \cdot (p-i)!$, so p must divide the binomial coefficient.

Now we prove Fermat's Little Theorem for *positive* integers x by induction on x. Certainly $1^p = 1 \bmod p$. Now suppose that we know that

$$x^p = x \bmod p$$

Then

$$(x+1)^p = \sum_{0 \le i \le p} \binom{p}{i} x^i\, 1^{p-i} = x^p + \sum_{0 < i < p} \binom{p}{i} x^i + 1$$

All the coefficients in the sum in the middle of the last expression are divisible by p, so

$$(x+1)^p = x^p + 0 + 1 = x + 1 \bmod p$$

This proves the theorem for positive x. ///

1.4.2 Example: Let p be a prime with $p = 3 \bmod 4$. Suppose that a is a **square modulo** p, in the sense that there exists an *integer* b such that

$$b^2 = a \bmod p$$

Such b is a **square root modulo** p of a. Then we claim that $a^{(p+1)/4}$ is a square root of a mod p. Indeed,

$$\left(a^{(p+1)/4}\right)^2 = \left((b^2)^{(p+1)/4}\right)^2 = b^{p+1} = b^p \cdot b = b \cdot b \bmod p$$

by Fermat. Then this is $a \bmod p$. ///

1.4.3 Example: Somewhat more generally, let q be a prime, and let p be another prime with $p = 1 \bmod q$ but $p \ne 1 \bmod q^2$.

$$r = q^{-1} \bmod \frac{p-1}{q}$$

Then when a is a qth power modulo p, a qth root of a mod p is given by the formula

$$q^{th} \text{ root of } a \bmod p = a^r \bmod p$$

If a is *not* a qth power mod p then this formula does *not* product a qth root.

1.4.4 Remark: For prime q and prime $p \neq 1 \bmod q$ there is an even simpler formula for q^{th} roots, namely let

$$r = q^{-1} \bmod p - 1$$

and then

$$q^{th} \text{ root of } a \bmod p = a^r \bmod p$$

Further, as can be seen from the even-easier proof of this formula, *everything* mod such p is a qth power.

For a positive integer n, the **Euler phi-function** $\varphi(n)$ is the number of integers b so that $1 \leq b \leq n$ and $\gcd(b, n) = 1$. Note that

$$\varphi(n) = \text{cardinality of } (\mathbb{Z}/n)^\times$$

1.4.5 Theorem: *(Euler)* For x relatively prime to a positive integer n,

$$x^{\varphi(n)} = 1 \bmod n$$

1.4.6 Remark: The special case that n is prime is Fermat's Little Theorem.

Proof: Let $G = (\mathbb{Z}/m)^\times$, for brevity. First note that the product

$$P = \prod_{g \in G} g = \text{product of all elements of } G$$

is again in G. Thus, P has a multiplicative inverse mod n, although we do not try to identify it. Let x be an element of G. Then we claim that the map $f : G \longrightarrow G$ defined by

$$f(g) = xg$$

is a bijection of G to itself. First, check that f really maps G to itself: for x and g both invertible mod n,

$$(xg)(g^{-1}x^{-1}) = 1 \bmod n$$

Next, injectivity: if $f(g) = f(h)$, then $xg = xh \bmod n$. Multiply this equality by $x^{-1} \bmod n$ to obtain $g = h \bmod n$. Last, surjectivity: given $g \in G$, note that $f(x^{-1}g) = g$.

Then

$$P = \prod_{g \in G} g = \prod_{g \in G} f(g)$$

since the map f merely permutes the elements of G. Then

$$P = \prod_{g \in G} f(g) = \prod_{g \in G} xg = x^{\varphi(n)} \prod_{g \in G} g = x^{\varphi(n)} \cdot P$$

Since P is invertible mod n, multiply through by $P^{-1} \bmod n$ to obtain

$$1 = x^{\varphi(n)} \bmod n$$

This proves Euler's Theorem. ///

1.4.7 Remark: This proof of Euler's theorem, while subsuming Fermat's Little Theorem as a special case, strangely uses fewer specifics. There is no mention of binomial coefficients, for example.

1.4.8 Remark: The argument above is a prototype example for the basic Lagrange's Theorem in basic group theory.

1.5 *Sun-Ze's theorem*

The result of this section is sometimes known as the **Chinese Remainder Theorem**. Indeed, the earliest results (including and following Sun-Ze's) were obtained in China, but such sloppy attribution is not good. Sun-Ze's result was obtained before 850, and the statement below was obtained by Chin Chiu Shao about 1250. Such results, with virtually the same proofs, apply much more generally.

1.5.1 Theorem: *(Sun-Ze)* Let m and n be relatively prime positive integers. Let r and s be integers such that
$$rm + sn = 1$$
Then the function
$$f : \mathbb{Z}/m \times \mathbb{Z}/n \longrightarrow \mathbb{Z}/mn$$
defined by
$$f(x,y) = y \cdot rm + x \cdot sn$$
is a bijection. The inverse map
$$f^{-1} : \mathbb{Z}/mn \longrightarrow \mathbb{Z}/m \times \mathbb{Z}/n$$
is
$$f^{-1}(z) = (x\text{-mod-}m, y\text{-mod-}n)$$

Proof: First, the peculiar characterization of $\gcd(m,n)$ as the smallest positive integer expressible in the form $rm + sn$ assures (since here $\gcd(m,n) = 1$) that integers r and s exist such that $rm + sn = 1$. Second, the function f is well-defined, that is, if $x' = x + am$ and $y' = y + bn$ for integers a and b, then still
$$f(x',y') = f(x,y)$$
Indeed,
$$f(x',y') = y'rm + x'sn = (y + an)rm + (x + am)sn$$
$$= yrm + xsn + mn(ar + bs) = f(x,y) \bmod mn$$
proving the well-definedness.

To prove surjectivity of f, for any integer z, let $x = z$ and $y = z$. Then
$$f(x,y) = zrm + zsn = z(rm + sn) = z \cdot 1 \bmod mn$$

(To prove injectivity, we *could* use the fact that $\mathbb{Z}/m \times \mathbb{Z}/n$ and $\mathbb{Z}/mn$ are finite sets of the same size, so a surjective function is necessarily injective, but a more direct argument is more instructive.) Suppose
$$f(x',y') = f(x,y)$$

Then modulo m the yrm and $y'rm$ are 0, so

$$xsn = x'sn \bmod m$$

From $rm + sn = 1 \bmod mn$ we obtain $sn = 1 \bmod m$, so

$$x = x' \bmod m$$

Symmetrically,
$$y = y' \bmod n$$

giving injectivity.

Finally, by the same reasoning,

$$f(x,y) = yrm + xsn = y \cdot 0 + x \cdot 1 \bmod m = x \bmod m$$

and similarly
$$f(x,y) = yrm + xsn = y \cdot 1 + x \cdot 0 \bmod n = y \bmod n$$

This completes the argument. ///

1.5.2 Remark: The above result is the simplest prototype for a very general result.

1.6 *Worked examples*

1.6.1 Example: Let D be an integer that is not the square of an integer. Prove that there is no $\sqrt{D}$ in $\mathbb{Q}$.

Suppose that a, b were integers ($b \neq 0$) such that $(a/b)^2 = D$. The fact/principle we intend to invoke here is that fractions can be put in *lowest terms*, in the sense that the numerator and denominator have greatest common divisor 1. This follows from *existence* of the *gcd*, and from the fact that, if $\gcd(a,b) > 1$, then let $c = a/\gcd(a,b)$ and $d = b/\gcd(a,b)$ and we have $c/d = a/b$. Thus, still $c^2/d^2 = D$. One way to proceed is to prove that c^2/d^2 is still in lowest terms, and thus cannot be an integer unless $d = \pm 1$. Indeed, if $\gcd(c^2, d^2) > 1$, this *gcd* would have a prime factor p. Then $p|c^2$ implies $p|c$, and $p|d^2$ implies $p|d$, by the critical proven property of primes. Thus, $\gcd(c,d) > 1$, contradiction.

1.6.2 Example: Let p be prime, $n > 1$ an integer. Show (directly) that the equation $x^n - px + p = 0$ has no rational root (where $n > 1$).

Suppose there were a rational root a/b, without loss of generality in lowest terms. Then, substituting and multiplying through by b^n, one has

$$a^n - pb^{n-1}a + pb^n = 0$$

Then $p|a^n$, so $p|a$ by the property of primes. But then p^2 divides the first two terms, so must divide pb^n, so $p|b^n$. But then $p|b$, by the property of primes, contradicting the lowest-common-terms hypothesis.

1.6.3 Example: Let p be prime, b an integer not divisible by p. Show (directly) that the equation $x^p - x + b = 0$ has no rational root.

Suppose there were a rational root c/d, without loss of generality in lowest terms. Then, substituting and multiplying through by d^p, one has

$$c^p - d^{p-1}c + bd^p = 0$$

If $d \neq \pm 1$, then some prime q divides d. From the equation, $q|c^p$, and then $q|c$, contradiction to the lowest-terms hypothesis. So $d = 1$, and the equation is

$$c^p - c + b = 0$$

By Fermat's Little Theorem, $p|c^p - c$, so $p|b$, contradiction.

1.6.4 Example: Let r be a positive integer, and p a prime such that $\gcd(r, p-1) = 1$. Show that every b in $\mathbb{Z}/p$ has a unique r^{th} root c, given by the formula

$$c = b^s \bmod p$$

where $rs = 1 \bmod (p-1)$. [*Corollary of Fermat's Little Theorem.*]

1.6.5 Example: Show that $R = \mathbb{Z}[\sqrt{-2}]$ and $\mathbb{Z}[\frac{1+\sqrt{-7}}{2}]$ are Euclidean.

First, we consider $R = \mathbb{Z}[\sqrt{-D}]$ for $D = 1, 2, \ldots$. Let $\omega = \sqrt{-D}$. To prove Euclidean-ness, note that the Euclidean condition that, given $\alpha \in \mathbb{Z}[\omega]$ and non-zero $\delta \in \mathbb{Z}[\omega]$, there exists $q \in \mathbb{Z}[\omega]$ such that

$$|\alpha - q \cdot \delta| < |\delta|$$

is equivalent to

$$|\alpha/\delta - q| < |1| = 1$$

Thus, it suffices to show that, given a complex number α, there is $q \in \mathbb{Z}[\omega]$ such that

$$|\alpha - q| < 1$$

Every complex number α can be written as $x + y\omega$ with real x and y. The simplest approach to analysis of this condition is the following. Let m, n be integers such that $|x - m| \leq 1/2$ and $|y - n| \leq 1/2$. Let $q = m + n\omega$. Then $\alpha - q$ is of the form $r + s\omega$ with $|r| \leq 1/2$ and $|s| \leq 1/2$. And, then,

$$|\alpha - q|^2 = r^2 + Ds^2 \leq \frac{1}{4} + \frac{D}{4} = \frac{1+D}{4}$$

For this to be strictly less than 1, it suffices that $1 + D < 4$, or $D < 3$. This leaves us with $\mathbb{Z}[\sqrt{-1}]$ and $\mathbb{Z}[\sqrt{-2}]$.

In the second case, consider $Z[\omega]$ where $\omega = (1 + \sqrt{-D})/2$ and $D = 3 \bmod 4$. (The latter condition assures that $\mathbb{Z}[x]$ works the way we hope, namely that everything in it is expressible as $a + b\omega$ with $a, b \in \mathbb{Z}$.) For D=3 (the Eisenstein integers) the previous approach still works, but fails for $D = 7$ and for $D = 11$. Slightly more cleverly, realize that first, given complex α, integer n can be chosen such that

$$-\sqrt{D}/4 \leq \text{imaginary part}(\alpha - n\omega) \leq +\sqrt{D}/4$$

since the imaginary part of ω is $\sqrt{D}/2$. *Then* choose integer m such that

$$-1/2 \leq \text{real part}(\alpha - n\omega - m) \leq 1/2$$

Then take $q = m + n\omega$. We have chosen q such that $\alpha - q$ is in the *rectangular* box of complex numbers $r + s\sqrt{-7}$ with

$$|r| \le 1/2 \quad \text{and} \quad |s| \le 1/4$$

Yes, $1/4$, not $1/2$. Thus, the size of $\alpha - q$ is at most

$$1/4 + D/16$$

The condition that this be strictly less than 1 is that $4 + D < 16$, or $D < 12$ (and $D = 1 \bmod 4$). This gives $D = 3, 7, 11$.

1.6.6 Example: Let $f : X \longrightarrow Y$ be a function from a set X to a set Y. Show that f has a left inverse if and only if it is injective. Show that f has a right inverse if and only if it is surjective. (Note where, if anywhere, the Axiom of Choice is needed.)

1.6.7 Example: Let $h : A \longrightarrow B$, $g : B \longrightarrow C$, $f : C \longrightarrow D$. Prove the associativity

$$(f \circ g) \circ h = f \circ (g \circ h)$$

Two functions are equal if and only if their values (for the same inputs) are the same. Thus, it suffices to evaluate the two sides at $a \in A$, using the definition of composite:

$$((f \circ g) \circ h)(a) = (f \circ g)(h(a)) = f(g((h(a))) = f((g \circ h)(a)) = (f \circ (g \circ h))(a)$$

1.6.8 Example: Show that a set is infinite if and only if there is an injection of it to a proper subset of itself. Do not set this up so as to trivialize the question.

The other definition of *finite* we'll take is that a set S is finite if there is a surjection to it from one of the sets

$$\{\}, \ \{1\}, \ \{1,2\}, \ \{1,2,3\}, \ \ldots$$

And a set is *infinite* if it has no such surjection.

We find a denumerable subset of an infinite set S, as follows. For infinite S, since S is not empty (or there'd be a surjection to it from $\{\}$), there is an element s_1. Define

$$f_1 : \{1\} \longrightarrow S$$

by $f(1) = s_1$. This cannot be surjective, so there is $s_2 \ne s_1$. Define

$$f_2 : \{1,2\} \longrightarrow S$$

by $f(1) = s_1$, $f(2) = s_2$. By induction, for each natural number n we obtain an injection $f_n : \{1, \ldots\} \longrightarrow S$, and distinct elements $s_1, s_2, \ldots$. Let S' be the complement to $\{s_1, s_2, \ldots\}$ in S. Then define $F : S \longrightarrow S$ by

$$F(s_i) = s_{i+1} \quad F(s') = s' \ (\text{for } s' \in S')$$

This is an injection to the proper subset $S - \{s_1\}$.

On the other hand, we claim that no set $\{1, \ldots, n\}$ admits an injection to a proper subset of itself. If there were such, by Well-Ordering there would be a least n such that this could happen. Let f be an injection of $S = \{1, \ldots, n\}$ to a proper subset of itself.

By hypothesis, f restricted to $S' = \{1, 2, \ldots, n-1\}$ does *not* map S' to a proper subset of itself. The restriction of an injective function is still injective. Thus, either $f(i) = n$ for some $1 \le i < n$, or $f(S')$ is the *whole* set S'. In the former case, let j be the least element not in the image $f(S)$. (Since $f(i) = n$, $j \neq n$, but this doesn't matter.) Replace f by $\pi \circ f$ where π is the permutation of $\{1, \ldots, n\}$ that interchanges j and n and leaves everything else fixed. Since permutations are bijections, this $\pi \circ f$ is still an injection of S to a proper subset. Thus, we have reduced to the second case, that $f(S') = S'$. By injectivity, $f(n)$ can't be in S', but then $f(n) = n$, and the image $f(S)$ is not a proper subset of S after all, contradiction.   ///

In a similar vein, one can *prove* the Pigeon-Hole Principle, namely, that for $m < n$ a function
$$f : \{1, \ldots, n\} \longrightarrow \{1, \ldots, m\}$$
cannot be injective. Suppose this is false. Let n be the smallest such that there is $m < n$ with an injective map as above. The restriction of an injective map is still injective, so f on $\{1, \ldots, n-1\}$ is still injective. By the minimality of n, it must be that $n-1 = m$, and that f restricted to $\{1, \ldots, m\}$ is a bijection of that set to itself. But then there is no possibility for $f(n)$ in $\{1, \ldots, m\}$ without violating the injectivity. Contradiction. Thus, there is no such injection to a smaller set.

Exercises

1.1 Let $f(x) = x^n + a_{n-1}x^{n-1} + \ldots + a_1 x + a_0$ be a polynomial with integer coefficients a_i. Show that if $f(x) = 0$ has a root in $\mathbb{Q}$, then this root is an integer dividing a_0.

1.2 Show that $x^2 - y^2 = 102$ has no solutions in integers.

1.3 Show that $x^3 - y^3 = 3$ has no solutions in integers.

1.4 Show that $x^3 + y^3 - z^3 = 4$ has no solutions in integers.

1.5 Show that $x^2 + 3y^2 + 6z^3 - 9w^5 = 2$ has no solutions in integers.

1.6 The defining property of *ordered pair* (a, b) is that $(a, b) = (a', b')$ if and only if $a = a'$ and $b = b'$. Show that the set-theoretic construction $(a, b) = \{\{a\}, \{a, b\}\}$ succeeds in making an object that behaves as an ordered pair is intended. (*Hint:* Beware: if $x = y$, then $\{x, y\} = \{x\}$.)

1.7 Let p be a prime, and q a positive integer power of p. Show that p divides the binomial coefficients $\binom{q}{i} = q!/i!(q-i)!$ for $0 < i < q$.

1.8 Show that the greatest common divisor of non-zero integers x, y, z is the smallest positive integer expressible as $ax + by + cz$ for integers a, b, c.

1.9 Let m, n be relatively prime integers. Without using factorizations, prove that $m|N$ and $n|N$ implies $mn|N$.

1.10 (*A warm-up to Hensel's lemma*) Let $p > 2$ be a prime. Suppose that b is an integer not divisible by p such that there is a solution y to the equation $y^2 = b \bmod p$. Show (by induction on n) that for $n \ge 1$ there is a unique $x \bmod p^n$ such that $x = b \bmod p$ and
$$x^p = b \bmod p^n$$

1.11 (*Another warm-up to Hensel's lemma*) Let $p > 2$ be a prime. Let y be an integer such that $y \equiv 1 \bmod p$. Show (by induction on n) that for $n \geq 1$ there is a unique $x \bmod p^n$ so that

$$x^p = y \bmod p^n$$

1.12 Let φ be Euler's phi-function, equal to the number of integers ℓ such that $1 \leq \ell < n$ with ℓ relatively prime to n. Show that for a positive integer n

$$n = \sum_{d | n, \, d > 0} \varphi(d)$$

2.   Groups I

2.1 *Groups*

The simplest, but not most immediately intuitive, object in abstract algebra is a *group*. Once introduced, one can see this structure nearly everywhere in mathematics. [1]

By definition, a **group** G is a set with an **operation** $g*h$ (formally, a function $G \times G \longrightarrow G$), with a special element e called **the identity**, and with properties:

• *The property of the identity*: for all $g \in G$, $e*g = g*e = g$.
• *Existence of* **inverses**: for all $g \in G$ there is $h \in G$ (the **inverse** of g) such that $h*g = g*h = e$.
• *Associativity*: for all $x, y, z \in G$, $x*(y*z) = (x*y)*z$.

---

[1] Further, the notion of group proves to be more than a mere *descriptive* apparatus. It provides unification and synthesis for arguments and concepts which otherwise would need individual development. Even more, abstract structure theorems for groups provide *predictive* indications, in the sense that we know something in advance about groups we've not yet seen.

If the operation $g * h$ is **commutative**, that is, if

$$g * h = h * g$$

then the group is said to be **abelian**. [2]   In that case, often, but not necessarily, the operation is written as *addition*. And when the operation is written as addition, the identity is often written as 0 instead of e.

In many cases the group operation is written as multiplication or simply as juxtaposition

$$g * h = g \cdot h = gh$$

This does not *preclude* the operation being abelian, but only denies the *presumption* that the operation is abelian. If the group operation is written as multiplication, then often the identity is denoted as 1 rather than e. Unless written additively, the **inverse** [3]   of an element g in the group is denoted

$$\text{inverse of } g = g^{-1}$$

If the group operation is written as *addition*, then the inverse is denoted

$$\text{inverse of } g = -g$$

Many standard mathematical items with natural operations are groups: The set $\mathbb{Z}$ of integers $\mathbb{Z}$ with addition $+$ is an abelian group. The set $n\mathbb{Z}$ of multiples of an integer n, with addition, is an abelian group. The set $\mathbb{Z}/m$ of integers mod m, with addition mod m as the operation is an abelian group. The set $\mathbb{Z}/m^\times$ of integers mod m *relatively prime to m*, with multiplication mod m as the operation is an abelian group.

The set $\mathbb{Z}$ of integers with operation being *multiplication* is *not* a group, because there are no inverses. [4]   The closest we can come is the set $\{1, -1\}$ with multiplication.

Other things which we'll define formally only a bit later are groups: vector spaces with vector addition are abelian groups. The set $GL(2, \mathbb{R})$ of invertible 2-by-2 real matrices, with group law matrix multiplication, is a non-abelian group. Here the identity is the matrix

$$1_2 = \begin{pmatrix} 1 & 0 \\ 0 & 1 \end{pmatrix}$$

The existence of inverses is part of the definition. The *associativity* of matrix multiplication is not entirely obvious from the definition, but can either be checked by hand or inferred from the fact that composition of functions is associative.

A more abstract example of a group is the set S_n of **permutations** of a set with n elements (n an integer), where *permutation* means *bijection to itself*. Here the operation is

---

[2] After N.H. Abel, who in his investigation of the solvability by radicals of algebraic equations came to recognize the significance of commutativity many decades before the notion of *group* was formalized.

[3] once we prove its uniqueness!

[4] The fact that there are multiplicative inverses in the larger set $\mathbb{Q}^\times$ of non-zero rational numbers is beside the point, since these inverses are not inside the given set $\mathbb{Z}$.

composition (as functions) of permutations. If there are more than two things in the set, S_n is non-abelian.

Some nearly trivial uniqueness issues should be checked: [5]
- *(Uniqueness of identity)* If $f \in G$ and $f * g = g * f = g$ for all g in G, then $f = e$.
- *(Uniqueness of inverses)* For given $g \in G$, if $a * g = e$ and $g * b = e$, then $a = b$.

Proof: For the first assertion,

$$\begin{aligned} f &= f * e & \text{(property of } e) \\ &= e & \text{(assumed property of } e) \end{aligned}$$

which was claimed. For the second, similarly,

$$a = a * e = a * (g * b) = (a * g) * b = e * b = b$$

where we use, successively, the property of the identity, the defining property of b, associativity, the defining property of a, and then the property of the identity again.
///

2.1.1 Remark: These uniqueness properties justify speaking of *the* inverse and *the* identity.

2.2 *Subgroups, Lagrange's theorem*

Subgroups are subsets of groups which are groups *in their own right*, in the following sense. A subset H of a group G is said to be a **subgroup** if, with the same operation and identity element as that used in G, it is a group.

That is, if H contains the identity element $e \in G$, if H contains inverses of all elements in it, and if H contains products of any two elements in it, then H is a subgroup.

Common terminology is that H is **closed under inverses** if for $h \in H$ the inverse h^{-1} is in H, and **closed under the group operation** if $h_1, h_2 \in H$ implies $h_1 * h_2$ is in H. [6]

Note that the associativity of the operation is assured since the operation was *assumed* associative for G itself to be a group.

The subset $\{e\}$ of a group G is always a subgroup, termed **trivial**. A subgroup of G other than the trivial subgroup and the group G itself is **proper**.

2.2.1 Proposition: The intersection $\bigcap_{H \in S} H$ of any collection of subgroups of a group G is again a subgroup of G.

Proof: Since the identity e of G lies in each H, it lies in their intersection. If h lies in H for every $H \in S$, then h^{-1} lies in H for every $H \in S$, so h^{-1} is in the intersection. Similarly,

---

[5] These are the sort of properties which, if they were *not* provable from the definition of group, would probably need to be added to the definition. We are fortunate that the innocent-looking definition does in fact yield these results.

[6] In reality, the very notion of *operation* includes the assertion that the output is again in the set. Nevertheless, the property is important enough that extra emphasis is worthwhile.

if h_1, h_2 are both in H for every $H \in S$, so is their product, and then the product is in the intersection.   ///

Given a set X of elements in a group G, the **subgroup generated by** [7] X is defined to be

$$\text{subgroup generated by } X \; = \langle X \rangle = \bigcap_{H \supset X} H$$

where H runs over *subgroups* of G containing X. The previous proposition ensures that this really is a subgroup. If $X = \{x_1, \ldots, x_n\}$ we may, by abuse of notation, write also

$$\langle X \rangle = \langle x_1, \ldots, x_n \rangle$$

and refer to the subgroup generated by $x_1, \ldots, x_n$ rather than by the subset X.

A **finite group** is a group which (as a set) is finite. The **order** of a finite group is the number of elements in it. Sometimes the order of a group G is written as $|G|$ or $o(G)$. The first real theorem in group theory is

2.2.2 Theorem: *(Lagrange)* [8]   Let G be a *finite* group. Let H be a subgroup of G. Then the order of H *divides* the order of G.

Proof: For $g \in G$, the **left coset** of H by g or **left translate** of H by g is

$$gH = \{gh : h \in H\}$$

(Similarly, the **right coset** of H by g or **right translate** of H by g is $Hg = \{hg : h \in H\}$.)

First, we will prove that the collection of all left cosets of H is a *partition* of G. Certainly $x = x \cdot e \in xH$, so every element of G lies in a left coset of H. Now suppose that $xH \cap yH \neq \phi$ for $x, y \in G$. Then for some $h_1, h_2 \in H$ we have $xh_1 = yh_2$. Multiply both sides of this equality on the right by h_2^{-1} to obtain

$$(xh_1)h_2^{-1} = (yh_2)h_2^{-1} = y$$

Let $z = h_1 h_2^{-1}$ for brevity. Since H is a *subgroup*, $z \in H$. Then

$$yH = \{yh : h \in H\} = \{(xz)h : h \in H\} = \{x(zh) : h \in H\}$$

Thus, $yH \subset xH$. Since the relationship between x and y is symmetrical, also $xH \subset yH$, and $xH = yH$. Thus, the left cosets of H in G partition G.

Next, show that the cardinalities of the left cosets of H are identical, by demonstrating a *bijection* from H to xH for any $x \in G$. Define

$$f(g) = xg$$

---

[7] Later we will see a constructive version of this notion. Interestingly, or, perhaps, disappointingly, the more constructive version is surprisingly complicated. Thus, the present quite non-constructive definition is useful, possibly essential.

[8] Since the notion of abstract group did not exist until about 1890, Lagrange, who worked in the late 18th and early 19th centuries, could not have proven the result as it is stated. However, his work in number theory repeatedly used results of this sort, as did Gauss's of about the same time. That is, Lagrange and Gauss recognized the principle without having a formal framework for it.

This maps H to yH, and if $f(g) = f(g')$, then

$$xg = xg'$$

from which left multiplication by x^{-1} gives $g = g'$. For *surjectivity*, note that the function f was arranged so that

$$f(h) = xh$$

Thus, all left cosets of H have the same number of elements as H.

So G is the disjoint union of the left cosets of H. From this, $|H|$ divides $|G|$. $///$

The **index** $[G : H]$ of a subgroup H in a group G is the number of disjoint (left or right) cosets of H in G. Thus, Lagrange's theorem says

$$|G| = [G : H] \cdot |H|$$

For a single element g of a group G, one can verify that

$$\langle g \rangle = \{g^n : n \in \mathbb{Z}\}$$

where $g^0 = e$, and

$$g^n = \begin{cases} \underbrace{g * g * \ldots * g}_{n} & (0 < n \in \mathbb{Z}) \\ \underbrace{g^{-1} * g^{-1} * \ldots * g^{-1}}_{|n|} & (0 > n \in \mathbb{Z}) \end{cases}$$

One might do the slightly tedious induction proof of the fact that, for all choices of sign of integers m, n,

$$g^{m+n} = g^m * g^n$$
$$(g^m)n = g^{mn}$$

That is, the so-called *Laws of Exponents* are provable properties. And, thus, $\langle g \rangle$ really is a subgroup. For various reasons, a (sub)group which can be generated by a single element is called a **cyclic subgroup**. Note that a cyclic group is necessarily abelian.

The smallest positive integer n (if it exists) such that

$$g^n = e$$

is the **order** or **exponent** of g, often denoted by $|g|$ or $o(g)$. If there is no such n, say that the order of g is *infinite*. [9]

2.2.3 Proposition: Let g be a finite-order element of a group G, with order n. Then the order of g (as group *element*) is equal to the order of $\langle g \rangle$ (as subgroup). In particular,

$$\langle g \rangle = \{g^0, g^1, g^2, \ldots, g^{n-1}\}$$

and, for arbitrary integers i, j,

$$g^i = g^j \quad \text{if and only if} \quad i = j \bmod n$$

---

[9] Yes, this use of the term *order* is in conflict with the use for subgroups, but we immediately prove their compatibility.

Proof: The last assertion implies the first two. On one hand, if $i = j \bmod n$, then write $i = j + \ell n$ and compute

$$g^i = g^{j+\ell n} = g^j \cdot (g^n)^\ell = g^j \cdot e^\ell = g^j \cdot e = g^j$$

On the other hand, suppose that $g^i = g^j$. Without loss of generality, $i \leq j$, and $g^i = g^j$ implies $e = g^{j-i}$. Let

$$j - i = q \cdot n + r$$

where $0 \leq r < n$. Then

$$e = g^{j-i} = g^{qn+r} = (g^n)^q \cdot g^r = e^q \cdot g^r = e \cdot g^r = g^r$$

Therefore, since n is the least such that $g^n = e$, necessarily $r = 0$. That is, $n | j - i$.   ///

2.2.4 Corollary: *(of Lagrange's theorem)* The order $|g|$ of an element g of a finite group G divides the order of G. [10]

Proof: We just proved that $|g| = |\langle g \rangle|$, which, by Lagrange's theorem, divides $|G|$.   ///

Now we can recover Euler's theorem as an example of the latter corollary of Lagrange's theorem:

2.2.5 Corollary: *(Euler's theorem, again)* Let n be a positive integer. For $x \in \mathbb{Z}$ relatively prime to n,

$$x^{\varphi(n)} = 1 \bmod n$$

Proof: The set $\mathbb{Z}/n^\times$ of integers mod n relatively prime to n is a group with $\varphi(n)$ elements. By Lagrange, the order k of $g \in \mathbb{Z}/n^\times$ divides $\varphi(n)$. Therefore, $\varphi(n)/k$ is an integer, and

$$g^{\varphi(n)} = (g^k)^{\varphi(n)/k} = e^{\varphi(n)/k} = e$$

as desired.   ///

The idea of Euler's theorem can be abstracted. For a group G, the smallest positive integer ℓ so that for every $g \in G$

$$g^\ell = e$$

is the **exponent** of the group G. It is not clear from the definition that there really is such a positive integer ℓ. Indeed, for *infinite* groups G there may not be. But for *finite* groups the mere finiteness allows us to characterize the exponent:

2.2.6 Corollary: *(of Lagrange's theorem)* Let G be a finite group. Then the exponent of G divides the order $|G|$ of G.

Proof: From the definition, the exponent is the least common multiple of the orders of the elements of G. From Lagrange's theorem, each such order is a divisor of $|G|$. The least common multiple of any collection of divisors of a fixed number is certainly a divisor of that number.   ///

---

[10] One can also imitate the direct proof of Euler's theorem, and produce a proof of this corollary at least for finite abelian groups.

2.3 *Homomorphisms, kernels, normal subgroups*

Group homomorphisms are the maps of interest among groups.

A *function* (or *map*)

$$f : G \longrightarrow H$$

from one group G to another H is a **(group) homomorphism** if the *group operation is preserved* in the sense that

$$f(g_1 g_2) = f(g_1)\, f(g_2)$$

for all $g_1, g_2 \in G$. Let e_G be the identity in G and e_H the identity in H. The **kernel** of a homomorphism f is

$$\text{kernel of } f \ = \ker f = \{g \in G : f(g) = e_H\}$$

The **image** of f is just like the image of any function:

$$\text{image of } f \ = \operatorname{im} f = \{h \in H : \text{ there is } g \in G \text{ so that } f(g) = h\}$$

2.3.1 Theorem: Let $f : G \longrightarrow H$ be a group homomorphism. Let e_G be the identity in G and let e_H be the identity in H. Then
• Necessarily f carries the identity of G to the identity of H: $f(e_G) = e_H$.
• For $g \in G$, $f(g^{-1}) = f(g)^{-1}$.
• The *kernel* of f is a subgroup of G.
• The *image* of f is a subgroup of H.
• Given a subgroup K of H, the *pre-image*

$$f^{-1}(K) = \{g \in G : f(g) \in K\}$$

of K under f is a subgroup of G.
• A group homomorphism $f : G \longrightarrow H$ is *injective* if and only if the kernel is *trivial* (that is, is the trivial subgroup $\{e_G\}$).

Proof: The image $f(e_G)$ has the property

$$f(e_G) = f(e_G \cdot e_G) = f(e_G) \cdot f(e_G)$$

Left multiplying by $f(e_G)^{-1}$ (whatever this may be),

$$f(e_G)^{-1} \cdot f(e_G) = f(e_G)^{-1} \cdot (f(e_G) \cdot f(e_G))$$

Simplifying,

$$e_H = (f(e_G)^{-1} \cdot f(e_G)) \cdot f(e_G) = e_H \cdot f(e_G) = f(e_G)$$

so the identity in G is mapped to the identity in H.

To check that the image of an inverse is the inverse of an image, compute

$$f(g^{-1}) \cdot f(g) = f(g^{-1} \cdot g) = f(e_G) = e_H$$

using the fact just proven that the identity in G is mapped to the identity in H.

Now prove that the kernel is a subgroup of G. The identity lies in the kernel since, as we just saw, it is mapped to the identity. If g is in the kernel, then g^{-1} is also, since, as just showed, $f(g^{-1}) = f(g)^{-1}$. Finally, suppose both x, y are in the kernel of f. Then

$$f(xy) = f(x) \cdot f(y) = e_H \cdot e_H = e_H$$

Let X be a subgroup of G. Let

$$f(X) = \{f(x) : x \in X\}$$

To show that $f(X)$ is a subgroup of H, we must check for presence of the identity, closure under taking inverses, and closure under products. Again, $f(e_G) = e_H$ was just proven. Also, we showed that $f(g)^{-1} = f(g^{-1})$, so the image of a subgroup is closed under inverses. And $f(xy) = f(x)f(y)$ by the defining property of a group homomorphism, so the image is closed under multiplication.

Let K be a subgroup of H. Let x, y be in the pre-image $f^{-1}(K)$. Then

$$f(xy) = f(x) \cdot f(y) \in K \cdot K = K$$

$$f(x^{-1}) = f(x)^{-1} \in K$$

And already $f(e_G) = e_H$, so the pre-image of a subgroup is a group.

Finally, we prove that a homomorphism $f : G \longrightarrow H$ is injective if and only if its kernel is trivial. First, if f is injective, then at most one element can be mapped to $e_H \in H$. Since we know that at least e_G is mapped to e_H by such a homomorphism, it must be that *only* e_G is mapped to e_H. Thus, the kernel is trivial. On the other hand, suppose that the kernel is trivial. We will suppose that $f(x) = f(y)$, and show that $x = y$. Left multiply $f(x) = f(y)$ by $f(x)^{-1}$ to obtain

$$e_H = f(x)^{-1} \cdot f(x) = f(x)^{-1} \cdot f(y)$$

By the homomorphism property,

$$e_H = f(x)^{-1} \cdot f(y) = f(x^{-1}y)$$

Thus, $x^{-1}y$ is in the kernel of f, so (by assumption) $x^{-1}y = e_G$. Left multiplying this equality by x and simplifying, we get $y = x$.   ///

If a group homomorphism $f : G \longrightarrow H$ is *surjective*, then H is said to be a **homomorphic image** of G. If a group homomorphism $f : G \longrightarrow H$ has an inverse homomorphism, then f is said to be an **isomorphism**, and G and H are said to be **isomorphic**, written

$$G \approx H$$

For groups, if a group homomorphism is a *bijection*, then it has an inverse which is a group homomorphism, so is an isomorphism.

2.3.2 Remark: Two groups that are *isomorphic* are considered to be 'the same', in the sense that any *intrinsic* group-theoretic assertion about one is also true of the other.

A subgroup N of a group G is **normal** [11] or **invariant** [12] if, for every $g \in G$,

$$gNg^{-1} = N$$

where the notation is

$$gNg^{-1} = \{gng^{-1} : n \in N\}$$

This is readily seen to be equivalent to the condition that

$$gN = Ng$$

for all $g \in G$. Evidently in an abelian group G *every* subgroup is normal. It is not hard to check that *intersections of normal subgroups are normal.*

2.3.3 Proposition: The kernel of a homomorphism $f : G \longrightarrow H$ is a normal subgroup.

Proof: For $n \in \ker f$, using things from just above,

$$f(gng^{-1}) = f(g)\,f(n)\,f(g^{-1}) = f(g)\,e_H\,f(g)^{-1} = f(g)\,f(g)^{-1} = e_H$$

as desired. ///

A group with no *proper* normal subgroups is **simple**. Sometimes this usage is restricted to apply only to groups *not* of orders which are prime numbers, since (by Lagrange) such groups have no proper subgroups whatsoever, much less normal ones.

2.4 *Cyclic groups*

Finite groups generated by a single element are easy to understand. The collections of all *subgroups* and of all *generators* can be completely understood in terms of elementary arithmetic, in light of the first point below. Recall that the set of integers modulo n is

$$\mathbb{Z}/n\mathbb{Z} = \mathbb{Z}/n = \{cosetsofn\mathbb{Z}in\mathbb{Z}\} = \{x + n\mathbb{Z} : x \in \mathbb{Z}\}$$

2.4.1 Proposition: Let $G = \langle g \rangle$, of order n. Then G is isomorphic to $\mathbb{Z}/n$ with addition, by the map

$$f(g^i) = i + n\mathbb{Z} \in \mathbb{Z}/n$$

Proof: The main point is the well-definedness of the map. That is, that $g^i = g^j$ implies $i = j \bmod n$, for $i, j \in \mathbb{Z}$. Suppose, without loss of generality, that $i < j$. Then $g^{j-i} = e$. Let

$$j - i = q \cdot n + r$$

with $0 \le r < n$. Then

$$e = e \cdot e = g^{j-i-qn} = g^r$$

---

[11] This is one of too many uses of this term, but it is irretrievably standard.

[12] The term *invariant* surely comes closer to suggesting the intent, but is unfortunately archaic.

and by the minimality of n we have $r = 0$. Thus, $n | j - i$, proving well-definedness of the map. The surjectivity and injectivity are then easy. The assertion that f is a homomorphism is just the well-definedness of addition modulo n together with properties of exponents:

$$f(g^i) + f(g^j) = (i + n\mathbb{Z}) + (j + n\mathbb{Z}) = (i + j) + n\mathbb{Z} = f(g^{i+j}) = f(g^i \cdot g^j)$$

This demonstrates the isomorphism.   ///

2.4.2 Corollary: Up to isomorphism, there is only one finite cyclic group of a given order.   ///

The following facts are immediate corollaries of the proposition and elementary properties of $\mathbb{Z}/n$.

- The *distinct* subgroups of G are exactly the subgroups $\langle g^d \rangle$ for all *divisors* d of N.
- For $d | N$ the order of the subgroup $\langle g^d \rangle$ is the order of g^d, which is N/d.
- The order of g^k with arbitrary integer $k \neq 0$ is $N/\gcd(k, N)$.
- For any integer n we have
$$\langle g^n \rangle = \langle g^{\gcd(n,N)} \rangle$$

- The distinct generators of G are the elements g^r where $1 \leq r < N$ and $\gcd(r, N) = 1$. Thus, there are $\varphi(N)$ of them, where φ is Euler's phi function.
- The number of elements of order n in a finite cyclic group of order N is 0 unless $n | N$, in which case it is N/n.

2.4.3 Proposition: A homomorphic image of a finite cyclic group is finite cyclic.

Proof: The image of a generator is a generator for the image.   ///

Using the isomorphism of a cyclic group to some $\mathbb{Z}/n$, it is possible to reach definitive conclusions about the solvability of the equation $x^r = y$.

2.4.4 Theorem: Let G be a cyclic group of order n with generator g. Fix an integer r, and define
$$f : G \longrightarrow G$$
by
$$f(x) = x^r$$
This map f is a group homomorphism of G to itself. If $\gcd(r, n) = 1$, then f is an *isomorphism*, and in that case every $y \in G$ has a unique r^{th} root. More generally,

$$\text{order of kernel of } f = \gcd(r, n)$$

$$\text{order of image of } f = n/\gcd(r, n)$$

If an element y has an rth root, then it has exactly $\gcd(r, n)$ of them. There are exactly $n/\gcd(r, n)$ r^{th} powers in G.

Proof: Since G is abelian the map f is a homomorphism. Use the fact that G is isomorphic to $\mathbb{Z}/n$. Converting to the additive notation for $\mathbb{Z}/n$-with-addition, f is

$$f(x) = r \cdot x$$

If $\gcd(r, n) = 1$ then there is a multiplicative inverse r^{-1} to r mod n. Thus, the function

$$g(x) = r^{-1} \cdot x$$

gives an inverse function to f, so f is an isomorphism.

For arbitrary r, consider the equation

$$r \cdot x = y \bmod n$$

for given y. This condition is

$$n | (rx - y)$$

Let $d = \gcd(r, n)$. Then certainly it is *necessary* that $d|y$ or this is impossible. On the other hand, suppose that $d|y$. Write $y = dy'$ with integer y'. We want to solve

$$r \cdot x = dy' \bmod n$$

Dividing through by the common divisor d, this congruence is

$$\frac{r}{d} \cdot x = y' \bmod \frac{n}{d}$$

The removal of the common divisor makes r/d prime to n/d, so there is an inverse $(r/d)^{-1}$ to r/d mod n/d, and

$$x = (r/d)^{-1} \cdot y' \bmod (n/d)$$

That is, any integer x meeting this condition is a solution to the original congruence. Letting x_0 be one such solution, the integers

$$x_0, \; x_0 + \frac{n}{d}, \; x_0 + 2 \cdot \frac{n}{d}, \; x_0 + 3 \cdot \frac{n}{d}, \; \ldots x_0 + (d-1) \cdot \frac{n}{d}$$

are also solutions, and are distinct mod n. That is, we have d distinct solutions mod n.

The kernel of f is the collection of x so that $rx = 0 \bmod n$. Taking out the common denominator $d = \gcd(r, n)$, this is $(r/d)x = 0 \bmod n/d$, or $(n/d)|(r/d)x$. Since r/d and n/d have no common factor, n/d divides x. Thus, mod n, there are d different solutions x. That is, the kernel of f has d elements.   ///

2.5 Quotient groups

Let G be a group and H a subgroup. The **quotient set** G/H of G by H is the set of H-cosets

$$G/H = \{xH : x \in G\}$$

in G. In general, there is *no* natural group structure on this set. [13]   But if H is *normal*, then we define a group operation $*$ on G/H by

$$xH * yH = (xy)H$$

---

[13] The key word is *natural*: of course any set can have several group structures put on it, but, reasonably enough, we are interested in group structures on G/H that have some connection with the original group structure on G.

Granting in advance that this works out, the **quotient map** $q : G \longrightarrow G/N$ defined by

$$q(g) = gN$$

will be a group homomorphism.

Of course, the same symbols can be written for non-normal H, *but will not give a well-defined operation.* That is, for well-definedness, one must verify that the operation does not depend upon the choice of coset representatives x, y in this formula. That is, one must show that if

$$xH = x'H \quad \text{and} \quad yH = y'H$$

then

$$(xy)H = (x'y')H$$

If H is *normal*, then $xH = Hx$ for all $x \in G$. Then, literally, as sets,

$$xH \cdot yH = x \cdot Hy \cdot H = x \cdot yH \cdot H = (xy)H \cdot H = (xy)H$$

That is, we can more directly define the group operation $*$ as

$$xH * yH = xH \cdot yH$$

2.5.1 Remark: If H is *not* normal, take $x \in G$ such that $Hx \neq xH$. That is, there is $h \in H$ such that $hx \notin xH$. Then $hxH \neq H$, and, if the same definition were to work, supposedly

$$hH * xH = (hx)H \neq xH$$

But, on the other hand, since $hH = eH$,

$$hH * xH = eH * xH = (ex)H = xH$$

That is, if H is not normal, this apparent definition is in fact not well-defined.

2.5.2 Proposition: *(Isomorphism Theorem)* Let $f : G \longrightarrow H$ be a *surjective* group homomorphism. Let $K = \ker f$. Then the map $\bar{f} : G/K \longrightarrow H$ by

$$\bar{f}(gK) = f(g)$$

is well-defined and is an isomorphism.

Proof: If $g'K = gK$, then $g' = gk$ with $k \in K$, and

$$f(g') = f(gk) = f(g)\, f(k) = f(g)\, e = f(g)$$

so the map $\bar{f}$ is well-defined. It is surjective because f is. For injectivity, if $\bar{f}(gK) = \bar{f}(g'K)$, then $f(g) = f(g')$, and

$$e_H = f(g)^{-1} \cdot f(g') = f(g^{-1}) \cdot f(g) = f(g^{-1}g')$$

Thus, $g^{-1}g' \in K$, so $g' \in gK$, and $g'K = gK$.   ///

In summary, the normal subgroups of a group are exactly the kernels of surjective homomorphisms.

As an instance of a counting principle, we have

2.5.3 Corollary: Let $f : G \longrightarrow H$ be a surjective homomorphism of finite groups. Let Y be a subgroup of H. Let

$$X = f^{-1}(Y) = \{x \in G : f(x) \in Y\}$$

be the **inverse image** of Y in G. Then

$$|X| = |\ker f| \cdot |Y|$$

Proof: By the isomorphism theorem, without loss of generality $Y = G/N$ where $N = \ker f$ is a normal subgroup in G. The quotient group is the set of cosets gN. Thus,

$$f^{-1}(Y) = \{xN : f(x) \in Y\}$$

That is, the inverse image is a disjoint union of cosets of N, and the number of cosets in the inverse image is $|Y|$. We proved earlier that X is a subgroup of G. ///

A variant of the previous corollary gives

2.5.4 Corollary: Given a normal subgroup N of a group G, and given any other subgroup H of G, let $q : G \longrightarrow G/N$ be the quotient map. Then

$$H \cdot N = \{hn : h \in H, n \in N\} = q^{-1}(q(H))$$

is a subgroup of G. If G is finite, the order of this group is

$$|H \cdot N| = \frac{|H| \cdot |N|}{|H \cap N|}$$

Further,

$$q(H) \approx H/(H \cap N)$$

Proof: By definition the inverse image $q^{-1}(q(H))$ is

$$\{g \in G : q(g) \in q(H)\} = \{g \in G : gN = hN \text{ for some } h \in H\}$$

$$= \{g \in G : g \in hN \text{ for some } h \in H\} = \{g \in G : g \in H \cdot N\} = H \cdot N$$

The previous corollary already showed that the inverse image of a subgroup is a subgroup. And if $hN = h'N$, then $N = h^{-1}h'N$, and $h^{-1}h' \in N$. Yet certainly $h^{-1}h' \in H$, so $h^{-1}h' \in H \cap N$. And, on the other hand, if $h^{-1}h' \in H \cap N$ then $hN = h'N$. Since $q(h) = hN$, this proves the isomorphism. From above, the inverse image $H \cdot N = q^{-1}(q(H))$ has cardinality

$$\operatorname{card} H \cdot N = |\ker q| \cdot |q(H)| = |N| \cdot |H/(H \cap N)| = \frac{|N| \cdot |H|}{|H \cap N|}$$

giving the counting assertion. ///

2.6 *Groups acting on sets*

Let G be a group and S a set. A map $G \times S \longrightarrow S$, denoted by juxaposition

$$g \times s \longrightarrow gs$$

is an **action** of the group on the set if
• $es = s$ for all $s \in S$
• *(Associativity)* $(gh)s = g(hs)$ for all $g, h \in G$ and $s \in S$.

These conditions assure that, for example, $gs = t$ for $s, t \in S$ and $g \in G$ implies that $g^{-1}t = s$. Indeed,

$$g^{-1}t = g^{-1}(gs) = (g^{-1}g)s = es = s$$

Sometimes a set with an action of a group G on it is called a G-**set**.

The action of G on a set is **transitive** if, for all $s, t \in S$, there is $g \in G$ such that $gs = t$. This definition admits obvious equivalent variants: for example, the seemingly weaker condition that there is $s_o \in S$ such that for every $t \in S$ there is $g \in G$ such that $gs_o = t$ implies transitivity. Indeed, given $s, t \in S$, let $g_s s_o = s$ and $g_t s_o = t$. Then

$$(g_t g_s^{-1})s = g_t(g_s^{-1}s) = g_t(s_o) = t$$

For G acting on a set S, a subset T of S such that $g(T) \subset T$ is G-**stable**.

2.6.1 Proposition: For a group G acting on a set S, and for a G-stable subset T of S, in fact $g(T) = T$ for all $g \in G$.

Proof: We have
$$T = eT = (gg^{-1})T = g(g^{-1}(T)) \subset g(T) \subset T$$

Thus, all the inclusions must be equalities.   ///

A single element $s_o \in S$ such that $gs_o = s_o$ for *all* $g \in G$ is G-**fixed**. Given an element $s_o \in S$, the **stabilizer** of s_o in G, often denoted G_{s_o}, is

$$G_{s_o} = \{g \in G : gs_o = s_o\}$$

More generally, for a subset T of S, the **stabilizer** of T in G is

$$\text{stabilizer of } T \text{ in } G = \{g \in G : g(T) = T\}$$

The **point-wise fixer** or **istropy subgroup** of a subset T is

$$\text{isotropy subgroup of } T = \text{point-wise fixer of } T \text{ in } G = \{g \in G : gt = t \text{ for all } t \in T\}$$

For a subgroup H of G, the **fixed points** S^H of H on S are the elements of the set

$$\text{fixed point set of } H = \{s \in S : hs = s \text{ for all } h \in H\}$$

2.6.2 Remark: In contrast to the situation of the previous proposition, if we attempt to define the stabilizer of a subset by the weaker condition $g(T) \subset T$, the following proposition can fail (for infinite sets S).

2.6.3 Proposition: Let G act on a set S, and let T be a subset of S. Then both the stabilizer and point-wise fixer of T in G are *subgroups* of G.

Proof: We only prove that the stabilizer of T is stable under inverses. Suppose $gT = T$. Then

$$g^{-1}T = g^{-1}(g(T)) = (g^{-1}g)(T) = e(T) = T$$

since $g(T) = T$. ///

With an action of G on the set S, a G-**orbit** in S is a *non-empty* G-stable subset of S, on which G is *transitive*.

2.6.4 Proposition: Let G act on a set S. For any element s_o in an orbit O of G on S,

$$O = G \cdot s_o = \{gs_o : g \in G\}$$

Conversely, for any $s_o \in S$, the set $G \cdot s_o$ is a G-orbit on S.

Proof: Since an orbit O is required to be non-empty, O contains an element s_o. Since O is G-stable, certainly $gs_o \in S$ for all $g \in G$. Since G is transitive on O, the collection of all images gs_o of s_o by elements $g \in G$ must be the whole orbit O. On the other hand, any set

$$Gs_o = \{gs_o : g \in G\}$$

is G-stable, since $h(gs_o) = (hg)s_o$. And certainly G is transitive on such a set. ///

Now we come to some consequences for counting problems. [14]

2.6.5 Proposition: Let G act transitively on a (non-empty) set S, and fix $s \in S$. Then S is in bijection with the set G/G_s of cosets gG_s of the isotropy group G_s of s in G, by

$$gs \longleftrightarrow gG_s$$

Thus,

$$\operatorname{card} S = [G : G_s]$$

Proof: If $hG_s = gG_s$, then there is $x \in G_s$ such that $h = gx$, and $hs = gxs = gs$. On the other hand, if $hs = gs$, then $g^{-1}hs = s$, so $g^{-1}h \in G_s$, and then $h \in gG_s$. ///

2.6.6 Corollary: *(Counting formula)* Let G be a finite group acting on a finite set S. Let X be the set of G-orbits in S. For $O \in X$ let $s_O \in O$. And

$$\operatorname{card} S = \sum_{O \in X} \operatorname{card} O = \sum_{O \in X} [G : G_{s_O}]$$

Proof: The set S is a disjoint union of the G-orbits in it, so the cardinality of S is the sum of the cardinalities of the orbits. The cardinality of each orbit is the index of the isotropy group of a chosen element in it, by the previous proposition. ///

---

[14] Yes, these look boring and innocent, in this abstraction.

Two fundamental examples of natural group actions are the following.

2.6.7 Example: A group G acts on itself (as a set) by **conjugation**: [15]   for $g, x \in G$,

$$\text{conjugate of } x \text{ by } g = gxg^{-1}$$

It is easy to verify that for fixed $g \in G$, the map

$$x \longrightarrow gxg^{-1}$$

is an isomorphism of G to itself. For x and y elements of G in the same G-orbit under this action, say that x and y **are conjugate**. The orbits of G on itself with the conjugation action are **conjugacy classes** (of elements). The **center** of a group G is the set of elements z whose orbit under conjugation is just $\{z\}$. That is,

$$\text{center of } G = \{z \in G : gz = zg \text{ for all } g \in G\}$$

Either directly or from general principles (above), the center Z of a group G is a *subgroup* of G. Further, it is *normal*:

$$gZg^{-1} = \{gzg^{-1} : z \in Z\} = \{z : z \in Z\} = Z$$

And of course the center is itself an *abelian* group.

2.6.8 Example: For a subgroup H of G and for $g \in G$, the **conjugate** subgroup gHg^{-1} is

$$gHg^{-1} = \{ghg^{-1} : h \in H\}$$

Thus, G acts on the set of its own subgroups by conjugation. [16]   As with the element-wise conjugation action, for H and K subgroups of G in the same G-orbit under this action, say that H and K **are conjugate**. The orbits of G on its subgroups with the conjugation action are **conjugacy classes** (of subgroups). The *fixed points* of G under the conjugation action on subgroups are just the *normal* subgroups. On the other hand, for a given subgroup H, the isotropy subgroup in G for the conjugation action is called the **normalizer** of H in G:

$$\text{normalizer of } H \text{ in } G = \{g \in G : gHg^{-1} = H\}$$

Either directly or from more general principles (above), the normalizer of H in G is a subgroup of G (containing H).

2.7 *The Sylow theorem*

There is not much that one can say about the subgroups of an arbitrary finite group. Lagrange's theorem is the simplest very general assertion. Sylow's theorem is perhaps the strongest and most useful relatively elementary result limiting the possibilities for subgroups and, therefore, for finite groups.

---

[15] It is obviously not wise to use the notation gh for ghg^{-1}.

[16] And, again, it is manifestly unwise to write gH for gH^{-1}.

Let p be a prime. A p-**group** is a finite group whose order is a power of the prime p. Let G be a finite group. Let p^e be the largest power of p dividing the order of G. A p-**Sylow** subgroup (if it exists) is a subgroup of G of order p^e.

2.7.1 Remark: By Lagrange's theorem, no larger power of p can divide the order of *any* subgroup of G.

2.7.2 Theorem: Let p be a prime. Let G be a finite group. Let p^e be the largest power of p dividing the order of G. Then
• G has p-Sylow subgroups.
• Every subgroup of G with order a power of p lies inside a p-Sylow subgroup of G.
• The number n_p of p-Sylow subgroups satisfies

$$n_p | \mathrm{order}(G) \qquad n_p = 1 \bmod p$$

• Any two p-Sylow subgroups P and Q are **conjugate**. [17]
• A group of order p^n has a non-trivial center.

It is convenient to prove a much weaker and simpler result first, which also illustrates a style of induction via subgroups and quotients:

2.7.3 Lemma: Let A be a finite *abelian* group, and let p be a prime dividing the order of A. Then there is an element a of A of order exactly p. Thus, there exists a subgroup of A of order p.

Proof: (*of lemma*) Use induction on the order of A. If the order is p exactly, then any non-identity element is of order p. Since a prime divides its order, A is not the trivial group, so we can choose a non-identity element g of A. By Lagrange, the order n of g divides the order of A. If p divides n, then $g^{n/p}$ is of order exactly p and we're done. So suppose that p does *not* divide the order of g. Then consider the quotient

$$q(A) = B = A/\langle g \rangle$$

of A by the cyclic subgroup generated by g. The order of B is still divisible by p (since $| < \langle g \rangle |$ is not), so by induction on order there is an element y in B of order exactly p. Let x be any element in A which maps to y under the quotient map $q : A \longrightarrow B$. Let N be the order of x. The prime p divides N, or else write $N = \ell p + r$ with $0 < r < p$, and

$$e_B = q(e_A) = q(x^N) = y^N = y^{\ell p + r} = y^r \neq e_B$$

contradiction. Then $x^{N/p}$ has order exactly p, and the cyclic subgroup generated by $x^{N/p}$ has order p. ///

Proof: Now prove the theorem. First, we prove *existence* of p-Sylow subgroups by induction on the exponent e of the power p^e of p dividing the order of G. (Really, the induction uses subgroups and quotient groups of G.) If $e = 0$ the p-Sylow subgroup is the trivial subgroup, and there is nothing to prove. For fixed $e > 1$, do induction on the order of the group G. If any proper subgroup H of G has order divisible by p^e, then invoke the theorem for H, and a p-Sylow subgroup of H is one for G. So suppose that *no* proper subgroup of G has

---

[17] This property is the sharpest and most surprising assertion here.

order divisible by p^e. Then for any subgroup H of G the prime p divides $[G : H]$. By the *counting formula* above, using the conjugation action of G on itself,

$$\operatorname{card} G = \sum_x [G : G_x]$$

where x is summed over (irredundant) representatives for conjugacy classes. Let Z be the center of G. Then Z consists of G-orbits each with a single element. We rewrite the counting formula as

$$\operatorname{card} G = \operatorname{card} Z + \sum_x^{\text{non-central}} [G : G_x]$$

where now x is summed over representatives *not* in the center. For non-central x the isotropy group G_x is a proper subgroup, so by assumption, p divides $[G : G_x]$ for all x. Since p divides the order of G, we conclude from the counting formula that p divides the order of the center Z (but p^e does not divide the order of Z). Using the lemma above, let A be a subgroup of Z of order p. Since A is inside the center it is still normal. Consider the quotient group $H = G/A$, with quotient map $q : A \longrightarrow B$. The power of p dividing the order of H is p^{e-1}, strictly smaller than p^e. By induction, let Q be a p-Sylow subgroup of H, and let $P = q^{-1}(Q)$ be the inverse image of Q under the quotient map q. Then

$$|P| = |q^{-1}(Q)| = |\ker q| \cdot |Q| = p \cdot p^{e-1} = p^e$$

from the counting corollary of the isomorphism theorem (above). Thus, G does have a p-Sylow theorem after all.

If it happens that $|G| = p^e$, looking at that same formula

$$\operatorname{card} G = \operatorname{card} Z + \sum_x^{\text{non-central}} [G : G_x]$$

the left-hand side is p^e, and all the summands corresponding to non-central conjugacy classes are divisible by p, so the order of the center is divisible by p. That is, p-power-order groups have non-trivial centers.

Let X be *any* G-conjugation stable set of p-Sylow subgroups. Fix a p-power-order subgroup Q not necessarily in X, and let Q act on the set X by conjugation. The counting formula gives

$$\operatorname{card} X = \sum_x [Q : Q_x]$$

where x runs over representatives for Q-conjugacy classes in X. If Q normalized a p-Sylow subgroup x not containing Q, then

$$H = Q \cdot x$$

would be a subgroup of order

$$|Q \cdot x| = \frac{|Q| \cdot |x|}{|Q \cap x|} > |x|$$

and would be a power of p, contradicting the maximality of x. Thus, the only p-Sylow subgroups normalized by any p-power-order subgroup Q are those containing Q. Thus, except for x containing Q, all the indices $[Q : Q_x]$ are divisible by p. Thus,

$$|X| = |\{x \in X : Q \subset x\}| + \sum_x^{Q \not\subset x} [Q : Q_x]$$

In the case that Q itself is a p-Sylow subgroup, and X is *all* p-Sylow subgroups in G,

$$|\{x \in X : Q \subset x\}| = |\{Q\}| = 1$$

so the number of *all* p-Sylow subgroups is 1 mod p.

Next, let X consist of a single G-conjugacy class of p-Sylow subgroups. Fix $x \in X$. Since X is a single orbit,

$$|X| = [G : G_x]$$

and the latter index is *not* divisible by p, since the normalizer G_x of x contains x. Let a p-power-subgroup Q act by conjugation on X. In the counting formula

$$|X| = |\{x \in X : Q \subset x\}| + \sum_{x}^{Q \not\subset x} [Q : Q_x]$$

all the indices $[Q : Q_x]$ are divisible by p, but $|X|$ is *not*, so

$$|\{x \in X : Q \subset x\}| \neq 0$$

That is, given a p-power-order subgroup Q, *every* G-conjugacy class of p-Sylow subgroups contains an x containing Q. This is only possible if there is a *unique* G-conjugacy class of p-Sylow subgroups. That is, the conjugation action of G is *transitive* on p-Sylow subgroups.

Further, since Q was not necessarily maximal in this last discussion, we have shown that every p-power-order subgroup of G lies inside at least one p-Sylow subgroup.

And, fixing a single p-Sylow subgroup x, using the transitivity, the number of *all* p-Sylow subgroups is

$$\text{number } p\text{-Sylow subgroups} = [G : G_x] = |G|/|G_x|$$

This proves the divisibility property. ///

2.7.4 Remark: For general integers d dividing the order of a finite group G, it is seldom the case that there is a subgroup of G of order d. By contrast, if G is *cyclic* there is a *unique* subgroup for every divisor of its order. If G is *abelian* there is *at least one* subgroup for every divisor of its order.

2.7.5 Remark: About the proof of the Sylow theorem: once one knows that the proof uses the conjugation action on elements and on subgroups, there are not so many possible directions the proof could go. Knowing these limitations on the proof methodology, one could hit on a correct proof after a relatively small amount of trial and error.

2.8 *Trying to classify finite groups, part I*

Lagrange's theorem and the Sylow theorem allow us to make non-trivial progress on the project of classifying finite groups whose orders have relatively few prime factors. That is, we can prove that there are not many non-isomorphic groups of such orders, sometimes a single isomorphism class for a given order. This sort of result, proving that an abstraction miraculously allows fewer instances than one might have imagined, is often a happy and useful result.

Groups of prime order: Let p be a prime, and suppose that G is a group with $|G| = p$. Then by Lagrange's theorem there are no proper subgroups of G. Thus, picking any element g of G other than the identity, the (cyclic) subgroup $\langle g \rangle$ generated by g is necessarily the whole group G. That is, for such groups G, choice of a non-identity element g yields

$$G = \langle g \rangle \approx \mathbb{Z}/p$$

Groups of order pq, **part I:** Let $p < q$ be primes, and suppose that G is a group with $|G| = pq$. Sylow's theorem assures that there exist subgroups P and Q of orders p and q, respectively. By Lagrange, the order of $P \cap Q$ divides both p and q, so is necessarily 1. Thus,

$$P \cap Q = \{e\}$$

Further, the number n_q of q-Sylow subgroups must be 1 mod q and also divide the order pq of the group. Since $q = 0$ mod q, the only possibilities (since p is prime) are that either $n_q = p$ or $n_q = 1$. But $p < q$ precludes the possibility that $n_q = p$, so $n_q = 1$. That is, with $p < q$, the q-Sylow subgroup is necessarily *normal.*

The same argument, apart from the final conclusion invoking $p < q$, shows that the number n_p of p-Sylow subgroups is either $n_p = 1$ or $n_p = q$, and (by Sylow) is $n_p = 1$ mod p. But now $p < q$ does *not* yield $n_p = 1$. There are two cases, $q = 1$ mod p and otherwise.

If $q \neq 1$ mod p, then we *can* reach the conclusion that $n_p = 1$, that is, that the p-Sylow subgroup is also normal. Thus, for $p < q$ and $q \neq 1$ mod p, we have a normal p-Sylow group P and a normal q-Sylow subgroup Q. Again, $P \cap Q = \{e\}$ from Lagrange.

How to reconstruct G from such facts about its subgroups?

We need to enrich our vocabulary: given two groups G and H, the **(direct) product** group $G \times H$ is the cartesian product with the operation

$$(g, h) \cdot (g', h') = (gg', hh')$$

(It is easy to verify the group properties.) For G and H abelian, with group operations written as addition, often the direct product is written instead as a **(direct) sum** [18]

$$G \oplus H$$

2.8.1 Proposition: Let A and B be normal [19] subgroups of a group G, such that $A \cap B = \{e\}$. Then

$$f : A \times B \longrightarrow A \cdot B = \{ab : a \in A, b \in B\}$$

by

$$f(a, b) = ab$$

is an isomorphism. The subgroup $A \cdot B \approx A \times B$ is a normal subgroup of G. In particular, $ab = ba$ for all $a \in A$ and $b \in B$.

---

[18] Eventually we will make some important distinctions between direct sums and direct products, but there is no need to do so just now.

[19] Unless at least one of the subgroups is normal, the set $A \cdot B$ may not even be a subgroup, much less normal.

Proof: The trick is to consider **commutator** expressions

$$aba^{-1}b^{-1} = aba^{-1} \cdot b^{-1} = a \cdot ba^{-1}b^{-1}$$

for $a \in A$ and $b \in B$. Since B is normal, the second expression is in B. Since A is normal, the third expression is in A. Thus, the commutator $aba^{-1}b^{-1}$ is in $A \cap B$, which is $\{e\}$. [20] Thus, right multiplying by b

$$aba^{-1} = b$$

or, right multiplying further by a,

$$ab = ba$$

The fact that $ab = ba$ for all $a \in A$ and all $in B$ allows one to easily show that f is a group homomorphism. Its kernel is trivial, since $ab = e$ implies

$$a = b^{-1} \in A \cap B = \{e\}$$

Thus, the map is injective, from earlier discussions. Now $|A \times B| = pq$, and the map is injective, so $f(A \times B)$ is a subgroup of G with order pq. Thus, the image is all of G. That is, f is an isomorphism. ///

2.8.2 Proposition: Let A and B by cyclic groups of relatively prime orders m and n. Then $A \times B$ is cyclic of order mn. In particular, for a a generator for A and b a generator for B, (a, b) is a generator for the product.

Proof: Let N be the least positive integer such that $N(a, b) = (e_A, e_B)$. Then $Na = e_A$, so $|a|$ divides N. Similarly, $|b|$ divides N. Since $|a|$ and $|b|$ are relatively prime, this implies that their product divides N. ///

2.8.3 Corollary: For $|G| = pq$ with $p < q$ and $q \neq 1 \bmod p$, G is cyclic of order pq. Hence, in particular, there is only *one* isomorphism class of groups of such orders pq.
///

2.8.4 Remark: Even without the condition $q \neq 1 \bmod p$, we do have the cyclic group $\mathbb{Z}/pq$ of order pq, but without that condition we cannot prove that there are no *other* groups of order pq. [21] We'll delay treatment of $|G| = pq$ with primes $p < q$ and $q = 1 \bmod p$ till after some simpler examples are treated.

2.8.5 Example: Groups of order $15 = 3 \cdot 5$, or order $35 = 5 \cdot 7$, of order $65 = 5 \cdot 13$, etc., are necessarily cyclic of that order. By contrast, we reach no such conclusion about groups of order $6 = 2 \cdot 4$, $21 = 3 \cdot 7$, $55 = 5 \cdot 11$, etc. [22]

Groups G of order pqr with distinct primes p, q, r: By Sylow, there is a p-Sylow subgroup P, a q-Sylow subgroup Q, and an r-Sylow subgroup R. Without any further assumptions, we cannot conclude anything about the normality of any of these Sylow

---

[20] By Lagrange, again. Very soon we will tire of explicit invocation of Lagrange's theorem, and let it go without saying.

[21] And, indeed, there *are* non-cyclic groups of those orders.

[22] And, again, there *are* non-cyclic groups of such orders.

subgroups, by contrast to the case where the order was pq, wherein the Sylow subgroup for the larger of the two primes was invariably normal.

One set of hypotheses which allows a simple conclusion is

$$q \neq 1 \bmod p \quad r \neq 1 \bmod p \quad qr \neq 1 \bmod p$$
$$p \neq 1 \bmod q \quad r \neq 1 \bmod q \quad pr \neq 1 \bmod q$$
$$p \neq 1 \bmod r \quad q \neq 1 \bmod r \quad pq \neq 1 \bmod r$$

These conditions would suffice to prove that all of P, Q, and R are normal. Then the little propositions above prove that $P \cdot Q$ is a normal cyclic subgroup of order pq, and then (since still pq and r are relatively prime) that $(PQ) \cdot R$ is a cyclic subgroup of order pqr, so must be the whole group G. That is, G is cyclic of order pqr.

Groups of order pq, **part II:** Let $p < q$ be primes, and now treat the case that $q = 1 \bmod p$, so that a group G of order pq need *not* be cyclic. Still, we know that the q-Sylow subgroup Q is *normal*. Thus, for each x in a fixed p-Sylow subgroup P, we have a map $a_x : Q \longrightarrow Q$ defined by

$$a_x(y) = xyx^{-1}$$

Once the normality of Q assures that this really does map back to Q, it is visibly an isomorphism of Q to itself. This introduces:

An isomorphism of a group to itself is an **automorphism**. [23] The **group of automorphisms** of a group G is

$$\mathrm{Aut}(G) = \mathrm{Aut}G = \{\text{isomorphisms } G \longrightarrow G\}$$

It is easy to check that $\mathrm{Aut}(G)$ is indeed a group, with operation being the composition of maps, and identity being the identity map 1_G defined by [24]

$$1_G(g) = g \quad \text{(for any } g \text{ in } G)$$

In general it is a non-trivial matter to determine in tangible terms the automorphism group of a given group, but we have a simple case:

2.8.6 Proposition:
$$\mathrm{Aut}(\mathbb{Z}/n) \approx (\mathbb{Z}/n)^\times$$

by defining, for each $z \in (\mathbb{Z}/n)^\times$, and for $x \in \mathbb{Z}/n$,

$$f_z(x) = zx$$

On the other hand, given an automorphism f, taking $z = f(1)$ gives $f_z = f$.

Proof: For z multiplicatively invertible mod n, since the addition and multiplication in $\mathbb{Z}/n$ enjoy a distributive property, f_z is an automorphism of $\mathbb{Z}/n$ to itself. On the other

---

[23] A homomorphism that is not necessarily an isomorphism of a group to itself is an *endomorphism*.

[24] This should be expected.

hand, given an automorphism f of $\mathbb{Z}/n$, let $z = f(1)$. Then, indeed, identifying x in $\mathbb{Z}/n$ with an ordinary integer,

$$f(x) = f(x \cdot 1) = f(\underbrace{1 + \ldots + 1}_{x}) = \underbrace{f(1) + \ldots + f(1)}_{x} = \underbrace{z + \ldots + z}_{x} = zx$$

That is, *every* automorphism is of this form. ///

Given two groups H and N, with a group homomorphism

$$f : H \longrightarrow \text{Aut}(N) \quad \text{denoted} \quad h \longrightarrow f_h$$

the **semi-direct product** group

$$H \times_f N$$

is the set $H \times N$ with group operation intending to express the idea that

$$hnh^{-1} = f_h(n)$$

But since H and N are not literally subgroups of anything yet, we must say, instead, that we want

$$(h, e_N)(e_H, n)(h^{-1}, e_N) = (e_H, f_h(n))$$

After some experimentation, one might decide upon the definition of the operation

$$(h, n) \cdot (h', n') = (hh', f_{h'^{-1}}(n)\, n')$$

Of course, when f is the trivial homomorphism (sending everything to the identity) the semi-direct product is simply the direct product of the two groups.

2.8.7 Proposition: With a group homomorphism $f : H \longrightarrow \text{Aut}(N)$, the semi-direct product $H \times_f N$ is a group. The maps $h \longrightarrow (h, e_N)$ and $n \longrightarrow (e_H, n)$ inject H and N, respectively, and the image of N is normal.

Proof: [25] The most annoying part of the argument would be proof of associativity. On one hand,

$$((h, n)(h', n'))\, (h'', n'') = (hh', f_{h'^{-1}}(n)n')\, (h'', n'') = (hh'h'', f_{h''^{-1}}\, (f_{h'^{-1}}(n)n')\, n'')$$

The H-component is uninteresting, so we only look at the N-component:

$$f_{h''^{-1}}\, (f_{h'^{-1}}(n)n')\, n'' = f_{h''^{-1}} \circ f_{h'^{-1}}(n) \cdot f_{h''^{-1}}(n') \cdot n'' = f_{(h'h'')^{-1}}(n) \cdot f_{h''^{-1}}(n') \cdot n''$$

which is the N-component which would arise from

$$(h, n)\, ((h', n')(h'', n''))$$

This proves the associativity. The other assertions are simpler. ///

———————————

[25] There is nothing surprising in this argument. It amounts to checking what must be checked, and there is no obstacle other than bookkeeping. It is surely best to go through it oneself rather than watch someone else do it, but we write it out here just to prove that it is possible to force oneself to carry out some of the details.

Thus, in the $|G| = pq$ situation, because Q is normal, we have a group homomorphism

$$\mathbb{Z}/p \approx P \longrightarrow \text{Aut}(Q) \approx (\mathbb{Z}/q)^\times$$

The latter is of order $q - 1$, so unless $p|(q - 1)$ this homomorphism must have trivial image, that is, P and Q commute, giving yet another approach to the case that $q \neq 1 \bmod p$. But for $p|(q - 1)$ there is at least one non-trivial homomorphism to the automorphism group: $(\mathbb{Z}/q)^\times$ is an abelian group of order divisible by p, so there exists [26] an element z of order p. Then take $f : \mathbb{Z}/p \longrightarrow \text{Aut}(\mathbb{Z}/q)$ by

$$f(x)(y) = z^x \cdot y \in \mathbb{Z}/q$$

This gives a semi-direct product of $\mathbb{Z}/p$ and $\mathbb{Z}/q$ which cannot be abelian, since elements of the copy P of $\mathbb{Z}/p$ and the copy Q of $\mathbb{Z}/q$ do not all commute with each other. That is, there is at least one non-abelian [27] group of order pq if $q = 1 \bmod p$.

How many different semi-direct products are there? [28]   Now we must use the non-trivial fact that $(\mathbb{Z}/q)^\times$ is *cyclic* for q prime. [29]   That is, granting this cyclicness, there are *exactly* $p - 1$ elements in $(\mathbb{Z}/q)^\times$ of order p. Thus, given a fixed element z of order p in $(\mathbb{Z}/q)^\times$, any other element of order p is a power of z.

Luckily, this means that, given a choice of isomorphism $i : \mathbb{Z}/p \approx P$ to the p-Sylow group P, and given non-trivial $f : P \longrightarrow \text{Aut}(Q)$, whatever the image $f(i(1))$ may be, we can alter the choice of i to achieve the effect that

$$f(i(1)) = z$$

Specifically, if at the outset

$$f(i(1)) = z'$$

with some other element z' of order p, use the cyclicness to find an integer ℓ (in the range $1, 2, \ldots, p - 1$) such that

$$z' = z^\ell$$

Since ℓ is prime to p, it has an inverse modulo $k \bmod p$. Then

$$f(i(k)) = f(k \cdot i(1)) = f(i(1))^k = (z')^k = (z^\ell)^k = z$$

since $\ell k = 1 \bmod p$ and z has order p.

In summary, with primes $q = 1 \bmod p$, up to isomorphism there is a *unique* non-abelian group of order pq, and it is a semi-direct product of $\mathbb{Z}/p$ and $\mathbb{Z}/q$. [30]

---

[26] Existence follows with or without use of the fact that there are primitive roots modulo primes. For small numerical examples this cyclicness can be verified directly, without necessarily appealing to any theorem that guarantees it.

[27] So surely non-cyclic.

[28] As usual in this context, *different* means *non-isomorphic*.

[29] This is the existence of *primitive roots* modulo primes.

[30] The argument that showed that seemingly different choices yield isomorphic groups is an ad hoc example of a wider problem, of classification up to isomorphism of *group extensions*.

Groups of order p^2, with prime p: A different aspect of the argument of the Sylow theorem is that a p-power-order group G necessarily has a non-trivial center Z. If $Z = G$ we have proven that G is abelian. Suppose Z is proper. Then [31] it is of order p, thus [32] necessarily *cyclic*, with generator z. Let x be any other group element *not* in Z. It cannot be that the order of x is p^2, or else $G = \langle x \rangle$ and $G = Z$, contrary to hypothesis. Thus, [33] the order of x is p, and [34]

$$\langle x \rangle \cap \langle z \rangle = \{e\}$$

Abstracting the situation just slightly:

2.8.8 Proposition: [35] Let G be a finite group with center Z and a subgroup A of Z. Let B be another abelian subgroup of G, such that $A \cap B = \{e\}$ and $A \cdot B = G$. Then the map

$$f : A \times B \longrightarrow A \cdot B$$

by

$$a \times b \longrightarrow ab$$

is an isomorphism, and $A \cdot B$ is abelian.

Proof: If $a \times b$ were in the kernel of f, then $ab = e$, and

$$a = b^{-1} \in A \cap B = \{e\}$$

And

$$f((a, b) \cdot (a', b')) = f(aa', bb') = aa'bb'$$

while

$$f(a, b) \cdot f(a', b') = ab \cdot a'b' = aa'bb'$$

because $ba' = a'b$, because elements of A commute with everything. That is, f is a homomorphism. Since both A and B are abelian, certainly the product is. ///

That is, any group G of order p^2 (with p prime) is abelian. So our supposition that the center of such G is of order only p is false.

Starting over, but knowing that G of order p^2 is abelian, if there is no element of order p^2 in G (so G is not cyclic), then in any case there is an element z of order p. [36] And take x not in $\langle z \rangle$. Necessarily x is of order p. By the same clichéd sort of argument as in the last proposition, $\langle x \rangle \cap \langle z \rangle = \{e\}$ and

$$\mathbb{Z}/p \times \mathbb{Z}/p \approx \langle x \rangle \times \langle z \rangle \approx \langle x \rangle \cdot \langle z \rangle = G$$

That is, *non-cyclic* groups of order p^2 are isomorphic to $\mathbb{Z}/p \times \mathbb{Z}/p$.

---

[31] Lagrange

[32] Lagrange

[33] Lagrange

[34] Lagrange

[35] This is yet another of an endless stream of variations on a theme.

[36] For example, this follows from the lemma preparatory to the Sylow theorem.

Automorphisms of groups of order q^2: Anticipating that we'll look at groups of order pq^2 with normal subgroups of order q^2, to understand semi-direct products $P \times_f Q$ with P of order p and Q of order q^2 we must have *some* understanding of the automorphism groups of groups of order q^2, since $f : P \longrightarrow \operatorname{Aut} Q$ determines the group structure. For the moment we will focus on merely the *order* of these automorphism groups. [37]

For Q cyclic of order q^2, we know that $Q \approx \mathbb{Z}/q^2$, and from above

$$\operatorname{Aut} Q \approx \operatorname{Aut}(\mathbb{Z}/q^2) \approx (\mathbb{Z}/q^2)^\times$$

In particular, the *order* is [38]

$$|\operatorname{Aut} Q| = \operatorname{card}(\mathbb{Z}/q^2)^\times = \varphi(q^2) = q(q-1)$$

This is easy.

For Q non-cyclic of order q^2, we saw that

$$Q \approx \mathbb{Z}/q \oplus \mathbb{Z}/q$$

where we write direct sum to emphasize the abelian-ness of Q. [39]   For the moment we only aim to *count* these automorphisms. Observe that [40]

$$(\bar{x}, \bar{y}) = x \cdot (\bar{1}, \bar{0}) + y \cdot (\bar{0}, \bar{1})$$

for any $x, y \in \mathbb{Z}$, where for the moment the bars denotes residue classes modulo q. Thus, for any automorphism α of Q

$$\alpha(\bar{x}, \bar{y}) = x \cdot \alpha(\bar{1}, \bar{0}) + y \cdot \alpha(\bar{0}, \bar{1})$$

where multiplication by an integer is repeated addition. Thus, the images of $(\bar{1}, \bar{0})$ and $(\bar{0}, \bar{1})$ determine α completely. And, similarly, *any* choice of the two images gives a group homomorphism of Q to itself. The only issue is to avoid having a proper kernel. To achieve this, $\alpha(\bar{1}, \bar{0})$ certainly must not be $e \in Q$, so there remain $q^2 - 1$ possible choices for the image of $(\bar{1}, \bar{0})$. Slightly more subtly, $\alpha(\bar{0}, \bar{1})$ must not lie in the cyclic subgroup generated by $\alpha(\bar{1}, \bar{0})$, which excludes exactly q possibilities, leaving $q^2 - q$ possibilities for $\alpha(\bar{0}, \bar{1})$ for each choice of $\alpha(\bar{1}, \bar{0})$. Thus, altogether,

$$\operatorname{card} \operatorname{Aut}(\mathbb{Z}/q \oplus \mathbb{Z}/q) = (q^2 - 1)(q^2 - q)$$

We will pursue this later.

Groups of order pq^2: As in the simpler examples, the game is to find some mild hypotheses that combine with the Sylow theorem to limit the possibilities for the arrangement of Sylow

---

[37] After some further preparation concerning finite fields and linear algebra we can say more definitive structural things.

[38] Using Euler's totient function φ.

[39] Thus, in fact, Q is a two-dimensional vector space over the finite field $\mathbb{Z}/q$. We will more systematically pursue this viewpoint shortly.

[40] Yes, this is linear algebra.

subgroups, and then to look at direct product or semi-direct product structures that can arise. Let G be a group of order pq^2 with p, q distinct primes.

As a preliminary remark, if G is assumed abelian, then G is necessarily the direct product of a p-Sylow subgroup and a q-Sylow subgroup (both of which are necessarily norml), and by the classification of order q^2 groups this gives possibilities

$$G = P \cdot Q \approx P \times Q \approx \mathbb{Z}/p \times Q \approx \begin{cases} \mathbb{Z}/p \oplus \mathbb{Z}/q^2 & \approx & \mathbb{Z}/pq^2 \\ \mathbb{Z}/p \oplus \mathbb{Z}/q \oplus \mathbb{Z}/q & \approx & \mathbb{Z}/q \oplus \mathbb{Z}/pq \end{cases}$$

in the two cases for Q, writing direct sums to emphasize the abelian-ness. So now we consider non-abelian possibilities, and/or hypotheses which force a return to the abelian situation.

The first and simplest case is that neither $p = 1 \bmod q$ nor $q^2 = 1 \bmod p$. Then, by Sylow, there is a unique q-Sylow subgroup Q and a unique p-Sylow subgroup P, both necessarily normal. We just saw that the group Q of order q^2 is necessarily [41] abelian. Since both subgroups are normal, elements of Q commute with elements of P. [42] This returns us to the abelian case, above.

A second case is that $p|(q-1)$. This implies that $p < q$. The number n_q of q-Sylow subgroups is 1 mod q and divides pq^2, so is either 1 or p, but $p < q$, so necessarily $n_q = 1$. That is, the q-Sylow subgroup Q is normal. But this does not follow for the p-Sylow subgroup, since now $p|(q-1)$. The Sylow theorem would seemingly allow the number n_p of p-Sylow subgroups to be 1, q, or q^2. Thus, we should consider the possible semi-direct products

$$\mathbb{Z}/p \times_f Q$$

for

$$f : \mathbb{Z}/p \longrightarrow \operatorname{Aut} Q$$

If f is the trivial homomorphism, then we obtain a direct product, returning to the abelian case. For $Q \approx \mathbb{Z}/q^2$ its automorphism group has order $q(q-1)$, which is divisible by p (by hypothesis), so [43] has an element of order p. That is, there does exist a non-trivial homomorphism

$$f : P \approx \mathbb{Z}/p \longrightarrow \operatorname{Aut} \mathbb{Z}/q^2$$

That is, there do exist non-abelian semi-direct products

$$\mathbb{Z}/p \times_f \mathbb{Z}/q^2$$

Distinguishing the isomorphism classes among these is similar to the case of groups of order pq with $p|(q-1)$, and we would find just a single non-abelian isomorphism class. For $\mathbb{Q} \approx \mathbb{Z}/q \oplus \mathbb{Z}/q$, we saw above that

$$|\operatorname{Aut}(\mathbb{Z}/q \oplus \mathbb{Z}/q)| = (q^2 - 1)(q^2 - q) = (q-1)^2 q (q+1)$$

[41] As a different sort of corollary of Sylow.

[42] The earlier argument is worth repeating: for a in one and b in another of two normal subgroups with trivial intersection, $(aba^{-1})b^{-1} = a(ba^{-1})b^{-1}$ must lie in both, so is e. Then $ab = ba$.

[43] By the lemma preceding the Sylow theorem, for example. In fact, all primes q the group $(\mathbb{Z}/q^2)^\times$ is *cyclic*, so will have *exactly one* subgroup of order p.

Thus, there is at least one non-abelian semi-direct product

$$\mathbb{Z}/p \times_f (\mathbb{Z}/q \oplus \mathbb{Z}/q)$$

Attempting to count the isomorphism classes would require that we have more information on the automorphism group, which we'll obtain a little later.

A third case is that $p|(q+1)$. This again implies that $p < q$, except in the case that $q = 2$ and $p = 3$, which we'll ignore for the moment. The number n_q of q-Sylow subgroups is 1 mod q and divides pq^2, so is either 1 or p, but $p < q$, so necessarily $n_q = 1$. That is, the q-Sylow subgroup Q is normal. But this does not follow for the p-Sylow subgroup, since now $p|(q+1)$. The Sylow theorem would seemingly allow the number n_p of p-Sylow subgroups to be 1 or q^2. Thus, we should consider the possible semi-direct products

$$\mathbb{Z}/p \times_f Q$$

for

$$f : \mathbb{Z}/p \longrightarrow \operatorname{Aut} Q$$

If f is the trivial homomorphism, then we obtain a direct product, returning to the abelian case. For $Q \approx \mathbb{Z}/q^2$ its automorphism group has order $q(q-1)$, which is not divisible by p, since $p|(q+1)$. That is, for $Q \approx \mathbb{Z}/q^2$ and $p|(q+1)$ and $q > 2$, there is *no* non-trivial homomorphism

$$f : P \approx \mathbb{Z}/p \longrightarrow \operatorname{Aut} \mathbb{Z}/q^2$$

That is, in this case there is *no* non-abelian semi-direct product

$$\mathbb{Z}/p \times_f \mathbb{Z}/q^2$$

For $\mathbb{Q} \approx \mathbb{Z}/q \oplus \mathbb{Z}/q$, we saw above that

$$|\operatorname{Aut}(\mathbb{Z}/q \oplus \mathbb{Z}/q)| = (q^2 - 1)(q^2 - q) = (q-1)^2 \, q \, (q+1)$$

Thus, there is at least one non-abelian semi-direct product

$$\mathbb{Z}/p \times_f (\mathbb{Z}/q \oplus \mathbb{Z}/q)$$

Again, attempting to count the isomorphism classes would require that we have more information on the automorphism group.

A fourth case is that $q|(p-1)$. Thus, by the same arguments as above, the p-Sylow subgroup P is normal, but the q-Sylow subgroup Q might not be. There are non-trivial homomorphisms in both cases

$$f : \begin{cases} \mathbb{Z}/p \oplus \mathbb{Z}/q^2 & \approx & \mathbb{Z}/pq^2 \\ \mathbb{Z}/p \oplus \mathbb{Z}/q \oplus \mathbb{Z}/q & \approx & \mathbb{Z}/q \oplus \mathbb{Z}/pq \end{cases} \longrightarrow \operatorname{Aut} \mathbb{Z}/p \approx (\mathbb{Z}/p)^\times$$

so either type of q-Sylow subgroup Q of order q^2 can give non-trivial automorphisms of the normal p-Sylow group P. Again, determining isomorphism classes in the first case is not hard, but in the second requires more information about the structure of $\operatorname{Aut}(\mathbb{Z}/q \oplus \mathbb{Z}/q)$.

2.8.9 Remark: The above discussion is fast approaching the limit of what we can deduce about finite groups based only on the prime factorization of their order. The cases of prime order and order pq with distinct primes are frequently genuinely useful, the others less so.

2.9 *Worked examples*

2.9.1 Example: Let G, H be finite groups with relatively prime orders. Show that any group homomorphism $f : G \longrightarrow H$ is necessarily trivial (that is, sends every element of G to the identity in H.)

The isomorphism theorem implies that

$$|G| = |\ker f| \cdot |f(G)|$$

In particular, $|f(G)|$ divides $|G|$. Since $f(G)$ is a subgroup of H, its order must also divide $|H|$. These two orders are relatively prime, so $|f(G)| = 1$.

2.9.2 Example: Let m and n be integers. Give a formula for an isomorphism of abelian groups

$$\frac{\mathbb{Z}}{m} \oplus \frac{\mathbb{Z}}{n} \longrightarrow \frac{\mathbb{Z}}{\gcd(m,n)} \oplus \frac{\mathbb{Z}}{\mathrm{lcm}(m,n)}$$

Let r, s be integers such that $rm + sn = \gcd(m, n)$. Let $m' = m/\gcd(m, n)$ and $n' = n/\gcd(m, n)$. Then $rm' + sn' = 1$. We claim that

$$f(a + m\mathbb{Z}, b + n\mathbb{Z}) = ((a - b) + \gcd(m, n)\mathbb{Z},\ (b \cdot rm' + a \cdot sn') + \mathrm{lcm}(m, n)\mathbb{Z})$$

is such an isomorphism. To see that it is well-defined, observe that

$$(a + m\mathbb{Z}) - (b + n\mathbb{Z}) = (a - b) + \gcd(m, n)\mathbb{Z}$$

since

$$m\mathbb{Z} + n\mathbb{Z} = \gcd(m, n)\mathbb{Z}$$

which itself follows from the facts that

$$\gcd(m, n) = rm + sn \in m\mathbb{Z} + n\mathbb{Z}$$

and (by definition) $m\mathbb{Z} \subset \gcd(m, n)\mathbb{Z}$ and $n\mathbb{Z} \subset \gcd(m, n)\mathbb{Z}$. And, similarly

$$sn' \cdot m\mathbb{Z} + rm' \cdot n\mathbb{Z} = \mathrm{lcm}(m, n)\mathbb{Z}$$

so the second component of the map is also well-defined.

Now since these things are finite, it suffices to show that the kernel is trivial. That is, suppose $b = a + k\gcd(m, n)$ for some integer k, and consider

$$b \cdot rm' + a \cdot sn'$$

The latter is

$$(a + k\gcd(m, n))rm' + a \cdot sn' = a \cdot rm' + a \cdot sn' = a \bmod m$$

since $\gcd(m, n)m' = m$ and $rm' + sn' = 1$. Symmetrically, it is $b \bmod n$. Thus, if it is 0 mod $\mathrm{lcm}(m, n)$, $a = 0 \bmod m$ and $b = 0 \bmod n$. This proves that the kernel is trivial, so the map is injective, and, because of finiteness, surjective as well.

2.9.3 Remark: I leave you the fun of guessing where the $a - b$ expression (above) comes from.

2.9.4 Example: Show that every group of order $5 \cdot 13$ is cyclic.

Invoke the Sylow theorem: the number of 5-Sylow subgroups is 1 mod 5 and also divides the order $5 \cdot 13$, so must be 1 (since 13 is not 1 mod 5). Thus, the 5-Sylow subgroup is normal. Similarly, even more easily, the 13-Sylow subgroup is normal. The intersection of the two is trivial, by Lagrange. Thus, we have two normal subgroups with trivial intersection and the product of whose orders is the order of the whole group, and conclude that the whole group is isomorphic to the (direct) product of the two, namely $\mathbb{Z}/5 \oplus \mathbb{Z}/13$. Further, this is isomorphic to $\mathbb{Z}/65$.

2.9.5 Example: Show that every group of order $5 \cdot 7^2$ is abelian.

From the classification of groups of prime-squared order, we know that there are only two (isomorphism classes of) groups of order 7^2, $\mathbb{Z}/49$ and $\mathbb{Z}/7 \oplus \mathbb{Z}/7$. From the Sylow theorem, since the number of 7-Sylow subgroups is 1 mod 7 and also divides the group order, the 7-Sylow subgroup is normal. For the same reason the 5-Sylow subgroup is normal. The intersection of the two is trivial (Lagrange). Thus, again, we have two normal subgroups with trivial intersection the product of whose orders is the group order, so the group is the direct product. Since the factor groups are abelian, so is the whole.

2.9.6 Example: Exhibit a non-abelian group of order $3 \cdot 7$.

We can construct this as a semi-direct product, since there exists a non-trivial homomorphism of $\mathbb{Z}/3$ to $\text{Aut}(\mathbb{Z}/7)$, since the latter automorphism group is isomorphic to $(\mathbb{Z}/7)^\times$, of order 6. Note that we are assured of the *existence* of a subgroup of order 3 of the latter, whether or not we demonstrate an explicit element.

2.9.7 Example: Exhibit a non-abelian group of order $5 \cdot 19^2$.

We can construct this as a semi-direct product, since there exists a non-trivial homomorphism of $\mathbb{Z}/5$ to $\text{Aut}(\mathbb{Z}/19 \oplus \mathbb{Z}/19)$, since the latter automorphism group has order $(19^2 - 1)(19^2 - 19)$, which is divisible by 5. Note that we are assured of the *existence* of a subgroup of order 5 of the latter, whether or not we demonstrate an explicit element.

2.9.8 Example: Show that every group of order $3 \cdot 5 \cdot 17$ is cyclic.

Again, the usual divisibility trick from the Sylow theorem proves that the 17-group is normal. Further, since neither 3 nor 5 divides $17 - 1 = |\text{Aut}(\mathbb{Z}/17)|$, the 17-group is *central*. But, since $3 \cdot 17 = 1$ mod 5, and $5 \cdot 17 = 1$ mod 3, we cannot immediately reach the same sort of conclusion about the 3-group and 5-group. But if *both* the 3-group and 5-group were *not* normal, then we'd have at least

$$1 + (17 - 1) + (5 - 1) \cdot 3 \cdot 17 + (3 - 1) \cdot 5 \cdot 17 = 391 > 3 \cdot 5 \cdot 17 = 255$$

elements in the group. So at least one of the two is normal. If the 5-group is normal, then the 3-group acts trivially on it by automorphisms, since 3 does not divide $5 - 1 = |\text{Aut}(\mathbb{Z}/5)|$. Then we'd have a *central* subgroup of order $5 \cdot 17$ group, and the whole group is abelian, so is cyclic by the type of arguments given earlier. Or, if the 3-group is normal, then for the same reason it is is central, so we have a central (cyclic) group of order $3 \cdot 17$, and again the whole group is cyclic.

2.9.9 Example: Do there exist 4 primes p, q, r, s such that every group of order $pqrs$ is necessarily abelian?

We want to arrange that all of the p, q, r, s Sylow subgroups P, Q, R, S are normal. Then, because the primes are distinct, still

$$P \cap Q = \{e\}$$

$$P \cdot Q \cap R = \{e\}$$

$$P \cdot Q \cdot R \cap S = \{e\}$$

(and all other combinations) so these subgroups commute with each other. And then, as usual, the whole group is the direct product of these Sylow subgroups.

One way to arrange that all the Sylow subgroups are normal is that, mod p, none of q, r, s, qr, qs, rs, qrs is 1, and symmetrically for the other primes. Further, with none of q, r, s dividing $p-1$ the p-group is *central*. For example, after some trial and error, plausible $p < q < r < s$ has $p = 17$. Take q, r, s mod 11 $= 2, 3, 5$ respectively. Take $q = 13$, so $p = -2$ mod 13, and require $r, s = 2, 5$ mod q. Then $r = 3$ mod 11 and $r = 2$ mod 13 is 80 mod 143, and 223 is the first prime in this class. With $s = 5$ mod 223, none of the 7 quantities is 1 mod r.. Then $s = 5$ mod $11 \cdot 13 \cdot 223$ and the first prime of this form is

$$s = 5 + 6 \cdot 11 \cdot 13 \cdot 223 = 191339$$

By this point, we know that the p, q, and r-sylow groups are central, so the whole thing is cyclic.

Exercises

2.1 Classify groups of order 7 or less, up to isomorphism.

2.2 Find two different non-abelian groups of order 8.

2.3 Classify groups of order 9 and 10.

2.4 Classify groups of order 12.

2.5 Classify groups of order 21.

2.6 Classify groups of order 27.

2.7 Classify groups of order 30.

2.8 Classify groups of order 77.

2.9 Let G be a group with just two subgroups, G and $\{e\}$. Prove that either $G = \{e\}$ or G is cyclic of prime order.

2.10 Let N be a normal subgroup of a group G, and let H be a subgroup of G such that $G = H \cdot N$, that is, such that the collection of all products $h \cdot n$ with $h \in H$ and $n \in N$ is the whole group G. Show that $G/N \approx H/(H \cap N)$.

2.11 (*Cayley's theorem*) Show that every finite group is isomorphic to a subgroup of a permutation group. (*Hint*: let G act on the *set G* by left multiplication.)

2.12 Let G be a group in which $g^2 = 1$ for every $g \in G$. Show that G is abelian.

2.13 Let H be a subgroup of index 2 in a finite group G. Show that H is normal.

2.14 Let p be the smallest prime dividing the order of a finite group G. Let H be a subgroup of index p in G. Show that H is normal.

2.15 Let H be a subgroup of finite index in a (not necessarily finite) group G. Show that there is a *normal* subgroup N of G such that $N \subset H$ and N is of finite index in G.

2.16 Find the automorphism group $\mathrm{Aut}\,\mathbb{Z}/n$ of the additive group $\mathbb{Z}/n$.

3. The players: rings, fields, etc.

Here we introduce some basic terminology, and give a sample of a modern construction of a *universal object*, namely a polynomial ring in one variable.

3.1 *Rings, fields*

The idea of **ring** generalizes the idea of *collection of numbers*, among other things, so maybe it is a little more intuitive than the idea of **group**. A **ring** R is a set with two operations, $+$ and $\cdot$, and with a special element 0 (**additive identity**) with most of the usual properties we expect or demand of *addition* and *multiplication*:
- R with its addition and with 0 is an abelian group. [1]
- The multiplication is *associative*: $a(bc) = (ab)c$ for all $a, b, c \in R$.
- The multiplication and addition have left and right **distributive** properties: $a(b + c) = ab + ac$ and $(b + c)a = ba + ca$ for all $a, b, c \in R$.

Often the multiplication is written just as juxtaposition

$$ab = a \cdot b$$

Very often, a particular ring has some additional special features or properties:

---

[1] This is a compressed way to say that 0 behaves as an additive identity, that there are additive inverses, and that addition is associative.

• If there is an element 1 in a ring with $1 \cdot a = a \cdot 1$ for all $a \in R$, then 1 is said to be **the (multiplicative) identity** or **the unit** [2] in the ring, and the ring is said to **have an identity** or **have a unit** or be a **ring with unit**. [3]

• If $ab = ba$ for all a, b in a ring R, then the ring is a **commutative ring**. That is, a ring is called *commutative* if and only if the *multiplication* is commutative.

• In a ring R with 1, for a given element $a \in R$, if there is $a^{-1} \in R$ so that $a \cdot a^{-1} = 1$ and $a^{-1} \cdot a = 1$, then a^{-1} is said to be a **multiplicative inverse** for a. If $a \in R$ *has* a multiplicative inverse, then a is called **a unit** [4] in R. The collection of all units in a ring R is denoted $R^{\times}$ and is called **the group of units in** R. [5]

• A commutative ring in which every nonzero element is a *unit* is a **field**.

• A not-necessarily commutative ring in which every nonzero element is a unit is a **division ring**.

• In a ring R an element r so that $r \cdot s = 0$ or $s \cdot r = 0$ for some nonzero $s \in R$ is called a **zero divisor**. [6] A commutative ring *without* nonzero zero-divisors is an **integral domain**.

• A commutative ring R has the **cancellation property** if, for any $r \neq 0$ in R, if $rx = ry$ for $x, y \in R$, then $x = y$.

If we take a ring R with 0 and with its addition, forgetting the multiplication in R, then we get an abelian group, called **the additive group of** R. And the group of units $R^{\times}$ is a (possibly non-abelian) group.

3.1.1 Example: The integers $\mathbb{Z}$ with usual addition and multiplication form a ring. This ring is certainly *commutative* and has a multiplicative identity 1. The group of units $\mathbb{Z}^{\times}$ is just $\{\pm 1\}$. This ring is an integral domain. The *even* integers $2\mathbb{Z}$ with the usual addition and multiplication form a commutative ring *without* unit. Just as this example suggests, sometimes the lack of a unit in a ring is somewhat artificial, because there is a larger ring it sits inside which *does* have a unit. There are no units in $2\mathbb{Z}$.

---

[2] Sometimes the word *unity* is used in place of *unit* for the special element 1, but this cannot be relied upon, and in any case does not fully succeed in disambiguating the terminology.

[3] We also demand that $1 \neq 0$ in a ring, if there is a 1.

[4] Yes, this usage is partly in conflict with the terminology for a special element 1.

[5] It is almost immediate that $R^{\times}$ truly is a group.

[6] The question of whether or not 0 should by convention be counted as a zero divisor has no clear answer.

3.1.2 Example: The integers mod m, denoted $\mathbb{Z}/m$, form a commutative ring with identity. It is not hard to verify that addition and multiplication are well-defined. *As the notation suggests,* the group of units is $\mathbb{Z}/m^{\times}$. [7]

3.1.3 Example: The ring $\mathbb{Z}/p$ of integers mod p is a *field* for p prime, since all non-zero residue classes have multiplicative inverses. [8] The group of units is $(\mathbb{Z}/p)^{\times}$. For n non-prime, $\mathbb{Z}/n$ is definitely not a field, because a proper factorization $n = ab$ exhibits non-zero zero divisors.

3.1.4 Example: Generally, a finite field with q elements is denoted $\mathbb{F}_q$. We will see later that, up to isomorphism, there is at most one finite field with a given number of elements, and, in fact, none unless that number is the power of a prime.

3.1.5 Example: The collection of n-by-n matrices (for fixed n) with entries in a ring R is a ring, with the usual matrix addition and multiplication. [9] Except for the silly case $n = 1$, rings of matrices over commutative rings R are *non-commutative.* The group of units, meaning matrices with an inverse of the same form, is the group $GL(n, R)$, the **general linear group** of size n over R.

3.1.6 Example: The rational numbers $\mathbb{Q}$, the real numbers $\mathbb{R}$, and the complex numbers $\mathbb{C}$ are all examples of *fields*, because all their nonzero elements have multiplicative inverses. The integers $\mathbb{Z}$ do not form a field.

There are some things about the behavior of rings which we might accidentally take for granted.

Let R be a ring.
• *Uniqueness of* 0 *additive identity:* From the analogous discussion at the beginning of group theory, we know that there is exactly one element $z = 0$ with the property that $r + z = r$ for all r in R. And there is exactly one additive inverse to any $r \in R$. And for $r \in R$, we have $-(-r) = r$. Similarly, if R has a unit 1, then, using the group $R^{\times}$, we deduce uniqueness of 1, and uniqueness of multiplicative inverses.

The following items are slightly subtler than the things above, involving the interaction of the multiplication and addition. Still, there are no surprises. [10]

Let R be a ring.
• For any $r \in R$, $0 \cdot r = r \cdot 0 = 0$. [11]
• Suppose that there is a 1 in R. Let -1 be the additive inverse of 1. Then for any $r \in R$ we have $(-1) \cdot r = r \cdot (-1) = -r$, where as usual $-r$ denotes the additive inverse of r.
• Let $-x, -y$ be the additive inverses of $x, y \in R$. Then $(-x) \cdot (-y) = xy$.

---

[7] Yes, we used the group-of-units notation in this case before we had introduced the terminology.

[8] Again, for a residue class represented by x relatively prime to p, there are integers r, s such that $rx + yp = \gcd(x, p) = 1$, and then the residue class of r is a multiplicative inverse to the residue class of x.

[9] Verification of the ring axioms is not terrifically interesting, but is worth doing once.

[10] No surprises except perhaps that these things do follow from the innocent-seeming ring axioms.

[11] One can easily take the viewpoint that this universal assertion has very little *semantic content*.

Proof: Let $r \in R$. Then

$$
\begin{aligned}
0 \cdot r &= (0 + 0) \cdot r &&\text{(since } 0 + 0 = 0) \\
&= 0 \cdot r + 0 \cdot r &&\text{(distributivity)}
\end{aligned}
$$

Then, adding $-(0 \cdot r)$ to both sides, we have

$$
0 = 0 \cdot r - 0 \cdot r = 0 \cdot r + 0 \cdot r - 0 \cdot r = 0 \cdot r + 0 = 0 \cdot r
$$

That is, $0 \cdot r$. The proof that $r \cdot 0 = 0$ is identical.

To show that $(-1)r$ is the additive inverse of r, which by now we know is unique, we check that

$$
r + (-1)r = 0
$$

We have

$$
r + (-1)r = 1 \cdot r + (-1) \cdot r = (1 - 1) \cdot r = 0 \cdot r = 0
$$

using the result we just $0 \cdot r = 0$.

To show that $(-x)(-y) = xy$, prove that $(-x)(-y) = -(-(xy))$, since $-(-r) = r$. We claim that $-(xy) = (-x)y$: this follows from

$$
(-x)y + xy = (-x + x)y = 0 \cdot y = 0
$$

Thus, we want to show

$$
(-x)(-y) + (-x)y = 0
$$

Indeed,

$$
(-x)(-y) + (-x)y = (-x)(-y + y) = (-x) \cdot 0 = 0
$$

using $r \cdot 0 = 0$ verified above. Thus, $(-x)(-y) = xy$.   ///

An **idempotent** element of a ring R is an element e such that

$$
e^2 = e
$$

A **nilpotent** element is an element z such that for some positive integer n

$$
z^n = 0_R
$$

3.2 *Ring homomorphisms*

Ring homomorphisms are maps from one ring to another which respect the ring structures.

Precisely, a **ring homomorphism** $f : R \to S$ from one ring R to another ring S is a map such that for all r, r' in R

$$
\begin{aligned}
f(r + r') &= f(r) + f(r') \\
f(rr') &= f(r)\,f(r')
\end{aligned}
$$

That is, f *preserves* or *respects* both addition and multiplication. [12]   A ring homomorphism which has a two-sided inverse homomorphism is an **isomorphism**. If a ring homomorphism is a bijection, it is an isomorphism. [13]

The **kernel** of a ring homomorphism $f : R \to S$ is

$$\ker f = \{r \in R : f(r) = 0\}$$

3.2.1 Example: The most basic worthwhile example of a ring homomorphism is

$$f : \mathbb{Z} \longrightarrow \mathbb{Z}/n$$

given by

$$f(x) = x + n\mathbb{Z}$$

The assertion that this f is a ring homomorphism is the combination of the two assertions

$$(x + n\mathbb{Z}) + (y + n\mathbb{Z}) = (x + y) + n\mathbb{Z}$$

and

$$(x + n\mathbb{Z}) \cdot (y + n\mathbb{Z}) + n\mathbb{Z} = (x \cdot y) + n\mathbb{Z}$$

Even though it is slightly misleading, this homomorphism is called the **reduction mod** m **homomorphism.**

3.2.2 Proposition: Let $f : R \to S$ be a ring homomorphism. Let $0_R, 0_S$ be the additive identities in R, S, respectively. Then $f(0_R) = 0_S$.

Proof: This is a corollary of the analogous result for groups.   ///

3.2.3 Proposition: Let $f : R \longrightarrow S$ be a *surjective* ring homomorphism. Suppose that R has a multiplicative identity 1_R. Then S has a multiplicative identity 1_S and

$$f(1_R) = 1_S$$

3.2.4 Remark: Notice that, unlike the discussion about the additive identity, now we need the further hypothesis of surjectivity.

---

[12] We do not make an attempt to use different notations for the addition and multiplication in the two different rings R and S in this definition, or in subsequent discussions. Context should suffice to distinguish the two operations.

[13] Since a bijective ring homomorphism has an inverse map which is a ring homomorphism, one could *define* an isomorphism to be a bijective homomorphism. However, in some other scenarios bijectivity of certain types of homomorphisms is *not* sufficient to assure that there is an inverse map of the same sort. The easiest example of such failure may be among continuous maps among topological spaces. For example, let $X = \{0, 1\}$ with the *indiscrete topology*, in which only the whole set and the empty set are open. Let $Y = \{0, 1\}$ with the *discrete* topology, in which all subsets are open. Then the identity map $X \longrightarrow Y$ is continuous, but its inverse is not. That is, the map is a continuous bijection, but its inverse is not continuous.

Proof: Given $s \in S$, let $r \in R$ be such that $f(r) = s$. Then

$$f(1_R) \cdot s = f(1_R) \cdot f(r) = f(1_R \cdot r) = f(r) = s$$

Thus, $f(1_R)$ behaves like a unit in S. By the *uniqueness* of units, $f(1_R) = 1_S$.   ///

3.2.5 Example: The image of a multiplicative identity 1_R under a ring homomorphism $f : R \to S$ is not necessarily the multiplicative identity 1_S of S. For example, define a ring homomorphism

$$f : \mathbb{Q} \to S$$

from the rational numbers $\mathbb{Q}$ to the ring S of 2-by-2 rational matrices by

$$f(x) = \begin{pmatrix} x & 0 \\ 0 & 0 \end{pmatrix}$$

Then the image of 1 is

$$\begin{pmatrix} 1 & 0 \\ 0 & 0 \end{pmatrix}$$

which is not the the multiplicative identity

$$\begin{pmatrix} 1 & 0 \\ 0 & 1 \end{pmatrix}$$

in S. As another example, let $R = \mathbb{Z}/3$ and $S = \mathbb{Z}/6$, and define $f : R \longrightarrow S$ by

$$f(r \bmod 3) = 4r \bmod 6$$

(This is well-defined, and is a homomorphism.) The essential feature is that

$$4 \cdot 4 = 4 \bmod 6$$

Then

$$f(x \cdot y) = 4(x \cdot y) = (4 \cdot 4)(x \cdot y) = (4x) \cdot (4y) = f(x) \cdot f(y)$$

But $f(1) = 4 \neq 1 \bmod 6$.

3.3 *Vector spaces, modules, algebras*

Let k be a field. A k-**vectorspace** V is an abelian group V (with operation written additively, referred to as **vector addition**) and a **scalar multiplication**

$$k \times V \longrightarrow V$$

written

$$\alpha \times v \longrightarrow \alpha \cdot v = \alpha v$$

such that, for $\alpha, \beta \in k$ and $v, v' \in V$,

$$
\begin{array}{llcl}
\text{(Distributivity)} & \alpha \cdot (v + v') & = & \alpha \cdot v + \alpha \cdot v' \\
\text{(Distributivity)} & (\alpha + \beta) \cdot v & = & \alpha \cdot v + \beta \cdot v \\
\text{(Associativity)} & (\alpha \cdot \beta) \cdot v & = & \alpha \cdot (\beta \cdot v) \\
& 1 \cdot v & = & v
\end{array}
$$

3.3.1 Remark: The requirement that $1 \cdot v = v$ does not follow from the other requirements. [14] By contrast, the zero element 0 in a field does reliably annihilate any vectorspace element v:

$$0_V = -(0 \cdot v) + 0 \cdot v = -(0 \cdot v) + (0 + 0) \cdot v = -(0 \cdot v) + 0 \cdot v + 0 \cdot v = 0 \cdot v$$

A k-vector **subspace** W of a k-vectorspace V is an additive subgroup closed under scalar multiplication.

A k-**linear combination** of vectors $v_1, \ldots, v_n$ in a k-vectorspace V is any vector of the form

$$\alpha_1 v_1 + \ldots + \alpha_n v_n$$

with $\alpha_i \in k$ and $v_i \in V$. Vectors $v_1, \ldots, v_n$ are **linearly dependent** if there is a linear combination of them which is 0, yet not all coefficients are 0. They are **linearly independent** if they are not linearly *dependent*. [15]

The pedestrian example of a vector space is, for fixed natural number n, the collection k^n of ordered n-tuples of elements of k with component-wise vector addition and component-wise scalar multiplication.

A k-**linear map** $T : V \longrightarrow W$ from one k-vectorspace V to another W is a homomorphism of abelian groups $T : V \longrightarrow W$

$$T(v + v') = Tv + Tv'$$

also respecting the scalar multiplication: for $\alpha \in k$

$$T(\alpha \cdot v) = \alpha \cdot Tv$$

The collection of all k-linear maps from V to W is denoted

$$\mathrm{Hom}_k(V, W) = \{ \text{ all } k\text{-linear maps from } V \text{ to } W \}$$

When $V = W$, write

$$\mathrm{End}_k(V, V) = \mathrm{End}_k(V)$$

This is the **ring of k-linear endomorphisms** of V.

The **kernel** $\ker T$ of a k-linear map $T : V \longrightarrow W$ is

$$\ker T = \{ v \in V : Tv = 0 \}$$

Let R be a ring. An R-**module** [16] M is an abelian group M (with operation written additively) and a **multiplication**

$$R \times M \longrightarrow M$$

---

[14] Sometimes the requirement that $1 \cdot v = v$ is given an unfortunate name, such as *unitary* property (in conflict with other usage), or *unital* property, which conjures up no clear image. The point is that the terminology is unpredictable.

[15] We will certainly continue this discussion of elementary linear algebra shortly, discussing the usual standard notions.

[16] In some older sources the word was *modul*, which is now obsolete. And, in some older sources, *module* was used for what we now call the multiplicative identity 1, as well as other things whose present names are otherwise.

written
$$r \times m \longrightarrow r \cdot m = rm$$
such that, for $r, r' \in R$ and $m, m' \in M$

$$
\begin{array}{rrcl}
\text{(Distributivity)} & r \cdot (m + m') & = & r \cdot m + e \cdot m' \\
\text{(Distributivity)} & (r + r') \cdot m & = & r \cdot m + r' \cdot m \\
\text{(Associativity)} & (r \cdot r') \cdot m & = & r \cdot (r' \cdot m)
\end{array}
$$

The notion of module-over-ring obviously subsumes the notion of vectorspace-over-field.

A R-**linear combination** of elements $m_1, \ldots, m_n$ in a R module M is any module element of the form
$$r_1 m_1 + \ldots + r_n m_n$$
with $r_i \in R$ and $m_i \in M$. [17]

We specifically do *not* universally require that $1_R \cdot m = m$ for all m in an R-module M when the ring R contains a unit 1_R. Nevertheless, on many occasions we *do* require this, but, therefore, must say so explicitly to be clear.

An R-**submodule** N of an R-module M is an additive subgroup which is closed under scalar multiplication.

An R-**linear map** $T : M \longrightarrow N$ from one T-module M to another N is a homomorphism of abelian groups $T : M \longrightarrow N$

$$T(m + m') = Tm + Tm'$$

also respecting the scalar multiplication: for $r \in R$

$$T(r \cdot m) = r \cdot Tm$$

The collection of all R-linear maps from M to N is denoted

$$\text{Hom}_R(M, N) = \{ \text{ all } R\text{-linear maps from } M \text{ to } N \}$$

When $M = N$, write
$$\text{End}_R(M, M) = \text{End}_R(M)$$

This is the **ring of R-linear endomorphisms** of M.

The **kernel** $\ker T$ of an R-linear map $T : M \longrightarrow N$ is

$$\ker T = \{ m \in M : Tm = 0 \}$$

3.3.2 Example: *Abelian groups are $\mathbb{Z}$-modules*: for $a \in A$ in an abelian group A, define the scalar multiplication by integers by

$$
n \cdot a =
\begin{cases}
0_A & \text{(for } n = 0) \\
\underbrace{a + \ldots + a}_{n} & \text{(for } n > 0) \\
-(\underbrace{a + \ldots + a}_{|n|}) & \text{(for } n < 0)
\end{cases}
$$

---

[17] While one should think of linear algebra over fields as a prototype for some of the phenomena concerning modules more generally, one should at the same time be prepared for deviation from the simpler reasonable expectations.

Observe that a homomorphism of abelian groups is inevitably $\mathbb{Z}$-linear.

3.3.3 Example: A (left) **ideal** I in a ring R is an additive subgroup I of R which is also closed under left multiplication by R: for $i \in I$ and $r \in R$, $r \cdot i \in I$. It is immediate that the collection of left ideals in R is identical to the collection of R-submodules of R (with left multiplication).

Let R be a *commutative* ring. [18] Let A be a not-necessarily commutative ring which is a left R-module. If, in addition to the requirements of a module, we have the *associativity*

$$r \cdot (a \cdot b) = (r \cdot a) \cdot b$$

for $r \in R$ and $a, b \in A$, then say A is an R-**algebra**. Often additional requirements are imposed. [19]

A ring homomorphism $f : A \longrightarrow B$ of R-algebras is an R-**algebra homomorphism** if it is also an R-module homomorphism.

3.4 *Polynomial rings I*

We should not be content to speak of *indeterminate x* or *variable x* to construct polynomial rings. Instead, we describe in precise terms the fundamental property that a polynomial ring is meant to have, namely, in colloquial terms, the *indeterminate* can be replaced by *any value*, or that any value can be *substituted for* the indeterminate.

Fix a commutative ring R, and let A be a *commutative* R-algebra with a distinguished element a_o. Say that A, or, more properly, the *pair* (A, a_o), is a **free (commutative) algebra** on one **generator** a_o if, for every commutative R-algebra B and chosen element $b_o \in B$ there is a *unique* R-algebra homomorphism

$$f_{B,b_o} : A \longrightarrow B$$

such that

$$f(a_o) = b_o$$

This condition is an example of a **universal mapping property**, and the polynomial ring (once we show that it is this object) is thus a **universal object** with respect to this property.

3.4.1 Remark: In different words, a_o can be mapped *anywhere*, and specifying the image of a_o completely determines the homomorphism.

3.4.2 Remark: We are to imagine that $A = R[x]$, $a_o = x$, and that the R-algebra homomorphism *is* the substitution of b_o for x.

---

[18] The requirement that R be commutative is not at all necessary to give a definition of R-algebra, but without that hypothesis it is much less clear what is best to offer as a first and supposedly general definition.

[19] One might require the commutativity $(ra)b = a(rb)$, for example. One might require that R have a unit 1 and that $1 \cdot a = a$ for all $a \in A$. However, not all useful examples meet these additional requirements.

The following uniqueness result is typical of what can be said when an object is characterized by universal mapping properties.

3.4.3 Proposition: Up to isomorphism, there is *at most one* free commutative R-algebra on one generator. That is, given two such things (A, a_o) and (A', a'_o), there is a unique isomorphism

$$i : A \longrightarrow A'$$

sending a_o to a'_o and such that, given a commutative R-algebra B with distinguished element b_o, the corresponding maps (as above)

$$f_{B,b_o} : A \longrightarrow B$$
$$f'_{B,b_o} : A' \longrightarrow B$$

satisfy

$$f = f' \circ i$$

3.4.4 Remark: Despite the possible unpalatableness of the definition and the proposition, this setup does what we want, and the proposition asserts the *essential uniqueness* of what will turn out to be recognizable as the polynomial ring $R[x]$.

Proof: This proof is typical of proving that there is at most one thing characterized by a universal property. First, take $B = A$ and $b_o = a_o$. Then there is a *unique R-algebra homomorphism $A \longrightarrow A$ taking a_o to a_o. Since the identity map on A does this, apparently *only* the identity has this property among all endomorphisms of A.

Next, let $B = A'$ and $b = a'_o$, and

$$f_{A',a'_o} : A \longrightarrow A' \quad \text{(with } a_o \longrightarrow a'_o\text{)}$$

the unique R-algebra homomorphism postulated. Reversing the roles of A and A', we have another *unique*

$$f'_{A,a_o} : A' \longrightarrow A \quad \text{(with } a'_o \longrightarrow a_o\text{)}$$

Consider $g = f' \circ f'$. It sends a_o to a_o, so, by our first observation, must be the identity map on A. Similarly, $f \circ f'$ is the identity map on A'. Thus, f and f' are mutual inverses, and A and A' are isomorphic, by a *unique* isomorphism, [20] and a_o is mapped to a'_o by this map.   ///

This slick uniqueness argument does not prove existence. Indeed, there seems to be no comparably magical way to prove existence, *but* the uniqueness result assures us that, whatever pathetic *ad hoc* device we do hit upon to construct the free algebra, the thing we make is *inevitably* isomorphic (and by a *unique* isomorphism) to what any *other* construction might yield. That is, the uniqueness result shows that particular choice of construction does not matter. [21]

---

[20] The virtues of there being a *unique* isomorphism may not be apparent at the moment, but already played a role in the uniqueness proof, and do play significant roles later.

[21] An elementary example of a construction whose internals are eventually ignored in favor of operational properties is *ordered pair*: in elementary set theory, the ordered pair (a, b) is defined as $\{\{a\}, \{a, b\}\}$, and the expected properties are verified. After that, this set-theoretic definition is forgotten. And, indeed, one should probably not consider this to be *correct* in any sense of providing further information about what an ordered pair truly is. Rather, it is an *ad hoc* construction which thereafter entitles us to do certain things.

How to construct the thing? On one hand, since the possible images $f(a_o)$ can be *anything* in another R-algebra B, a_o ought not satisfy any relations such as $a_o^3 = a_o$ since a homomorphism would carry such a relation forward into the R-algebra B, and we have no reason to believe that $b_o^3 = b_o$ for every element b_o of every R-algebra B. [22] On the other hand, since the image of a_o under an R-algebra homomorphism is intended to determine the homomorphism completely, the free algebra A should not contain more than R-linear combinations of powers of a_o.

For fixed commutative ring R with identity 1, let S be the set [23] of R-valued functions P on the set $\{0, 1, 2, \ldots\}$ such that, for each P, there is an index n such that for $i > n$ we have $P(i) = 0$. [24] Introduce an addition which is simply componentwise: for $P, Q \in S$,

$$(P + Q)(i) = P(i) + Q(i)$$

And there is the value-wise R-module structure with scalar multiplication

$$(r \cdot P)(i) = r \cdot P(i)$$

All this is obligatory, simply to have an R-module. We take the distinguished element to be

$$a_o = \text{ the function } P_1 \text{ such that } P_1(1) = 1 \text{ and } P_1(i) = 0 \text{ for } i \neq 1$$

A misleadingly glib way of attempting to define the multiplication is [25] as

$$(P \cdot Q)(i) = \sum_{j+k=i} P(j) \, Q(k)$$

using the idea that a function is completely described by its values. Thus, since R is commutative,

$$P \cdot Q = Q \cdot P$$

For P, Q, T in S, associativity

$$(P \cdot Q) \cdot T = P \cdot (Q \cdot T)$$

---

[22] While it is certainly true that we should doubt that this a_o satisfies any relations, in other situations specification of universal objects *can* entail unexpected relations. In others, the fact that there *are* no relations apart from obvious ones may be non-trivial to prove. An example of this is the Poincaré-Birkhoff-Witt theorem concerning universal enveloping algebras. We may give this as an example later.

[23] This construction presumes that sets and functions are legitimate primitive objects. Thus, we tolerate possibly artificial-seeming constructions for their validation, while clinging to the uniqueness result above to rest assured that any peculiarities of a construction do not harm the object we create.

[24] We would say that P is *eventually zero*. The intent here is that $P(i)$ is the coefficient of x^i in a polynomial.

[25] If one is prepared to describe polynomial multiplication by telling the coefficients of the product then perhaps this is not surprising. But traditional informal discussions of polynomials often to treat them more as strings of symbols, *expressions*, rather than giving them set-theoretic substance.

follows from rewriting the left-hand side into a symmetrical form

$$((P \cdot Q) \cdot T)(i) = \sum_{j+k=i} (P \cdot Q)(j) \, T(k)$$

$$= \sum_{j+k=i} \sum_{m+n=j} P(m)Q(n)T(k) = \sum_{m+n+k=i} P(m)Q(n)T(k)$$

Distributivity of the addition and multiplication in S follows from that in R:

$$(P \cdot (Q+T))(i) = \sum_{j+k=i} P(j) \cdot (Q+T)(k) = \sum_{j+k=i} (P(j)Q(k) + P(j)T(k))$$

$$= \sum_{j+k=i} P(j)Q(k) + \sum_{j+k=i} P(j)T(k) = (P \cdot Q)(i) + (P \cdot T)(i)$$

The associativity

$$r \cdot (P \cdot Q) = (rP) \cdot Q$$

is easy. Note that, by an easy induction

$$P_1^i(j) = \begin{cases} 1 & (\text{if } j = i) \\ 0 & (\text{if } j \neq i) \end{cases}$$

So far, we have managed to make a commutative R-algebra S with a distinguished element P_1. With the above-defined multiplication, we claim that

$$\sum_i r_i \, P_1^i$$

(with coefficients r_i in R) is 0 (that is, the zero function in S) if and only if all coefficients are 0. Indeed, the value of this function at j is r_j. Thus, as a consequence, if

$$\sum_i r_i \, P_1^i = \sum_j r_j' \, P_1^j$$

then subtract one side from the other, so see that $r_i = r_i'$ for all indices i. That is, there is only one way to express an element of S in this form.

Given another R-algebra B and element $b_o \in B$, we would like to define

$$f(\sum_i r_i P_1^i) = \sum_i r_i b_o^i$$

At least this is *well-defined*, since there is only one expression for elements of S as R-linear combinations of powers of P_1^i. The R-linearity of this f is easy. The fact that it respects the multiplication of S is perhaps less obvious, but not difficult: [26]

$$f((\sum_i r_i P_1^i) \cdot (\sum_j r_j' P_1^j)) = f(\sum_{i,j} r_i r_j' P_1^{i+j}) = \sum_{i,j} r_i r_j' b_o^{i+j} = (\sum_i r_i b_o^i) \cdot (\sum_j r_j' b_o^j))$$

---

[26] The formulas for multiplication of these finite sums with many summands could be proven by induction if deemed necessary.

Thus, this f is an R-algebra homomorphism which sends $a_o = P_1$ to b_o.

Finally, there is no *other* R-algebra homomorphism of S to B sending $a_o = P_1$ to b_o, since every element of S is expressible as $\sum r_i P_1^i$, and the R-algebra homomorphism property yields

$$f\left(\sum_i r_i P_1^i\right) = \sum_i f(r_i P_1^i) = \sum_i r_i f(P_1^i) = \sum_i r_i f(P_1)^i$$

There is no further choice possible. [27]

3.4.5 Remark: This tedious construction (or something equivalent to it) is necessary. The uniqueness result assures us that no matter which of the several choices we make for (this tiresome) construction, the resulting thing is the same.

Exercises

3.1 Let r be nilpotent in a commutative ring. Show that $1 + r$ is a unit.

3.2 Give an example of an integer n such that $\mathbb{Z}/n$ has at least 4 different idempotent elements.

3.3 Give an example of an integer n such that $\mathbb{Z}/n$ has at least 4 different idempotent elements.

3.4 Let $f \in k[x]$ for a field k. For indeterminates x, y, show that we have a Taylor-Maclaurin series expansion of the form

$$f(x + y) \;=\; f(x) + \sum_{i=1}^{n} f_i(x)\, y^i$$

for some polynomials $f_i(x)$. For k of characteristic 0, show that

$$f_i(x) \;=\; \left(\frac{\partial}{\partial x}\right)^i f(x)/i!$$

3.5 Show that a **local ring** R (that is, a ring having a unique *maximal* proper ideal) has no idempotent elements other than 0 and 1.

3.6 Let $p > 2$ be a prime. Show that for $\ell \geq 1 \geq \frac{p}{p-1}$ the power of p dividing $(p^\ell)^n$ is larger than or equal to the power of p dividing $n!$.

3.7 *The exponential map modulo p^n:* Let $p > 2$ be prime. Make sense of the map

$$E : p\mathbb{Z}/p^n \to 1 + p\mathbb{Z} \bmod p^n$$

defined by the dubious formula

$$E(px) = 1 + \frac{px}{1!} + \frac{(px)^2}{2!} + \frac{(px)^3}{3!} + \dots$$

---

[27] One may paraphrase this by saying that if g were another such map, then $f - g$ evaluated on any such element of S is 0.

(*Hint:* cancel powers of p before trying to make sense of the fractions. And only finitely-many of the summands are non-zero mod p^n, so this is a *finite* sum.)

3.8 With the exponential map of the previous exercise, show that $E(px + py) = E(px) \cdot E(py)$ modulo p^n, for $x, y \in \mathbb{Z}/p^{n-1}$. That is, prove that E is a group homomorphism from $p\mathbb{Z}/p^n\mathbb{Z}$ to the subgroup of $(\mathbb{Z}/p^n)^\times$ consisting of $a = 1$ mod p.

3.9 Prove that $(\mathbb{Z}/p^n)^\times$ is *cyclic* for $p > 2$ prime.

3.10 Figure out the correct analogue of the exponential map for $p = 2$.

3.11 Figure out the correct analogue of the exponential maps modulo primes for the Gaussian integers $\mathbb{Z}[i]$.

3.12 For which ideals I of $\mathbb{Z}[i]$ is the multiplicative group $\mathbb{Z}[i]/I^\times$ of the quotient ring $\mathbb{Z}[i]/I$ *cyclic*?

3.13 Show that there are no *proper* two-sided ideals in the ring R of 2-by-2 rational matrices.

3.14 (*Hamiltonian quaternions*) Define the **quaternions** $\mathfrak{H}$ to be an $\mathbb{R}$-algebra generated by $1 \in \mathbb{R}$ and by elements i, j, k such that $i^2 = j^2 = k^2 = -1$, $ij = k$, $jk = i$, and $ki = j$. Define the *quaternion conjugation* $\alpha \longrightarrow \alpha^*$ by

$$(a + bi + cj + dk)^* = a - bi - cj - dk$$

Show that $*$ is an *anti-automorphism*, meaning that

$$(\alpha \cdot \beta)^* = \beta^* \cdot \alpha^*$$

for quaternions α, β. Show that $\mathfrak{H}$ is a *division ring*.

3.15 Provide a *construction* of the quaternions, by showing that

$$a + bi + cj + dk \longrightarrow \begin{pmatrix} a + bi & c + di \\ c - di & a - bi \end{pmatrix}$$

is a ring homomorphism from the quaternions to a subring of the 2-by-2 complex matrices.

4. Commutative rings I

Throughout this section the rings in question will be commutative, and will have a unit 1.

4.1 *Divisibility and ideals*

Many of the primitive ideas about divisibility we bring from the ordinary integers $\mathbb{Z}$, though few of the conclusions are as simple in any generality.

Let R be a commutative [1]   ring with unit [2]   1. Let $\mathbb{R}^{\times}$ be the group of units in R.

Say d **divides** m, equivalently, that m is a **multiple** of d, if there exists a $q \in R$ such that $m = qd$. Write $d|m$ if d divides m. It is easy to prove, from the definition, that if $d|x$ and

---

[1] Divisibility and ideals can certainly be discussed without the assumption of commutativity, but the peripheral complications obscure simpler issues.

[2] And, certainly, one can contemplate divisibility in rings without units, but this leads to needlessly counterintuitive situations.

$d|y$ then $d|(ax + by)$ for any $x, y, a, b \in R$: let $x = rd$ and $y = sd$, and

$$ax + by = a(rd) + b(sd) = d \cdot (ar + bs)$$

A ring element d is a **common divisor** of ring elements $n_1, \ldots, n_m$ if d divides each n_i. A ring element N is a **common multiple** of ring elements $n_1, \ldots, n_m$ if N is a multiple of each.

A divisor d of n is **proper** if it is not a unit multiple of n and is not a unit itself. A ring element is **irreducible** if it has no proper factors. A ring element p is **prime** if $p|ab$ implies $p|a$ or $p|b$ and p is not a unit and is not 0. [3]   If two prime elements p and p' are related by $p = up'$ with a unit u, say that p and p' are **associate**. We view associate primes as being essentially identical. [4]   Recall that an **integral domain** [5] is a commutative ring in which $cd = 0$ implies either c or d is 0. [6]

4.1.1 Proposition: Prime elements of an integral domain R are irreducible. [7]

Proof: Let p be a prime element of R, and suppose that $p = ab$. Then $p|ab$, so $p|a$ or $p|b$. Suppose $a = a'p$. Then $p = p \cdot a'b$, and $p \cdot (1 - a'b) = 0$. Since the ring is an integral domain, either $p = 0$ or $a'b = 1$, but $p \neq 0$. Thus, $a'b = 1$, and b is a unit. This proves that any factorization of the prime p is non-proper.   ///

An integral domain R with 1 is a **unique factorization domain** (**UFD**) if every element $r \in R$ has a unique (up to ordering of factors and changing primes by units) expression

$$r = up_1 \ldots p_\ell$$

with unit u and primes p_i.

4.1.2 Remark: The ordinary integers are the primary example of a UFD. The second important example is the ring of polynomials in one variable over a field, treated in the next section.

----

[3] Yes, the definition of *prime* rewrites what was a theorem for the ordinary integers as the definition in general, while demoting the lack of proper factors to a slightly more obscure classification, of *irreducibility*.

[4] In the case of the ordinary integers, $\pm p$ are associate, for prime p. We naturally distinguish the positive one of the two. But in more general situations there is not a reliable special choice among associates.

[5] Some sources have attempted to popularize the term *entire* for a ring with no proper zero divisors, but this has not caught on.

[6] If one insisted, one could say that an integral domain is a commutative ring in which 0 is prime, but for practical reasons we want our convention not to include 0 when we speak of prime *elements*. Likewise by convention we do not want *units* to be included when we speak of primes.

[7] The converse is not generally true.

4.2 *Polynomials in one variable over a field*

We will prove that the ring $k[x]$ of polynomials in one variable with coefficients in a field k is *Euclidean*, and thus has unique factorization. This example is comparable in importance to $\mathbb{Z}$ and its elementary properties.

As usual, the **degree** of a polynomial $\sum_i c_i x^i$ is the highest index i such that $c_i \neq 0$.

4.2.1 Proposition: For polynomials P, Q with coefficients in a field [8] k, the degree of the product is the sum of the degrees:

$$\deg(P \cdot Q) = \deg P + \deg Q$$

4.2.2 Remark: To make this correct even when one of the two polynomials is the 0 polynomial, the 0 polynomial is by convention given degree $-\infty$.

Proof: The result is clear if either polynomial is the zero polynomial, so suppose that both are non-zero. Let

$$P(x) = a_m x^m + a_{m-1} x^{m-1} + \ldots + a_2 x^2 + a_1 x + a_0$$

$$Q(x) = b_n x^n + b_{n-1} x^{n-1} + \ldots + b_2 x^2 + b_1 x + b_0$$

where the (apparent) highest-degree coefficients a_m and b_n non-zero. Then in $P \cdot Q$ the highest-degree term is $a_m b_n x^{m+n}$. Since the product of non-zero elements of a field is non-zero, [9] the coefficient of x^{m+n} is non-zero. ///

4.2.3 Corollary: *(Cancellation property)* For polynomials in $k[x]$, with a field k, let $A \cdot P = B \cdot P$ for a non-zero polynomial P. Then $A = B$.

Proof: The equality $AP = BP$ gives $(A - B)P = 0$. Because the degree of the product is the sum of the degrees of the factors,

$$\deg(A - B) + \deg P = \deg 0 = -\infty$$

Since P is non-zero, $\deg P \geq 0$. Then $\deg(A - B) = -\infty$, so $A - B = 0$, and $A = B$. ///

4.2.4 Corollary: The group of units $k[x]^\times$ in the polynomial ring in one variable over a field k is just the group of units $k^\times$ in k. [10]

---

[8] The proof only uses the fact that a product of non-zero elements is necessarily non-zero. Thus, the same conclusion can be reached if the coefficients of the polynomials are merely in an *integral domain*.

[9] In case this is not clear: let a, b be elements of a field k with $ab = 0$ and $a \neq 0$. Since non-zero elements have inverses, there is a^{-1}, and $a^{-1} ab = a^{-1} \cdot 0 = 0$, but also $a^{-1} ab = b$. Thus, $b = 0$.

[10] We identify the scalars with degree-zero polynomials, as usual. If one is a bit worried about the legitimacy of this, the free-algebra definition of the polynomial ring can be invoked to prove this more formally.

Proof: Suppose that $P \cdot Q = 1$. Then $\deg P + \deg Q = 0$, so both degrees are 0, that is, P and Q are in k. ///

A polynomial is **monic** if its highest degree coefficient is 1. Since elements of $k^\times$ are units in $k[x]$, any polynomial can be multiplied by a unit to make it monic.

4.2.5 Proposition: (*Euclidean property*) Let k be a field and M a non-zero polynomial in $k[x]$. Let H be any other polynomial in $k[x]$. Then there are unique polynomials Q and R in $k[x]$ such that $\deg R < \deg M$ and

$$H = Q \cdot M + R$$

Proof: [11] Let X be the set of polynomials expressible in the form $H - S \cdot M$ for some polynomial S. Let $R = H - Q \cdot M$ be an element of X of minimal degree. Claim that $\deg R < \deg M$. If not, let a be the highest-degree coefficient of R, let b be the highest-degree coefficient of M, and define

$$G = (ab^{-1}) \cdot x^{\deg R - \deg M}$$

Then

$$R - G \cdot M$$

removes the highest-degree term of R, and

$$\deg(R - G \cdot M) < \deg R$$

But $R - GM$ is still in X, since

$$R - G \cdot M = (H - Q \cdot M) - G \cdot M = H - (Q + G) \cdot M$$

By choice of R this is impossible, so, in fact, $\deg R < \deg M$. For uniqueness, suppose

$$H = Q \cdot M + R = Q' \cdot M + R'$$

Subtract to obtain

$$R - R' = (Q' - Q) \cdot M$$

Since the degree of a product is the sum of the degrees, and since the degrees of R, R' are less than the degree of M, this is impossible unless $Q' - Q = 0$, in which case $R - R' = 0$. ///

Compatibly with general terminology, a non-zero polynomial is **irreducible** if it has no proper divisors.

The **greatest common divisor** of two polynomials A, B is the *monic* polynomial g of highest degree dividing both A and B.

4.2.6 Theorem: For polynomials f, g in $k[x]$, the monic polynomial of the form $sf + tg$ (for $s, t \in k[x]$) of smallest degree is the *gcd* of f, g. In particular, greatest common divisors exist.

---

[11] This argument is identical to that for the ordinary integers, as are many of the other proofs here.

Proof: Among the non-negative integer values $\deg(sf + tg)$ there is at least one which is minimal. Let $h = sf + tg$ be such, and multiply through by the inverse of the highest-degree coefficient to make h monic. First, show that $h|f$ and $h|g$. We have

$$f = q(sf + tg) + r$$

with $\deg r < \deg(sf + tg)$. Rearranging,

$$r = (1 - qs)f + (-qt)g$$

So r itself is $s'f + t'g$ with $s', t' \in k[x]$. Since $sf + tg$ had the smallest non-negative degree of any such expression, and $\deg r < \deg(sf + tg)$, it must be that $r = 0$. So $sf + tg$ divides f. Similarly, $sf + tg$ divides g, so $sf + tg$ is a divisor of both f and g. On the other hand, if $d|f$ and $d|g$, then certainly $d|sf + tg$.   ///

4.2.7 Corollary: Let P be an irreducible polynomial. For two other polynomials A, B, if $P|AB$ then $P|A$ or $P|B$. Generally, if an irreducible P divides a product $A_1 \ldots A_n$ of polynomials then P must divide one of the factors A_i.

Proof: It suffices to prove that if $P|AB$ and $P \nmid A$ then $P|B$. Since $P \nmid A$, and since P is irreducible, the *gcd* of P and A is just 1. Therefore, there are $s, t \in k[x]$ so that

$$1 = sA + tP$$

Then

$$B = B \cdot 1 = B \cdot (sA + tP) = s(AB) + (Bt)P$$

Since $P|AB$, surely P divides the right-hand side. Therefore, $P|B$, as claimed.

4.2.8 Corollary: Irreducible polynomials P in $k[x]$ are prime, in the sense that $P|AB$ implies $P|A$ or $P|B$ for polynomials A and B.

Proof: Let $AB = M \cdot P$, and suppose that P does not divide A. Since P is irreducible, any proper factor is a unit, hence a non-zero constant. Thus, $\gcd(P, A) = 1$, and there are polynomials R, S such that $RP + SA = 1$. Then

$$B = B \cdot 1 = B \cdot (RP + SA) = P \cdot BR + S \cdot AB = P \cdot BR + S \cdot M \cdot P = P \cdot (BR + SM)$$

so B is a multiple of P.   ///

4.2.9 Corollary: Any polynomial M in $k[x]$ has a unique factorization (up to ordering of factors) as

$$M = u \cdot P_1^{e_1} \ldots P_\ell^{e_\ell}$$

where $u \in k^\times$ is a unit in $k[x]$, the P_i are distinct primes, and the exponents are positive integers.

Proof: [12] First prove existence by induction on degree. [13] Suppose some polynomial F admitted no such factorization. Then F is not irreducible (or the non-factorization is

---

[12] It bears emphasizing that the argument here proves unique factorization from the propery of primes that $p|ab$ implies $p|a$ or $p|b$, which comes from the Euclidean property. There are many examples in which a unique factorization result does hold without Euclidean-ness, such as polynomial rings $k[x_1, \ldots, x_n]$ in *several* variables over a field, but the argument is more difficult. See *Gauss' Lemma*.

[13] An induction on size completely analogous to the induction on size for the ordinary integers.

the factorization), so $R = A \cdot B$ with both of A, B of lower degree but not degree 0. By induction, both A and B have factorizations into primes.

Uniqueness is a sharper result, proven via the property that $P|AB$ implies $P|A$ or $P|B$ for prime P. As in the case of integers, given two alleged prime factorizations, any prime in one of them must be equal to a prime in the other, and by cancelling we do an induction to prove that all the primes are the same. ///

4.2.10 Proposition: *(Testing for linear factors)* A polynomial $f(x)$ with coefficients in a field k has a linear factor $x - a$ (with $a \in k$) if and only if $F(a) = 0$.

Proof: If $x - a$ is a factor, clearly $f(a) = 0$. On the other hand, suppose that $f(a) = 0$. Use the division algorithm to write

$$f(x) = Q(x) \cdot (x - a) + R$$

Since $\deg R < \deg(x - a) = 1$, R is a constant. Evaluate both sides at a to obtain

$$0 = f(a) = Q(a) \cdot (a - a) + R = Q(a) \cdot 0 + R = R$$

Therefore, $R = 0$ and $x - a$ divides $f(x)$. ///

4.2.11 Corollary: For polynomial P in $k[x]$, the equation $P(a) = 0$ has no more roots a than the degree of P.

Proof: By the proposition, a root gives a monic linear factor, and by unique factorization there cannot be more of these than the degree. ///

4.2.12 Example: With coefficients not in a field, the intuition that a polynomial equation has no more roots than its degree is inaccurate. For example, with coefficients in $\mathbb{Z}/15$, the equation

$$a^2 - 1 = 0$$

has the obvious roots ± 1, but also the roots 6 and 10. And there are two different factorizations in $(\mathbb{Z}/15)[x]$

$$x^2 - 1 = (x - 1)(x + 1) = (x - 6)(x - 10)$$

4.3 *Ideals*

Let R be a commutative ring with unit 1. An **ideal** in R is an additive subgroup I of R such that $R \cdot I \subset I$. That is, I is an R-submodule of R with (left) multiplication.

4.3.1 Example: One archetype is the following. In the ring $\mathbb{Z}$, for any fixed n, the set $n \cdot \mathbb{Z}$ of multiples of n is an ideal.

4.3.2 Example: Let $R = k[x]$ be the ring of polynomials in one variable x with coefficients in a field k. Fix a polynomial $P(x)$, and let $I \subset R$ be the set of polynomial multiples $M(x) \cdot P(x)$ of $P(x)$.

4.3.3 Example: Abstracting the previous two examples: fix $n \in R$. The set $I = n \cdot R = \{rn : r \in R\}$ of multiples of m is an ideal, the **principal ideal generated by** n. A convenient lighter notation is to write

$$\langle n \rangle = R \cdot n = \text{principal ideal generated by } n$$

4.3.4 Example: In any ring, the **trivial ideal** is $I = \{0\}$. An ideal is **proper** if it is neither the trivial ideal $\{0\}$ nor the whole ring R (which is also an ideal).

4.3.5 Example: If an ideal I contains a unit u in R, then $I = R$. Indeed, for any $r \in R$,

$$r = r \cdot 1 = r \cdot (u^{-1} \cdot u) \in r \cdot u^{-1} \cdot I \subset I$$

For two subsets X, Y of a *ring* R, write

$$X + Y = \{x + y : x \in X, \ y \in Y\}$$

and [14]

$$X \cdot Y = \{\text{finite sums } \sum_i x_i \, y_i : x_i \in X, \ y_i \in Y\}$$

In this notation, for an ideal I in a commutative ring R with 1 we have $R \cdot I = I$.

An integral domain in which every ideal is principal is a **principal ideal domain**. [15]

4.3.6 Corollary: [16] Every ideal I in $\mathbb{Z}$ is principal, that is, of the form $I = n \cdot \mathbb{Z}$. In particular, unless $I = \{0\}$, the integer n is the least positive element of I.

Proof: Suppose I is non-zero. Since I is closed under additive inverses, if I contains $x < 0$ then it also contains $-x > 0$. Let n be the least element of I. Let $x \in I$, take $q, r \in \mathbb{Z}$ with $0 \le r < n$ such that

$$x = q \cdot n + r$$

Certainly qn is in I, and $-qn \in I$ also. Since $r = x - qn$, $r \in I$. Since n was the smallest positive element of I, $r = 0$. Thus, $x = qn \in n \cdot \mathbb{Z}$, as desired.   ///

4.3.7 Corollary: [17] Let k be a field. Let $R = k[x]$ be the ring of polynomials in one variable x with coefficients in k. Then every ideal I in R is principal, that is, is of the form $I = k[x] \cdot P(x)$ for some polynomial P. In particular, $P(x)$ is the monic polynomial of smallest degree in I, unless $I = \{0\}$, in which case $P(x) = 0$.

Proof: If $I = \{0\}$, then certainly $I = k[x] \cdot 0$, and we're done. So suppose I is non-zero. Suppose that $Q(x) = a_n x^n + \ldots + a_0$ lies in I with $a_n \ne 0$. Since k is a field, there is an inverse a_n^{-1}. Then, since I is an ideal, the polynomial

$$P(x) = a_n^{-1} \cdot Q(x) = x^n + a_n^{-1} a_{n-1} x^{n-1} + \ldots + a_n^{-1} a_0$$

---

[14] Note that here the notation $X \cdot Y$ has a different meaning than it does in group theory, since in the present context it is implied that we take all finite sums of products, not just products.

[15] If we do not assume that the ring is a domain, then we certainly may form the notion of *principal ideal ring*. However, the presence of zero divisors is a distraction.

[16] This is a corollary of the Euclidean-ness of $\mathbb{Z}$.

[17] This is a corollary of the Euclidean-ness of $\mathbb{Z}$.

also lies in I. That is, there is indeed a *monic* polynomial of lowest degree of any element of the ideal. Let $x \in I$, and use the Division Algorithm to get $Q, R \in k[x]$ with $\deg R < \deg P$ and

$$x = Q \cdot P + R$$

Certainly $Q \cdot P$ is still in I, and then $-Q \cdot P \in I$ also. Since $R = x - Q \cdot P$, we conclude that $R \in I$. Since P was the monic polynomial in I of smallest degree, it must be that $R = 0$. Thus, $x = Q \cdot P \in n \cdot k[x]$, as desired.   ///

4.3.8 Remark: The proofs of these two propositions can be abstracted to prove that every ideal in a Euclidean ring is principal.

4.3.9 Example: Let R be a commutative ring with unit 1, and fix two elements $x, y \in R$. Then

$$I = R \cdot x + R \cdot y = \{rx + sy : r, s \in R\}$$

is an ideal in R. The two elements x, y are the **generators** of I.

4.3.10 Example: Similarly, for fixed elements $x_1, \ldots, x_n$ of a commutative ring R, we can form an ideal

$$I = R \cdot x_1 + \ldots + R \cdot x_n$$

4.3.11 Example: To construct new, larger ideals from old, smaller ideals proceed as follows. Let I be an ideal in a commutative ring R. Let x be an element of R. Then let

$$J = R \cdot x + I = \{rx + i : r \in R, \ i \in I\}$$

Let's check that J is an ideal. First

$$0 = 0 \cdot x + 0$$

so 0 lies in J. Second,

$$-(rx + i) = (-r)x + (-i)$$

so J is closed under inverses. Third, for two elements $rx + i$ and $r'x + i'$ in J (with $r, r' \in R$ and $i, i' \in I$) we have

$$(rx + i) + (r'x + i') = (r + r')x + (i + i')$$

so J is closed under addition. Finally, for $rx + i \in J$ with $r \in R$, $i \in I$, and for $r' \in R$,

$$r' \cdot (rx + i) = (r'r)x + (r'i)$$

so $R \cdot J \subset J$ as required. Thus, this type of set J is indeed an ideal.

4.3.12 Remark: In the case of rings such as $\mathbb{Z}$, where we know that every ideal is principal, the previous construction does not yield any more general type of ideal.

4.3.13 Example: In some rings R, *not* every ideal is principal. We return to an example used earlier to illustrate a failure of unique factorization. Let

$$R = \{a + b\sqrt{-5} : a, b \in \mathbb{Z}\}$$

Let
$$I = \{x \cdot 2 + y \cdot (1 + \sqrt{-5}) : x, y \in R\}$$

These phenomena are not of immediate relevance, but did provide considerable motivation in the historical development of algebraic number theory.

4.4 *Ideals and quotient rings*

Here is a construction of new rings from old in a manner that includes as a special case the construction of $\mathbb{Z}/n$ from $\mathbb{Z}$.

Let R be a commutative ring with unit 1. Let I be an ideal in R. The **quotient ring** R/I is the set of cosets
$$r + I = \{r + i : i \in I\}$$
with operations of addition and multiplication on R/I by
$$(r + I) + (s + I) = (r + s) + I$$
$$(r + I) \cdot (s + I) = (r \cdot s) + I$$
The zero in the quotient is $0_{R/I} = 0 + I$, and the unit is $1_{R/I} = 1 + I$.

4.4.1 Example: The basic example is that $\mathbb{Z}/n$ is the quotient ring $\mathbb{Z}/I$ where $I = n \cdot \mathbb{Z}$.

4.4.2 Remark: It's tedious, but someone should check that the operations of addition and multiplication in $\mathbb{Z}/n$ are *well-defined*: we want the alleged addition and multiplication operations not to depend on the way the coset is *named*, but only on *what it is*. So suppose $r + I = r' + I$ and $s + I = s' + I$. We need to check that
$$(r + s) + I = (r' + s') + I$$
and to prove well-definedness of multiplication check that
$$(r \cdot s) + I = (r' \cdot s') + I$$

Since $r' + I = r + I$, in particular $r' = r' + 0 \in r + I$, so r' can be written as $r' = r + i$ for some $i \in I$. Likewise, $s' = s + j$ for some $j \in I$. Then
$$(r' + s') + I = (r + i + s + j) + I = (r + s) + (i + j + I)$$

The sum $k = i + j$ is an element of I. We claim that for any $k \in I$ we have $k + I = I$. Certainly since I is closed under addition, $k + I \subset I$. On the other hand, for any $x \in I$ we can write
$$x = k + (x - k)$$
with $x - k \in I$, so also $k + I \supset I$. Thus, indeed, $k + I = I$. Thus,
$$(r' + s') + I = (r + s) + I$$

which proves the well-definedness of addition in the quotient ring. Likewise, looking at multiplication:
$$(r' \cdot s') + I = (r + i) \cdot (s + j) + I = (r \cdot s) + (rj + si + I)$$

Since I is an ideal, rj and si are again in I, and then $rj+si \in I$. Therefore, as just observed in the discussion of addition, $rj + si + I = I$. Thus,

$$(r' \cdot s') + I = (r \cdot s) + I$$

and multiplication is well-defined. The proofs that $0 + I$ is the zero and $1 + I$ is the unit are similar.

The **quotient homomorphism**

$$q : R \longrightarrow R/I$$

is the natural map

$$q(r) = r + I$$

The definition and discussion above proves

4.4.3 Proposition: For a commutative ring R and ideal I, the quotient map $R \longrightarrow R/I$ is a (surjective) ring homomorphism.   ///

4.5 *Maximal ideals and fields*

Now we see how to make fields by taking quotients of commutative rings by *maximal ideals* (defined just below). This is a fundamental construction.

Let R be a commutative ring with unit 1. [18]   An ideal M in R is **maximal** if $M \neq R$ and if for any other ideal I with $I \supset M$ it must be that $I = R$. That is, M is a maximal ideal if there is no ideal strictly larger than M (containing M) except R itself.

4.5.1 Proposition: For a commutative ring R with unit, and for an ideal I, the quotient ring R/I is a *field* if and only if I is a *maximal* ideal.

Proof: Let $x + I$ be a non-zero element of R/I. Then $x + I \neq I$, so $x \notin I$. Note that the ideal $Rx + I$ is therefore strictly larger than I. Since I was already maximal, it must be that $Rx + I = R$. Therefore, there are $r \in R$ and $i \in I$ so that $rx + i = 1$. Looking at this last equation modulo I, we have $rx \equiv 1 \bmod I$. That is, $r + I$ is the multiplicative inverse to $x + I$. Thus, R/I is a field.

On the other hand, suppose that R/I is a field. Let $x \in R$ but $x \notin I$. Then $x + I \neq 0 + I$ in R/I. Therefore, $x + I$ has a multiplicative inverse $r + I$ in R/I. That is,

$$(r + I) \cdot (x + I) = 1 + I$$

From the definition of the multiplication in the quotient, this is $rx + I = 1 + I$, or $1 \in rx + I$, which implies that the ideal $Rx + I$ is R. But $Rx + I$ is the smallest ideal containing I and x. Thus, there cannot be any proper ideal strictly larger than I, so I is maximal.   ///

---

[18] The commutativity allows us to avoid several technical worries which are not the current point, and the presence of 1 likewise skirts some less-than-primary problems. The applications we have in mind of the results of this section do not demand that we worry about those possibilities.

4.6 *Prime ideals and integral domains*

Let R be a commutative ring with unit 1. An ideal P in R is **prime** if $ab \in P$ implies either $a \in P$ or $b \in P$. [19]

4.6.1 Proposition: For a commutative ring R with unit, and for an ideal I, the quotient ring R/I is an *integral domain* [20] if and only if I is a *prime* ideal.

Proof: Let I be prime. Suppose that

$$(x + I) \cdot (y + I) = 0 + I$$

Recall that the product in the quotient is not defined exactly as the set of products of elements from the factors, but, rather, in effect,

$$(x + I) \cdot (y + I) = xy + I$$

Then $(x + I)(y + I) = 0 + I$ implies that $xy \in I$. By the prime-ness of I, either x or y is in I, so either $x + I = 0 + I$ or $y + I = 0 + I$.

On the other hand, suppose that R/I is an integral domain. Suppose that $xy \in I$. The definition $(x + I)(y + I) = xy + I$ then says that $(x + I)(y + I) = 0 + I$. Since R/I is an integral domain, either $x + I = I$ or $y + I = I$. That is, either $x \in I$ or $y \in I$, and I is prime. ///

4.6.2 Corollary: Maximal ideals are prime. [21]

Proof: If I is a maximal ideal in a ring R, then R/I is a field, from above. Fields are certainly integral domains, so I is prime, from above. ///

4.6.3 Remark: Not all prime ideals are maximal.

4.6.4 Example: Let $R = \mathbb{Z}[x]$ be the polynomial ring in one variable with integer coefficients. Consider the ideal $I = \mathbb{Z}[x] \cdot x$ generated by x. We claim that this ideal is prime, but *not* maximal. Indeed,

$$R/I = \mathbb{Z}[x]/x\mathbb{Z}[x] \approx \mathbb{Z}$$

via the homomorphism

$$P(x) + I \longrightarrow P(0)$$

(One might verify that this map is indeed well-defined, and is a homomorphism.) [22] Since $\mathbb{Z} \approx \mathbb{Z}[x]/I$ is an integral domain, I is prime, but since $\mathbb{Z}$ is not a field, I is not maximal.

---

[19] Yes, by this point the property *proven* for prime numbers is taken to be the *definition*.

[20] Again, an integral domain has no zero divisors.

[21] ... in commutative rings with identity, at least.

[22] This can be verified in different styles. One style is the following. The universality of the polynomial ring assures us that there is a unique $\mathbb{Z}$-algebra homomorphism $e : \mathbb{Z}[x] \longrightarrow \mathbb{Z}$ which sends $x \longrightarrow 0$. Implicit in the $\mathbb{Z}$-algebra homomorphism property is that $n \longrightarrow n$ for $n \in \mathbb{Z}$, so no non-zero integers lie in the kernel of this *evaluation homomorphism* e. Thus, this homomorphism is a surjection to $\mathbb{Z}$.

4.6.5 Example: Let $R = \mathbb{Z}[x]$ again and let $I = \mathbb{Z}[x] \cdot p$ be the ideal generated by a prime number p. Then

$$R/I = \mathbb{Z}[x]/p\mathbb{Z}[x] \approx (\mathbb{Z}/p)[x]$$

via the map

$$P(x) \longrightarrow (P \text{ with coefficients reduced mod } p)(x)$$

The ring $(\mathbb{Z}/p)[x]$ is an integral domain [23] but not a field, so the ideal is prime but not maximal.

4.6.6 Example: [24] Let k be a field, and consider the polynomial ring $R = k[x, y]$ in two variables. Let $I = k[x, y] \cdot x$ be the ideal generated by x. We claim that this ideal I is prime but not maximal. Indeed, the quotient R/I is [25] (naturally isomorphic to) $k[y]$ under the evaluation map

$$P(x, y) \longrightarrow P(0, y)$$

Since $k[xy]$ is an integral domain but not a field, we reach the conclusion, in light of the results just above.

4.7 *Fermat-Euler on sums of two squares*

4.7.1 Theorem: [26] A prime integer p is expressible as

$$p = a^2 + b^2$$

if and only if $p = 1 \bmod 4$ (or $p = 2$).

Proof: Parts of this are very easy. Certainly $2 = 1^2 + 1^2$. Also, if an odd [27] prime is

---

[23] Since p is prime, $\mathbb{Z}/p$ is a field, so this is a polynomial ring in one variable over a field, which we know is an integral domain.

[24] While the conclusion of this example is correct, the most natural full proof that such things are what they seem requires results we do not yet have in hand, such as Gauss' Lemma.

[25] Certainly if we have a polynomial of the form $xf(x, y)$, replacing x by 0 gives the 0 polynomial in y. On the other hand, it is less clear that $f(0, y) = 0$ implies that f is of the form $f(x, y) = xg(x, y)$ for some polynomial g. The conceptual proof of results of this sort would use the unique factorization property of $k[x, y]$, which follows from the one-variable case via Gauss' lemma. For the present case, with the special factor x (rather than a more general polynomial), a direct approach is still easy. Let $f(x, y) = xg(x, y) + h(x)$ where $h(y)$ is the collection of all monomials in $f(x, y)$ in which x does not appear. Then $f(0, y) = h(y)$. If this is the 0 polynomial in y, then $f(x, y) = xg(x, y)$.

[26] Fermat stated in correspondence that he knew this, roughly around 1650, but there was no recorded argument. About 100 years later Euler reconsidered this and many other unsupported statements of Fermat's, and gave a proof that was publicly available. In this and other cases, it is not clear that Fermat was sufficiently aware of all the things that might go wrong to enable us to be sure that he had a complete proof. It is plausible, but not clear.

[27] The phrase *odd prime* is a standard if slightly peculiar way to refer to prime integers other than 2. Sometimes the import of this is that the prime is *larger* than 2, and sometimes it really is that the prime is *odd*.

expressible as $p = a^2 + b^2$, then, since the squares modulo 4 are just 0 and 1, it must be that one of a, b is odd and one is even, and the sum of the squares is 1 modulo 4.

On the other hand, suppose that $p = 1 \bmod 4$. If p were expressible as $p = a^2 + b^2$ then

$$p = (a + bi)(a - bi)$$

where $i = \sqrt{-1}$ in $\mathbb{C}$. That is, p is expressible as a sum of two squares, if and only if p factors in a particular manner in $\mathbb{Z}[i]$. One might have at some point already observed that the only units in $\mathbb{Z}[i]$ are ± 1 and $\pm i$, so if neither of a, b is 0, then neither of $a \pm bi$ is a unit. We need to analyze the possible factorization of p in $\mathbb{Z}[i]$ a little more closely to understand the close connection to the present issue.

Let $N(a+bi) = a^2 + b^2$ be the usual (square-of) norm. One can check that the only elements of $\mathbb{Z}[i]$ with norm 1 are the 4 units, and norm 0 occurs only for 0. If $p = \alpha \cdot \beta$ is a proper factorization, then by the multiplicative property of N

$$p^2 = N(p) = N(\alpha) \cdot N(\beta)$$

Thus, since neither α nor β is a unit, it must be that

$$N(\alpha) = p = N(\beta)$$

Similarly, α and β must both be irreducibles in $\mathbb{Z}[i]$, since applying N to any proper factorization would give a contradiction. Also, since p is its own complex conjugate,

$$p = \alpha \cdot \beta$$

implies

$$p = \overline{p} = \overline{\alpha} \cdot \overline{\beta}$$

Since we know that the Gaussian integers $\mathbb{Z}[i]$ are Euclidean and, hence, have unique factorization, it must be that these two prime factors are the same *up to units*. [28]

Thus, either $\alpha = \pm \overline{\alpha}$ and $\beta = \pm \overline{\beta}$ (with matching signs), or $\alpha = \pm i \overline{\alpha}$ and $\beta = \mp i \overline{\beta}$, or $\alpha = u \overline{\beta}$ with u among $\pm 1, \pm i$. If $\alpha = \pm \overline{\alpha}$, then α is either purely imaginary or is real, and in either case its norm is a square, but no square divides p. If $\alpha = \pm i \overline{\alpha}$, then α is of the form $t \pm it$ for $t \in \mathbb{Z}$, and then $N(\alpha) \in 2\mathbb{Z}$, which is impossible.

Thus, $\alpha = u \overline{\beta}$ for some unit u, and $p = uN(\beta)$. Since $p > 0$, it must be that $u = 1$. Letting $\alpha = a + bi$, we have recovered an expression as (proper) sum of two squares

$$p = a^2 + b^2$$

Thus, a prime integer p is a (proper) sum of two squares if and only if it is *not prime* in $\mathbb{Z}[i]$. From above, this is equivalent to

$$\mathbb{Z}[i]/p\mathbb{Z}[x] \text{ is not an integral domain}$$

---

[28] This *up to units* issue is nearly trivial in $\mathbb{Z}$, since positivity and negativity give us a convenient handle. But in $\mathbb{Z}[i]$ and other rings with more units, greater alertness is required.

We grant that for $p = 1 \bmod 4$ there is an integer α such that $\alpha^2 = -1 \bmod p$. [29]   That is, (the image of) the polynomial $x^2 + 1$ factors in $\mathbb{Z}/p[x]$.

Note that we can rewrite $\mathbb{Z}[i]$ as

$$\mathbb{Z}[x]/(x^2 + 1)\mathbb{Z}[x]$$

We'll come back to this at the end of this discussion. Then [30]

$$\mathbb{Z}[i]/\langle p \rangle \approx \left(\mathbb{Z}[x]/\langle x^2 + 1 \rangle \right)/\langle p \rangle$$

$$\approx \left(\mathbb{Z}[x]/\langle p \rangle \right)/\langle x^2 + 1 \rangle \approx (\mathbb{Z}/p)[x]/\langle x^2 + 1 \rangle$$

and the latter is *not* an integral domain, since

$$x^2 + 1 = (x - \alpha)(x + \alpha)$$

is not irreducible in $(\mathbb{Z}/p)[x]$. That is, $\mathbb{Z}[i]/\langle p \rangle$ is not an integral domain when p is a prime with $p = 1 \bmod 4$. That is, p is not irreducible in $\mathbb{Z}[i]$, so factors properly in $\mathbb{Z}[i]$, thus, as observed above, p is a sum of two squares.   ///

4.7.2 Remark: Let's follow up on the isomorphism

$$\mathbb{Z}[x]/\langle x^2 + 1 \rangle \approx \mathbb{Z}[i]$$

Since $\mathbb{Z}[x]$ is the free $\mathbb{Z}$-algebra on the generator x, there is a unique $\mathbb{Z}$-algebra homomorphism $\mathbb{Z}[x] \longrightarrow \mathbb{Z}[i]$ taking x to i. We claim that the kernel is identifiable as the principal ideal generated by $x^2 + 1$, after which the obvious isomorphism theorem for rings would yield the desired isomorphism.

That is, we claim that if a polynomial $P(x)$ in $\mathbb{Z}[x]$ has the property that $P(i) = 0$, then P is a multiple (in $\mathbb{Z}[x]$) of $x^2 + 1$. This is less trivial than in the case of polynomials in one variable over a *field*, but the fact that $x^2 + 1$ is *monic* saves us. That is, we can claim that for a *monic* poly $M(x)$, given any other polynomial $P(x) \in \mathbb{Z}[x]$, there are $Q(x)$ and $R(x)$ in $\mathbb{Z}[x]$ with $\deg R < \deg M$, such that

$$P = Q \cdot M + R$$

Indeed, suppose not. Let

$$P(x) = a_n x^n + \ldots + a_0$$

---

[29] If we grant that there are primitive roots modulo primes, that is, that $(\mathbb{Z}/p)^\times$ is cyclic, then this assertion follows from basic and general properties of cyclic groups. Even without knowledge of primitive roots, we can still give a special argument in this limited case, as follows. Let $G = (\mathbb{Z}/p)^\times$. This group is abelian, and has order divisible by at least 2^2. Thus, for example by Sylow theorems, there is a 2-power-order subgroup A of order at least 4. By unique factorization in polynomial rings, the equation $x^2 - 1 = 0$ has only the solutions ± 1. Thus, there is only a *single* element in A of order 2, and the identity 1 of order 1. Other elements in A must have order a larger power of 2, and then one can arrange elements of order 4. Such things would be 4th roots of 1.

[30] A scrupulous reader should verify that the change in order of quotient-taking is legitimate. It is certainly a good trick, assuming that it works properly.

be the polynomial of least degree n which we cannot divide by M and obtain a smaller remainder. Let $m = \deg M$. Necessarily $n \geq m$ or P is itself already of lower degree than M. And, for $n \geq m$,

$$P - a_n \cdot x^{n-m} \cdot M$$

is of strictly lower degree than P, so is expressible as $QM + R$. Then

$$P = (Q + a_n x^{n-m}) \cdot M + R$$

Since the degree of R was of degree at most $n - 1$, which is strictly less than n, this contradicts the supposition that P had no such expression.

4.8 Worked examples

4.8.1 Example: Let $R = \mathbb{Z}/13$ and $S = \mathbb{Z}/221$. Show that the map

$$f : R \longrightarrow S$$

defined by $f(n) = 170 \cdot n$ is *well-defined* and is a ring homomorphism. (Observe that it does not map $1 \in R$ to $1 \in S$.)

The point is that $170 = 1 \bmod 13$ and $170 = 17 \cdot 10 = 0 \bmod 17$, and $221 = 13 \cdot 17$. Thus, for $n' = n + 13\ell$,

$$170 \cdot n' = 17 \cdot 10 \cdot n + 10 \cdot 17 \cdot 13 = 17 \cdot 10 \cdot n \bmod 13 \cdot 17$$

so the map is well-defined. Certainly the map respects addition, since

$$170(n + n') = 170n + 170n'$$

That it respects multiplication is slightly subtler, but we verify this separately modulo 13 and modulo 17, using unique factorization to know that if $13|N$ and $17|N$ then $(13 \cdot 17)|N$. Thus, since $170 = 1 \bmod 13$,

$$170(nn') = 1 \cdot (nn') = nn' = (170n) \cdot (170n') \bmod 13$$

And, since $17 = 0 \bmod 17$,

$$170(nn') = 0 \cdot (nn') = 0 = (170n) \cdot (170n') \bmod 17$$

Putting these together gives the multiplicativity.

4.8.2 Example: Let p and q be distinct prime numbers. Show directly that there is no field with pq elements.

There are several possible approaches. One is to suppose there exists such a field k, and first invoke Sylow (or even more elementary results) to know that there exist (non-zero!) elements x, y in k with (additive) orders p, q, respectively. That is, $p \cdot x = 0$ (where left multiplication by an ordinary integer means repeated addition). Then claim that $xy = 0$, contradicting the fact that a field (or even integral domain) has no proper zero divisors.

Indeed, since p and q are distinct primes, $\gcd(p,q)=1$, so there are integers r,s such that $rp+sq=1$. Then

$$xy = 1\cdot xy = (rp+sq)\cdot xy = ry\cdot px + sx\cdot qy = ry\cdot 0 + sx\cdot 0 = 0$$

4.8.3 Example: Find all the idempotent elements in $\mathbb{Z}/n$.

The idempotent condition $r^2=r$ becomes $r(r-1)=0$. For each prime p dividing n, let p^e be the exact power of p dividing n. For the image in $\mathbb{Z}/n$ of an ordinary integer b to be idempotent, it is necessary and sufficient that $p^e|b(b-1)$ for each prime p. Note that p cannot divide both b and $b-1$, since $b-(b-1)=1$. Thus, the condition is $p^e|b$ or $p^e|b-1$, for each prime p dividing n. Sun-Ze's theorem assures that we can choose either of these two conditions for each p as p various over primes dividing n, and be able to find a simultaneous solution for the resulting family of congruences. That is, let $p_1,\ldots,p_t$ be the distinct primes dividing n, and let $p_i^{e_i}$ be the exact power of p_i dividing n. For each p_i choose $\varepsilon_i\in\{0,1\}$. Given a sequence $\varepsilon=(\varepsilon_1,\ldots,\varepsilon_t)$ of 0s and 1s, consider the collection of congruences $p_i^{e_i}|(b-\varepsilon_i)$, for $i=1,\ldots,t$. Sun-Ze guarantees that there is a solution, and that it is unique mod n. Thus, each of the 2^t choices of sequences of 0s and 1s gives an idempotent.

4.8.4 Example: Find all the nilpotent elements in $\mathbb{Z}/n$.

For each prime p dividing n, let p^e be the exact power of p dividing n. For the image in $\mathbb{Z}/n$ of an ordinary integer b to be nilpotent, it is necessary and sufficient that for some n sufficiently large $p^e|b^n$ for each prime p. Then surely $p|b^n$, and since p is prime $p|b$. And, indeed, if every prime dividing n divides b, then a sufficiently large power of b will be 0 modulo p^e, hence (by unique factorization, etc.) modulo n. That is, for b to be nilpotent it is necessary and sufficient that every prime dividing n divides b.

4.8.5 Example: Let $R=\mathbb{Q}[x]/(x^2-1)$. Find e and f in R, neither one 0, such that

$$e^2=e \quad f^2=f \quad ef=0 \quad e+f=1$$

(Such e and f are **orthogonal** idempotents.) Show that the maps $p_e(r)=re$ and $p_f(r)=rf$ are ring homomorphisms of R to itself.

Let ξ be the image of x in the quotient. Then $(\xi-1)(\xi+1)=0$. Also note that

$$(\xi-1)^2 = \xi^2-2\xi+1 = (\xi^2-1)-2\xi+2 = -2\xi+2$$

so

$$\left(\frac{\xi-1}{2}\right)^2 = \frac{\xi^2-2\xi+1}{4} = \frac{(\xi^2-1)-2\xi+2}{4} = \frac{-\xi+1}{2}$$

Similarly,

$$\left(\frac{\xi+1}{2}\right)^2 = \frac{\xi^2+2\xi+1}{4} = \frac{(\xi^2-1)+2\xi+2}{4} = \frac{\xi+1}{2}$$

Thus, $e=(-\xi+1)/2$ and $f=(\xi+1)/2$ are the desired orthogonal idempotents.

4.8.6 Example: Prove that in $(\mathbb{Z}/p)[x]$ we have the factorization

$$x^p - x = \prod_{a\in\mathbb{Z}/p}(x-a)$$

By Fermat's Little Theorem, the left-hand side is 0 when x is replaced by any of $0, 1, 2, \ldots, p-1$. Thus, by unique factorization in $k[x]$ for k a field (which applies to $\mathbb{Z}/p$ since p is prime), all the factors $x-0$, $x-1$, $x-2$, $\ldots$, $x-(p-1)$ divide the left-hand side, and (because these are mutually relatively prime) so does their product. Their product is the right-hand side, which thus at least *divides* the left-hand side. Since degrees add in products, we see that the right-hand side and left-hand side could differ at most by a unit (a polynomial of degree 0), but both are *monic*, so they are identical, as claimed.

4.8.7 Example: Let $\omega = (-1+\sqrt{-3})/2$. Prove that

$$\mathbb{Z}[\omega]/p\mathbb{Z}[\omega] \approx (\mathbb{Z}/p)[x]/(x^2+x+1)(\mathbb{Z}/p)[x]$$

and, as a consequence, that a prime p in $\mathbb{Z}$ is expressible as x^2+xy+y^2 with integers x, y if and only if $p = 1 \bmod 3$ (apart from the single anomalous case $p = 3$).

If a prime is expressible as $p = a^2 + ab + b^2$, then, modulo 3, the possibilities for p modulo 3 can be enumerated by considering $a = 0, \pm 1$ and $b = 0, \pm 1 \bmod 3$. Noting the symmetry that $(a, b) \longrightarrow (-a, -b)$ does not change the output (nor does $(a, b) \longrightarrow (b, a)$) we reduce from $3 \cdot 3 = 9$ cases to a smaller number:

$$p = a^2 + ab + b^2 = \begin{cases} 0^2 + 0\cdot 0 + 0^2 & = 1 \quad \bmod 3 \\ 1^2 + 1\cdot 1 + 1^2 & = 0 \quad \bmod 3 \\ 1^2 + 1\cdot(-1) + (-1)^2 & = 1 \quad \bmod 3 \end{cases}$$

Thus, any prime p expressible as $p = a^2 + ab + b^2$ is either 3 or is 1 mod 3.

On the other hand, suppose that $p = 1 \bmod 3$. If p were expressible as $p = a^2+ab+b^2$ then

$$p = (a+b\omega)(a+b\overline{\omega})$$

where $\omega = (-1+\sqrt{-3})/2$. That is, p is expressible as $a^2 + ab + b^2$ if and only if p factors in a particular manner in $\mathbb{Z}[\omega]$.

Let $N(a+b\omega) = a^2+ab+b^2$ be the usual (square-of) norm. To determine the units in $\mathbb{Z}[\omega]$, note that $\alpha \cdot \beta = 1$ implies that

$$1 = N(\alpha) \cdot N(\beta)$$

and these norms from $\mathbb{Z}[\omega]$ are integers, so units have norm 1. By looking at the equation $a^2+ab+b^2 = 1$ with integers a, b, a little fooling around shows that the only units in $\mathbb{Z}[\omega]$ are ± 1, $\pm\omega$ and $\pm\omega^2$. And norm 0 occurs only for 0.

If $p = \alpha \cdot \beta$ is a proper factorization, then by the multiplicative property of N

$$p^2 = N(p) = N(\alpha) \cdot N(\beta)$$

Thus, since neither α nor β is a unit, it must be that

$$N(\alpha) = p = N(\beta)$$

Similarly, α and β must both be irreducibles in $\mathbb{Z}[\omega]$, since applying N to any proper factorization would give a contradiction. Also, since p is its own complex conjugate,

$$p = \alpha \cdot \beta$$

implies

$$p = \bar{p} = \bar{\alpha} \cdot \bar{\beta}$$

Since we know that the (Eisenstein) integers $\mathbb{Z}[\omega]$ are Euclidean and, hence, have unique factorization, it must be that these two prime factors are the same *up to units*.

Thus, either $\alpha = \pm\bar{\alpha}$ and $\beta = \pm\bar{\beta}$ (with matching signs), or $\alpha = \pm\omega\bar{\alpha}$ and $\beta = \pm\omega^2\bar{\beta}$, or $\alpha = \pm\omega^2\bar{\alpha}$ and $\beta = \pm\omega\bar{\beta}$, or $\alpha = u\bar{\beta}$ with u among $\pm 1, \pm\omega, \pm\omega^2$. If $\alpha = \pm\bar{\alpha}$, then α is either in $\mathbb{Z}$ or of the form $t \cdot \sqrt{-3}$ with $t \in \mathbb{Z}$. In the former case its norm is a square, and in the latter its norm is divisible by 3, neither of which can occur. If $\bar{\alpha} = \omega\alpha$, then $\alpha = t \cdot \omega$ for some $t \in \mathbb{Z}$, and its norm is a square, contradiction. Similarly for $\alpha = \pm\omega^2\bar{\alpha}$.

Thus, $\alpha = u\bar{\beta}$ for some unit u, and $p = uN(\beta)$. Since $p > 0$, it must be that $u = 1$. Letting $\alpha = a + b\omega$, we have recovered an expression

$$p = a^2 + ab + b^2$$

with neither a nor b zero.

Thus, a prime integer $p > 3$ is expressible (properly) as $a^2 + ab + b^2$ of two squares if and only if it is *not prime* in $\mathbb{Z}[\omega]$. From above, this is equivalent to

$$\mathbb{Z}[\omega]/\langle p \rangle \text{ is not an integral domain}$$

We grant that for $p = 1 \bmod 3$ there is an integer α such that $\alpha^2 + alf + 1 = 0 \bmod p$. [31] That is, (the image of) the polynomial $x^2 + x + 1$ factors in $(\mathbb{Z}/p)[x]$.

Note that we can rewrite $\mathbb{Z}[\omega]$ as

$$\mathbb{Z}[x]/\langle x^2 + x + 1 \rangle$$

Then

$$\mathbb{Z}[\omega]/\langle p \rangle \approx \left(\mathbb{Z}[x]/\langle x^2 + 1 \rangle\right)/\langle p \rangle \approx \left(\mathbb{Z}[x]/\langle p \rangle\right)/\langle x^2 + 1 \rangle \approx (\mathbb{Z}/p)[x]/\langle x^2 + 1 \rangle$$

and the latter is *not* an integral domain, since

$$x^2 + x + 1 = (x - \alpha)(x - \alpha^2)$$

is not irreducible in $(\mathbb{Z}/p)[x]$. That is, $\mathbb{Z}[\omega]/\langle p \rangle$ is not an integral domain when p is a prime with $p = 1 \bmod 3$. That is, p is not irreducible in $\mathbb{Z}[\omega]$, so factors properly in $\mathbb{Z}[\omega]$, thus, as observed above, p is expressible as $a^2 + ab + b^2$.   ///

---

[31] If we grant that there are primitive roots modulo primes, that is, that $(\mathbb{Z}/p)^\times$ is cyclic, then this assertion follows from basic and general properties of cyclic groups. Even without knowledge of primitive roots, we can still give a special argument in this limited case, as follows. Let $G = (\mathbb{Z}/p)^\times$. This group is abelian, and has order divisible by 3. Thus, for example by Sylow theorems, there is a 3-power-order subgroup A, and, thus, at least one element of order exactly 3.

Exercises

4.1 Show that in a commutative ring the set of nilpotent elements is an ideal (the **nilradical** of R). Give an example to show that the set of nilpotent elements may fail to be an ideal in a non-commutative ring.

4.2 Let R be a commutative ring with unit, such that for every $r \in R$ there is an integer $n > 1$ (possibly depending upon r) such that $r^n = r$. Show that every prime ideal in R is maximal.

4.3 Let k be a field. Let P, Q be two polynomials in $k[x]$. Let K be an extension field of k. Show that, if P divides Q in $K[x]$, then P divides Q in $k[x]$.

4.4 Let R be a commutative ring with unit. Show that the set of prime ideals in R has minimal elements under the ordering by inclusion. (*Hint:* You may want to use Zorn's lemma or some other equivalent of the Axiom of Choice.)

4.5 The **radical** of an ideal I in a commutative ring R with unit is

$$\mathrm{rad}\, I = \{r \in R : r^n \in I \text{ for some } n\}$$

Show that a proper ideal I of a ring is equal to its own radical if and only if it is an intersection of prime ideals.

4.6 Let R be a commutative ring with unit. Check that the **nilradical** N of R, defined to be the set of all nilpotent elements, is

$$\mathrm{nilrad}\, R = \mathrm{rad}\, \{0\}$$

Show that R has a *unique* prime ideal if and only if every element of R is either nilpotent or a unit, if and only if R/N is a field.

4.7 Show that a prime p in $\mathbb{Z}$ is expressible as $p = m^2 + 2n^2$ with integers m, n if and only if -2 is a square mod p.

4.8 Let R be a commutative ring with unit. Suppose R contains an *idempotent* element r other than 0 or 1. (That is, $r^2 = r$.) Show that every prime ideal in R contains an idempotent other than 0 or 1.

5. Linear algebra I: dimension

5.1 *Some simple results*

Several observations should be made. Once stated explicitly, the proofs are easy. [1]

- The intersection of a (non-empty) set of subspaces of a vector space V is a subspace.

Proof: Let $\{W_i : i \in I\}$ be a set of subspaces of V. For w in every W_i, the additive inverse $-w$ is in W_i. Thus, $-w$ lies in the intersection. The same argument proves the other properties of subspaces. ///

The **subspace spanned** by a set X of vectors in a vector space V is the intersection of all subspaces containing X. From above, this intersection is a subspace.

- The subspace spanned by a set X in a vector space V is the collection of all linear combinations of vectors from X.

---

[1] At the beginning of the abstract form of this and other topics, there are several results which have little informational content, but, rather, only serve to assure us that the definitions/axioms have not included phenomena too violently in opposition to our expectations. This is not surprising, considering that the definitions have endured several decades of revision exactly to address foundational and other potential problems.

Proof: Certainly every linear combination of vectors taken from X is in any subspace containing X. On the other hand, we must show that any vector in the intersection of subspaces containing X is a linear combination of vectors in X. Now it is not hard to check that the collection of such linear combinations is *itself* a subspace of V, and contains X. Therefore, the intersection is no larger than this set of linear combinations.   ///

A *linearly independent* set of vectors *spanning* a subspace W of V is a **basis** for W.

5.1.1 Proposition: Given a basis $e_1, \ldots, e_n$ for a vector space V, there is *exactly one* expression for an arbitrary vector $v \in V$ as a linear combination of $e_1, \ldots, e_n$.

Proof: That there is *at least* one expression follows from the spanning property. On the other hand, if

$$\sum_i a_i e_i = v = \sum_i b_i e_i$$

are two expressions for v, then subtract to obtain

$$\sum_i (a_i - b_i) e_i = 0$$

Since the e_i are linearly independent, $a_i = b_i$ for all indices i.   ///

5.2 *Bases and dimension*

The argument in the proof of the following fundamental theorem is the *Lagrange replacement principle*. This is the first non-trivial result in linear algebra.

5.2.1 Theorem: Let $v_1, \ldots, v_m$ be a linearly independent set of vectors in a vector space V, and let $w_1, \ldots, w_n$ be a basis for V. Then $m \leq n$, and (renumbering the vectors w_i if necessary) the vectors

$$v_1, \ldots, v_m, w_{m+1}, w_{m+2}, \ldots, w_n$$

are a basis for V.

Proof: Since the w_i's are a basis, we may express v_1 as a linear combination

$$v_1 = c_1 w_1 + \ldots + c_n w_n$$

Not all coefficients can be 0, since v_1 is not 0. Renumbering the w_i's if necessary, we can assume that $c_1 \neq 0$. Since the scalars k are a *field*, we can express w_1 in terms of v_1 and $w_2, \ldots, w_n$

$$w_1 = c_1^{-1} v_1 + (-c_1^{-1} c_2) w_2 + \ldots + (-c_1^{-1} c_2) w_n$$

Replacing w_1 by v_1, the vectors $v_1, w_2, w_3, \ldots, w_n$ span V. They are still linearly independent, since if v_1 were a linear combination of $w_2, \ldots, w_n$ then the expression for w_1 in terms of $v_1, w_2, \ldots, w_n$ would show that w_1 was a linear combination of $w_2, \ldots, w_n$, contradicting the linear independence of $w_1, \ldots, w_n$.

Suppose inductively that $v_1, \ldots, v_i, w_{i+1}, \ldots, w_n$ are a basis for V, with $i < n$. Express v_{i+1} as a linear combination

$$v_{i+1} = a_1 v_1 + \ldots + a_i v_i + b_{i+1} w_{i+1} + \ldots + b_n w_n$$

Some b_j is non-zero, or else v_i is a linear combination of $v_1, \ldots, v_i$, contradicting the linear independence of the v_j's. By renumbering the w_j's if necessary, assume that $b_{i+1} \neq 0$. Rewrite this to express w_{i+1} as a linear combination of $v_1, \ldots, v_i, w_{i+1}, \ldots, w_n$

$$w_{i+1} = (-b_{i+1}^{-1}a_1)v_1 + \ldots + (-b_{i+1}^{-1}a_i)v_i + (b_{i+1}^{-1})v_{i+1}$$

$$+ (-b_{i+1}^{-1}b_{i+2})w_{i+2} + \ldots + (-b_{i+1}^{-1}b_n)w_n$$

Thus, $v_1, \ldots, v_{i+1}, w_{i+2}, \ldots, w_n$ span V. Claim that these vectors are linearly independent: if for some coefficients a_j, b_j

$$a_1 v_1 + \ldots + a_{i+1}v_{i+1} + b_{i+2}w_{i+2} + \ldots + b_n w_n = 0$$

then some a_{i+1} is non-zero, because of the linear independence of $v_1, \ldots, v_i, w_{i+1}, \ldots, w_n$. Thus, rearrange to express v_{i+1} as a linear combination of $v_1, \ldots, v_i, w_{i+2}, \ldots, w_n$. The expression for w_{i+1} in terms of $v_1, \ldots, v_i, v_{i+1}, w_{i+2}, \ldots, w_n$ becomes an expression for w_{i+1} as a linear combination of $v_1, \ldots, v_i, w_{i+2}, \ldots, w_n$. But this would contradict the (inductively assumed) linear independence of $v_1, \ldots, v_i, w_{i+1}, w_{i+2}, \ldots, w_n$.

Consider the possibility that $m > n$. Then, by the previous argument, $v_1, \ldots, v_n$ is a basis for V. Thus, v_{n+1} is a linear combination of $v_1, \ldots, v_n$, contradicting their linear independence. Thus, $m \leq n$, and $v_1, \ldots, v_m, w_{m+1}, \ldots, w_n$ is a basis for V, as claimed.
///

Now define the $(k\text{-})$**dimension** [2] of a vector space (over field k) as the number of elements in a $(k\text{-})$basis. The theorem says that this number is well-defined. Write

$$\dim V = \text{ dimension of } V$$

A vector space is **finite-dimensional** if it has a finite basis. [3]

5.2.2 Corollary: A linearly independent set of vectors in a finite-dimensional vector space can be augmented to be a basis.

Proof: Let $v_1, \ldots, v_m$ be as linearly independent set of vectors, let $w_1, \ldots, w_n$ be a basis, and apply the theorem.
///

5.2.3 Corollary: The dimension of a *proper* subspace of a finite-dimensional vector space is strictly less than the dimension of the whole space.

Proof: Let $w_1, \ldots, w_m$ be a basis for the subspace. By the theorem, it can be extended to a basis $w_1, \ldots, w_m, v_{m+1}, \ldots, v_n$ of the whole space. It must be that $n > m$, or else the subspace is the whole space.
///

---

[2] This is an instance of terminology that is nearly too suggestive. That is, a naive person might all too easily accidentally assume that there is a connection to the colloquial sense of the word *dimension*, or that there is an appeal to physical or visual intuition. Or one might assume that it is somehow *obvious* that *dimension* is a well-defined invariant.

[3] We proved only the finite-dimensional case of the well-definedness of dimension. The infinite-dimensional case needs *transfinite induction* or an equivalent.

5.2.4 Corollary: The dimension of k^n is n. The vectors

$$\begin{aligned} e_1 &= (1, 0, 0, \ldots, 0, 0) \\ e_2 &= (0, 1, 0, \ldots, 0, 0) \\ e_3 &= (0, 0, 1, \ldots, 0, 0) \\ &\cdots \\ e_n &= (0, 0, 0, \ldots, 0, 1) \end{aligned}$$

are a basis (the **standard basis**).

Proof: Those vectors *span* k^n, since

$$(c_1, \ldots, c_n) = c_1 e_1 + \ldots + c_n e_n$$

On the other hand, a linear dependence relation

$$0 = c_1 e_1 + \ldots + c_n e_n$$

gives

$$(c_1, \ldots, c_n) = (0, \ldots, 0)$$

from which each c_i is 0. Thus, these vectors are a basis for k^n.   ///

5.3 *Homomorphisms and dimension*

Now we see how dimension behaves under homomorphisms.

Again, a vector space **homomorphism** [4] $f : V \longrightarrow W$ from a vector space V over a field k to a vector space W over the same field k is a function f such that

$$\begin{aligned} f(v_1 + v_2) &= f(v_1) + f(v_2) \quad \text{(for all } v_1, v_2 \in V) \\ f(\alpha \cdot v) &= \alpha \cdot f(v) \quad \text{(for all } \alpha \in k, \, v \in V) \end{aligned}$$

The **kernel** of f is

$$\ker f = \{v \in V : f(v) = 0\}$$

and the **image** of f is

$$\operatorname{Im} f = \{f(v) : v \in V\}$$

A homomorphism is an **isomorphism** if it has a two-sided inverse homomorphism. For vector spaces, a homomorphism that is a bijection is an isomorphism. [5]

• A vector space homomorphism $f : V \longrightarrow W$ sends 0 (in V) to 0 (in W, and, for $v \in V$, $f(-v) = -f(v)$. [6]

---

[4] Or **linear map** or **linear operator**.

[5] In most of the situations we will encounter, bijectivity of various sorts of homomorphisms is sufficient (and certainly necessary) to assure that there is an inverse map of the same sort, justifying this description of *isomorphism*.

[6] This follows from the analogous result for groups, since V with its additive structure is an abelian group.

5.3.1 Proposition: The kernel and image of a vector space homomorphism $f : V \longrightarrow W$ are vector subspaces of V and W, respectively.

Proof: Regarding the kernel, the previous proposition shows that it contains 0. The last bulleted point was that additive inverses of elements in the kernel are again in the kernel. For $x, y \in \ker f$

$$f(x + y) = f(x) + f(y) = 0 + 0 = 0$$

so $\ker f$ is closed under addition. For $\alpha \in k$ and $v \in V$

$$f(\alpha \cdot v) = \alpha \cdot f(v) = \alpha \cdot 0 = 0$$

so $\ker f$ is closed under scalar multiplication. Thus, the kernel is a vector subspace.

Similarly, $f(0) = 0$ shows that 0 is in the image of f. For $w = f(v)$ in the image of f and $\alpha \in k$

$$\alpha \cdot w = \alpha \cdot f(v) = f(\alpha v) \in \operatorname{Im} f$$

For $x = f(u)$ and $y = f(v)$ both in the image of f,

$$x + y = f(u) + f(v) = f(u + v) \in \operatorname{Im} f$$

And from above

$$f(-v) = -f(v)$$

so the image is a vector subspace. ///

5.3.2 Corollary: A linear map $f : V \longrightarrow W$ is injective if and only if its kernel is the trivial subspace $\{0\}$.

Proof: This follows from the analogous assertion for groups. ///

5.3.3 Corollary: Let $f : V \longrightarrow W$ be a vector space homomorphism, with V finite-dimensional. Then

$$\dim \ker f + \dim \operatorname{Im} f = \dim V$$

Proof: Let $v_1, \ldots, v_m$ be a basis for $\ker f$, and, invoking the theorem, let $w_{m+1}, \ldots, w_n$ be vectors in V such that $v_1, \ldots, v_m, w_{m+1}, \ldots, w_n$ form a basis for V. We claim that the images $f(w_{m+1}), \ldots, f(w_n)$ are a basis for $\operatorname{Im} f$. First, show that these vectors *span*. For $f(v) = w$, express v as a linear combination

$$v = a_1 v_1 + \ldots + a_m v_m + b_{m+1} w_{m+1} + \ldots + b_n w_n$$

and apply f

$$w = a_1 f(v_1) + \ldots + a_m f(v_m) + b_{m+1} f(w_{m+1}) + \ldots + b_n f(w_n)$$

$$= a_1 \cdot 0 + \ldots + a_m \cdot 0(v_m) + b_{m+1} f(w_{m+1}) + \ldots + b_n f(w_n)$$

$$= b_{m+1} f(w_{m+1}) + \ldots + b_n f(w_n)$$

since the v_is are in the kernel. Thus, the $f(w_j)$'s *span* the image. For linear independence, suppose

$$0 = b_{m+1} f(w_{m+1}) + \ldots + b_n f(w_n)$$

Then

$$0 = f(b_{m+1}w_{m+1} + \ldots + b_n w_n)$$

Then, $b_{m+1}w_{m+1} + \ldots + b_n w_n$ would be in the kernel of f, so would be a linear combination of the v_i's, contradicting the fact that $v_1, \ldots, v_m, w_{m+1}, \ldots, w_n$ is a basis, unless all the b_j's were 0. Thus, the $f(w_j)$ are linearly independent, so are a basis for $\operatorname{Im} f$. ///

Exercises

5.1 For subspaces V, W of a vector space over a field k, show that

$$\dim_k V + \dim_k W = \dim_k(V + W) + \dim_k(V \cap W)$$

5.2 Given two bases $e_1, \ldots, e_n$ and $f_1, \ldots, f_n$ for a vector space V over a field k, show that there is a unique k-linear map $T : V \longrightarrow V$ such that $T(e_i) = f_i$.

5.3 Given a basis $e_1, \ldots, e_n$ of a k-vectorspace V, and given arbitrary vectors $w_1, \ldots, w_n$ in a k-vectorspace W, show that there is a unique k-linear map $T : V \longrightarrow W$ such that $Te_i = w_i$ for all indices i.

5.4 The space $\operatorname{Hom}_k(V, W)$ of k-linear maps from one k-vectorspace V to another, W, is a k-vectorspace under the operation

$$(\alpha \dot{T})(v) = \alpha \cdot (T(v))$$

for $\alpha \in k$ and $T \in \operatorname{Hom}_k(V, W)$. Show that

$$\dim_k \operatorname{Hom}_k(V, W) = \dim_k V \cdot \dim_k W$$

5.5 A **flag** $V_1 \subset \ldots \subset V_\ell$ of subspaces of a k-vectorspace V is simply a collection of subspaces satisfying the indicated inclusions. The **type** of the flag is the list of *dimensions* of the subspaces V_i. Let W be a k-vectorspace, with a flag $W_1 \subset \ldots \subset W_\ell$ of the same *type* as the flag in V. Show that there exists a k-linear map $T : V \longrightarrow W$ such that T restricted to V_i is an isomorphism $V_i \longrightarrow W_i$.

5.6 Let $V_1 \subset V_\ell$ be a flag of subspace inside a finite-dimensional k-vectorspace V, and $W_1 \subset \ldots \subset W_\ell$ a flag inside another finite-dimensional k-vectorspace W. We do not suppose that the two flags are of the same type. Compute the dimension of the space of k-linear homomorphisms $T : V \longrightarrow W$ such that $TV_i \subset W_i$.

6. Fields I

6.1 *Adjoining things*

The general *intention* of *adjoining* a new element α to a field k is arguably clear: k itself does *not* contain a root of an equation, and we want to enlarge k so that it *does* include such a root. The possibility or legitimacy of doing so may seem to depend upon one's philosophical outlook, but the situation is more robust than that. [1]

Let k be a field. Let $k \subset K$ where K is a bigger field. For $\alpha \in K$, define the **field extension (in K) over k generated by** α [2]

$$k(\alpha) = \bigcap_{\text{fields } E \subset K,\ E \supset k,\ \alpha \in E} E$$

---

[1] In the 19th century there was widespread confusion or at least concern over issues of existence of *quantities* having various properties. Widespread belief in the legitimacy of the complex numbers was not in place until well into that century, and ironically was abetted by pictorial emphasis on complex numbers as *two-dimensional* things. The advent of the Hamiltonian quaternions in mid-century made the complex numbers seem innocent by comparison.

[2] The notation here is uncomfortably fragile: exchanging the parentheses for any other delimiters alters the meaning.

It is easy to check that the intersection of subfields of a common field is a field, so this intersection is a field. Rather than a single element, one could as well adjoin any subset of the over-field K. [3]

Before studying $k(\alpha)$ in more detail, consider a different procedure of *adjoining* something: for a commutative ring R with 1 which is a subring of an R-algebra A, for $\alpha \in A$, one might *attempt* to define [4]

$$R[\alpha] = \{\text{ polynomials in } \alpha\}$$

One probably understands the intent, that this is

$$R[\alpha] = \{c_0 + c_1\alpha + \ldots + c_n\alpha^n : c_i \in R\}$$

More precisely, a proper definition would be

$$R[\alpha] = \text{the image in } A \text{ of the unique } R\text{-algebra homomorphism sending } x \text{ to } \alpha$$

where we invoke the universal mapping property of $R[x]$.

Specialize R again to be a field k, and let A be a (not necessarily commutative) k-algebra, $\alpha \in A$. Then the natural homomorphism

$$\varphi : k[x] \longrightarrow k[\alpha] \quad (\text{by } x \longrightarrow \alpha)$$

has a kernel which is a principal ideal $\langle f \rangle$. [5] By the usual Isomorphism Theorem the map φ *descends* to the quotient by the kernel, giving an isomorphism

$$\overline{\varphi} : k[x]/\langle f \rangle \approx k[\alpha]$$

If $f = 0$, that is, if the kernel is trivial, then $k[\alpha]$ of course inherits properties of the polynomial ring. [6]

At this point we need to begin using the fact that a k-algebra A is a k-vectorspace. [7] The **degree** of A over k is

$$[A : k] = \text{degree of } A \text{ over } k = \text{ dimension of } A \text{ as } k\text{-vectorspace}$$

If $k[\alpha] \approx k[x]$, then, for example, the various powers of α are linearly independent over k, and $k[\alpha]$ is infinite-dimensional as a k-vectorspace. And there is *no* polynomial $P(x) \in k[x]$

---

[3] This definition does *not* give a good computational handle on such field extensions. On the other hand, it is unambiguous and well-defined.

[4] Note again the fragility of the notation: $k(\alpha)$ is generally quite different from $k[\alpha]$, although in some useful cases (as below) the two can coincide.

[5] ... since $k[x]$ is a principal ideal domain for k a field. For more general commutative rings R the corresponding discussion is more complicated, though not impossible.

[6] ... to which it is *isomorphic* by the just-demonstrated isomorphism!

[7] By *forgetting* the multiplication in A, if one insists.

such that $P(\alpha) = 0$. Especially in the simple situation that the k-algebra A is a *field*, such elements α with $k[\alpha] \approx k[x]$ are **transcendental** over k. [8]

On the other hand, a perhaps more interesting situation is that in which the kernel of the natural

$$k[x] \longrightarrow k[\alpha]$$

has non-zero kernel $\langle f \rangle$, with f monic without loss of generality. This f is the **minimal polynomial** of α (in A) over k.

Although our immediate concern is field extensions, there is at least one other useful application of this viewpoint, as follows. Let V be a k-vectorspace, and let A be the k-algebra

$$A = \mathrm{End}_k V$$

of k-linear maps (i.e., **endomorphisms**) of V to itself. For $T : V \longrightarrow V$ a k-linear map, we can consider the natural k-algebra map

$$k[x] \longrightarrow \mathrm{End}_k V \quad (\text{by } x \longrightarrow T)$$

We give $\mathrm{End}_k V$ a k-vectorspace structure value-wise by

$$(\alpha \cdot T)(v) = \alpha \cdot (Tv)$$

for $v \in V$ and $\alpha \in k$. If V is finite-dimensional, then $\mathrm{End}_k V$ is also finite-dimensional. [9] In particular, the kernel of the natural map from $k[x]$ cannot be just 0. Let f be the non-zero monic generator for the kernel. Again, [10] this monic is the **minimal polynomial** for T. The general construction shows that for any $P(x) \in k[x]$,

$$P(T) = 0 \in \mathrm{End}_k V \quad \text{if and only if } f \text{ divides } P$$

In particular, if the polynomial equation $f(x) = 0$ has a *root* λ in k, then [11] we can prove that T has **eigenvalue** λ. That is, there is a non-zero vector $v \in V$ (the λ-**eigenvector**) such that

$$Tv = \lambda \cdot v$$

Indeed, let $f(x) = (x - \lambda) \cdot g(x)$ for some $g(x) \in k[x]$. Since g is not the minimal polynomial for T, then there is a vector $w \in V$ such that $g(T) \cdot w \neq 0$. We claim that $v = g(T)w$ is a λ-eigenvector. Indeed,

$$0 = f(T) \cdot w = (T - \lambda) \cdot g(T)w = (T - \lambda) \cdot v$$

---

[8] This is an essentially *negative* definition: there are no relations.

[9] This is not hard to prove: let $e_1, \ldots, e_n$ be a k-basis for V. Then the k-linearity $T(\sum_i c_i e_i) = \sum_i c_i T(e_i)$ shows that T is determined completely by the collection of images Te_i. And $Te_i = \sum_j T_{ij} e_j$ for some collection of n^2 elements T_{ij} of k. Thus, if V is n-dimensional then its endomorphism algebra is n^2-dimensional.

[10] This is terminology completely consistent with linear algebra usage.

[11] From the fact that roots correspond perfectly to linear factors, for polynomials in one variable with coefficients in a field.

and by the previous comment $v = g(T)w$ is not 0. [12]

Returning to field extensions: let K be a field containing a smaller field k, $\alpha \in K$, and let f be the generator for the kernel of the natural map $k[x] \longrightarrow k[\alpha]$. We do assume that f is non-zero, so we can make f monic, without loss of generality. Since f is non-zero, we do call it the **minimal polynomial** of α over k, and, since α *has* a minimal polynomial over k, we say that α is **algebraic** over k. [13]   If every element α of a field extension K of k is algebraic over k, then say that the field extension K itself is **algebraic** over k.

Once again, given any polynomial $P(x)$, there are *unique* $Q(x)$ and $R(x)$ with $\deg R < \deg f$ such that

$$P = Q \cdot f + R$$

and

$$P(\alpha) = Q(\alpha) \cdot f(\alpha) + R(\alpha) = Q(\alpha) \cdot 0 + R(\alpha) = R(\alpha)$$

Letting $n = \deg f$, this implies that $1, \alpha, \alpha^2, \ldots, \alpha^{n-1}$ are a k-basis for $k[\alpha]$. [14]

6.1.1 Proposition: For α algebraic over k (all inside K), the ring $k[\alpha]$ is a *field*. [15] That is, for α *algebraic*, $k(\alpha) = k[\alpha]$. The minimal polynomial f of α over k is *irreducible* in $k[x]$. And the degree (dimension) of $k(\alpha)$ over k is

$$[k(\alpha) : k] = \dim_k k(\alpha) = \deg f$$

Proof: First, from above,

$$k[\alpha] \approx k[x]/\langle f \rangle$$

To prove irreducibility, suppose we can write $f = g \cdot h$ with $g, h \in k[x]$ with proper factors. By minimality of f, neither $g(\alpha)$ nor $h(\alpha)$ is 0. But $f(\alpha) = 0$, so $g(\alpha$ and $h(\alpha)$ are zero-divisors, contradiction. [16]

Since $k(\alpha)$ is the smallest field inside the ambient field K containing α and k, certainly $k[\alpha] \subset k(\alpha)$. To prove equality, it would suffice to show that non-zero elements of $k[\alpha]$ have multiplicative inverses in $k[\alpha]$. For polynomial $g(x) \in k[x]$, $g(\alpha) \neq 0$ if and only if the minimal polynomial $f(x)$ of α over k does not divide $g(x)$. Since f is irreducible and does not divide g, there are polynomials r, s in $k[x]$ such that

$$1 = \gcd(f, g) = r(x) \cdot f(x) + s(x) \cdot g(x)$$

---

[12] Even this brief discussion of minimal polynomials and linear operators should suggest, and correctly so, that use of determinants and invocation of the Cayley-Hamilton theorem, concerning the *characteristic polynomial* of a linear operator, is not exactly to the point.

[13] Again, this situation, where $f(\alpha) = 0$ with a non-zero polynomial f, is in contrast to the case where α satisfies *no* algebraic equation with coefficients in k.

[14] Indeed, the identity $P = Qf + R$ shows that any polynomial in α is expressible as a polynomial of degree $< n$. This proves spanning. On the other hand, a linear dependence relation $\sum_i c_i \alpha^i = 0$ with coefficient c_i in k is nothing other than a polynomial relation, and our hypothesis is that any such is a (polynomial) multiple of f. Thus, the monomials of degrees less than $\deg f$ are linearly independent.

[15] This should be a little surprising.

[16] Everything is taking place inside the larger field K.

so, mapping x to α,

$$1 = r(\alpha) \cdot f(\alpha) + s(\alpha) \cdot g(\alpha) = r(\alpha) \cdot 0 + s(\alpha) \cdot g(\alpha) = s(\alpha) \cdot g(\alpha)$$

That is, $s(\alpha)$ is a multiplicative inverse to $g(\alpha)$, and $k[\alpha]$ is a field. The degree is as asserted, since the polynomials of degree $< \deg f$ are irredundant representatives for the equivalence classes of $k[x]/\langle f \rangle$.   ///

6.2 *Fields of fractions, fields of rational functions*

For $k \subset K$ fields and $\alpha \in K$ transcendental over k, it is not true that $k[\alpha] \approx k(\alpha)$, in complete contrast to the case that α is algebraic, discussed just above. [17]

But from elementary mathematics we have the idea that for *indeterminate* [sic] x

$$k(x) = \text{ field of rational functions in } x \ = \{\frac{g(x)}{h(x)} : g, h \in k[x], \ h \neq 0\}$$

We can reconcile this primitive idea with our present viewpoint.

Let R be an *integral domain* (with unit 1) [18]   and define the **field of fractions** Q of R to be the collection of ordered pairs (r, s) with $r, s \in R$ and $s \neq 0$, modulo the equivalence relation [19]

$$(r, s) \sim (r', s') \qquad \text{if} \qquad rs' = sr'$$

The addition is suggested by the usual addition of fractions, namely that

$$(r, s) + (r', s') = (rs' + r's, ss')$$

and the multiplication is the more obvious

$$(r, s) \cdot (r', s') = (rr', ss')$$

One should verify that these operations are well-defined on the quotient Q by that equivalence relation, that Q is a commutative ring with unit (the equivalence class of)$(1, 1)$, that $r \longrightarrow (r, 1)$ *injects* R to Q. This constructs the field of fractions.

The latter construction is *internal,* in the sense that it constructs a concrete thing in set-theoretic terms, given the original ring R. On the other hand, we can characterize the field of fractions *externally,* by properties of its mappings to other rings or fields. In particular, we have

6.2.1 Proposition: For an integral domain R with unit 1, its field of fractions Q, with the natural inclusion $i : R \longrightarrow Q$, is the unique field (and inclusion of R into it) such that, for any *injective* ring homomorphism $\varphi : R \longrightarrow K$ with a field K, there is a unique $\tilde{\varphi} : Q \longrightarrow K$ such that

$$\varphi \circ i = \tilde{\varphi}$$

Specifically, $\tilde{\varphi}(r, s) = \varphi(r)/\varphi(s)$. [20]

---

[17] In particular, since $k[\alpha] \approx k[x]$, $k[\alpha]$ is not a field at all.

[18] A definition can be made for more general commutative rings, but the more general definition has more complicated features which are not of interest at the moment.

[19] This would be the usual requirement that two fractions r/s and r'/s' be equal.

[20] Implicitly we must claim that this is well-defined.

Proof: Indeed, try to define [21]

$$\tilde{\varphi}(r,s) = \varphi(r)/\varphi(s)$$

where the quotient on the right-hand side is in the field K, and the injectivity of φ assure that $s \neq 0$ implies that $\varphi(s) \neq 0$. This is certainly compatible with φ on R, since

$$\tilde{\varphi}(r,1) = \varphi(r)/\varphi(1) = \varphi(r)$$

and the smallest subfield of K containing R certainly must contain all such quotients. The main thing to check is that this definition really is well-defined, namely that if $(r,s) \sim (r',s')$, then

$$\tilde{\varphi}(r,s) = \tilde{\varphi}(r',s')$$

Do this as follows. The equivalence relation is that $rs' = r's$. Applying φ on R gives

$$\varphi(r)\varphi(s') = \varphi(r')\varphi(s)$$

Since φ is injective, for s,s' nonzero in R their images are nonzero in K, so we can divide, to obtain

$$\varphi(r)/\varphi(s) = \varphi(r')/\varphi(s')$$

This proves the well-definedness. That multiplication is preserved is easy, and that addition is preserved is straightforward. ///

To practice categorical arguments, we can also prove, without using formulas or explicit constructions:

6.2.2 Proposition: Let Q' be a field with inclusion $i' : R \longrightarrow Q'$ such that, for every injective homomorphism $\varphi : R \longrightarrow K$ with a field K, there is a unique $\tilde{\varphi} : Q' \longrightarrow K$ such that

$$\varphi \circ i' = \tilde{\varphi}$$

Then there is a unique isomorphism $j : Q \longrightarrow Q'$ of the field of fractions Q of R (with inclusion $i : R \longrightarrow Q$) to Q' such that

$$i' = j \circ i$$

That is, up to unique isomorphism, there is only one field of fractions of an integral domain.

Proof: First prove that any field map $f : Q \longrightarrow Q$ such that $f \circ i = i$ must be the identity on Q. Indeed, taking $K = Q$ and $f = i : R \longrightarrow K$ in the defining property, we see that the identity map id_Q on Q has the property $\mathrm{id}_K \circ i = i$. The *uniqueness* property assures that any other f with this property must be id_K.

Then let Q' and $i' : R \longrightarrow Q'$ be another pair satisfying the universal mapping condition. Taking $\varphi = i' : R \longrightarrow Q'$ yields $\tilde{\varphi} : Q \longrightarrow Q'$ with $\varphi = \tilde{\varphi} \circ i$. Reversing the roles, taking $\varphi' = i : R \longrightarrow Q$ yields $\tilde{\varphi}' : Q' \longrightarrow Q$ with $\varphi' = \tilde{\varphi}' \circ i'$. Then (by the previous paragraph) $\tilde{\varphi} \circ \tilde{\varphi}' : Q \longrightarrow Q$ must be the identity on Q, and, similarly, $\tilde{\varphi}' \circ \tilde{\varphi} : Q' \longrightarrow Q$; must be the

[21] What else could it be?

identity on Q'. Thus, $\tilde{\varphi}$ and $\tilde{\varphi}'$ are mutual inverses. This proves the isomorphism of the two objects. [22] ///

Thus, without having a larger field in which the polynomial ring $k[x]$ sits, we simply form the field of fractions of this integral domain, and denote it [23]

$$k(x) = \text{field of fractions of } k[x] = \text{rational functions in } x$$

Despite having this construction available, it still may be true that for fields $k \subset K$, there is α in K *transcendental* over k, in the sense (above) that α satisfies no polynomial relation with coefficients in k. [24] In that case, we have the more general definition of $k(\alpha)$ as the intersection of all subfields of K containing k and containing α.

For notational consistency, we should check that $k(\alpha)$ is isomorphic to the field of fractions of $k[\alpha]$. And, indeed, since $k[x]$ injects to $k[\alpha]$ (taking x to α), by the mapping property characterization the field of fractions $k(x)$ of $k[x]$ has a unique injection j to the field $k(\alpha)$ extending the given map. Certainly $k(\alpha) \subset j(k(x))$, since $k(\alpha)$ is the intersection of all subfields of K containin k and α. Thus, the image of the injective map j is exactly $k(\alpha)$, and j is an isomorphism of $k(x)$ to $k(\alpha)$.

6.3 *Characteristics, finite fields*

The linear algebra viewpoint is decisive in understanding many elementary features of fields, for example, the result below on possible cardinalities of finite fields.

First, observe that any ring R is a $\mathbb{Z}$-algebra in a canonical [25] manner, with the action

$$n \cdot r = \begin{cases} \underbrace{r + \ldots + r}_{n} & (n > 0) \\ 0_R & (n = 0) \\ -\underbrace{(r + \ldots + r)}_{|n|} & (n < 0) \end{cases}$$

An easy but tedious induction proves that this $\mathbb{Z}$-algebra structure deserves the name. [26] As evidence for the naturality of this $\mathbb{Z}$-structure, notice that if $f : R \longrightarrow S$ is any ring

---

[22] The *uniqueness* of the isomorphism also follows from discussion, since if there were two isomorphisms h and h' from Q to Q', then $h' \circ h^{-1} : Q \longrightarrow Q$ would be a non-identity map with the desired property, but only the identity on Q has the universal mapping property.

[23] To say that these are rational *functions* is a bit of a misnomer, but no worse than to refer to polynomial *functions*, which is also misleading but popular.

[24] Again, more precisely, the condition that α be transcendental is that the natural map $k[x] \longrightarrow k[\alpha]$ by $x \longrightarrow \alpha$ has trivial kernel.

[25] This sort of use of *canonical* is meant for the moment to insinuate that there is no whimsical choice involved. A more precise formulation of what *canonical* could mean would require a category-theoretical set-up. We may do this later.

[26] The arguments to prove this are of the same genre as those proving the so-called Laws of Exponents. Here, one must show that $(m + n)r = mr + nr$ and $(mn)r = m(nr)$ for $m, n \in \mathbb{Z}$ and $r \in R$, and $m(rs) = (mr)s$ for $s \in R$.

homomorphism, then f is a $\mathbb{Z}$-algebra homomorphism when the above $\mathbb{Z}$-algebra structures are put on R and S.

When a ring R has an identity 1_R, there is a canonical $\mathbb{Z}$-algebra homomorphism $i : \mathbb{Z} \longrightarrow R$ by

$$i : n \longrightarrow n \cdot 1_R$$

Granting that the $\mathbb{Z}$-algebra structure on R works as claimed, the proof that this is a homomorphism is nearly trivial:

$$i(m + n) = (m + n) \cdot 1_R = m \cdot 1_R + n \cdot 1_R = i(m) + i(n)$$

$$i(m \cdot n) = (m \cdot n) \cdot 1_R = m \cdot (n \cdot 1_R) = m \cdot (1_R \cdot (n \cdot 1_R)) = (m \cdot 1_R) \cdot (n \cdot 1_R) = i(m) \cdot i(n)$$

Now consider the canonical $\mathbb{Z}$-algebra homomorphism $i : \mathbb{Z} \longrightarrow k$ for a field k. [27] If i is injective, then it extends to an injection of the field of fractions $\mathbb{Q}$ of $\mathbb{Z}$ into k. In this case, say k is of **characteristic zero**, and this canonical copy of $\mathbb{Q}$ inside k is the **prime field** inside k. If i is not injective, its kernel is a principal ideal in $\mathbb{Z}$, say $p\mathbb{Z}$ with $p > 0$. Since the image $i(\mathbb{Z})$ is inside a field, it is an integral domain, so $p\mathbb{Z}$ is a (non-zero) prime ideal, which implies that p is prime. This integer p is the **characteristic** of k. We know that $\mathbb{Z}/\langle p \rangle$ is a *field*. Then we see that (by the Isomorphism Theorem for rings) the homomorphism $i : \mathbb{Z} \longrightarrow k$ with kernel $p\mathbb{Z}$ induces an isomorphism

$$\mathbb{Z}/p \approx i(\mathbb{Z}) \subset k$$

This canonical copy of $\mathbb{Z}/p$ inside k is the **prime field** inside k.

A finite field with q elements is often denoted $\mathbb{F}_q$ or $GF(q)$. [28]

6.3.1 Theorem: A finite field K has p^n elements for some prime p and integer n. [29] In particular, let $n = [K : \mathbb{F}_p]$ be the degree of K over its prime field $\mathbb{F}_p \approx \mathbb{Z}/p$ with prime p. Then

$$|K| = p^n$$

Proof: Let $\mathbb{F}_p$ be the prime field in K. Let $e_1, \ldots, e_n$ be a $\mathbb{F}_p$-basis for the $\mathbb{F}_p$-vectorspace K. Then there are p^n choices of coefficients $c_i \in \mathbb{F}_p$ to form linear combinations

$$\alpha = \sum_{i=1}^{n} c_i \, e_i \in K$$

so K has p^n elements.   ///

---

[27] It is no coincidence that we begin our study of fields by considering homomorphisms of the two simplest interesting rings, $k[x]$ for a field k, and $\mathbb{Z}$, into rings and fields.

[28] This notation begs the question of *uniqueness* (up to isomorphism) of a finite field once its cardinality is specified. We address this shortly.

[29] We will prove *existence* and uniqueness results for finite fields a bit later.

6.4 *Algebraic field extensions*

The first of the following two examples is amenable to *ad hoc* manipulation, but the second is designed to frustrate naive explicit computation.

6.4.1 Example: Let γ be a root (in some field k of characteristic 0, thus containing the prime field $\mathbb{Q}$) of the equation

$$x^2 - \sqrt{2}x + \sqrt{3} = 0$$

Is γ a root of a polynomial equation with rational coefficients?

In the same spirit as *completing the square*, we can manipulate the equation $x^2 - \sqrt{2}x + \sqrt{3} = 0$ to make the square roots disappear, as follows. Move the x^2 to the opposite side and square both sides, to obtain

$$2x^2 - 2\sqrt{6}x + 3 = x^4$$

Then move everything but the remaining square root to the right-hand side

$$-2\sqrt{6}\,x = x^4 - 2x^2 - 3$$

and square again

$$24x^2 = x^8 - 4x^6 - 2x^4 + 6x^2 + 9$$

and then we find that γ is a root of

$$0 = x^8 - 4x^6 - 2x^4 - 18x^2 + 9$$

It is not so obvious that the original [30]

$$\gamma = \frac{\sqrt{2} \pm \sqrt{2 - 4\sqrt{3}}}{2}$$

are roots. [31]

6.4.2 Example: Let α be a root of the equation

$$x^5 - x + 1 = 0$$

and let β be a root of the equation

$$x^7 - x + 1 = 0$$

Then let γ be a root of the equation

$$x^6 - \alpha x + \beta = 0$$

---

[30] Solving the original quadratic equation directly, by completing the square, for example.

[31] For that matter, it appears that the original equation has exactly two roots, while a degree 8 equation might have 8. Thus, we seem to have introduced 6 *spurious* roots in this process. Of course, an explanation for this is that there are two different square roots of 2 and two different square roots of 3 in k, so really $2 \cdot 2 = 4$ versions of the original quadratic equation, each with perhaps 2 roots in k.

Is γ a root of a polynomial equation with rational coefficients?

In this second example manipulations at the level of the first example fail. [32]   But one might speculate that in answering an existential question it might be possible to avoid explicit computations entirely, as in the proofs of the following results.

6.4.3 Proposition: Let $k \subset K \subset L$ be fields, with $[K : k] < \infty$ and $[L : K] < \infty$. Then

$$[L : k] = [L : K] \cdot [K : k] < \infty$$

In particular, for a K-basis $\{E_i\}$ of L, and for a k-basis e_j of K, the set $\{E_i e_j\}$ is a k-basis for L. [33]

Proof: On one hand, any linear relation

$$\sum_{ij} A_{ij} E_i e_j = 0$$

with $A_{ij} \in k$ gives

$$\sum_i \left(\sum_j A_{ij} e_j \right) E_i = 0$$

so for each i we have $\sum_j A_{ij} e_j = 0$, by the linear independence of the E_i. And by the linear independence of the e_j we find that $A_{ij} = 0$ for all indices. On the other hand, given

$$\beta = \sum_i b_i E_i \in L$$

with $b_i \in K$, write $b_i = \sum_j a_{ij} e_j$ with $a_{ij} \in k$, and then

$$\beta = \sum_i \left(\sum_j a_{ij} e_j \right) E_i = \sum_{ij} a_{ij} E_i e_j$$

which proves the spanning property. Thus, the elements $E_i e_j$ are a k-basis for L.   ///

A field extension K of a field k is **finite** if the degree $[K : k]$ is finite. *Finite* field extensions can be built up by adjoining elements. To economize on parentheses and brackets, [34] write

$$k(\alpha_1, \ldots, \alpha_n) \quad \text{for} \quad k(\alpha_1)(\alpha_2) \ldots (\alpha_n)$$

and

$$k[\alpha_1, \ldots, \alpha_n] \quad \text{for} \quad k[\alpha_1][\alpha_2] \ldots [\alpha_n]$$

---

[32] The provable limitations of familiar algebraic operations are packaged up in *Galois theory*, a bit later.

[33] The first assertion is merely a qualitative version of the last. Note that this proposition does not mention field *elements* explicitly, but rather emphasizes the vector space structures.

[34] On might worry that this notation glosses over potential issues. But, for example, one can prove that a polynomial ring in two variables really is *naturally* isomorphic to a polynomial ring in one variable over a polynomial ring in one variable.

6.4.4 Proposition: Let K be a field containing k, and suppose that $[K : k] < \infty$. Then any element α in K is algebraic over k, and there are finitely-many $\alpha_1, \ldots, \alpha_n$ such that

$$K = k(\alpha_1, \ldots, \alpha_n) = k[\alpha_1, \ldots, \alpha_n]$$

In particular, *finite* extensions K are necessarily *algebraic*. [35]

Proof: Given $\alpha \in K$, the countably many powers $1, \alpha, \alpha^2, \ldots$ cannot be linearly independent over k, since the whole K is finite-dimensional over k. A linear dependence relation among these powers is a polynomial equation satisfied by α. [36] If K is strictly larger than k, take $\alpha_1 \in K$ but not in k. Then $[k(\alpha_1) : k] > 1$, and the multiplicativity

$$[K : k] = [K : k(\alpha_1)] \cdot [k(\alpha_1) : k]$$

with $[K : k] < \infty$ implies that

$$[K : k(\alpha_1)] < [K : k]$$

If K is still larger than $k(\alpha_1)$, take α_2 in K not in $k(\alpha_1)$. Again,

$$[K : k(\alpha_1, \alpha_2)] < [K : k(\alpha_1)] < [K : k]$$

These degrees are positive integers, so a decreasing sequence must reach 1 in finitely-many steps (by Well-Ordering). The fact [37] that $k(\alpha) = k[\alpha]$ for α algebraic over k was proven earlier. ///

Let K and L be subfields of a larger field E. The **compositum** $K \cdot L$ of K and L is the smallest subfield of E containing both K and L. [38]

6.4.5 Proposition: Let $k \subset E$ be fields. Let K, L be subfields of K containing k. Suppose that $[K : k] < \infty$ and $[L : k] < \infty$. Then

$$[K \cdot L : k] \leq [K : k] \cdot [L : k] < \infty$$

In particular, if

$$K = k(\alpha_1, \ldots, \alpha_m) = k[\alpha_1, \ldots, \alpha_m]$$

$$L = k(\beta_1, \ldots, \beta_n) = k[\beta_1, \ldots, \beta_n]$$

then

$$K \cdot L = k(\alpha_1, \ldots, \alpha_m, \beta_1, \ldots, \beta_n) = k[\alpha_1, \ldots, \alpha_m, \beta_1, \ldots, \beta_n]$$

---

[35] The converse is not true. That is, some fields k admit extensions K with the property that every element in K is algebraic over k, but K is infinite-dimensional over k. The rational numbers $\mathbb{Q}$ can be proven to have this property, as do the p-adic numbers $\mathbb{Q}_p$ discussed later. It is not completely trivial to prove this.

[36] A more elegant argument is to map $k[x]$ to K by $x \longrightarrow \alpha$, and note that the kernel must be non-zero, since otherwise the image would be infinite-dimensional over k.

[37] Potentially disorienting and quite substantial.

[38] As with many of these constructions, the notion of *compositum* does not make sense, or at least is not well-defined, unless the two fields lie in a common larger field.

Proof: From the previous proposition, there do exist the α_i and β_j expressing K and L as k with finitely many elements adjoined as in the statement of the proposition. Recall that these mean that

$$K = \text{ intersection of subfields of } E \text{ containing } k \text{ and all } \alpha_i$$

$$L = \text{ intersection of subfields of } E \text{ containing } k \text{ and all } \beta_i$$

On one hand, $K \cdot L$ contains all the α_i and β_j. On the other hand, since these elements are algebraic over k, we do have

$$k(\alpha_1, \ldots, \alpha_m, \beta_1, \ldots, \beta_n) = k[\alpha_1, \ldots, \alpha_m, \beta_1, \ldots, \beta_n]$$

The left-hand side is a field, by definition, namely the smallest subfield [39] of E containing all the α_i and β_j. Thus, it contains K, and contains L. Thus, we have equality.   ///

6.4.6 Proposition: Let $k \subset E$ be fields, and K, L fields between k and E. Let $\alpha \in L$ be algebraic over k. Then

$$[k(\alpha) : k] \geq [K(\alpha) : K]$$

Proof: Since α is algebraic over k, $k(\alpha) = k[\alpha]$, and the degree $[k(\alpha) : k]$ is the degree of the minimal polynomial of α over k. This degree cannot increase when we replace k by the larger field K, and we obtain the indicated inequality.   ///

6.4.7 Proposition: Let k be a field, K a field algebraic over k, and L a field containing K. Let $\beta \in L$ be algebraic over K. Then β is algebraic over k.

Proof: Let $M(x)$ be the monic irreducible in $K[x]$ which is the minimal polynomial for β over K. Let $\{\alpha_0, \ldots, \alpha_{n-1}\}$ be the *finite* set (inside K) of coefficients of $M(x)$. Each field $k(\alpha_i)$ is of finite degree over k, so by the previous proposition their compositum $k(\alpha_1, \ldots, \alpha_n)$ is finite over k. The polynomial $M(x)$ is in $k(\alpha_1, \ldots, \alpha_n)[x]$, so β is algebraic over $k(\alpha_1, \ldots, \alpha_n)$. From above, the degree of $k(\alpha_1, \ldots, \alpha_n)(\beta)$ over k is the product

$$[k(\alpha_1, \ldots, \alpha_n)(\beta) : k] = [k(\alpha_1, \ldots, \alpha_n)(\beta) : k(\alpha_1, \ldots, \alpha_n)] \cdot [k(\alpha_1, \ldots, \alpha_n) : k] < \infty$$

Thus, $k(\alpha_1, \ldots, \alpha_n)(\beta)$ is finite over k, and in particular β is algebraic over k.

                                                                                        ///

6.4.8 Corollary: Let $k \subset K \subset L$ be fields, with K algebraic over k and L algebraic over K. Then L is algebraic over k.

Proof: This is an element-wise assertion, and for each β in L the previous proposition proves the algebraicity.   ///

---

[39] This discussion would appear to depend perhaps too much upon the larger ambient field E. In one sense, this is true, in that *some* larger ambient field is necessary. On the other hand, if K and L are both contained in a smaller subfield E' of E, we can replace E by E' for this discussion. One may reflect upon the degree to which the outcome genuinely depends upon any difference between E' and E, and how to avoid this concern.

6.4.9 Remark: An arrangement of fields of the form $k \subset K \subset L$ is sometimes called a **tower** of fields, with a corresponding picture

The situation that K and L are intermediate fields between k and E, with compositum KL, is depicted as

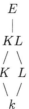

6.5 *Algebraic closures*

A field K is **algebraically closed** if every non-constant polynomial $f(x) \in k[x]$ has at least one root $\alpha \in k$, that is,

$$f(\alpha) = 0$$

Upon division, this algebraic closure property implies that any polynomial in $K[x]$ factors into linear factors in $K[x]$.

Given a field k, a larger field K which is algebraically closed [40] *and* such that every element of K is algebraic over k, is an **algebraic closure of** k. [41]

6.5.1 Theorem: Any field k has an algebraic closure $\overline{k}$, unique up to isomorphism. Any algebraic field extension E of k has at least one injection to $\overline{k}$ (which restricts to the identity on k).

Proof: (Artin) Let S be the set of monic irreducibles in $k[x]$, for each $s \in S$ let x_s be an indeterminate, and consider the polynomial ring

$$R = k[\ldots, x_s, \ldots] \quad (s \in S)$$

---

[40] Note that not only polynomials with coefficients in k must have roots in K, but polynomials with coefficients in K. Thus, one can perhaps imagine a different universe in which one makes a large enough field K such that all polynomials with coefficients in k have roots, but polynomials with coefficients in K need a larger field for their roots. That this does not happen, and that the process of constructing algebraic closures terminates, is the content of the theorem below.

[41] The second requirement is desirable, since we do not want to have algebraic closures be needlessly large. That is, an algebraic closure of k should not contain elements transcendental over k.

in S-many variables. [42]   We claim that there is at least one maximal proper ideal M in R containing every $f(x_f)$ for $f \in S$. First, one must be sure that the ideal F generated by all $f(x_f)$ is *proper* in R. If F were not proper, there would be elements $r_i \in R$ and irreducibles f_i such that (a finite sum)

$$\sum_{i=1}^{n} r_i f_i(x_{f_i}) = 1$$

Make a finite field extension E of k such that all the finitely-many f_i have roots α_i in E, inductively, as follows. First, let $k_1 = k[x]/\langle f_1 \rangle$. Then let F_2 be an *irreducible* factor of f_2 in k_1, and let $k_2 = k_1[x]/\langle F_2 \rangle$. And so on, obtaining $E = k_n$. Using the universal mapping property of polynomial rings, we can send x_{f_i} to $\alpha_i \in E$, thus sending $f_i(x_{f_i})$ to 0. [43] Then the relation becomes

$$0 = 1$$

Thus, there is no such relation, and the ideal F is proper.

Next, we claim that F lies in a *maximal* proper ideal M in R. This needs an equivalent of the Axiom of Choice, such as Hausdorff Maximality or Zorn's Lemma. In particular, among all chains of *proper* ideals containing F

$$F \subset \ldots \subset I \subset \ldots$$

there exists a maximal chain. [44]   The union of an ascending chain of proper ideals cannot contain 1, or else one of the ideals in the chain would contain 1, and would not be proper. Thus, the union of the ideals in a maximal chain is still proper. If it were not a *maximal* proper ideal then there would be a further (proper) ideal that could be added to the chain, contrary to assumption. Thus, we have a maximal ideal M in R. Thus, $K = R/M$ is a field.

By construction, for monic irreducible (non-constant) f the equation $f(Y) = 0$ has a root in K, namely the image of x_f under the quotient map, since $f(x_f) \in M$ for all irreducibles f. This proves that all non-constant polynomials in $k[x]$ have roots in K.

Now we prove that every element in $\overline{k}$ is algebraic over k. Let α_f be the image of x_f in *kbar*. Since α_f is a zero of f it is algebraic over k. An element β of $\overline{k}$ is a polynomial in finitely-many of the α_fs, say $\alpha_{f_1}, \ldots, \alpha f_n$. That is, $\beta \in k[\alpha_1, \ldots, \alpha_n]$, which is a field since each α_i is algebraic over k. Since (for example) the compositum (inside $\overline{k}$) of the algebraic extensions $k(\alpha_{f_i}) = k[\alpha_{f_i}]$ is algebraic, β is algebraic over k.

Next, we prove that non-constant $F(x) \in \overline{k}[x]$ has a zero in $\overline{k}$ (hence, it has *all* zeros in $\overline{k}$). The coefficients of F involve some finite list $\alpha_{f_1}, \ldots, \alpha_{f_n}$ out of all α_f, and $F(x)$ has a zero in $\overline{k}(\alpha_{f_1}, \ldots, \alpha_{f_n})[x]/\langle F \rangle$. Thus, since β is algebraic over an algebraic extension of k, it is algebraic over k, and, thus, is a root of a polynomial in $k[x]$.

---

[42] This ostentatiously extravagant construction would not have been taken seriously prior to Bourbaki's influence on mathematics. It turns out that once one sacrifices a *little* finiteness, one may as well fill things out symmetrically and accept a *lot* of non-finiteness. Such extravagance will reappear in our modern treatment of tensor products, for example.

[43] No, we have no idea what happens to the r_i, but we don't care.

[44] Maximal in the sense that there is no other proper ideal J containing F that either contains or is contained in every element of the (maximal) chain.

Now consider an algebraic extension E of k, and show that it admits an imbedding into $\overline{k}$. First, if $\alpha \in E$, let f be the minimal polynomial of α over k, and let β be a zero of f in $\overline{k}$. Map $k[x] \longrightarrow \overline{k}$ by sending $x \longrightarrow \beta$. The kernel is exactly the ideal generated by f, so (by an isomorphism theorem) the homomorphism $k[x] \longrightarrow \overline{k}$ descends to an injection $k[\alpha] \longrightarrow \overline{k}$. This argument can be repeated to extend the inclusion $k \subset \overline{k}$ to any extension $E = k(\alpha_1, \ldots, \alpha_n)$ with α_i algebraic over k. We use an equivalent of the Axiom of Choice to complete the argument: consider the collection of ascending chains of fields E_i (containing k) inside E admitting families of injections $\psi_i : E_i \longrightarrow \overline{k}$ with the compatibility condition that

$$\psi_j|_{E_i}| = \psi_i \quad \text{for} \quad E_i \subset E_j$$

We can conclude that there is a *maximal* chain. Let E' be the union of the fields in this maximal chain. The field E' imbeds in $\overline{k}$ by ψ_i on E_i, and the compatibility condition assures us that this is well-defined. We claim that $E' = E$. Indeed, if not, there is $\alpha \in E$ that is not in E'. But then the first argument shows that $E'(\alpha)$ does admit an imbedding to $\overline{k}$ extending the given one of E', contradiction. Thus, $E' = E$ and we have, in fact, imbedded the whole algebraic extension E to $\overline{k}$.

Last, we prove that any other algebraic closure K of k is isomorphic to $\overline{k}$. [45] Indeed, since K and $\overline{k}$ are algebraic over k, we have at least one injection $K \longrightarrow \overline{k}$, and at least one injection $\overline{k} \longrightarrow K$, but there is no reason to think that our capricious construction assures that these are mutual inverses. A different mechanism comes into play. Consider K imbedded into $\overline{k}$. Our claim is that K is necessarily all of $\overline{k}$. Indeed, any element of $\overline{k}$ is algebraic over k, so is the zero of a polynomial f in $k[x]$, say of degree n, which has all n roots in the subfield K of $\overline{k}$ because K is algebraically closed. That is, every element of the overfield $\overline{k}$ is actually in the subfield K, so the two are equal. ///

Exercises

6.1 Let γ be a root of the equation $x^2 + \sqrt{5}x + \sqrt{2} = 0$ in an algebraic closure of $\mathbb{Q}$. Find an equation with *rational* coefficients having root γ.

6.2 Let γ be a root of the equation $x^2 + \sqrt{5}x + \sqrt[3]{2} = 0$ in an algebraic closure of $\mathbb{Q}$. Find an equation with *rational* coefficients having root γ.

6.3 Find a polynomial with rational coefficients having a root $\sqrt{2} + \sqrt{3}$.

6.4 Find a polynomial with rational coefficients having a root $\sqrt{2} + \sqrt[3]{5}$.

6.5 Let γ be a root of $x^5 - x + 1 = 0$ in an algebraic closure of $\mathbb{Q}$. Find a polynomial with rational coefficients of which $\gamma + \sqrt{2}$ is a root.

6.6 Show that the field obtained by adjoining $\sqrt{2}$, $\sqrt[4]{2}$, $\sqrt[8]{2}$, $\sqrt[16]{2}$, ..., $\sqrt[2^n]{2}$, ..., to $\mathbb{Q}$ is *not* of finite degree over $\mathbb{Q}$.

---

[45] Note that we do not claim uniqueness of the isomorphism. Indeed, typically there are many different maps of a given algebraic closure $\overline{k}$ to itself that fix the underlying field k.

7. Some irreducible polynomials

Linear factors $x - \alpha$ of a polynomial $P(x)$ with coefficients in a field k correspond precisely to roots $\alpha \in k$ of the equation $P(x) = 0$. This follows from unique factorization in the ring $k[x]$. [1]   Here we also look at some special higher-degree polynomials, over *finite* fields, where we useful structural interpretation of the polynomials. [2]

Here we take for granted the existence of an algebraic closure $\overline{k}$ of a given field, as a fixed universe in which to consider roots of polynomial equations.

7.1 *Irreducibles over a finite field*

7.1.1 Proposition: Let (non-constant) $M(x)$ be an irreducible in $k[x]$, with field k. Let I be the ideal generated in $k[x]$ by $M(x)$. Let α be the image of x in the field $K = k[x]/I$. Then α is a root of the equation $M(x) = 0$. [3]

Proof: The salient aspects of the ring structure in the quotient can be summarized by the point that the quotient map $k[x] \longrightarrow k[x]/I$ is a ring homomorphism, in fact, a k-algebra

---

[1] And this unique factorization follows from the *Euclidean*-ness of the polynomial ring.

[2] All these are *cyclotomic* polynomials, that is, divisors of $x^n - 1$ for some n. A systematic investigation of these polynomials is best done with a little more preparation. But they do provide accessible examples immediately.

[3] This is immediate, when one looks at the proof, but deserves complete explicitness.

homomorphism. Thus, for any polynomial f,

$$f(x) + I = f(x + I)$$

In particular,

$$M(x + I) = M(x) + I = I = 0 + I$$

which shows that $x + I$ is a root of the equations.   ///

7.1.2 Proposition: [4] Let $P(x)$ be a polynomial in $k[x]$ for a field k. The equation $P(x) = 0$ has a root α generating [5] a degree d extension K of k if and only if $P(x)$ has a degree d irreducible factor $f(x)$ in $k[x]$.

Proof: Let α be a root of $P(x) = 0$ generating a degree d extension [6] $k(\alpha) = k[\alpha]$ over k. Let $M(x)$ be the minimal polynomial for α over k. Let

$$P = Q \cdot M + R$$

in $k[x]$ with $\deg R < \deg M$. Then, evaluating these polynomials at α, $R(\alpha) = 0$, but the minimality of the degree of M with this property assures that $R = 0$. That is, M divides P.

On the other hand, for an irreducible (monic, without loss of generality) $M(x)$ dividing $P(x)$, the quotient $K = k[x]/\langle M(x)\rangle$ is a field containing (a canonical copy of) k, and the image α of x in that extension is a root of $M(x) = 0$. Letting $P = Q \cdot M$,

$$P(\alpha) = Q(\alpha) \cdot M(\alpha) = Q(\alpha) \cdot 0 = 0$$

showing that $P(x) = 0$ has root α.   ///

The first two examples use only the correspondence between linear factors and roots in the ground field.

7.1.3 Example: $x^2 + 1$ is irreducible over $k = \mathbb{Z}/p$ for any prime $p = 3 \bmod 4$.

Indeed, if $x^2 + 1$ had a linear factor then the equation $x^2 + 1 = 0$ would have a root α in k. This alleged root would have the property that $\alpha^2 = -1$. Thus, $\alpha \neq 1$, $\alpha \neq -1$, but $\alpha^4 = 1$. That is, the order of α in $k^\times$ is 4. But the order of $(\mathbb{Z}/p)^\times$ is $p-1$. The hypothesis $p = 3 \bmod 4$ was exactly designed to deny the existence of an element of order 4 in $(\mathbb{Z}/p)^\times$. Thus, $x^2 + 1$ is irreducible in such $k[x]$.

7.1.4 Example: $x^2 + x + 1$ is irreducible over $k = \mathbb{Z}/p$ for any prime $p = 2 \bmod 3$.

---

[4] This assertion should not be surprising, when one looks at the technique of the proof, which is nearly identical to the proof that linear factors correspond to roots in the base field.

[5] As earlier, the field extension $k(\alpha)$ *generated by* α makes sense only inside a fixed larger field. Throughout the present discussion we fix an algebraic closure of any ground field k and consider extensions inside that algebraic closure.

[6] Since the degree of the extension is finite, it is equal to polynomials in α over k, as we saw earlier.

If $x^2 + x + 1$ had a linear factor then $x^2 + x + 1 = 0$ would have a root α in k, and, since

$$x^3 - 1 = (x-1)(x^2 + x + 1)$$

$\alpha^3 = 1$ but $\alpha \neq 1$ since $1 + 1 + 1 \neq 0$ in k. That is, the order of α in $k^\times$ is 3. But the order of $(\mathbb{Z}/p)^\times$ is $p - 1$, and the hypothesis $p = 2 \bmod 3$ exactly precludes any element of order 3 in $(\mathbb{Z}/p)^\times$. Thus, $x^2 + x + 1$ is irreducible in such $k[x]$.

7.1.5 Example: $P(x) = x^4 + x^3 + x^2 + x + 1$ is irreducible over $k = \mathbb{Z}/p$ for prime $p \neq \pm 1 \bmod 5$ and $p \neq 5$. Note that

$$x^5 - 1 = (x-1)(x^4 + x^3 + x^2 + x + 1)$$

Thus, any root of $P(x) = 0$ has order [7] 5 or 1 (in whatever field it lies). The only element of order 1 is the identity element 1. If $P(x)$ had a linear factor in $k[x]$, then $P(x) = 0$ would have a root in k. Since $1 + 1 + 1 + 1 + 1 \neq 0$ in k, 1 is not a root, so any possible root must have order 5. [8] But the order of $k^\times = (\mathbb{Z}/p)^\times$ is $p - 1$, which is not divisible by 5, so there is no root in the base field k.

If $P(x)$ had an irreducible *quadratic* factor $q(x)$ in $k[x]$, then $P(x) = 0$ would have a root in a quadratic extension K of k. Since $[K : k] = 2$, the field K has p^2 elements, and

$$K^\times = p^2 - 1 = (p-1)(p+1)$$

By Lagrange, the order of any element of $K^\times$ is a divisor of $p^2 - 1$, but 5 does not divide $p^2 - 1$, so there is no element in K of order 5. That is, there is no quadratic irreducible factor.

By additivity of degrees in products, lack of factors up to half the degree of a polynomial assures that the polynomial is irreducible. Thus, since the quartic $x^4 + x^3 + x^2 + x + 1$ has no linear or quadratic factors, it is irreducible.

7.1.6 Example: $P(x) = x^6 + x^5 + x^4 + x^3 + x^2 + x + 1$ is irreducible over $k = \mathbb{Z}/p$ for prime $p = 3 \bmod 7$ or $p = 5 \bmod 7$.

Note that
$$x^7 - 1 = (x-1)(x^6 + x^5 + x^4 + x^3 + x^2 + x + 1)$$

Thus, any root of $P(x) = 0$ has order 7 or 1 (in whatever field it lies). The only element of order 1 is the identity element 1. If $P(x)$ had a linear factor in $k[x]$, then $P(x) = 0$ would have a root in k. Since $1 + 1 + 1 + 1 + 1 + 1 + 1 \neq 0$ in k, 1 is not a root, so any possible root must have order 7. But the order of $k^\times = (\mathbb{Z}/p)^\times$ is $p - 1$, which is not divisible by 7, so there is no root in the base field k.

If $P(x)$ had an irreducible *quadratic* factor $q(x)$ in $k[x]$, then $P(x) = 0$ would have a root in a quadratic extension K of k. Since $[K : k] = 2$, the field K has p^2 elements, and

$$|K^\times| = p^2 - 1 = (p-1)(p+1)$$

---

[7] By Lagrange.

[8] The only other positive divisor of 5, thinking of Lagrange.

By Lagrange, the order of any element of $K^\times$ is a divisor of $p^2 - 1$, but 7 divides neither $3^2 - 1 = 8 = 1 \bmod 7$ nor $5^2 - 1 = 24 = 3 \bmod 7$, so there is no element in K of order 7. That is, there is no quadratic irreducible factor.

If $P(x)$ had an irreducible *cubic* factor $q(x)$ in $k[x]$, then $P(x) = 0$ would have a root in a cubic extension K of k. Since $[K : k] = 3$, the field K has p^3 elements, and

$$|K^\times| = p^3 - 1$$

By Lagrange, the order of any element of $K^\times$ is a divisor of $p^3 - 1$, but 7 divides neither $3^3 - 1 = 26 = 5 \bmod 7$ nor $5^3 - 1 = -8 = -1 \bmod 7$, so there is no element in K of order 7. That is, there is no cubic irreducible factor.

By additivity of degrees in products, lack of factors up to half the degree of a polynomial assures that the polynomial is irreducible. Thus, since the sextic $x^6 + x^5 + x^4 + x^3 + x^2 + x + 1$ has no linear, quadratic, or cubic factors, it is irreducible.

7.1.7 Example: $P(x) = (x^{11} - 1)/(x - 1)$ is irreducible over $k = \mathbb{Z}/p$ for prime p of order 10 (multiplicatively) mod 11. That is, modulo $p = 2, 6, 7, 8 \bmod 11$ this polynomial is irreducible. [9]

Again, any root of $P(x) = 0$ has order 11 or 1 (in whatever field it lies). The only element of order 1 is the identity element 1. If $P(x)$ had a linear factor in $k[x]$, then $P(x) = 0$ would have a root in k. Since $11 \neq 0$ in k, 1 is not a root, so any possible root must have order 11. But the order of $k^\times = (\mathbb{Z}/p)^\times$ is $p - 1$, which is not divisible by 11, so there is no root in the base field k.

If $P(x)$ had an irreducible degree d factor $q(x)$ in $k[x]$, then $P(x) = 0$ would have a root in a degree d extension K of k. The field K has p^d elements, so

$$|K^\times| = p^d - 1$$

By Lagrange, the order of any element of $K^\times$ is a divisor of $p^d - 1$, but 11 divides none of $p - 1$, $p^2 - 1$, $p^3 - 1$, $p^4 - 1$, ..., $p^9 - 1$, by design.

7.2 Worked examples

7.2.1 Example: *(Lagrange interpolation)* Let $\alpha_1, \ldots, \alpha_n$ be *distinct* elements in a field k, and let $\beta_1, \ldots, \beta_n$ be any elements of k. Prove that there is a unique polynomial $P(x)$ of degree $< n$ in $k[x]$ such that, for all indices i,

$$P(\alpha_i) = \beta_i$$

Indeed, letting

$$Q(x) = \prod_{i=1}^{n} (x - \alpha_i)$$

---

[9] By this point, one might have guessed that the irreducibility will be assured by taking primes p such that $p^d \neq 1$ for $d < 10$. The fact that there are such primes can be verified in an *ad hoc* fashion by simply looking for them, and Dirichlet's theorem on primes in arithmetic progressions assures that there are infinitely many such. The presence of primitive roots $2, 6, 7, 8$ (that is, generators for the cyclic group $(\mathbb{Z}/11)^\times$) modulo 11 is yet another issue, when we replace 11 by a different prime.

show that

$$P(x) = \sum_{i=1}^{n} \frac{Q(x)}{(x - \alpha_i) \cdot Q'(\alpha_i)} \cdot \beta_i$$

Since the α_i are distinct,

$$Q'(\alpha_i) = \prod_{j \neq i} (\alpha_i - \alpha_j) \neq 0$$

(One could say more about purely algebraic notions of derivative, but maybe not just now.) Evaluating $P(x)$ at $x \longrightarrow \alpha_i$,

$$\frac{Q(x)}{(x - \alpha_j)} \text{ evaluated at } x \longrightarrow \alpha_i = \begin{cases} 1 & (\text{for } j = i) \\ 0 & (\text{for } j = i) \end{cases}$$

Thus, all terms but the ith vanish in the sum, and the ith one, by design, gives β_i. For uniqueness, suppose $R(x)$ were another polynomial of degree $< n$ taking the same values at n distinct points α_i as does $Q(x)$. Then $Q - R$ is of degree $< n$ and vanishes at n points. A non-zero degree ℓ polynomial has at most ℓ zeros, so it must be that $Q - R$ is the 0 polynomial.

7.2.2 Example: *(Simple case of partial fractions)* Let $\alpha_1, \ldots, \alpha_n$ be *distinct* elements in a field k. Let $R(x)$ be any polynomial in $k[x]$ of degree $< n$. Show that there exist unique constants $c_i \in k$ such that in the field of rational functions $k(x)$

$$\frac{R(x)}{(x - \alpha_1) \ldots (x - \alpha_n)} = \frac{c_1}{x - \alpha_1} + \ldots + \frac{c_n}{x - \alpha_n}$$

In particular, let

$$Q(x) = \prod_{i=1}^{n} (x - \alpha_i)$$

and show that

$$c_i = \frac{R(\alpha_i)}{Q'(\alpha_i)}$$

We might emphasize that the field of rational functions $k(x)$ is most precisely the *field of fractions* of the polynomial ring $k[x]$. Thus, in particular, equality $r/s = r'/s'$ is exactly equivalent to the equality $rs' = r's$ (as in elementary school). Thus, to test whether or not the indicated expression performs as claimed, we test whether or not

$$R(x) = \sum_{i} \left(\frac{R(\alpha_i)}{Q'(\alpha_i)} \cdot \frac{Q(x)}{x - \alpha_i} \right)$$

One might notice that this is the previous problem, in case $\beta_i = R(\alpha_i)$, so its correctness is just a special case of that, as is the uniqueness (since $\deg R < n$).

7.2.3 Example: Show that the ideal I generated in $\mathbb{Z}[x]$ by $x^2 + 1$ and 5 is *not* maximal.

We will show that the quotient is not a field, which implies (by the standard result proven above) that the ideal is not maximal (proper).

First, let us make absolutely clear that the quotient of a ring R by an ideal $I = Rx + Ry$ generated by two elements can be expressed as a two-step quotient, namely

$$(R/\langle x \rangle)/\langle \bar{y} \rangle \approx R/(Rx + Ry)$$

where the $\langle \bar{y} \rangle$ is the principal ideal generated by the *image* $\bar{y}$ of y in the quotient $R/\langle x \rangle$. The principal ideal generated by y in the quotient $R/\langle x \rangle$ is the set of cosets

$$\langle \bar{y} \rangle = \{(r + Rx) \cdot (y + Rx) : r \in R\} = \{ry + Rx : r \in R\}$$

noting that the multiplication of cosets in the quotient ring is *not* just the element-wise multiplication of the cosets. With this explication, the natural map is

$$r + \langle x \rangle = r + \langle x \rangle \longrightarrow r + \langle x \rangle + \langle y \rangle' = r + (Rx + Rx)$$

which is visibly the same as taking the quotient in a single step.

Thus, first

$$\mathbb{Z}[x]/\langle 5 \rangle \approx (\mathbb{Z}/5)[x]$$

by the map which reduces the coefficients of a polynomial modulo 5. In $(\mathbb{Z}/5)[x]$, the polynomial $x^2 + 1$ *does* factor, as

$$x^2 + 1 = (x - 2)(x + 2)$$

(where these 2s are in $\mathbb{Z}/5$, not in $\mathbb{Z}$). Thus, the quotient $(\mathbb{Z}/5)[x]/\langle x^2 + 1 \rangle$ has proper zero divisors $\bar{x} - 2$ and $\bar{x} + 2$, where $\bar{x}$ is the image of x in the quotient. Thus, it's not even an integral domain, much less a field.

7.2.4 Example: Show that the ideal I generated in $\mathbb{Z}[x]$ by $x^2 + x + 1$ and 7 is *not* maximal.

As in the previous problem, we compute the quotient in two steps. First,

$$\mathbb{Z}[x]/\langle 7 \rangle \approx (\mathbb{Z}/7)[x]$$

by the map which reduces the coefficients of a polynomial modulo 7. In $(\mathbb{Z}/7)[x]$, the polynomial $x^2 + x + 1$ *does* factor, as

$$x^2 + x + 1 = (x - 2)(x - 4)$$

(where 2 and 4 are in $\mathbb{Z}/7$). Thus, the quotient $(\mathbb{Z}/7)[x]/\langle x^2 + x + 1 \rangle$ has proper zero divisors $\bar{x} - 2$ and $\bar{x} - 4$, where $\bar{x}$ is the image of x in the quotient. Thus, it's not even an integral domain, so certainly not a field.

Exercises

7.1 Show that $x^2 + x + 1$ is irreducible in $\mathbb{F}_5[x]$, and in $\mathbb{F}_{29}[x]$.

7.2 Show that $x^3 - a$ is irreducible in $\mathbb{F}_7[x]$ unless $a = 0$ or ± 1.

7.3 Determine how $x^5 + 1$ factors into irreducibles in $\mathbb{F}_2[x]$.

7.4 Exhibit an irreducible quintic in $\mathbb{F}_{11}[x]$.

7.5 Show that the ideal generated by $x^2 - x + 1$ and 13 in $\mathbb{Z}[x]$ is *not* maximal.

7.6 Show that the ideal generated by $x^2 - x + 1$ and 17 in $\mathbb{Z}[x]$ *is* maximal.

7.7 Let $\alpha_1,\ldots,\alpha_n$ be distinct elements of a field k not of characteristic 2. Show that there are elements $a_1, b_1, \ldots, a_n, b_n$ in k such that

$$\frac{1}{(x-\alpha_1)^2 \ldots (x-\alpha_n)^2} = \frac{a_1}{x-\alpha_1} + \frac{b_1}{(x-\alpha_1)^2} + \ldots + \frac{a_n}{x-\alpha_n} + \frac{b_n}{(x-\alpha_n)^2}$$

8. Cyclotomic polynomials

8.1 *Multiple factors in polynomials*

There is a simple device to detect repeated occurrence of a factor in a polynomial with coefficients in a field.

Let k be a *field*. For a polynomial

$$f(x) = c_n x^n + \ldots + c_1 x + c_0$$

with coefficients c_i in k, *define* the (**algebraic**) **derivative** [1] $Df(x)$ of $f(x)$ by

$$Df(x) = nc_n x^{n-1} + (n-1)c_{n-1}x^{n-2} + \ldots + 3c_3 x^2 + 2c_2 x + c_1$$

Better said, D is by definition a k-linear map

$$D : k[x] \longrightarrow k[x]$$

---

[1] Just as in the calculus of polynomials and rational functions one is able to evaluate all limits algebraically, one can readily prove (without reference to any limit-taking processes) that the notion of *derivative* given by this formula has the usual properties.

defined on the k-basis $\{x^n\}$ by

$$D(x^n) = nx^{n-1}$$

8.1.1 Lemma: For f, g in $k[x]$,

$$D(fg) = Df \cdot g + f \cdot Dg$$

8.1.2 Remark: Any k-linear map T of a k-algebra R to itself, with the property that

$$T(rs) = T(r) \cdot s + r \cdot T(s)$$

is a k-linear **derivation** on R.

Proof: Granting the k-linearity of T, to prove the derivation property of D is suffices to consider basis elements x^m, x^n of $k[x]$. On one hand,

$$D(x^m \cdot x^n) = Dx^{m+n} = (m+n)x^{m+n-1}$$

On the other hand,

$$Df \cdot g + f \cdot Dg = mx^{m-1} \cdot x^n + x^m \cdot nx^{n-1} = (m+n)x^{m+n-1}$$

yielding the product rule for monomials.   ///

A field k is **perfect** if either the characteristic of k is 0 [2] or if, in characteristic $p > 0$, there is a pth root $a^{1/p}$ in k for every $a \in k$. [3]

8.1.3 Proposition: Let $f(x) \in k[x]$ with a field k, and P an irreducible polynomial in $k[x]$. If P^e divides f then P divides $\gcd(f, Df)$. If k is *perfect* and $e - 1 \neq 0$ in k, there is a converse: [4] if P^{e-1} divides both f and Df then P^e divides f.

Proof: On one hand, suppose $f = P^e \cdot g$ with ≥ 2. By the product rule,

$$Df = eP^{e-1}DP \cdot g + P^e \cdot Dg$$

is a multiple of P^{e-1}. [5] This was the easy half.

On the other hand, for the harder half of the assertion, suppose P^{e-1} divides both f and Df. Write

$$f/P^{e-1} = Q \cdot P + R$$

with $\deg R < \deg P$. Then $f = QP^e + RP^{e-1}$. Differentiating,

$$Df = DQ\,P^e + eQP^{e-1}DP + DR\,P^{e-1} + R(e-1)P^{e-2}\,DP$$

---

[2] as for $\mathbb{Q}$, $\mathbb{R}$, and $\mathbb{C}$

[3] As is the case for finite fields such as $\mathbb{Z}/p$, by Fermat's Little Theorem.

[4] In particular, this converse holds if the characteristic of k is 0.

[5] This half does not need the irreducibility of P.

By hypothesis P^{e-1} divides Df. All terms on the right-hand side except possibly $R(e-1)P^{e-2}DP$ are divisible by P^{e-1}, so P divides $R(e-1)P^{e-2}DP$. Since P is irreducible, either $e-1=0$ in k, or P divides R, or P divides DP. If P divides R, P^e divides f, and we're done.

If P does not divide R then P divides DP. Since $\deg DP < \deg P$, if P divides DP then $DP = 0$. This would require that all the exponents of x occurring with non-zero coefficient are divisible by the characteristic p, which must be positive. So P is of the form

$$P(x) = a_{pm}x^{pm} + a_{p(m-1)}x^{p(m-1)} + a_{p(m-2)}x^{p(m-2)} + \ldots + a_{2p}x^{2p} + a_p x^p + a_0$$

Using the perfect-ness of the field k, each a_i has a pth root b_i in k. Because the characteristic is $p > 0$,

$$(A + B)^p = A^p + B^p$$

Thus, $P(x)$ is the pth power of

$$b_{pm}x^n + b_{p(m-1)}x^{(m-1)} + b_{p(m-2)}x^{(m-2)} + \ldots + b_{2p}x^2 + b_p x + b_0$$

If P is a pth power it is not irreducible. Therefore, for P irreducible DP is not the zero polynomial. Therefore, $R = 0$, which is to say that P^e divides f, as claimed. ///

8.2 Cyclotomic polynomials

For $b \neq 0$ in a field k, the **exponent** of b is the smallest positive integer n (if it exists) such that $b^n = 1$. That is, b is a root of $x^n - 1$ but not of $x^d - 1$ for any smaller d. We construct polynomials $\Phi_n(x) \in \mathbb{Z}[x]$ such that

$$\Phi_n(b) = 0 \text{ if and only if } b \text{ is of exponent } n$$

These polynomials Φ_n are **cyclotomic polynomials**.

8.2.1 Corollary: The polynomial $x^n - 1$ has no repeated factors in $k[x]$ if the field k has characteristic *not* dividing n.

Proof: It suffices to check that $x^n - 1$ and its derivative nx^{n-1} have no common factor. Since the characteristic of the field does not to divide n, $n \cdot 1_k \neq 0$ in k, so has a multiplicative inverse t in k, and

$$(x^n - 1) - (tx) \cdot (nx^{n-1}) = -1$$

and $\gcd(x^n - 1, nx^{n-1}) = 1$. ///

Define the nth **cyclotomic polynomial** $\Phi_n(x)$ by

$$\Phi_1(x) = x - 1$$

and for $n > 1$, inductively,

$$\Phi_n(x) = \frac{x^n - 1}{lcm \text{ of all } x^d - 1 \text{ with } 0 < d < n, d \text{ dividing } n}$$

with the least common multiple *monic*.

8.2.2 Theorem:

- Φ_n is a monic polynomial with integer coefficients. [6]
- For α in the field k, $\Phi_n(\alpha) = 0$ if and only if $\alpha^n = 1$ and $\alpha^t \neq 1$ for all $0 < t < n$.
- $\gcd(\Phi_m(x), \Phi_n(x)) = 1$ for $m < n$ with neither m nor n divisible by the characteristic of the field k.
- The degree of $\Phi_n(x)$ is $\varphi(n)$ (Euler's phi-function)
- Another description of $\Phi_n(x)$:

$$\Phi_n(x) = \frac{x^n - 1}{\prod_{1 \leq d < n, d|n} \Phi_d(x)}$$

- $x^n - 1$ factors as

$$x^n - 1 = \prod_{1 \leq d \leq n, d|n} \Phi_d(x)$$

Proof: We know that $d|n$ (and $d > 0$) implies that $x^d - 1$ divides $x^n - 1$. Therefore, by unique factorization, the least common multiple of a collection of things each dividing $x^n - 1$ also divides $x^n - 1$. Thus, the indicated *lcm* does divide $x^n - 1$.

For α in k, $x - \alpha$ divides $\Phi_n(x)$ if and only if $\Phi_n(\alpha) = 0$. And $\alpha^t = 1$ if and only if $x - \alpha$ divides $x^t - 1$. The definition

$$\Phi_n(x) = \frac{x^n - 1}{lcm \text{ of all } x^d - 1 \text{ with } 0 < d < n, \, d \text{ dividing } n}$$

shows first that $\Phi_n(\alpha) = 0$ implies $\alpha^n = 1$. Second, if $\alpha^t = 1$ for any proper divisor t of n then $x - \alpha$ divides $x^t - 1$, and thus $x - \alpha$ divides the denominator. But $x^n - 1$ has no repeated factors, so $x - \alpha$ dividing the denominator would prevent $x - \alpha$ dividing $\Phi_n(x)$, contradiction. That is, $\Phi_n(\alpha) = 0$ if and only if α is of order n.

To determine the *gcd* of Φ_m and Φ_n for neither m nor n divisible by the characteristic of k, note that Φ_m divides $x^m - 1$ and Φ_n divides $x^n - 1$, so

$$\gcd(\Phi_m, \Phi_n) \quad \text{divides} \quad \gcd(x^m - 1, x^n - 1)$$

We claim that for m, n two integers (divisible by the characteristic or not)

$$\gcd(x^m - 1, x^n - 1) = x^{\gcd(m,n)} - 1$$

Prove this claim by induction on the maximum of m and n. Reduce to the case $m > n$, wherein

$$x^m - 1 - x^{m-n} \cdot (x^n - 1) = x^{m-n} - 1$$

For g a polynomial dividing both $x^m - 1$ and $x^n - 1$, g divides $x^{m-n} - 1$. By induction,

$$\gcd(x^{m-n} - 1, x^n - 1) = x^{\gcd(m-n,n)} - 1$$

---

[6] More properly, if the ambient field k is of characteristic 0, then the coefficients lie in the copy of $\mathbb{Z}$ inside the prime field $\mathbb{Q}$ inside k. If the ambient field is of positive characteristic, then the coefficients lie inside the prime field (which is the natural image of $\mathbb{Z}$ in k). It would have been more elegant to consider the cyclotomic polynomials as polynomials in $\mathbb{Z}[x]$, but this would have required that we wait longer.

But

$$\gcd(m, n) = \gcd(m - n, n)$$

and

$$x^m - 1 = x^{m-n} \cdot (x^n - 1) + x^{m-n} - 1$$

so

$$\gcd(x^m - 1, x^n - 1) = \gcd(x^{m-n} - 1, x^n - 1)$$

and induction works. Thus,

$$\gcd(x^m - 1, x^n - 1) = x^{\gcd(m,n)} - 1$$

Since

$$d \le m < n$$

d is a *proper* divisor of n. Thus, from

$$\Phi_n(x) = \frac{x^n - 1}{lcm \text{ of all } x^d - 1 \text{ with } 0 < d < n, \, d \text{ dividing } n}$$

we see that $\Phi_n(x)$ divides $(x^n - 1)/(x^d - 1)$. Since $x^n - 1$ has no repeated factors, $\Phi_n(x)$ has no factors in common with $x^d - 1$. Thus, $\gcd(\Phi_m, \Phi_n) = 1$.

Next, use induction to prove that

$$x^n - 1 = \prod_{1 \le d \le n, \, d|n} \Phi_d(x)$$

For $n = 1$ the assertion is true. From the definition of Φ_n,

$$x^n - 1 = \Phi_n(x) \cdot lcm\{x^d - 1 : d|n, 0 < d < n\}$$

By induction, for $d < n$

$$x^d - 1 = \prod_{0 < e \le d, \, e|d} \Phi_e(x)$$

Since for $m < n$ the *gcd* of Φ_m and Φ_n is 1,

$$lcm\{x^d - 1 : d|n, 0 < d < n\} = \prod_{d|n, d<n} \Phi_d(x)$$

Thus,

$$x^n - 1 = \Phi_n(x) \cdot \prod_{d|n, d<n} \Phi_d(x)$$

as claimed.

Inductively, since all lower-index cyclotomic polynomials have integer coefficients [7] and are monic, and $x^n - 1$ is monic with integer coefficients, the quotient of $x^n - 1$ by the product of the lower ones is monic with integer coefficients.

---

[7] Or, more properly, coefficients in the canonical image of $\mathbb{Z}$ in the field k.

The assertion about the degree of Φ_n follows from the identity (see below) for Euler's phi-function

$$\sum_{d|n,d>0} \varphi(d) = n$$

This completes the proof of the theorem. ///

8.2.3 Proposition: Let $\varphi(x)$ be Euler's phi function

$$\varphi(x) = \sum_{1\leq\ell\leq x;\gcd(\ell,x)=1} 1$$

Then for m and n relatively prime

$$\varphi(mn) = \varphi(m) \cdot \varphi(n) \quad \text{(weak multiplicativity)}$$

For p prime and ℓ a positive integer

$$\varphi(p^\ell) = (p-1) \cdot p^{\ell-1}$$

And

$$\sum_{d|n,d>0} \varphi(d) = n$$

Proof: By unique factorization, for $\gcd(m,n)=1$,

$$\gcd(t,mn) = \gcd(t,m) \cdot \gcd(t,n)$$

so, t is prime to mn if and only if t is prime to both m and n. The gcd of m and n is the smallest positive integer of the form $rm + sn$. By Sun-Ze,

$$f : \mathbb{Z}/m \oplus \mathbb{Z}/n \longrightarrow \mathbb{Z}/mn$$

by

$$f : (x,y) \longrightarrow rmy + snx$$

is a *bijection*, since m and n are coprime. From $rm+yn=1$, $rm = 1$ mod n so rm is prime to n, and $sn = 1$ mod m so sn is prime to m. Thus, $rmy + snx$ has a common factor with m if and only if x does, and $rmy + snx$ has a common factor with n if and only if y does. Thus, f gives a bijection

$$\{x : 1 \leq x < m, \gcd(x,m) = 1\} \times \{y : 1 \leq y < n, \gcd(y,n) = 1\}$$

$$\longrightarrow \{z : 1 \leq z < mn, \gcd(z,mn) = 1\}$$

and $\varphi(mn) = \varphi(m) \cdot \varphi(n)$. This reduces calculation of $\varphi()$ to calculation for prime powers p^e. An integer x in $1 \leq x < p^e$ is prime to p^e if and only if it is not divisible by p, so there are

$$\varphi(p^e) = p^e - p^{e-1} = (p-1)p^{e-1}$$

such x, as claimed.

To prove

$$\sum_{d|n,d>0} \varphi(d) = n$$

start with n a prime power p^e, in which case

$$\sum_{d|p^e} \varphi(d) = \sum_{0 \leq k \leq e} \varphi(p^k) = 1 + \sum_{1 \leq k \leq e} (p-1)p^{k-1} = 1 + (p-1)(p^e-1)/(p-1) = p^e$$

Let $n = p_1^{e_1} \dots p_t^{e_t}$ with distinct primes p_i. Then

$$\sum_{d|n} \varphi(d) = \prod_{i=1,\dots,t} \left(\sum_{d|p_i^{e_i}} \varphi(d) \right) = \prod_{i=1,\dots,t} \varphi(p_i^{e_i}) = \varphi(\prod_i p_i^{e_i}) = \varphi(n)$$

This proves the desired identity for φ.   ///

8.3 *Examples*

For prime p, the factorization of $x^p - 1$ into cyclotomic polynomials is boring

$$\Phi_p(x) = \frac{x^p - 1}{x - 1} = x^{p-1} + x^{p-2} + \dots + x^2 + x + 1$$

For $n = 2p$ with odd prime p

$$\Phi_{2p}(x) = \frac{x^{2p} - 1}{\Phi_1(x)\,\Phi_2(x)\,\Phi_p(x)} = \frac{x^{2p} - 1}{\Phi_2(x)\,(x^p - 1)} = \frac{x^p + 1}{x + 1}$$

$$= x^{p-1} - x^{p-2} + x^{p-3} - \dots + x^2 - x + 1$$

For $n = p^2$ with p prime,

$$\Phi_{p^2}(x) = \frac{x^{p^2} - 1}{\Phi_1(x)\,\Phi_p(x)} = \frac{x^{p^2} - 1}{x^p - 1} = x^{p(p-1)} + x^{p(p-2)} + \dots + x^p + 1$$

Generally, one observes that for $n = p^e$ a prime power

$$\Phi_{p^e}(x) = \Phi_p(x^{p^{e-1}}) = x^{p^{e-1}(p-1)} + x^{p^{e-1}(p-2)} + \dots + x^{2p^{e-1}} + x^{2p^{e-1}} + 1$$

For $n = 2 \cdot m$ (with odd $m > 1$) we claim that

$$\backslash Phi_{2m}(x) = \Phi_m(-x)$$

Note that, anomalously, $\Phi_2(-x) = -x + 1 = -\Phi_1(x)$. Prove this by induction:

$$\Phi_{2m}(x) = \frac{x^{2m} - 1}{\prod_{d|m} \Phi_d(x) \cdot \prod_{d|m,\,d<m} \Phi_{2d}(x)} = \frac{x^{2m} - 1}{(x^m - 1) \prod_{d|m,\,d<m} \Phi_d(-x)}$$

$$= \frac{x^m + 1}{\prod_{d|m,\,d<m} \Phi_d(-x)} = \frac{(x^m + 1)\,\Phi_m(-x)}{((-x)^m - 1) \cdot (-1)} = \Phi_m(-x)$$

by induction, where the extra -1 in the denominator was for $\Phi_2(-x) = -\Phi_1(x)$, and $(-1)^m = -1$ because m is odd.

Thus,

$$\Phi_3(x) = \frac{x^3 - 1}{x - 1} = x^2 + x + 1$$

$$\Phi_9(x) = \Phi_3(x^3) = x^6 + x^3 + 1$$

$$\Phi_{18}(x) = \Phi_9(-x) = x^6 - x^3 + 1$$

For $n = pq$ with distinct primes p, q some unfamiliar examples appear.

$$\Phi_{15}(x) = \frac{x^{15} - 1}{\Phi_1(x)\Phi_3(x)\Phi_5(x)} = \frac{x^{15} - 1}{\Phi_1(x)\Phi_3(x)\Phi_5(x)} = \frac{x^{15} - 1}{\Phi_3(x)(x^5 - 1)}$$

$$= \frac{x^{10} + x^5 + 1}{x^2 + x + 1} = x^8 - x^7 + x^5 - x^4 + x^3 - x + 1$$

by direct division [8] at the last step. And then

$$\Phi_{30}(x) = \Phi_{15}(-x) = x^8 + x^7 - x^5 - x^4 - x^3 + x + 1$$

8.3.1 Remark: Based on a few hand calculations, one might speculate that all coefficients of all cyclotomic polynomials are either $+1$, -1, or 0, but this is not true. It *is* true for n prime, and for n having at most 2 distinct prime factors, but not generally. The smallest n where $\Phi_n(x)$ has an exotic coefficient seems to be $n = 105 = 3 \cdot 5 \cdot 7$.

$$\Phi_{105}(x) = \frac{x^{105} - 1}{\Phi_1(x)\Phi_3(x)\Phi_5(x)\Phi_7(x)\Phi_{15}(x)\Phi_{21}(x)\Phi_{35}(x)}$$

$$= \frac{x^{105} - 1}{\Phi_3(x)\Phi_{15}(x)\Phi_{21}(x)(x^{35} - 1)} = \frac{x^{70} + x^{35} + 1}{\Phi_3(x)\Phi_{15}(x)\Phi_{21}(x)} = \frac{(x^{70} + x^{35} + 1)(x^7 - 1)}{\Phi_{15}(x)(x^{21} - 1)}$$

$$= \frac{(x^{70} + x^{35} + 1)(x^7 - 1)\Phi_1(x)\Phi_3(x)\Phi_5(x)}{(x^{15} - 1)(x^{21} - 1)}$$

$$= \frac{(x^{70} + x^{35} + 1)(x^7 - 1)(x^5 - 1)\Phi_3(x)}{(x^{15} - 1)(x^{21} - 1)}$$

Instead of direct polynomial computations, we do *power series* [9] computations, imagining that $|x| < 1$, for example. Thus,

$$\frac{-1}{x^{21} - 1} = \frac{1}{1 - x^{21}} = 1 + x^{21} + x^{42} + x^{63} + \dots$$

---

[8] Only mildly painful. Any lesson to be learned here?

[9] In fact, one is not obliged to worry about convergence, since one can do computations in a *formal power series* ring. Just as polynomials can be precisely defined by their *finite* sequences of coefficients, with the obvious addition and multiplication mirroring our intent, formal power series are not-necessarily-finite sequences with the same addition and multiplication, noting that the multiplication does not require any infinite sums. The *formal* adjective here merely indicates that convergence is irrelevant.

We anticipate that the degree of $\Phi_{105}(x)$ is $(3-1)(5-1)(7-1) = 48$. We also observe that the coefficients of all cyclotomic polynomials are the same back-to-front as front-to-back (why?). Thus, we'll use power series in x and ignore terms of degree above 24. Thus,

$$\Phi_{105}(x) = \frac{(x^{70} + x^{35} + 1)(x^7 - 1)(x^5 - 1)(x^2 + x + 1)}{(x^{15} - 1)(x^{21} - 1)}$$

$$= (1 + x + x^2)(1 - x^7)(1 - x^5)(1 + x^{15})(1 + x^{21})$$

$$= (1 + x + x^2) \times (1 - x^5 - x^7 + x^{12} + x^{15} - x^{20} + x^{21} - x^{22})$$

$$= 1 + x + x^2 - x^5 - x^6 - x^7 - x^7 - x^8 - x^9 + x^{12} + x^{13} + x^{14} + x^{15} + x^{16} + x^{17}$$

$$- x^{20} - x^{21} - x^{22} + x^{21} + x^{22} + x^{23} - x^{22} - x^{23} - x^{24}$$

$$= 1 + x + x^2 - x^5 - x^6 - 2x^7 - x^8 - x^9 + x^{12} + x^{13} + x^{14}$$

$$+ x^{15} + x^{16} + x^{17} - x^{20} - x^{22} - x^{24}$$

Looking closely, we have a $-2x^7$.

Less well known are *Lucas-Aurifeullian-LeLasseur* factorizations such as

$$x^4 + 4 = (x^4 + 4x^2 + 4) - (2x)^2 = (x^2 + 2x + 2)(x^2 - 2x + 2)$$

More exotic are

$$\frac{x^6 + 27}{x^2 + 3} = (x^2 + 3x + 3)(x^2 - 3x + 3)$$

$$\frac{x^{10} - 5^5}{x^2 - 5} = (x^4 + 5x^3 + 15x^2 + 25x + 25) \times (x^4 - 5x^3 + 15x^2 - 25x + 25)$$

and

$$\frac{x^{12} + 6^6}{x^4 + 36} = (x^4 + 6x^3 + 18x + 36x + 36) \times (x^4 - 6x^3 + 18x - 36x + 36)$$

and further

$$\frac{x^{14} + 7^7}{x^2 + 7} = (x^6 + 7x^5 + 21x^4 + 49x^3 + 147x^2 + 343x + 343)$$

$$\times (x^6 - 7x^5 + 21x^4 - 49x^3 + 147x^2 - 343x + 343)$$

The possibility and nature of these factorizations are best explained by Galois theory.

8.4 *Finite subgroups of fields*

Now we can prove that the multiplicative group $k^\times$ of a finite field k is a *cyclic group*. When k is *finite*, a generator of $k^\times$ is a **primitive root** for k.

8.4.1 Theorem: Let G be a finite subgroup of $k^\times$ for a field k. Then G is cyclic.

8.4.2 Corollary: For a finite field k, the multiplicative group $k^\times$ is cyclic.   ///

Proof: Let n be the order of G. Then [10]   any element of G is a root of the polynomial $f(x) = x^n - 1$. We know that a polynomial with coefficients in a field k has at most as many roots (in k) as its degree, so this polynomial has at most n roots in k. Therefore, it has *exactly* n roots in k, namely the elements of the subgroup G.

The characteristic p of k cannot divide n, since if it did then the derivative of $f(x) = x^n - 1$ would be zero, and $\gcd(f, f') = f$ and f would have multiple roots. Thus,

$$x^n - 1 = \prod_{d|n} \Phi_d(x)$$

Since $x^n - 1$ has n roots in k, and since the Φ_d's here are relatively prime to each other, each Φ_d with $d|n$ must have a number of roots (in k) equal to its degree. Thus, Φ_d for $d|q-1$ has $\varphi(d) > 0$ roots in k (Euler's phi-function).

The roots of $\Phi_n(x)$ are $b \in k^\times$ such that $b^n = 1$ and no smaller positive power than n has this property.

Any root of $\Phi_n(x) = 0$ in $k^\times$ would be a generator of the (therefore cyclic) group G. The cyclotomic polynomial Φ_n has $\varphi(n) > 0$ zeros, so G has a generator, and is cyclic.   ///

8.5 *Infinitude of primes* $p = 1 \bmod n$

This is a very special case of Dirichlet's theorem that, given a modulus n and a fixed integer a relatively prime to n, there are infinitely-many primes $p = a \bmod n$. We only treat the case $a = 1$.

8.5.1 Corollary: Fix $1 < n \in \mathbb{Z}$. There are infinitely many primes $p = 1 \bmod n$.

Proof: Recall that the nth cyclotomic polynomial $\Phi_n(x)$ is monic (by definition), has integer coefficients, and has constant coefficient ± 1. [11]   And $\Phi_n(x)$ is not constant. Suppose there were only finitely-many primes $p_1, \ldots, p_t$ equal to 1 mod n. Then for large-enough positive integer ℓ,

$$N = \Phi_n(\ell \cdot np_1 \ldots p_t) > 1$$

and N is an integer. Since $\Phi_n(x)$ has integer coefficients and has constant term ± 1, for each p_i we have $N = \pm 1 \bmod p_i$, so in particular no p_i divides N. But since $N > 1$ it does have some prime factor p. Further, since the constant term is ± 1, $N = \pm 1 \bmod n$, so p is relatively prime to n. Then

$$\Phi_n(\ell \cdot np_1 \ldots p_t) = N = 0 \bmod p$$

Thus, $\ell \cdot np_1 \ldots p_t$ has order n in $\mathbb{F}_p^\times$. By Lagrange, n divides $|\mathbb{F}_p^\times| = p - 1$, so $p = 1 \bmod n$. Contradiction to the finiteness assumption, [12]   so there are infinitely-many primes $p = 1 \bmod n$.   ///

---

[10] Lagrange, again.

[11] The assertion about the constant coefficient follows from the fact that $\Phi_n(x)$ is monic, together with the fact that $\Phi(x^{-1}) = \pm\Phi(x)$, which is readily proven by induction.

[12] Mildly ironic that we have a contradiction, considering that we seem to have just succeeded in proving that there is one more prime of the type that we want. Perhaps this suggests that it is needlessly inefficient to couch this argument as proof by contradiction.

8.6 *Worked examples*

8.6.1 Example: Gracefully verify that the octic $x^8 + x^7 + x^6 + x^5 + x^4 + x^3 + x^2 + x + 1$ factors properly in $\mathbb{Q}[x]$.

This octic is

$$\frac{x^9 - 1}{x - 1} = \frac{x^3 - 1)(x^6 + x^3 + 1)}{x - 1} = (x^2 + x + 1)\,(x^6 + x^3 + 1)$$

for example. We might anticipate this reducibility by realizing that

$$x^9 - 1 = \Phi_1(x)\,\Phi_3(x)\,\Phi_9(x)$$

where Φ_n is the nth cyclotomic polynomial, and the given octic is just $(x^9 - 1)/\Phi_1(x)$, so what is left *at least* factors as $\Phi_3(x)\,\Phi_9(x)$.

8.6.2 Example: Gracefully verify that the quartic $x^4 + x^3 + x^2 + x + 1$ is irreducible in $\mathbb{F}_2[x]$.

Use the recursive definition of cyclotomic polynomials

$$\Phi_n(x) = \frac{x^n - 1}{\prod_{d|n,\ d<n}\ \Phi_d(x)}$$

Thus, the given quartic is $\Phi_5(x)$. And use the fact that for the characteristic of the field k not dividing n, $\Phi_n(\alpha) = 0$ if and only if α is of order n in $k^\times$. If it had a linear factor $x - \alpha$ with $\alpha \in \mathbb{F}_2$, then $\Phi_4(\alpha) = 0$, and α would be of order 5 in $\mathbb{F}_2^\times$. But $\mathbb{F}_2^\times$ is of order 1, so has no elements of order 5 (by Lagrange). (We saw earlier that) existence of an irreducible quadratic factor of $\Phi_4(x)$ in $\mathbb{F}_2[x]$ is equivalent to existence of an element α of order 5 in $\mathbb{F}_{2^2}^\times$, but $|\mathbb{F}_{2^2}^\times| = 2^2 - 1 = 3$, which is not divisible by 5, so (Lagrange) has no element of order 5. The same sort of argument would show that there is no irreducible cubic factor, but we already know this since if there were any proper factorization then there would be a proper factor of at most half the degree of the quartic. But there is no linear or quadratic factor, so the quartic is irreducible.

8.6.3 Example: Gracefully verify that the sextic $x^6 + x^5 + x^4 + x^3 + x^2 + x + 1$ is irreducible in $\mathbb{F}_3[x]$.

Use the recursive definition of cyclotomic polynomials

$$\Phi_n(x) = \frac{x^n - 1}{\prod_{d|n,\ d<n}\ \Phi_d(x)}$$

Thus, the given sextic is $\Phi_7(x)$. And use the fact that for the characteristic of the field k not dividing n, $\Phi_n(\alpha) = 0$ if and only if α is of order n in $k^\times$. If it had a linear factor $x - \alpha$ with $\alpha \in \mathbb{F}_3$, then $\Phi_7(\alpha) = 0$, and α would be of order 7 in $\mathbb{F}_2^\times$. But $\mathbb{F}_3^\times$ is of order 2, so has no elements of order 7 (Lagrange). Existence of an (irreducible) quadratic factor of $\Phi_7(x)$ in $\mathbb{F}_3[x]$ is equivalent to existence of an element α of order 7 in $\mathbb{F}_{3^2}^\times$, but $|\mathbb{F}_{3^2}^\times| = 3^2 - 1 = 8$, which is not divisible by 7, so (Lagrange) has no element of order 5. Similarly, if there were an (irreducible) cubic factor, then there would be a root in a cubic extension $\mathbb{F}_{3^3}$ of $\mathbb{F}_3$, but $\mathbb{F}_{3^3}^\times$ has order $3^3 - 1 = 26$ which is not divisible by 7, so there is no such element. If there were any proper factorization then there would be a proper factor of at most half the degree of the sextic. But there is no linear, quadratic, or cubic factor, so the sextic is irreducible.

8.6.4 Example: Gracefully verify that the quartic $x^4 + x^3 + x^2 + x + 1$ in factors into two irreducible quadratics in $\mathbb{F}_{19}[x]$.

As above, we see that the quartic is the 5th cyclotomic polynomial. If it had a linear factor in $\mathbb{F}_{19}[x]$ then (since the characteristic 19 does not divide the index 5) there would be an element of order 5 in $\mathbb{F}_{19}^\times$, but the latter group has order $19 - 1$ not divisible by 5, so (Lagrange) there is no such element. But the quadratic extension $\mathbb{F}_{19^2}$ of $\mathbb{F}_{19}$ has multiplicative group with order $19^2 - 1 = 360$ which is divisible by 5, so there is an element α of order 5 there.

Since $\alpha \in \mathbb{F}_{19^2} - \mathbb{F}_{19}$, the minimal polynomial $M(x)$ of α over $\mathbb{F}_{19}$ is quadratic. We have shown that in this circumstance the polynomial M divides the quartic. (Again, the proof is as follows: Let

$$x^4 + x^3 + x^2 + x + 1 = Q(x) \cdot M(x) + R(x)$$

with $Q, R \in \mathbb{F}_{19}[x]$ and $\deg R < \deg M$. Evaluating at α gives $R(\alpha) = 0$, which (by minimality of M) implies R is the 0 polynomial. Thus, M divides the quartic.) The quotient of the quartic by M is quadratic, and (as we've already seen) has no linear factor in $\mathbb{F}_{19}[x]$, so is irreducible.

8.6.5 Example: Let $f(x) = x^6 - x^3 + 1$. Find primes p with each of the following behaviors: f is irreducible in $\mathbb{F}_p[x]$, f factors into irreducible quadratic factors in $\mathbb{F}_p[x]$, f factors into irreducible cubic factors in $\mathbb{F}_p[x]$, f factors into linear factors in $\mathbb{F}_p[x]$.

By the recursive definition and properties of cyclotomic polynomials, we recognize $f(x)$ as the 18th cyclotomic polynomial $\Phi_{18}(x)$. For a prime p not dividing 18, zeros of Φ_{18} are exactly elements of order 18. Thus, if $p^d - 1 = 0 \bmod 18$ but no smaller exponent than d achieves this effect, then $\mathbb{F}_{p^d}^\times$ (proven *cyclic* by now) has an element of order 18, whose minimal polynomial divides $\Phi_{18}(x)$.

We might observe that $(\mathbb{Z}/18)^\times$ is itself *cyclic*, of order $\varphi(18) = \varphi(2)\varphi(3^2) = (3-1)3 = 6$, so has elements of all possible orders, namely $1, 2, 3, 6$.

For $p = 1 \bmod 18$, for example $p = 19$, already $p - 1 = 0 \bmod 18$, so $f(x)$ has a *linear* factor in $\mathbb{F}_{19}[x]$. This is the case of order 1 element in $(\mathbb{Z}/18)^\times$.

A moment's thought might allow a person to realize that $17 = -1$ is an element (and the only element) of order 2 in $(\mathbb{Z}/18)^\times$. So any prime $p = 17 \bmod 18$ (for example $p = 17$ itself, by coincidence prime) will have the property that $\mathbb{F}_{p^2}^\times$ has elements of order 18. Indeed, by properties of cyclic groups, it will have $\varphi(18) = 6$ elements of order 18 there, each of whose minimal polynomial is quadratic. Thus (since a quadratic has at most two zeros) there are at least 3 irreducible quadratics dividing the sextic $\Phi_{18}(x)$ in $\mathbb{F}_p[x]$. Thus, since degrees add in products, these three quadratics are *all* the factors of the sextic.

After a bit of trial and error, one will find an element of order 3 in $(\mathbb{Z}/18)^\times$, such as 7. Thus, for $p = 7 \bmod 18$ (such as 7 itself, which by coincidence is prime), there is no element of order 18 in $\mathbb{F}_p$ or in $\mathbb{F}_{p^2}$, but there *is* one in $\mathbb{F}_{p^3}$, whose minimal polynomial over $\mathbb{F}_p$ is therefore cubic and divides Φ_{18}. Again, by properties of cyclic groups, there are exactly $\varphi(18) = 6$ such elements in $\mathbb{F}_{p^3}$, with cubic minimal polynomials, so there are at least (and, thus, exactly) two different irreducible cubics in $\mathbb{F}_p[x]$ dividing $\Phi_{18}(x)$ for such p.

After a bit more trial and error, one finds an element of order 6 in $(\mathbb{Z}/18)^\times$, such as 5. (The other is 11.) Thus, for $p = 5 \mod 18$ (such as 5 itself, which by coincidence is prime), there is no element of order 18 in $\mathbb{F}_p$ or in $\mathbb{F}_{p^2}$, or $\mathbb{F}_{p^3}$, but there is one in $\mathbb{F}_{p^6}$. (By Lagrange, the only possible orders of p in $(\mathbb{Z}/18)^\times$ are $1, 2, 3, 6$, so we need not worry about p^4 or p^5). The minimal polynomial of such an element is $\Phi_{18}(x)$, which is (thus, necessarily) irreducible in $\mathbb{F}_p[x]$.

8.6.6 Example: Explain why $x^4 + 1$ properly factors in $\mathbb{F}_p[x]$ for any prime p.

As in the previous problems, we observe that $x^4 + 1$ is the 8th cyclotomic polynomial. If $p|8$, namely $p = 2$, then this factors as $(x - 1)^4$. For odd p, if $p = 1 \mod 8$ then $\mathbb{F}_p^\times$, which we now know to be *cyclic*, has an element of order 8, so $x^4 + 1$ has a linear factor. If $p \neq 1 \mod 8$, write $p = 2m + 1$, and note that

$$p^2 - 1 = (2m+1)^2 - 1 = 4m^2 + 4m = m(m+1) \cdot 4$$

so, if m is odd, $m + 1$ is even and $p^2 - 1 = 0 \mod 8$, and if m is even, the same conclusion holds. That is, for odd p, $p^2 - 1$ is invariably divisible by 8. That is, (using the cyclicness of any finite field) there is an element of order 8 in $\mathbb{F}_{p^2}$. The minimal polynomial of this element, which is quadratic, divides $x^4 + 1$ (as proven in class, with argument recalled above in another example).

8.6.7 Example: Explain why $x^8 - x^7 + x^5 - x^4 + x^3 - x + 1$ properly factors in $\mathbb{F}_p[x]$ for any prime p. (*Hint:* It factors either into linear factors, irreducible quadratics, or irreducible quartics.)

The well-read person will recognize this octic as $\Phi_{15}(x)$, the fifteenth cyclotomic polynomial. For a prime p not dividing 15, zeros of Φ_{15} in a field $\mathbb{F}_{p^d}$ are elements of order 15, which happens if and only if $p^d - 1 = 0 \mod 15$, since we have shown that $\mathbb{F}_{p^d}^\times$ is cyclic. The smallest d such that $p^d = 1 \mod 15$ is the order of p in $(\mathbb{Z}/15)^\times$. After some experimentation, one may realize that $(\mathbb{Z}/15)^\times$ is *not* cyclic. In particular, every element is of order 1, 2, or 4. (How to see this?) Granting this, for any p other than 3 or 5, the minimal polynomial of an order 15 element is linear, quadratic, or quartic, and divides Φ_{15}.

For $p = 3$, there is some degeneration, namely $x^3 - 1 = (x - 1)^3$. Thus, in the (universal) expression

$$\Phi_{15}(x) = \frac{x^{15} - 1}{\Phi_1(x)\,\Phi_3(x)\,\Phi_5(x)}$$

we actually have

$$\Phi_{15}(x) = \frac{(x^5 - 1)^3}{(x - 1)^2\,(x^5 - 1)} = \frac{(x^5 - 1)^2}{(x - 1)^2} = (x^4 + x^3 + x^2 + 1)^2$$

For $p = 5$, similarly, $x^5 - 1 = (x - 1)^5$, and

$$\Phi_{15}(x) = \frac{x^{15} - 1}{\Phi_1(x)\,\Phi_3(x)\,\Phi_5(x)} = \frac{(x^3 - 1)^5}{(x^3 - 1)\,(x - 1)^4} = \frac{(x^3 - 1)^4}{(x - 1)^4} = (x^2 + x + 1)^4$$

8.6.8 Example: Why is $x^4 - 2$ irreducible in $\mathbb{F}_5[x]$?

A zero of this polynomial would be a fourth root of 2. In $\mathbb{F}_5^\times$, one verifies by brute force that 2 is of order 4, so is a generator for that (cyclic) group, so is not a square in $\mathbb{F}_5^\times$, much less a fourth power. Thus, there is no linear factor of $x^4 - 2$ in $\mathbb{F}_5[x]$.

The group $\mathbb{F}_{5^2}^\times$ is cyclic of order 24. If 2 were a fourth power in $\mathbb{F}_{5^2}$, then $2 = \alpha^4$, and $2^4 = 1$ gives $\alpha^{16} = 1$. Also, $\alpha^{24} = 1$ (Lagrange). Claim that $\alpha^8 = 1$: let $r, s \in \mathbb{Z}$ be such that $r \cdot 16 + s \cdot 24 = 8$, since 8 is the greatest common divisor. Then

$$\alpha^8 = \alpha^{16r+24s} = (\alpha^{16})^r \cdot (\alpha^{24})^s = 1$$

This would imply

$$2^2 = (\alpha^4)^2 = \alpha^8 = 1$$

which is false. Thus, 2 is not a fourth power in $\mathbb{F}_{5^2}$, so the polynomial $x^4 - 2$ has no quadratic factors.

A quartic with no linear or quadratic factors is irreducible (since any proper factorization of a polynomial P must involve a factor of degree at most half the degree of P). Thus, $x^4 - 2$ is irreducible in $\mathbb{F}_5[x]$.

8.6.9 Example: Why is $x^5 - 2$ irreducible in $\mathbb{F}_{11}[x]$?

As usual, to prove irreducibility of a quintic it suffices to show that there are no linear or quadratic factors. To show the latter it suffices to show that there is no zero in the underlying field (for linear factors) or in a quadratic extension (for irreducible quadratic factors).

First determine the order of 2 in $\mathbb{F}_{11}$: since $|\mathbb{F}_{11}^\times| = 10$, it is either $1, 2, 5,$ or 10. Since $2 \neq 1 \bmod 11$, and $2^2 - 1 = 3 \neq 0 \bmod 11$, and $2^5 - 1 = 31 \neq 0 \bmod 11$, the order is 10. Thus, in $\mathbb{F}_{11}$ it cannot be that 2 is a fifth power.

The order of $\mathbb{F}_{11^2}^\times$ is $11^2 - 1 = 120$. If there were a fifth root α of 2 there, then $\alpha^5 = 2$ and $2^{10} = 1$ imply $\alpha^{50} = 1$. Also, (Lagrange) $\alpha^{120} = 1$. Thus, (as in the previous problem) α has order dividing the *gcd* of 50 and 120, namely 10. Thus, if there were such α, then

$$2^2 = (\alpha^5)^2 = \alpha^{10} = 1$$

But $2^2 \neq 1$, so there is no such α.

Exercises

8.1 Determine the coefficients of the 12th cyclotomic polynomial.

8.2 Gracefully verify that $(x^{15} - 1)/(x^5 - 1)$ factors properly in $\mathbb{Q}[x]$.

8.3 Find a prime p such that the 35th cyclotomic polynomial has an irreducible 12th-degree factor in $\mathbb{F}_p[x]$.

8.4 Determine the factorization into irreducibles of $(x^7 - 1)/(x - 1)$ in $\mathbb{F}_2[x]$.

8.5 Explain why the 12th cyclotomic polynomial factors properly in $\mathbb{F}_p[x]$ for any prime p.

8.6 Explain why the thirty-fifth cyclotomic polynomial factors properly in $\mathbb{F}_p[x]$ for any prime p.

8.7 Show that a finite field extension of $\mathbb{Q}$ contains only finitely-many roots of unity.

8.8 Let p be a prime and $n \geq 1$. Let φ_m be the mth cyclotomic polynomial. Show that

$$\varphi_{pn}(x) = \begin{cases} \varphi_n(x^p) & \text{(for } p|n\text{)} \\ \dfrac{\varphi_n(x^p)}{\varphi_n(x)} & \text{(otherwise)} \end{cases}$$

8.9 Let $n = 2^a\, p^b\, q^c$ for primes p, q. Show that the coefficients of the cyclotomic polynomial φ_n are in the set $\{-1, 0, 1\}$.

8.10 Suppose n is divisible by p^2 for some prime p. Show that the sum of the primitive n^{th} roots of unity is 0.

9. Finite fields

9.1 *Uniqueness*

Among other things, the following result justifies speaking of *the* field with p^n elements (for prime p and integer n), since, we prove, these parameters completely determine the isomorphism class.

9.1.1 Theorem: Given a prime p and an integer n, there is exactly one (up to isomorphism) finite field $\mathbb{F}_{p^n}$ with p^n elements. Inside a fixed algebraic closure of $\mathbb{F}_p$, the field $\mathbb{F}_{p^m}$ lies inside $\mathbb{F}_{p^n}$ if and only if $m|n$. In particular, $\mathbb{F}_{p^n}$ is the set of solutions of

$$x^{p^n} - x = 0$$

inside an algebraic closure of $\mathbb{F}_p$.

Proof: Let E be an algebraic closure of $\mathbb{F}_p$. Let $F(x) = x^{p^n} - x$ in $\mathbb{F}_p[x]$. The algebraic derivative of F is -1, so $\gcd(F, F') = 1$, and F has no repeated factors. Let $K = \mathbb{F}_p(\alpha_1, \ldots, \alpha_{p^n})$ be the subfield of E generated over $\mathbb{F}_p$ by the roots of $F(x) = 0$, which we know are exactly the p^n distinct α_is occuring as linear factors $x - \alpha_i$ in $F(x)$. [1]

Perhaps unsurprisingly, we claim that K is exactly the set of all the roots of $F(x) = 0$. Naturally we use the fact [2] that binomial coefficients $\binom{p}{i}$ are 0 in characteristic p, for

---

[1] Later we would say that K is a **splitting field** for F since F factors into linear factors in K.

[2] As in the most pedestrian proof of Fermat's Little Theorem.

$0 < i < p$. Thus,

$$(\alpha + \beta)^{p^n} = (\ldots((\alpha + \beta)^p)^p \ldots)^p = \alpha^{p^n} + \beta^{p^n}$$

In particular, if $\alpha^{p^n} = \alpha$ and $\beta^{p^n} = \beta$, then $\alpha + \beta$ has the same property. And even more obviously

$$(\alpha \cdot \beta)^{p^n} = \alpha^{p^n} \cdot \beta^{p^n} = \alpha \cdot \beta$$

Additive inverses of roots of $F(x) = 0$ are present in the collection of roots, because $\alpha + \beta = 0$ implies $\alpha^{p^n} + \beta^{p^n} = 0$. Far more simply, certainly non-zero roots have multiplicative inverses among the roots. And 0 is among the roots. Finally, because $\alpha^p = \alpha$ for $\alpha \in \mathbb{F}_p$, certainly $\mathbb{F}_p$ is a subset of the set of roots.

In summary, the smallest subfield K (of some algebraic closure E of $\mathbb{F}_p$) containing the roots of $x^{p^n} - x = 0$ is exactly the set of all roots, and K contains $\mathbb{F}_p$. Thus, K has exactly p^n elements. This proves *existence* of a field with p^n elements.

For *uniqueness* (up to isomorphism) of a field with p^n elements, it suffices to prove that inside a given algebraic closure E of $\mathbb{F}_p$ there is exactly one such field, since[3] any algebraic extension L of $\mathbb{F}_p$ can be mapped injectively to E (by an injection that is the identity on $\mathbb{F}_p$). For L of degree n over $\mathbb{F}_p$, necessarily $L^\times$ is of order $p^n - 1$. That is, the non-zero elements of $L^\times$ all satisfy $x^{p^n-1} - 1 = 0$. [4] Thus, adding a factor of x, all elements of L are roots of $x^{p^n} - x = 0$. Thus, with L sitting inside the fixed algebraic closure E of $\mathbb{F}_p$, since a degree p^n equation has at most p^n roots in E, the elements of L must be just the field K constructed earlier. [5] This proves uniqueness (up to isomorphism). [6]

Inside a fixed algebraic closure of $\mathbb{F}_p$, if $\mathbb{F}_{p^m} \subset \mathbb{F}_{p^n}$ then the larger field is a vector space over the smaller. Given a basis $e_1, \ldots, e_t$, every element of the larger field is uniquely expressible as $\sum_i c_i e_i$ with c_i in the smaller field, so there are $(p^m)^t$ elements in the larger field. That is, $n = mt$, so $m | n$. Conversely, if $m | n$, then the roots of $x^{p^m-1} - 1 = 0$ are among those of $x^{p^n-1} - 1 = 0$. We have identified $\mathbb{F}_{p^m}^\times$ as the set of roots of $x^{p^m-1} - 1 = 0$ inside a fixed algebraic closure, and similarly for $\mathbb{F}_{p^n}^\times$, so $\mathbb{F}_{p^m} \subset \mathbb{F}_{p^n}$.   ///

---

[3] By part of the main theorem on algebraic closures.

[4] By Lagrange. In fact, we know that the multiplicative group is cyclic, but this is not used.

[5] For non-finite fields, we will not be able to so simply or completely identify all the extensions of the prime field.

[6] Note that we do not at all assert any uniqueness of the isomorphism between any two such fields. To the contrary, there will be several different isomorphisms. This is clarified just below, in discussion of the Frobenius automorphisms.

9.2 *Frobenius automorphisms*

Let q be a power of a prime p, and let E be an algebraic closure of $\mathbb{F}_q$. [7] For $\alpha \in E$, the **Frobenius automorphism** (depending on q) is

$$F(\alpha) = \alpha^q$$

9.2.1 Proposition: For fixed prime power q and algebraic closure E of finite field $\mathbb{F}_q$, the Frobenius map $F : \alpha \longrightarrow \alpha^q$ is the identity map on $\mathbb{F}_q$, and stabilizes any overfield K of $\mathbb{F}_q$ inside E. Further, if $\beta \in E$ has the property that $F\beta = \beta$, then $\beta \in \mathbb{F}_q$. Generally, the fixed points of $\alpha \longrightarrow \alpha^{q^n}$ make up the field $\mathbb{F}_{q^n}$ inside E.

Proof: Certainly $F(\alpha\beta) = F(\alpha)F(\beta)$. Since the characteristic is P, also $(\alpha+\beta)^p = \alpha^p + \beta^p$, and F truly is a field homomorphism of E to itself.

Since any subfield K of E is stable under taking powers, certainly F maps K to itself.

By now we know that $\mathbb{F}_{q^n}^\times$ is cyclic, and consists exactly of the roots of $x^{q^n - 1} - 1 = 0$ in E. That is, $\mathbb{F}_{q^n}$ is exactly the roots of $x^{q^n} - x = 0$. That is, the fixed points of F^n are exactly $\mathbb{F}_{q^n}$, as claimed. ///

9.2.2 Proposition: Let $f(x)$ be a polynomial with coefficients in $\mathbb{F}_q$. Let $\alpha \in K$ be a root (in a fixed algebraic closure E of $\mathbb{F}_q$) of the equation $f(x) = 0$. Then $F(\alpha) = \alpha^q$, $F^2(\alpha) = F(F(\alpha)) = \alpha^{q^2}$, ... are also roots of the equation.

Proof: Let f have coefficients

$$f(x) = c_n x^n + c_{n-1} x^{n-1} + \ldots + c_2 x^2 + c_1 x + c_0$$

with all the c_i's in $\mathbb{F}_q$. Apply the Frobenius map to both sides of the equation

$$0 = c_n \alpha^n + c_{n-1} \alpha^{n-1} + \ldots + c_2 \alpha^2 + c_1 \alpha + c_0$$

to obtain

$$F(0) = F(c_n)F(\alpha)^n + F(c_{n-1})F(\alpha)^{n-1} + \ldots + F(c_2)F(\alpha)^2 + F(c_1)F(\alpha) + F(c_0)$$

since F is a field homomorphism. The coefficients c_i are in $\mathbb{F}_q$, as is the 0 on the left-hand side, so F does not change them. Thus,

$$0 = c_n F(\alpha)^n + c_{n-1} F(\alpha)^{n-1} + \ldots + c_2 F(\alpha)^2 + c_1 F(\alpha) + c_0$$

That is,

$$0 = f(F(\alpha))$$

and $F(\alpha)$ is a root of $P(x) = 0$ if α is. ///

---

[7] We take the liberty of considering not only $\mathbb{F}_p$ but any finite field $\mathbb{F}_q$ to be at the *bottom* of whatever towers of fields we consider. This is a simple case of *Galois theory*, which studies automorphisms of general fields.

9.2.3 Proposition: Let
$$A = \{\alpha_1, \ldots, \alpha_t\}$$

be a set of (t distinct) elements of and algebraic closure E of $\mathbb{F}_q$, with the property that for any α in A, $F(\alpha)$ is again in A. Then the polynomial

$$(x - \alpha_1)(x - \alpha_2) \ldots (x - \alpha_t)$$

(when multiplied out) has coefficients in k.

Proof: For a polynomial

$$f(x) = c_n x^n + c_{n-1} x^{n-1} + \ldots + c_2 x^2 + c_1 x + c_0$$

with coefficients in E, define a new polynomial $F(f)$ by letting the Frobenius F act on the coefficients

$$F(f)(x) = F(c_n) x^n + F(c_{n-1}) x^{n-1} + \ldots + F(c_2) x^2 + F(c_1) x + F(c_0)$$

This action gives a $\mathbb{F}_q$-algebra homomorphism $\mathbb{F}_q[x] \longrightarrow \mathbb{F}_q[x]$. Applying F to the product

$$(x - \alpha_1)(x - \alpha_2) \ldots (x - \alpha_t)$$

merely permutes the factors, by the hypothesis that F permutes the elements of A. Thus,

$$c_n x^n + c_{n-1} x^{n-1} + \ldots + c_1 x + c_0 = (x - \alpha_1)(x - \alpha_2) \ldots (x - \alpha_t)$$

$$= (x - F\alpha_1)(x - F\alpha_2) \ldots (x - F\alpha_t) = F(c_n) x^n + F(c_{n-1}) x^{n-1} + \ldots + F(c_1) x + F(c_0)$$

Equality of polynomials is coefficient-wise equality, so $F(c_i) = c_i$ for all indices i.   ///

9.2.4 Corollary: Let α be an element of an algebraic closure E of $\mathbb{F}_q$. Suppose that $[\mathbb{F}_q(\alpha) : \mathbb{F}_q] = n$. Then the minimal polynomial $M(x)$ of α is

$$M(x) = (x - \alpha)(x - F(\alpha))(x - F^2(\alpha)) \ldots (x - F^{n-1}(\alpha))$$

Proof: By definition of the minimal polynomial, M is the unique monic polynomial in $\mathbb{F}_q[x]$ such that any other polynomial in $\mathbb{F}_q[x]$ of which α is a zero is a polynomial multiple of M. Since α generates a degree n extension of $\mathbb{F}_q$, from above $F^n \alpha = \alpha$. Thus, the set α, $F\alpha$, $F^2\alpha$, ..., $F^{n-1}\alpha$ is F-stable, and the right-hand side product (when multiplied out) has coefficients in $\mathbb{F}_q$. Thus, it is a polynomial multiple of M. Since it is monic and has degree n (as does M), it must be M itself.   ////

Given ground field $\mathbb{F}_q$ and α in an algebraic extension E of $\mathbb{F}_q$, the images

$$\alpha, \ \alpha^q, \ \alpha^{q^2}, \ \ldots$$

of α under the Frobenius are the **(Galois) conjugates** of α over $\mathbb{F}_q$. Indeed, the notion of *Frobenius* automorphism is relative to the ground field $\mathbb{F}_q$. Two elements α, β in an algebraic extension E of $\mathbb{F}_q$ are **conjugate** if

$$\beta = \alpha^{q^t}$$

for some power F^t of the Frobenius over $\mathbb{F}_q$.

9.2.5 Proposition: Inside a given algebraic extension E of $\mathbb{F}_q$, the property of being conjugate is an equivalence relation. ///

9.2.6 Corollary: Given α in an algebraic field extension E of $\mathbb{F}_q$, the *number* of distinct conjugates of α over $\mathbb{F}_q$ is equal to the degree $[\mathbb{F}_q(\alpha) : \mathbb{F}_q]$. ///

9.2.7 Corollary: Let $f(x) \in \mathbb{F}_q[x]$ be irreducible, of degree n. Then $f(x)$ factors into linear factors in $\mathbb{F}_{q^n}$, (up to isomorphism) the unique extension of $\mathbb{F}_q$ of degree n. ///

Fix a prime power q, and an integer n. The set

$$= \mathrm{Aut}_{\mathbb{F}_q} \mathbb{F}_{q^n} = \{ \text{ automorphisms } h : \mathbb{F}_{q^n} \longrightarrow \mathbb{F}_{q^n} \text{ trivial on } \mathbb{F}_q \}$$

is a *group*, with operation *composition*. [8]

9.2.8 Theorem: The group $G = \mathrm{Aut}_{\mathbb{F}_q} \mathbb{F}_{q^n}$ of automorphisms of $\mathbb{F}_{q^n}$ trivial on $\mathbb{F}_q$ is cyclic of order n, generated by the Frobenius element $F(\alpha) = \alpha^q$.

Proof: First, we check that the Frobenius map is a field automorphism. It certainly preserves multiplication. Let p be the prime of which q is a power. Then p divides all the inner binomial coefficients $\binom{q}{i}$ with $0 < i < q$, essentially because p divides all the inner binomial coefficients $\binom{p}{i}$ with $0 < i < p$. Thus, for $\alpha, \beta \in \mathbb{F}_{q^n}$, by the binomial expansion,

$$(\alpha + \beta)^q = \alpha^q + \sum_{0<i<q} \binom{q}{i} \alpha^i \beta^{q-i} + \beta^q = \alpha^q + \beta^q$$

We should show that Frobenious does fix $\mathbb{F}_q$ pointwise. Since $\mathbb{F}_q^\times$ has order $q-1$, every element has order dividing $q-1$, by Lagrange. Thus, for $\beta \in \mathbb{F}_q$,

$$\beta^q = \beta^{q-1} \cdot \beta = 1 \cdot \beta = \beta$$

Certainly 0 is mapped to itself by Frobenius, so Frobenius fixes $\mathbb{F}_q$ pointwise, and, therefore, is a field automorphism of $\mathbb{F}_{q^n}$ over $\mathbb{F}_q$. Last, note that F^n fixes $\mathbb{F}_{q^n}$ pointwise, by the same argument that just showed that F fixes $\mathbb{F}_q$ pointwise. That is, F^n is the identity automorphism of $\mathbb{F}_{q^n}$. We note that F is invertible on $\mathbb{F}_{q^n}$, for any one of several reasons. One argument is that F^n is the identity.

The powers of the Frobenius element clearly form a subgroup of the automorphism group G, so the question is whether *every* automorphism is a power of Frobenius. There are many ways to approach this, but one straightforward way is as follows. We have seen that the multiplicative group $\mathbb{F}_{q^n}^\times$ is *cyclic*. Let α be a generator. Any field automorphism σ of $\mathbb{F}_{q^n}^\times$ is completely determined by $\sigma\alpha$, since a field map preserves multiplication, and, therefore,

$$\sigma(\alpha^n) = \left(\sigma(\alpha)\right)^n$$

[8] As usual, an *automorphism* of a thing is an isomorphism of it to itself, of whatever sort is currently under discussion. Here, we are concerned with field isomorphisms of $\mathbb{F}_{q^n}$ to itself which fix $\mathbb{F}_q$ pointwise. In general, with some further hypotheses to avoid various problems, roughly speaking the automorphism group of one field over another is a *Galois group*.

And we know that the only possible images of $\sigma\alpha$ are the other roots in $\mathbb{F}_{q^n}$ of the monic irreducible $f(x)$ of α in $\mathbb{F}_q[x]$, which is of degree n, since we know that

$$\mathbb{F}_{q^n} \approx \mathbb{F}_q[x]/f$$

That is, there are at most n possible images $\sigma\alpha$ of α, including α itself. Let's count the number of distinct images of α under powers of Frobenious. First, for $i < j$, using the invertibility of F, $F^i\alpha = F^j\alpha$ is equivalent to $\alpha = F^{j-i}\alpha$. Thus, it suffices to determine the smallest positive exponent j such that $F^j\alpha = \alpha$. In fact, being the generator of the cyclic group $\mathbb{F}_{q^n}^\times$, α has order exactly $q^n - 1$. Thus, the positive powers of α of orders less than $q^n - 1$ are distinct. Thus, $\alpha^{q^\ell} = \alpha$ implies $\alpha^{q^\ell - 1} = 1$, and then

$$q^n - 1 \text{ divides } q^\ell - 1$$

Thus, it must be that $\ell = n$. This shows that $\alpha, F\alpha, F^2\alpha, \ldots, F^{n-1}\alpha$ are distinct, and therefore are *all* the possible images of α by automorphisms. We noted that the image of α by an automorphism determines that automorphism completely, so $1, F, F^2, \ldots, F^{n-1}$ are all the automorphisms of $\mathbb{F}_{q^n}$ over $\mathbb{F}_q$. ///

9.3 *Counting irreducibles*

By now we might anticipate that counting irreducible polynomials $f(x) \in \mathbb{F}_q[x]$ of degree n is intimately connected with elements α [9] such that $[\mathbb{F}_q(\alpha) : \mathbb{F}_q] = n$, by taking roots α of $f(x) = 0$.

9.3.1 Proposition: The collection of monic irreducible polynomials $f(x)$ of degree n in $\mathbb{F}_q[x]$ is in bijection with sets of n mutually conjugate generators of $\mathbb{F}_{q^n}$ over $\mathbb{F}_q$, by

$$\alpha, \alpha^q, \ldots, \alpha^{q^{n-1}} \quad \longleftrightarrow \quad (x-\alpha)(x-\alpha^q)\ldots(x-\alpha^{q^{n-1}})$$

Proof: On one hand, a degree n monic irreducible f has a root α in $\mathbb{F}_q[x]/\langle f\rangle$, which is a degree n field extension of $\mathbb{F}_q$. In particular, $\mathbb{F}_q(\alpha) = \mathbb{F}_q[\alpha]$ is of degree n over $\mathbb{F}_q$. And (from just above)

$$f(x) = (x-\alpha)(x-\alpha^q)(x-\alpha^{q^2})\ldots(x-\alpha^{q^{n-1}})$$

We have noted that the n distinct images α^{q^i} are an equivalence class under the equivalence relation of being conjugate, and any one of these roots generates the same degree n extension as does α.

On the other hand, let α generate the unique degree n extension of $\mathbb{F}_q$ inside a fixed algebraic closure. That is, $\mathbb{F}_q(\alpha) = \mathbb{F}_q[\alpha]$ is of degree n over $\mathbb{F}_q$, which implies that the minimal polynomial f of α over $\mathbb{F}_q$ is of degree n. From above, the other roots of $f(x) = 0$ are exactly the conjugates of α over $\mathbb{F}_q$. ///

Let $\mu(n)$ be the Möbius function

$$\mu(n) = \begin{cases} 0 & \text{(if the square of any prime divides } n) \\ (-1)^t & \text{(otherwise, where distinct primes divide } n, \text{ but no square does)} \end{cases}$$

---

[9] In a fixed algebraic closure of $\mathbb{F}_q$, for example.

9.3.2 Corollary: The number of irreducible degree n polynomials in $\mathbb{F}_q[x]$ is

$$\text{number irreducibles degree } n \;=\; \frac{1}{n} \cdot \left(\sum_{d|n} \mu(d)\, q^{n/d} \right)$$

Proof: We want to remove from $\mathbb{F}_{q^n}$ the elements which generate (over $\mathbb{F}_q$) proper subfields of $\mathbb{F}_{q^n}$, and then divide by n, the number of conjugates of a given generator of $\mathbb{F}_{q^n}$ over $\mathbb{F}_q$. Above we showed that $\mathbb{F}_{q^m} \subset \mathbb{F}_{q^n}$ if and only if $m|n$. Thus, the maximal proper subfields of $\mathbb{F}_{q^n}$ are the fields $\mathbb{F}_{q^{n/r}}$ with r a prime dividing n. But the attempted count $q^n - \sum_{r|n} q^{n/r}$ over-counts the intersections of subfields $\mathbb{F}_{q^{n/r_1}}$ and $\mathbb{F}_{q^{n/r_2}}$, for primes $r_1 \neq r_2$. Thus, typically, we put back $q^{n/r_1 r_2}$, but we have put back too much, and must subtract the common triple intersections, and so on. After this inclusion-exclusion process, we divide by n so that we count equivalence classes of mutually conjugate generators of the degree n extension, rather than the individual generators. ///

Exercises

9.1 Show that any root α of $x^3 + x + 1 = 0$ in an algebraic closure of the finite field $\mathbb{F}_2$ with 2 elements is a generator for the multiplicative group $\mathbb{F}_{2^3}^\times$.

9.2 Find the irreducible quartic equation with coefficients in $\mathbb{F}_2$ satisfied by a generator for the cyclic group $\mathbb{F}_{2^4}^\times$.

9.3 Let f be an irreducible polynomial of degree n in $\mathbb{F}_q[x]$, where $\mathbb{F}_q$ is a field with q elements. Show that $f(x)$ divides $x^{q^n} - x$ if and only if $\deg f$ divides n.

9.4 Show that the *general linear group* $GL_n(\mathbb{F}_q)$ of invertible matrices with entries in the finite field $\mathbb{F}_q$ has an element of order $q^n - 1$.

9.5 Let k be a finite field. Show that $k[x]$ contains irreducibles of every positive integer degree.

9.6 For a power q of a prime p, find a p-Sylow subgroup of $GL_n(\mathbb{F}_q)$.

9.7 For q a power of an odd prime p, find a 2-Sylow subgroup of $GL_2(\mathbb{F}_q)$.

9.8 For q a power of an odd prime p, find a 2-Sylow subgroup of $GL_3(\mathbb{F}_q)$.

9.9 Find a 3-Sylow subgroup of $GL_3(\mathbb{F}_7)$.

9.10 (*Artin-Schreier polynomials*) Let q be a power of a prime p. Take $a \neq 0$ in $\mathbb{F}_q$. Show that if α is a root of $x^p - x + a = 0$ then so is $\alpha + i$ for $i = 1, 2, \ldots, p - 1$.

9.11 Show that Artin-Schreier polynomials are irreducible in $\mathbb{F}_q[x]$.

10. Modules over PIDs

The structure theorem for finitely-generated abelian groups and Jordan canonical form for endomorphisms of finite-dimensional vector spaces are example corollaries of a common idea.

10.1 *The structure theorem*

Let R be a **principal ideal domain**, that is, a commutative ring with identity such that every ideal I in R is **principal**, that is, the ideal can be expressed as

$$I = R \cdot x = \{r \cdot x : r \in R\}$$

for some $x \in R$. An R-module M is **finitely-generated** if there are finitely-many $m_1, \ldots, m_n$ in M such that every element m in M is expressible in at least one way as

$$m = r_1 \cdot m_1 + \ldots + r_n \cdot m_n$$

with $r_i \in R$.

A basic construction of new R-modules from old is as **direct sums**: given R-modules $M_1, \ldots, M_n$, the direct sum R-module

$$M_1 \oplus \ldots \oplus M_n$$

is the collection of n-tuples $(m_1, \ldots, m_n)$ with $m_i \in M_i$, with component-wise operation[1]

$$(m_1, \ldots, m_n) + (m'_1, \ldots, m'_n) = (m_1 + m'_1, \ldots, m_n + m'_n)$$

and the multiplication[2] by elements $r \in R$ by

$$r \cdot (m_1, \ldots, m_n) = (rm_1, \ldots, rm_n)$$

10.1.1 Theorem: Let M be a finitely-generated module over a PID R. Then there are uniquely determined ideals

$$I_1 \supset I_2 \supset \ldots \supset I_t$$

such that

$$M \approx R/I_1 \oplus R/I_2 \oplus \ldots \oplus R/I_t$$

The ideals I_i are the **elementary divisors** of M, and this expression is the **elementary divisor form** of M. [3]

Proof: (next chapter)   ///

10.2 *Variations*

The following proposition (which holds in more general circumstances) suggests variations on the form of the structure theorem above.

10.2.1 Proposition: [4] Let I and J be ideals of a commutative ring R with identity 1 such that

$$I + J = R$$

Take $r \in I$ and $s \in J$ such that $r + s = 1$. Then

$$R/I \oplus R/J \approx R/IJ$$

by [5]

$$(x + I, y + J) \longrightarrow sx + ry + IJ$$

Proof: First, prove well-definedness. Note that $r + s = 1$ implies that $1 - r = s \in J$, and, similarly, $1 - s = r \in I$. If $x - x' \in I$ and $y - y' \in I$, then

$$(sx + ry) - (sx' + ry') = s(x - x') + r(y - y') \in JI + IJ = IJ$$

---

[1] This certainly is the obvious generalization, to modules, of vector addition in vector spaces written as ordered n-tuples.

[2] Obviously generalizing the scalar multiplication in vector spaces.

[3] Other sources call the I_i's the **invariant factors** of the module.

[4] Yes, this is simply an abstracted form of Sun-Ze's theorem. The proof is exactly the same.

[5] Yes, the element $s \in J$ is the coefficient of x, and the element $r \in I$ is the coefficient of y.

This proves well-definedness. [6]   Next, show that the kernel is trivial. Indeed, if $sx + ry \in IJ$, then $(1-r)x = sx \in I$. Thus, as $rx \in I$, $x \in I$. Similarly $y \in J$, and we have injectivity. For surjectivity, take any $z \in R$, and compute that

$$(z+I,\; z+J) \longrightarrow sz + rz + IJ = (r+s)z + IJ = z + IJ$$

since $r + s = 1$. ///

Returning to the situation that R is a PID, let $I = R \cdot x$. Factor the generator x into prime element powers $p_i^{e_i}$ and a unit u in R

$$x = u \cdot p_1^{e_1} \dots p_t^{e_t}$$

Then, iterating the result of the previous proposition,

$$R/I = R/\langle x \rangle = R/\langle p_1^{e_1} \rangle \oplus \dots \oplus R/\langle p_t^{e_t} \rangle$$

10.2.2 Remark: Depending on the circumstances, it may be interesting that the left-hand side is expressible as the right-hand side, or, at other moments, that the right-hand side is expressible as the left-hand side.

Now if we have a direct sum

$$R/I_1 \oplus \dots \oplus R/I_n$$

we can do the same further prime-wise decomposition, if we want. That is, let

$$I_i = \langle p_1^{e_{i1}} \dots p_t^{e_{it}} \rangle$$

(with a common set of primes p_j in R), with non-negative integer exponents, [7]   the divisibility condition is

$$e_{1j} \leq e_{2j} \leq \dots \leq e_{nj}$$

and

$$R/I_1 \oplus \dots \oplus R/I_n \approx (R/\langle p_1^{e_{11}} \rangle \oplus \dots \oplus R/\langle p_1^{e_{n1}} \rangle) \oplus \dots \oplus (R/\langle p_t^{e_{1t}} \rangle \oplus \dots \oplus R/\langle p_t^{e_{nt}} \rangle)$$

That is, for each prime p_i, we can extract a summand in elementary divisor form whose elementary divisor ideals are generated simply by powers of p_i. (If some $e_{ij} = 0$, then $R/\langle p_j^{e_{ij}} \rangle = \{0\}$.)

Conversely, a direct sum of direct sums corresponding to distinct (non-associate) primes in R can be reassembled in a *unique* manner to fit the conclusion of the structure theorem.

As an example of the re-assembly into canonical form, taking $R = \mathbb{Z}$, let

$$M = \left(\mathbb{Z}/2 \oplus \mathbb{Z}/4 \oplus \mathbb{Z}/8 \right) \oplus \left(\mathbb{Z}/9 \oplus \mathbb{Z}/27 \right)$$

---

[6] Keep in mind that in this context IJ is not merely the collection of product xy with $x \in I$ and $y \in J$, but is the set of finite sums $\sum_i x_i y_i$ with $x_i \in I$, $y_i \in J$.

[7] Allowing non-negative integer exponents keeps the notation from becoming even more ridiculous than it is, though, at the same time, it creates some potential for confusion.

It is important to realize that there is *unique* choice of how to put the summands together in the form of the conclusion of the Structure Theorem, here

$$M \approx \mathbb{Z}/2 \oplus \mathbb{Z}/36 \oplus \mathbb{Z}/216$$

It is *true* that this is *also* isomorphic (for example) to

$$\mathbb{Z}/18 \oplus \mathbb{Z}/4 \oplus \mathbb{Z}/216$$

but this is not in canonical form, and mere permutation of the summands is insufficient to put it into canonical form.

10.2.3 Remark: Even *without* the condition

$$e_1 \leq \ldots \leq e_n$$

for prime p in R any direct sum

$$R/\langle p^{e_1} \rangle \oplus \ldots \oplus R/\langle p^{e_t} \rangle$$

involving just the prime p at worst needs merely a permutation of its factors to be put into elementary divisor form.

10.3 *Finitely-generated abelian groups*

Surely a very popular choice of PID is the ring of integers $\mathbb{Z}$. Finitely-generated $\mathbb{Z}$-modules are exactly *abelian groups*, since any abelian group is a $\mathbb{Z}$-module with structure given as usual by

$$n \cdot m = \begin{cases} \underbrace{m + \ldots + m}_{n} & (n \geq 0) \\ -(\underbrace{m + \ldots + m}_{|n|}) & (n \leq 0) \end{cases}$$

Any ideal I in $\mathbb{Z}$ has a unique non-negative generator. Thus, the Structure Theorem becomes

10.3.1 Corollary: Let M be a finitely-generated $\mathbb{Z}$-module (that is, a finitely-generated abelian group). Then there are uniquely determined non-negative integers $d_1, \ldots, d_n$ such that [8]

$$d_1 | d_2 | \ldots | d_n$$

and

$$M \approx \mathbb{Z}/d_1 \oplus \mathbb{Z}/d_2 \oplus \ldots \oplus \mathbb{Z}/d_n$$

10.3.2 Corollary: Let $n = p_1^{e_1} \ldots p_t^{e_t}$ with $p_1, \ldots, p_t$ distinct primes in $\mathbb{Z}$. Then every abelian group of order n is uniquely expressible as a direct sum

$$A_{p_1^{e_1}} \oplus \ldots \oplus A_{p_t^{e_t}}$$

of abelian groups $A_{p_i^{e_i}}$ of orders $p_i^{e_i}$.

---

[8] Keep in mind that any integer divides 0, so it may happen that some of the d_i are 0. Of course, if $d_i = 0$, then $d_j = 0$ for $j \geq i$.

Proof: This second corollary comes from the observations on variations of the Structure Theorem that we can obtain by thinking in terms of Sun-Ze's theorem. ///

10.3.3 Corollary: The finite abelian groups of order p^n for prime p are

$$\mathbb{Z}/p^{e_1} \oplus \ldots \oplus \mathbb{Z}/p^{e_t}$$

for all sets of positive integers $e_1, \ldots, e_t$ (for varying t) with

$$e_1 \leq \ldots \leq e_t$$

and

$$e_1 + \ldots + e_t = n$$

Proof: The inequalities on the exponents are the conditions organizing the elementary divisors, and the last equality reflects the condition that the order of the whole group be as required. ///

10.3.4 Example: Count the number of abelian groups of order 1000.

A finite abelian group is certainly finitely generated. Since $100000 = 2^5 \cdot 5^5$ (and 2 and 5 are distinct primes in $\mathbb{Z}$), by observations above, every abelian group of order 100000 is uniquely expressible as a direct sum of an abelian group of order 2^5 and an abelian group of order 5^5. From the last corollary, the number of abelian groups of order p^5 for any prime p is the number of sums of non-decreasing sequences of positive integers which sum to the exponent, here 5. [9] For 5 the possibilities are

$$
\begin{aligned}
1+1+1+1+1 &= 5 \\
1+1+1+2 &= 5 \\
1+2+2 &= 5 \\
1+1+3 &= 5 \\
2+3 &= 5 \\
1+4 &= 5 \\
5 &= 5
\end{aligned}
$$

That is, the abelian groups of order p^5 for prime p are

$$\mathbb{Z}/p \oplus \mathbb{Z}/p \oplus \mathbb{Z}/p \oplus \mathbb{Z}/p \oplus \mathbb{Z}/p$$

$$\mathbb{Z}/p \oplus \mathbb{Z}/p \oplus \mathbb{Z}/p \oplus \mathbb{Z}/p^2$$

$$\mathbb{Z}/p \oplus \mathbb{Z}/p^2 \oplus \mathbb{Z}/p^2$$

$$\mathbb{Z}/p \oplus \mathbb{Z}/p \oplus \mathbb{Z}/p^3$$

$$\mathbb{Z}/p^2 \oplus \mathbb{Z}/p^3$$

$$\mathbb{Z}/p \oplus \mathbb{Z}/p^4$$

$$\mathbb{Z}/p^5$$

[9] The number of non-decreasing sequences of positive integers summing to n is the number of **partitions** of n. This number grows rapidly with n, and seems not to be expressible by any simple computationally useful formula.

Thus, there are 7 abelian groups of order 2^5, and 7 of order 5^5, and $7 \cdot 7 = 49$ abelian groups of order $2^5 \cdot 5^5 = 100000$.

A useful and commonly occurring manner of describing a finitely-generated $\mathbb{Z}$-module is as a *quotient* of

$$M = \mathbb{Z}^t = \underbrace{\mathbb{Z} \oplus \ldots \oplus \mathbb{Z}}_{t}$$

by a submodule N, which itself can be described as the image of some $\mathbb{Z}^r$. The **standard basis** of $\mathbb{Z}^t$ is the set of generators given by

$$e_i = (0, \ldots, 0, 1, 0, \ldots, 0) \qquad (1 \text{ at } i^{th} \text{ place})$$

In the following chapter we will prove a result giving as a special case

10.3.5 Corollary: Let M be a $\mathbb{Z}$-module generated by $m_1, \ldots, m_t$. Then there is a unique $\mathbb{Z}$-module homomorphism $f : \mathbb{Z}^t \longrightarrow M$ such that $f(e_i) = m_i$. The kernel of f is finitely generated on at most t generators, and in fact is isomorphic to $\mathbb{Z}^r$ for some $r \leq t$.

$$/\!/\!/$$

10.4 *Jordan canonical form*

In this section we make the other popular choice of PID, namely $k[x]$ for a field k.

Let k be a field, V a finite-dimensional vector space over k, and T a k-linear endomorphism of V. Let $k[x] \longrightarrow \mathrm{End}_k(V)$ be the unique k-algebra homomorphism which sends $x \longrightarrow T$. This makes V into a $k[x]$-module. To say that V is finite-dimensional is to say that it is finitely-generated as a k-module, so certainly is finitely-generated as a $k[x]$-module. Thus, by the Structure Theorem, and by the prime-wise further decompositions,

10.4.1 Corollary: Given k-vectorspace V and k-linear endomorphism T of V, there is a sequence of monic polynomials $d_1, \ldots, d_t$ with [10]

$$d_1 | d_2 | \ldots | d_t$$

such that, as a $k[x]$-module,

$$V \approx k[x]/\langle d_1 \rangle \oplus \ldots \oplus k[x]/\langle d_t \rangle$$

where x acts on V by T, and x acts on the right-hand side by multiplication by x. [11] Further, taking monic irreducibles $f_1, \ldots, f_r$ and exponents e_{ij} such that $d_i = \prod_j f_j^{e_{ij}}$, we have

$$V \approx \left(k[x]/\langle f_1^{e_{11}} \rangle \oplus \ldots \oplus k[x]/\langle f_1^{e_{t1}} \rangle \right) \oplus \ldots \oplus \left(k[x]/\langle f_r^{e_{1r}} \rangle \oplus \ldots \oplus k[x]/\langle f_r^{e_{tr}} \rangle \right)$$

---

[10] One must remember that divisibility of elements and inclusion of the corresponding principal ideals run in opposite directions, namely, $a|b$ if and only if $Ra \supset Rb$.

[11] What else could the action be on a sum of $k[x]$-modules of the form $k[x]/I$?

Though we have not chosen a basis nor written matrices, this $k[x]$-module decomposition of the original k-vectorspace with endomorphism T is a **Jordan canonical form** of V with respect to T. [12]   The monic polynomials d_i that occur are the **elementary divisors** of T on V. [13]

Breaking V up into $k[x]$-module summands of the form

$$N = k[x]/\langle f^e \rangle \quad (f \text{ irreducible monic in } k[x])$$

is the finest reasonable decomposition to expect.   Each such N is a k-vectorspace of dimension $\deg f$. [14]

10.4.2 Example: Let
$$V = k[x]/\langle (x-\lambda)^e \rangle$$
be a k-vectorspace with with operator T given by multiplication by x (on the quotient), with $\lambda \in k$. Then

$$x \cdot (x-\lambda)^{e-1} = (x-\lambda)(x-\lambda)^{e-1} + \lambda(x-\lambda)^{e-1} = \lambda \cdot (x-\lambda)^{e-1} \bmod (x-\lambda)^e$$

That is, $(x-\lambda)^{e-1}$ modulo $(x-\lambda)^e$ is a λ-**eigenvector** with **eigenvalue** λ for the operator T. [15]

10.4.3 Example: A $k[x]$-module of the form

$$V = k[x]/\langle F \rangle$$

with (not necessarily irreducible) monic

$$F(x) = x^n + a_{n-1}x^{n-1} + a_{n-2}x^{n-2} + \ldots + xa_2 x^2 + a_1 x + a_0$$

in $k[x]$ of degree n is called a **cyclic module** for $k[x]$, since it can be generated by a single element, as we shall see here. A reasonable choice of k-basis is

$$1, \, x, \, x^2, \, \ldots, \, x^{n-1}$$

Then the endomorphism T on V given by multiplication by x is, with respect to this basis,

$$T \cdot x^i = \begin{cases} x^{i+1} & (i < n-1) \\ -(a_0 + a_1 x + \ldots + a_{n-1}x^{n-1}) & (i = n-1) \end{cases}$$

---

[12] It is only *a* canonical form because there are typically many different $k[x]$-isomorphisms of V to such a direct sum.

[13] It is not hard to see that the *minimal polynomial* of T (that is, the monic generator for the kernel of the map $k[x] \longrightarrow \mathrm{End}_k(V)$ that sends $x \longrightarrow T$) is the largest d_t of the elementary divisors d_i of T.

[14] If the exponent e is strictly larger than 1, then there are yet smaller $k[x]$ submodules, but they will not appear in a direct sum decomposition. This is clarified in examples below.

[15] As usual, for a k-linear endomorphism T of a k-vectorspace V, a non-zero vector $v \in V$ is a T-eigenvector with eigenvalue $\lambda \in k$ if $Tv = \lambda v$. In some sources these are called *proper values* rather than eigenvalues, but this terminology seems to be no longer in use.

That is, with respect to this basis, T has the matrix

$$\begin{pmatrix} 0 & 0 & 0 & 0 & 0 & \cdots & -a_0 \\ 1 & 0 & 0 & 0 & 0 & & -a_1 \\ 0 & 1 & 0 & 0 & 0 & & -a_2 \\ \vdots & & \ddots & \ddots & & & \vdots \\ & & & 1 & 0 & & -a_{n-3} \\ & & & 0 & 1 & 0 & -a_{n-2} \\ 0 & & & \cdots & 0 & 1 & -a_{n-1} \end{pmatrix}$$

This is the **rational canonical form** of T. [16]

10.4.4 Remark: If we want to make a matrix T (viewed as an endomorphism of k^n, viewed as column vectors) such that with the $k[x]$-module structure on k^n created by $k[x] \longrightarrow \mathrm{End}_k k^n$ given by $x \longrightarrow T$,

$$k^n \approx k[x]/\langle F \rangle$$

as $k[x]$-modules, we simply take the matrix T as above, namely with sub-diagonal 1s and the coefficients of the desired polynomial arranged in the right column. [17]

10.4.5 Example: To make a 3-by-3 matrix T so that the associated $k[x]$-structure on k^3 gives a module isomorphic to

$$k[x]/\langle x^3 + 2x^2 + 5x + 7 \rangle$$

we take

$$T = \begin{pmatrix} 0 & 0 & -7 \\ 1 & 0 & -5 \\ 0 & 1 & -2 \end{pmatrix}$$

If k happens to be *algebraically closed*, then a monic irreducible is of the form $x - \lambda$ for some $\lambda \in k$. Thus, the simplest $k[x]$-modules we're looking at in the context of the Structure Theorem are k-vectorspaces V of the form

$$V = k[x]/\langle (x - \lambda)^e \rangle$$

The endomorphism T of V is multiplication by x. [18]   At this point we can choose k-bases with respect to which the matrix of T (multiplication by x) is of various simple sorts. One obvious choice [19]   is to take k-basis consisting of (the *images* of, in the quotient)

$$1, \, x, \, x^2, \, \ldots, \, x^{e-1}$$

---

[16]   Note that only on a *cyclic* $k[x]$-module (where x acts by a k-linear endomorphism T) is there such a rational canonical form of the endomorphism T. And, yes, the question of whether or not a k-vectorspace with distinguished endomorphism T is cyclic or not certainly does depend on the endomorphism T. If there is a vector such that $v, Tv, T^2 v, \ldots$ form a basis, the module is cyclic, and v is a **cyclic vector**.

[17]   No, this sort of construction does not give any idea about eigenvalues or eigenvectors. But, on some occasions, this is not the issue.

[18]   Now that we've forgotten the original T above, having replaced it by multiplication by x on a quotient of $k[x]$!

[19]   Which might have the appeal of not depending upon λ.

We have

$$T \cdot x^i = \begin{cases} x^{i+1} & \text{(for } i < e-1) \\ x^e - (x-\lambda)^e & \text{(for } i = e-1) \end{cases}$$

The slightly peculiar expression in the case $i = e-1$ is designed to be a polynomial of degree $< e$, hence, a linear combination of the specified basis elements $1, x, \ldots, x^{e-1}$. [20] The other obvious choice is

$$1, \ x - \lambda, \ (x-\lambda)^2, \ \ldots, \ (x-\lambda)^{e-1}$$

In this case, since

$$x(x-\lambda)^i = (x-\lambda)^{i+1} + \lambda(x-\lambda)^i$$

we have

$$T \cdot (x-\lambda)^i = \begin{cases} \lambda(x-\lambda)^i + (x-\lambda)^{i+1} & \text{(for } i < e-1) \\ \lambda(x-\lambda)^i & \text{(for } i = e-1) \end{cases}$$

The latter choice shows that (the image in the quotient of)

$$(x-\lambda)^{e-1} = \lambda - \text{eigenvalue of } T$$

Indeed, the matrix for T with respect to the latter basis is the **Jordan block**

$$\begin{pmatrix} \lambda & 0 & 0 & 0 & 0 & \cdots & 0 \\ 1 & \lambda & 0 & 0 & 0 & & \\ 0 & 1 & \lambda & 0 & 0 & & \\ \vdots & & & \ddots & & & \vdots \\ & & & 1 & \lambda & 0 & 0 \\ & & & 0 & 1 & \lambda & 0 \\ 0 & & \cdots & & 0 & 1 & \lambda \end{pmatrix}$$

Thus, concerning matrices, the Structure Theorem says

10.4.6 Corollary: For algebraically closed fields k, given an endomorphism T of a finite-dimensional k-vectorspace, there is a choice of basis such that the associated matrix is of the form

$$\begin{pmatrix} B_1 & & & \\ & B_2 & & \\ & & \ddots & \\ & & & B_t \end{pmatrix}$$

where each B_i on the diagonal is a **Jordan block**, and all other entries are 0. ///

10.4.7 Example: When k is not necessarily algebraically closed, there may be irreducibles in $k[x]$ of higher degree. For monic irreducible f in $k[x]$ consider the $k[x]$-module

$$V = k[x]/\langle f^e \rangle$$

with endomorphism T being multiplication by x (on the quotient). Choice of k-basis that illuminates the action of T is more ambiguous now. Still, there are not very many plausible

---

[20] The matrix arising from this choice of basis is, in some circumstances, for example as just above, called the *rational canonical form*, though one should not depend upon this.

natural choices. Let $d = \deg f$. Then take k-basis consisting of (the images in the quotient of)

$$1, x, x^2, \ldots, x^{d-1}, \ f, f \cdot x, f \cdot x^2 \ldots, f \cdot x^{d-1}, \ \ldots, \ f^{e-1}, f^{e-1}x, \ldots, f^{e-1} \cdot x^{d-1}$$

That is, we choose a basis $1, x, x^2, \ldots, x^{d-1}$ for (images of) polynomials of degrees less than f, and then multiply those by powers of f below the power f^e that is (by definition) 0 in the quotient. [21]   The endomorphism T is still multiplication by x in the quotient. For certain of the basis elements the effect is easy to describe in terms of the basis: [22]

$$T \cdot f^i \cdot x^j = f^i \cdot x^{j+1} \quad \text{(for } j < d - 1)$$

However the other cases are somewhat messier than before. Namely,

$$T \cdot f^i \cdot x^{d-1} = \begin{cases} f^i \cdot x^d & = & f^i \cdot (x^d - f) + f^{i+1} & (i < e - 1) \\ f^i \cdot x^d & = & f^i \cdot (x^d - f) & (i = e - 1) \end{cases}$$

Note that $x^d - f$ is a linear combination of monomials x^j with $0 \le j \le d - 1$. This is still called a **Jordan canonical form**.

10.4.8 Example: Let $k = \mathbb{R}$, $f(x) = x^2 + 1$, and consider

$$V = \mathbb{R}[x]/\langle (x^2 + 1)^3 \rangle$$

According to the prescription just given, we take basis

$$1, \ x, \ x^2 + 1, \ (x^2 + 1)x, \ (x^2 + 1)^2, \ (x^2 + 1)^2 x$$

Then the endomorphism T which is multiplication by x is, in terms of this basis,

$$
\begin{array}{rcccc}
T \cdot 1 & & & = & x \\
T \cdot x & = & x^2 & = & -1 + (x^2 + 1) \\
T \cdot x^2 + 1 & & & = & (x^2 + 1)x \\
T \cdot (x^2 + 1)x & = & (x^2 + 1)x^2 & = & -(x^2 + 1) + (x^2 + 1)^2 \\
T \cdot (x^2 + 1)^2 & & & = & (x^2 + 1)^2 x \\
T \cdot (x^2 + 1)^2 x & = & (x^2 + 1)^2 x^2 & = & -(x^2 + 1)^2
\end{array}
$$

since $(x^2 + 1)^3 = 0$ in the quotient. That is, with respect to this basis, the matrix is

$$
\begin{pmatrix}
0 & -1 & 0 & 0 & 0 & 0 \\
1 & 0 & 0 & 0 & 0 & 0 \\
0 & 1 & 0 & -1 & 0 & 0 \\
0 & 0 & 1 & 0 & 0 & 0 \\
0 & 0 & 0 & 1 & 0 & -1 \\
0 & 0 & 0 & 0 & 1 & 0
\end{pmatrix}
$$

Notice that there is no eigenvector or eigenvalue in the usual more elementary sense. But we still do have some understanding of what the endomorphism does.

---

[21] Note that there is no obvious choice to replace $1, x, x^2, \ldots, x^{d-1}$.

[22] Note that this easy case did not occur at all when the monic irreducible f was *linear*.

10.4.9 Remark: Note that the latter more complicated (because k need not be algebraically closed) version of Jordan canonical form incorporates both the simpler version of Jordan canonical form as well as the rational canonical form.

10.5 *Conjugacy versus $k[x]$-module isomorphism*

First, it is important to realize that conjugation of matrices

$$A \longrightarrow gAg^{-1}$$

(for invertible g) is exactly changing the basis with respect to which one computes the matrix of the underlying endomorphism.

Two n-by-n matrices A and B with entries in a field k are **conjugate** if there is an invertible n-by-n matrix g with entries in k such that

$$B = gAg^{-1}$$

Conjugacy is obviously an equivalence relation. The **conjugacy classes** of n-by-n matrices with entries in k are the corresponding equivalence classes with respect to this equivalence relation.

But it is somewhat misguided to fix upon matrices as descriptive apparatus for linear endomorphisms, [23] since, in effect, a matrix specifies not only the endomorphism but also a *basis* for the vector space, and a whimsically or accidentally chosen basis will not illuminate the structure of the endomorphism.

Thus, we will take the viewpoint that, yes, the set $V = k^n$ of size n column matrices with entries in k is a k-vectorspace, and, yes, n-by-n matrices give k-linear endomorphisms [24] of V by matrix multiplication, *but*, no, this is only one of several possible descriptions, and admittedly sub-optimal for certain purposes.

Thus, more properly, given two k-linear endomorphisms S and T of a finite-dimensional k-vectorspace V, say that S and T are **conjugate** if there is an *automorphism*[25] $g : V \longrightarrow V$ such that

$$g \circ S \circ g^{-1} = T$$

---

[23] Though, certainly, a matrix-oriented version of linear algebra is a reasonable developmental stage, probably *necessary*. And writing out a small numerical matrix is a compellingly direct description of an endomorphism, entirely adequate for many purposes.

[24] At least once in one's life one should check that matrix multiplication and composition of endomorphisms are compatible. Given a k-vectorspace V with k-basis $e_1, \ldots, e_n$, define a map φ from the ring $\text{End}_k(V)$ of k-linear endomorphisms to the ring of n-by-n matrices with entries in k, by defining the ij^{th} entry $\varphi(T)_{ij}$ of $\varphi(T)$ (for endomorphism T) by $Te_j = \sum_i \varphi(T)_{ij} e_i$. (Yes, there is another obvious possibility for indexing, but the present choice is what we want. One should check this, too.) Then φ is a ring homomorphism. The main point is multiplication: On one hand, for two endomorphisms S and T, $(S \circ T)e_k = \sum_j \varphi(S \circ T)_{ik} e_i$. On the other hand, using the linearity of S, $S(Te_k) = S(\sum_k \varphi(T)_{jk} e_j) = \sum_k \varphi(T)_{jk} \sum_k \varphi(S)_{ij} e_i$. Since a matrix product has ij^{th} entry $(\varphi(S)\varphi(T))_{ij} = \sum_\ell \varphi(S)_{i\ell}\varphi(T)_{\ell j}$, the two expressions are the same thing.

[25] As usual, an automorphism is an endomorphism that is an isomorphism of the thing to itself.

Emphatically, this includes conjugation of matrices if or when we write endomorphisms as matrices.

The following proposition, which is trivial to prove once laid out clearly, illustrates (among other things) that conjugacy of matrices is a special example of a more meaningful structural property.

10.5.1 Proposition: Let V be a finite-dimensional k-vectorspace. Let S and T be two k-linear endomorphisms of V. Let V_S be V with the $k[x]$-module structure in which x acts on $v \in V$ by $xv = Sv$, and let V_T be V with the $k[x]$-module structure in which x acts on $v \in V$ by $xv = Tv$. Then S and T are conjugate if and only if $V_S \approx V_T$ as $k[x]$-modules.

Proof: First, suppose that $V_S \approx V_T$ as $k[x]$-modules. Let $g : V_S \longrightarrow V_T$ be the k-vectorspace isomorphism that is also a $k[x]$-module isomorphism. The latter condition means exactly that (in addition to the vectorspace isomorphism aspect) for all v in V

$$g(x \cdot v) = x \cdot g(v)$$

That is, since in V_S the action of x is $xv = Sv$ and in V_T is $xv = Tv$

$$g(Sv) = T\, g(v)$$

Since this is true for all $v \in V$, we have an equality of endomorphisms

$$g \circ S = T \circ g$$

Since g is invertible,

$$g \circ S \circ g^{-1} = T$$

as claimed. It is clear that this argument is reversible, giving the opposite inclusion as well.   ///

10.5.2 Remark: The uniqueness part of the Structure Theorem says that the elementary divisors (ideals) $I_1 \supset \ldots \supset I_t$ in an expression

$$M \approx R/I_1 \oplus \ldots \oplus R/I_t$$

(with M a finitely-generated module over a PID R) are uniquely determined by the R-module isomorphism class of M, and, conversely, that choice of such ideals uniquely determines the isomorphism class. Thus,

10.5.3 Corollary: Conjugacy classes of endomorphisms of a finite-dimensional k-vectorspace V are in bijection with choices of (monic) elementary divisors $d_1 | \ldots | d_t$ in an expression
$$V \approx k[x]/\langle d_1 \rangle \oplus \ldots \oplus k[x]/\langle d_t \rangle$$
as $k[x]$-module.   ////

10.5.4 Remark: Further, using the unique factorization in the PID $k[x]$, one finds that in prime-wise decompositions (via Sun-Ze)

$$k[x]/\langle f_1^{e_1} \ldots f_r^{e_r} \rangle \approx k[x]/\langle f_1^{e_1} \rangle \oplus \ldots \oplus k[x]/\langle f_r^{e_r} \rangle$$

with f_is irreducible, the exponents e_i are uniquely determined by the isomorphism class, and vice-versa. Combining this uniqueness with the Structure Theorem's uniqueness gives the corresponding uniqueness for the general prime-wise expression of a module as a direct sum.

10.5.5 Remark: The extent to which different-appearing Jordan forms of *matrices* can be conjugate is (likewise) answered by the uniqueness assertion of the Structure Theorem. For example, in the case of algebraically closed field, two Jordan forms are conjugate if and only if the collection (counting repeats) of Jordan blocks of one is equal to that of the other, allowing only permutations among the blocks. That is, in matrix form, the only mutually conjugate Jordan forms are visually obvious.

10.5.6 Example: The following two Jordan forms are conjugate

$$\begin{pmatrix} 3 & 0 & 0 & 0 & 0 \\ 1 & 3 & 0 & 0 & 0 \\ 0 & 1 & 3 & 0 & 0 \\ 0 & 0 & 0 & 7 & 0 \\ 0 & 0 & 0 & 1 & 7 \end{pmatrix} \quad \text{is conjugate to} \quad \begin{pmatrix} 7 & 0 & 0 & 0 & 0 \\ 1 & 7 & 0 & 0 & 0 \\ 0 & 0 & 3 & 0 & 0 \\ 0 & 0 & 1 & 3 & 0 \\ 0 & 0 & 0 & 1 & 3 \end{pmatrix}$$

by

$$\begin{pmatrix} 3 & 0 & 0 & 0 & 0 \\ 1 & 3 & 0 & 0 & 0 \\ 0 & 1 & 3 & 0 & 0 \\ 0 & 0 & 0 & 7 & 0 \\ 0 & 0 & 0 & 1 & 7 \end{pmatrix} = \begin{pmatrix} 0 & 0 & 1 & 0 & 0 \\ 0 & 0 & 0 & 1 & 0 \\ 0 & 0 & 0 & 0 & 1 \\ 1 & 0 & 0 & 0 & 0 \\ 0 & 1 & 0 & 0 & 0 \end{pmatrix} \begin{pmatrix} 7 & 0 & 0 & 0 & 0 \\ 1 & 7 & 0 & 0 & 0 \\ 0 & 0 & 3 & 0 & 0 \\ 0 & 0 & 1 & 3 & 0 \\ 0 & 0 & 0 & 1 & 3 \end{pmatrix} \begin{pmatrix} 0 & 0 & 1 & 0 & 0 \\ 0 & 0 & 0 & 1 & 0 \\ 0 & 0 & 0 & 0 & 1 \\ 1 & 0 & 0 & 0 & 0 \\ 0 & 1 & 0 & 0 & 0 \end{pmatrix}^{-1}$$

10.5.7 Remark: Without further indications from context, it is not clear whether one would want to parametrize conjugacy classes by the decomposition given immediately by the Structure Theorem, namely sums of rational canonical forms, or by the further prime-wise decomposition (as in the Jordan decomposition), which do still involve some sort of rational canonical forms when the underlying field is not algebraically closed.

Generally, for a commutative ring R with identity 1, let

$$GL(n, R) = \{ \text{ invertible } n\text{-by-}n \text{ matrices, entries in } R \}$$

This is the **general linear group** of size n over R.

10.5.8 Example: Determine the conjugacy classes in $GL(2, k)$ for a field k. Note that $GL(2, k)$ is the set (group) of *invertible* endomorphisms of the two-dimensional k-vectorspace k^2. From the Structure Theorem, and from the observation above that conjugacy classes are in bijection with batches of elementary divisors, we can immediately say that $k[x]$-module structures

$$V \approx k[x]/\langle d_1 \rangle \oplus \ldots \oplus k[x]/\langle d_t \rangle$$

with $d_1 | \ldots | d_t$ and

$$\deg d_1 + \ldots + \deg d_t = 2 = \dim_k V$$

specify the conjugacy classes. [26]   Since the dimension is just 2, there are only two cases

$$V \approx \begin{cases} k[x]/\langle d_1 \rangle & (d_1 \text{ monic quadratic}) \\ k[x]/\langle x - \lambda \rangle \oplus k[x]/\langle x - \lambda \rangle & (\text{monic linear case}) \end{cases}$$

Yes, in the second case the linear monic is repeated, due to the divisibility requirement. Using rational canonical forms in the first case, and (in a degenerate sense) in the second case as well, we have corresponding irredundant conjugacy class representatives

$$\begin{pmatrix} 0 & -a_2 \\ 1 & -a_1 \end{pmatrix} \quad (a_1 \in k, \ a_2 \in k^\times)$$

$$\begin{pmatrix} \lambda & 0 \\ 0 & \lambda \end{pmatrix} \quad (\lambda \in k^\times)$$

One might object that in the first of the two cases we have no indication of eigenvalues/eigenvectors. Thus, we might consider the two cases of quadratic monics, namely, irreducible and not. In the irreducible case nothing further happens, but with reducible $d_1(x) = (x - \lambda)(x - \mu)$ if $\lambda \neq \mu$ we have

$$k[x]/\langle (x - \lambda)(x - \mu) \rangle \approx k[x]/\langle x - \lambda \rangle \oplus k[x]/\langle x - \mu \rangle$$

We still have no real recourse but to use a rational canonical form for the quadratic irreducible case, but the reducible case with distinct zeros is diagonalized, and the *repeated factor* (reducible) case gives a non-trivial Jordan block. The $k[x]$-module structures are, respectively,

$$V \approx \begin{cases} k[x]/\langle d_1 \rangle & (d_1 \text{ irreducible monic quadratic}) \\ k[x]/\langle x - \lambda \rangle \oplus k[x]/\langle x - \mu \rangle & (\lambda \neq \mu, \text{ both in } k^\times) \\ k[x]/\langle (x - \lambda)^2 \rangle & (\text{repeated root case}, \lambda \in k^\times) \\ k[x]/\langle x - \lambda \rangle \oplus k[x]/\langle x - \lambda \rangle & (\text{monic linear case}) \end{cases}$$

In matrix form, the irredundant representatives are, respectively,

$$\begin{pmatrix} 0 & -a_0 \\ 1 & -a_1 \end{pmatrix} \quad (x^2 + a_1 x + a_0 \text{ irreducible})$$

$$\begin{pmatrix} \lambda & 0 \\ 0 & \mu \end{pmatrix} \quad (\lambda \neq \mu, \text{ both in } k^\times)$$

$$\begin{pmatrix} \lambda & 0 \\ 1 & \lambda \end{pmatrix} \quad (\lambda \in k^\times)$$

$$\begin{pmatrix} \lambda & 0 \\ 0 & \lambda \end{pmatrix} \quad (\lambda \in k^\times)$$

---

[26] Again, the endomorphism T representing the conjugacy class is the one given by multiplication by x on the right-hand side. It is *transported* back to V via the k-vectorspace isomorphism.

10.5.9 Example: Determine the conjugacy classes in $GL(3,k)$ for a field k. From the Structure Theorem and the fact that conjugacy classes are in bijection with batches of elementary divisors, the $k[x]$-module structures

$$V \approx k[x]/\langle d_1 \rangle \oplus \ldots \oplus k[x]/\langle d_t \rangle$$

with $d_1 | \ldots | d_t$ and

$$\deg d_1 + \ldots + \deg d_t = \dim_k V = 3$$

specify the conjugacy classes. Since the dimension is 3, there are 3 cases

$$V \approx \begin{cases} k[x]/\langle C \rangle & (C \text{ monic cubic}) \\ k[x]/\langle x - \lambda \rangle \oplus k[x]/\langle (x-\lambda)(x-\mu) \rangle & (\lambda, \mu \in k^{\times}) \\ k[x]/\langle x-\lambda \rangle \oplus k[x]/\langle x-\lambda \rangle \oplus k[x]/\langle x-\lambda \rangle & (\lambda \in k^{\times}) \end{cases}$$

Yes, in the second case the linear monic is repeated, as even more so in the third, by the divisibility requirement. We can still use a rational canonical form in each of the cases, to write matrix versions of these

$$\begin{pmatrix} 0 & 0 & -a_0 \\ 1 & 0 & -a_1 \\ 0 & 1 & -a_2 \end{pmatrix} \quad (a_0 \in k^{\times})$$

$$\begin{pmatrix} \lambda & 0 & 0 \\ 0 & 0 & -\lambda\mu \\ 0 & 1 & -\lambda - \mu \end{pmatrix} \quad (\lambda, \mu \in k^{\times})$$

$$\begin{pmatrix} \lambda & 0 & 0 \\ 0 & \lambda & 0 \\ 0 & 0 & \lambda \end{pmatrix} \quad (\lambda \in k^{\times})$$

In the first two cases the eigenvalues/eigenvectors are not delineated. It breaks up into 3 subcases, namely, irreducible cubic, linear and irreducible quadratic, and three linear factors. The $k[x]$-module structures are, respectively,

$$V \approx \begin{cases} k[x]/\langle C \rangle & (C \text{ irred monic cubic}) \\ k[x]/\langle (x-\lambda)Q(x) \rangle & (\lambda \in k^{\times}, Q \text{ irred monic quadratic}) \\ k[x]/\langle (x-\lambda)(x-\mu)(x-\nu) \rangle & (\lambda, \mu, \nu \in k^{\times}) \\ k[x]/\langle x-\lambda \rangle \oplus k[x]/\langle (x-\lambda)(x-\mu) \rangle & (\lambda, \mu \in k^{\times}) \\ k[x]/\langle x-\lambda \rangle \oplus k[x]/\langle x-\lambda \rangle \oplus k[x]/\langle x-\lambda \rangle & (\lambda \in k^{\times}) \end{cases}$$

The third and fourth cases break up into subcases depending upon the confluence (or not) of the parameters. That is, in the third case there are three subcases, where all the $\lambda, \mu\, nu$ are the same, only two are the same, or all different. The fourth case has two subcases, $\lambda = \mu$ or not. [27] In the following display, it is assumed that λ, μ, ν are distinct and non-zero.

---

[27] If the field k is $\mathbb{F}_2$, then non-zero parameters in the ground field cannot be distinct.

The last-mentioned subcases are presented on the same line. And Q is an irreducible monic quadratic, C an irreducible monic cubic.

$$\frac{k[x]}{\langle C \rangle}$$

$$\frac{k[x]}{\langle (x-\lambda)Q(x) \rangle}$$

$$\frac{k[x]}{\langle x-\lambda \rangle} \oplus \frac{k[x]}{\langle x-\mu \rangle} \oplus \frac{k[x]}{\langle x-\nu \rangle} \qquad \frac{k[x]}{\langle (x-\lambda)^2 \rangle} \oplus \frac{k[x]}{\langle x-\mu \rangle} \qquad \frac{k[x]}{\langle (x-\lambda)^3 \rangle}$$

$$\frac{k[x]}{\langle x-\lambda \rangle} \oplus \frac{k[x]}{\langle x-\lambda \rangle} \oplus \frac{k[x]}{\langle x-\mu \rangle} \qquad \frac{k[x]}{\langle x-\lambda \rangle} \oplus \frac{k[x]}{\langle (x-\lambda)^2 \rangle}$$

$$\frac{k[x]}{\langle x-\lambda \rangle} \oplus \frac{k[x]}{\langle x-\lambda \rangle} \oplus \frac{k[x]}{\langle x-\lambda \rangle}$$

In matrix form, the irredundant representatives are, respectively, (with λ, μ, ν distinct and non-zero)

$$\begin{pmatrix} 0 & 0 & -a_0 \\ 1 & 0 & -a_1 \\ 0 & 1 & -a_2 \end{pmatrix} \quad (x^3 + a_2 x^2 + a_1 x + a_2 \text{ irreducible})$$

$$\begin{pmatrix} \lambda & 0 & 0 \\ 0 & 0 & -a_0 \\ 0 & 1 & -a_1 \end{pmatrix} \quad (x^2 + a_1 x + a_0 \text{ irreducible})$$

$$\begin{pmatrix} \lambda & 0 & 0 \\ 0 & \mu & 1 \\ 0 & 0 & \nu \end{pmatrix} \qquad \begin{pmatrix} \lambda & 0 & 0 \\ 1 & \lambda & 0 \\ 0 & 0 & \mu \end{pmatrix} \qquad \begin{pmatrix} \lambda & 0 & 0 \\ 1 & \lambda & 0 \\ 0 & 1 & \lambda \end{pmatrix}$$

$$\begin{pmatrix} \lambda & 0 & 0 \\ 0 & \lambda & 0 \\ 0 & 0 & \mu \end{pmatrix} \qquad \begin{pmatrix} \lambda & 0 & 0 \\ 0 & \lambda & 0 \\ 0 & 1 & \lambda \end{pmatrix}$$

$$\begin{pmatrix} \lambda & 0 & 0 \\ 0 & \lambda & 0 \\ 0 & 0 & \lambda \end{pmatrix}$$

10.5.10 Example: Determine the conjugacy classes in $GL(5, k)$ for field k. Use the Structure Theorem and the bijection of conjugacy classes with batches of elementary divisors. There are seven different patterns of degrees of elementary divisors, with

$\lambda, \mu, \nu \in k^{\times}$, and all polynomials *monic*

(1) $$\frac{k[x]}{\langle Q \rangle} \qquad (Q \text{ quintic})$$

(2) $$\frac{k[x]}{\langle x - \lambda \rangle} \oplus \frac{k[x]}{\langle (x - \lambda)C(x) \rangle} \qquad (C \text{ cubic})$$

(3) $$\frac{k[x]}{\langle x - \lambda \rangle} \oplus \frac{k[x]}{\langle x - \lambda \rangle} \oplus \frac{k[x]}{\langle (x - \lambda)Q(x) \rangle} \qquad (Q \text{ quadratic})$$

(4) $$\frac{k[x]}{\langle Q(x) \rangle} \oplus \frac{k[x]}{\langle (x - \lambda)Q(x) \rangle} \qquad (Q \text{ quadratic})$$

(5) $$\frac{k[x]}{\langle x - \lambda \rangle} \oplus \frac{k[x]}{\langle x - \lambda \rangle} \oplus \frac{k[x]}{\langle x - \lambda \rangle} \oplus \frac{k[x]}{\langle (x - \lambda)(x - \mu) \rangle}$$

(6) $$\frac{k[x]}{\langle x - \lambda \rangle} \oplus \frac{k[x]}{\langle x - \lambda \rangle} \oplus \frac{k[x]}{\langle x - \lambda \rangle} \oplus \frac{k[x]}{\langle x - \lambda \rangle} \oplus \frac{k[x]}{\langle x - \lambda \rangle}$$

There is no obvious graceful way to write rational canonical forms that indicate divisibility. Listing the divisors to save space, the reducibility subcases are (with $\lambda, \mu, \nu, \sigma, \tau$ non-zero)

(1a) $\qquad\qquad\qquad Q(x) \qquad\qquad\qquad\qquad\qquad (Q \text{ irred quintic})$

(1b) $\qquad\qquad\qquad Q(x)C(x) \qquad\qquad\qquad (Q \text{ irred quadratic}, C \text{ irred cubic})$

(1c) $\qquad\qquad (x - \lambda)Q_1(x)Q_2(2) \qquad\qquad\qquad (Q_1, Q_2 \text{ irred quadratic})$

(1d) $\qquad (x - \lambda)(x - \mu)(x - \nu)Q(2) \qquad\qquad\qquad (Q \text{ irred quadratic})$

(1e) $\qquad (x - \lambda)(x - \mu)(x - \nu)(x - \sigma)(x - \tau)$

(2a) $\qquad\qquad (x - \lambda), \; (x - \lambda)C(x) \qquad\qquad\qquad\qquad (C \text{ irred cubic})$

(2b) $\qquad\quad (x - \lambda), \; (x - \lambda)(x - \mu)Q(x) \qquad\qquad\qquad (Q \text{ irred quadratic})$

(2c) $\qquad (x - \lambda), \; (x - \lambda)(x - \mu)(x - \nu)(x - \tau)$

(3a) $\qquad\quad (x - \lambda), \; (x - \lambda), \; (x - \lambda)Q(x) \qquad\qquad\quad (Q \text{ irred quadratic})$

(3b) $\qquad (x - \lambda), \; (x - \lambda), \; (x - \lambda)(x - \mu)(x - \nu)$

(3c) $\qquad\qquad\quad Q(x), \; (x - \lambda)Q(x) \qquad\qquad\qquad\quad (Q \text{ irred quadratic})$

(3d) $\qquad (x - \mu)(x - \nu), \; (x - \lambda)(x - \mu)(x - \nu)$

(4) $\qquad (x - \lambda), \; (x - \lambda), \; (x - \lambda), \; (x - \lambda)(x - \mu)$

(5) $\qquad (x - \lambda), \; (x - \lambda), \; (x - \lambda), \; (x - \lambda), \; (x - \lambda)$

There still remains the sorting into subcases, depending upon confluence of parameters. The most novel case is the case denoted 1c above, where there is a single elementary divisor

$(x - \lambda)Q_1(x)Q_2(x)$, with irreducible monic quadratics [28] Q_i. If $Q_1 \neq Q_2$, the canonical form is merely a direct sum of previous (smaller) cases. But if $Q_1 = Q_2$, a new thing happens in the direct summand $k[x]/\langle Q(x)^2 \rangle$ in

$$k[x]/\langle (x - \lambda)\, Q(x)^2 \rangle \approx k[x]/\langle x - \lambda \rangle \oplus k[x]/\langle Q(x)^2 \rangle$$

As earlier, letting $Q(x) = x^2 + a_1 x + a_2$, we can choose a basis

$$1,\ x,\ Q(x),\ xQ(x) \mod Q(x)^2$$

for $k[x]/\langle Q(x)^2 \rangle$. Then the endomorphism given by multiplication by x has matrix

$$\begin{pmatrix} 0 & -a_0 & 0 & 0 \\ 1 & -a_1 & 0 & 0 \\ 0 & 1 & 0 & -a_0 \\ 0 & 0 & 1 & -a_1 \end{pmatrix}$$

10.5.11 Example: Determine the matrix canonical form for an endomorphism T of a 6-dimensional k-vectorspace V, where T has the single elementary divisor $C(x)^2$m where C is an irreducible monic cubic

$$C(x) = x^3 + a_2 x^2 + a_1 x + a_0$$

Take basis

$$1,\ x,\ x^2,\ C(x),\ xC(x),\ x^2 C(x) \mod C(x)^2$$

for $k[x]/\langle C(x)^2 \rangle$. Then the endomorphism T given by multiplication by x has matrix

$$\begin{pmatrix} 0 & 0 & -a_0 & 0 & 0 & 0 \\ 1 & 0 & -a_1 & 0 & 0 & 0 \\ 0 & 1 & -a_2 & 0 & 0 & 0 \\ 0 & 0 & 1 & 0 & 0 & -a_0 \\ 0 & 0 & 0 & 1 & 0 & -a_1 \\ 0 & 0 & 0 & 0 & 1 & -a_2 \end{pmatrix}$$

10.5.12 Example: If the single elementary divisor were $C(x)^3$ with a monic cubic $C(x) = x^3 + a^2 x^2 + a_1 x + a_0$, then the basis

$$1,\ x,\ x^2,\ C(x),\ xC(x),\ x^2 C(x),\ C(x)^@,\ xC(x)^2,\ x^2 C(x)^2 \mod C(x)^3$$

gives matrix

$$\begin{pmatrix} 0 & 0 & -a_0 & 0 & 0 & 0 & 0 & 0 & 0 \\ 1 & 0 & -a_1 & 0 & 0 & 0 & 0 & 0 & 0 \\ 0 & 1 & -a_2 & 0 & 0 & 0 & 0 & 0 & 0 \\ 0 & 0 & 1 & 0 & 0 & -a_0 & 0 & 0 & 0 \\ 0 & 0 & 0 & 1 & 0 & -a_1 & 0 & 0 & 0 \\ 0 & 0 & 0 & 0 & 1 & -a_2 & 0 & 0 & 0 \\ 0 & 0 & 0 & 0 & 0 & 1 & 0 & 0 & -a_0 \\ 0 & 0 & 0 & 0 & 0 & 0 & 1 & 0 & -a_1 \\ 0 & 0 & 0 & 0 & 0 & 0 & 0 & 1 & -a_2 \end{pmatrix}$$

---

[28] If the underlying field k is algebraically closed, this and more complicated situations do not arise.

10.6 *Worked examples*

10.6.1 Example: Given a 3-by-3 matrix M with integer entries, find A, B integer 3-by-3 matrices with determinant ± 1 such that AMB is diagonal.

Let's give an *algorithmic*, rather than *existential*, argument this time, saving the existential argument for later.

First, note that given two integers x, y, not both 0, there are integers r, s such that $g = \gcd(x, y)$ is expressible as $g = rx + sy$. That is,

$$(x \quad y) \begin{pmatrix} r & * \\ s & * \end{pmatrix} = (g \quad *)$$

What we want, further, is to figure out what other two entries will make the second entry 0, *and* will make that 2-by-2 matrix invertible (in $GL_2(\mathbb{Z})$). It's not hard to guess:

$$(x \quad y) \begin{pmatrix} r & -y/g \\ s & x/g \end{pmatrix} = (g \quad 0)$$

Thus, given $(x \ y \ z)$, there is an invertible 2-by-2 integer matrix $\begin{pmatrix} a & b \\ c & d \end{pmatrix}$ such that

$$(y \quad z) \begin{pmatrix} a & b \\ c & d \end{pmatrix} = (\gcd(y, z) \quad 0)$$

That is,

$$(x \quad y \quad z) \begin{pmatrix} 1 & 0 & 0 \\ 0 & a & b \\ 0 & c & d \end{pmatrix} = (x \quad \gcd(y, z) \quad 0)$$

Repeat this procedure, now applied to x and $\gcd(y, z)$: there is an invertible 2-by-2 integer matrix $\begin{pmatrix} a' & b' \\ c' & d' \end{pmatrix}$ such that

$$(x \quad \gcd(y, z)) \begin{pmatrix} a' & b' \\ c' & d' \end{pmatrix} = (\gcd(x, \gcd(y, z)) \quad 0)$$

That is,

$$(x \quad \gcd(y, z) \quad 0) \begin{pmatrix} a' & b' & 0 \\ c' & d' & 0 \\ 0 & 0 & 1 \end{pmatrix} = (\gcd(x, y, z) \quad 0 \quad 0)$$

since *gcds* can be computed iteratively. That is,

$$(x \quad y \quad z) \begin{pmatrix} 1 & 0 & 0 \\ 0 & a & b \\ 0 & c & d \end{pmatrix} \begin{pmatrix} a' & b' & 0 \\ c' & d' & 0 \\ 0 & 0 & 1 \end{pmatrix} = (\gcd(x, y, z) \quad 0 \quad 0)$$

Given a 3-by-3 matrix M, *right*-multiply by an element A_1 of $GL_3(\mathbb{Z})$ to put M into the form

$$MA_1 = \begin{pmatrix} g_1 & 0 & 0 \\ * & * & * \\ * & * & * \end{pmatrix}$$

where (necessarily!) g_1 is the *gcd* of the top row. Then *left*-multiply by an element $B_2 \in GL_3(\mathbb{Z})$ to put MA into the form

$$B_2 \cdot MA_1 = \begin{pmatrix} g_2 & * & * \\ 0 & * & * \\ 0 & * & * \end{pmatrix}$$

where (necessarily!) g_2 is the *gcd* of the left column entries of MA_1. Then right multiply by $A_3 \in GL_3(\mathbb{Z})$ such that

$$B_2 M A_1 \cdot A_3 = \begin{pmatrix} g_3 & 0 & 0 \\ * & * & * \\ * & * & * \end{pmatrix}$$

where g_3 is the *gcd* of the top row of $B_2 M A_1$. Continue. Since these *gcds* divide each other successively

$$\dots |g_3| g_2 |g_1 \neq 0$$

and since any such chain must be finite, after finitely-many iterations of this the upper-left entry ceases to change. That is, for some $A, B \in GL_3(\mathbb{Z})$ we have

$$BMA = \begin{pmatrix} g & * & * \\ 0 & x & y \\ 0 & * & * \end{pmatrix}$$

and also g divides the top row. That is,

$$u = \begin{pmatrix} 1 & -x/g & -y/g \\ 0 & 1 & 0 \\ 0 & 0 & 1 \end{pmatrix} \in GL_3(\mathbb{Z})$$

Then

$$BMA \cdot u = \begin{pmatrix} g & 0 & 0 \\ 0 & * & * \\ 0 & * & * \end{pmatrix}$$

Continue in the same fashion, operating on the lower right 2-by-2 block, to obtain a form

$$\begin{pmatrix} g & 0 & 0 \\ 0 & g_2 & 0 \\ 0 & 0 & g_3 \end{pmatrix}$$

Note that since the r, s such that $\gcd(x, y) = rx + sy$ can be found via Euclid, this whole procedure is *effective*. And it certainly applies to larger matrices, not necessarily square.

10.6.2 Example: Given a row vector $x = (x_1, \dots, x_n)$ of integers whose *gcd* is 1, prove that there exists an *n*-by-*n* integer matrix M with determinant ± 1 such that $xM = (0, \dots, 0, 1)$.

(*The iterative/algorithmic idea of the previous solution applies here, moving the gcd to the right end instead of the left.*)

10.6.3 Example: Given a row vector $x = (x_1, \dots, x_n)$ of integers whose *gcd* is 1, prove that there exists an *n*-by-*n* integer matrix M with determinant ± 1 whose bottom row is x.

This is a corollary of the previous exercise. Given A such that

$$xA = (0 \quad \ldots \quad 0 \quad \gcd(x_1, \ldots, x_n)) = (0 \quad \ldots \quad 0 \quad 1)$$

note that this is saying

$$\begin{pmatrix} * & \cdots & * \\ \vdots & & \vdots \\ * & \cdots & * \\ x_1 & \cdots & x_n \end{pmatrix} \cdot A = \begin{pmatrix} * & \cdots & * & * \\ \vdots & & \vdots & \vdots \\ * & \cdots & * & * \\ 0 & \cdots & 0 & 1 \end{pmatrix}$$

or

$$\begin{pmatrix} * & \cdots & * \\ \vdots & & \vdots \\ * & \cdots & * \\ x_1 & \cdots & x_n \end{pmatrix} = \begin{pmatrix} * & \cdots & * & * \\ \vdots & & \vdots & \vdots \\ * & \cdots & * & * \\ 0 & \cdots & 0 & 1 \end{pmatrix} \cdot A^{-1}$$

This says that x is the bottom row of the invertible A^{-1}, as desired.

10.6.4 Example: Show that $GL(2, \mathbb{F}_2)$ is isomorphic to the permutation group S_3 on three letters.

There are exactly 3 non-zero vectors in the space $\mathbb{F}_2^2$ of column vectors of size 2 with entries in $\mathbb{F}_2$. Left multiplication by elements of $GL_2(\mathbb{F}_2)$ permutes them, since the invertibility assures that no non-zero vector is mapped to zero. If $g \in GL_2(\mathbb{F}_2)$ is such that $gv = v$ for all non-zero vectors v, then $g = 1_2$. Thus, the map

$$\varphi : GL_2(\mathbb{F}_2) \longrightarrow \text{permutations of the set } N \text{ of non-zero vectors in } \mathbb{F}_2^2$$

is *injective*. It is a group homomorphism because of the associativity of matrix multiplication:

$$\varphi(gh)(v) = (gh)v = g(hv) = \varphi(g)(\varphi(h)(v))$$

Last, we can confirm that the injective group homomorphism φ is also surjective by showing that the order of $GL_2(\mathbb{F}_2)$ is the order of S_3, namely, 6, as follows. An element of $GL_2(\mathbb{F}_2)$ can send any basis for $\mathbb{F}_2^2$ to any other basis, and, conversely, is completely determined by telling what it does to a basis. Thus, for example, taking the first basis to be the standard basis $\{e_1, e_2\}$ (where e_i has a 1 at the ith position and 0s elsewhere), an element g can map e_1 to any non-zero vector, for which there are $2^2 - 1$ choices, counting *all* less 1 for the zero-vector. The image of e_2 under g must be linearly independent of e_1 for g to be invertible, and conversely, so there are $2^2 - 2$ choices for ge_2 (*all* less 1 for 0 and less 1 for ge_1). Thus,

$$|GL_2(\mathbb{F}_2)| = (2^2 - 1)(2^2 - 2) = 6$$

Thus, the map of $GL_2(\mathbb{F}_2)$ to permutations of non-zero vectors gives an isomorphism to S_3.

10.6.5 Example: Determine all conjugacy classes in $GL(2, \mathbb{F}_3)$.

First, $GL_2(\mathbb{F}_3)$ is simply the group of *invertible* k-linear endomorphisms of the $\mathbb{F}_3$-vectorspace $\mathbb{F}_3^2$. As observed earlier, conjugacy classes of endomorphisms are in bijection with $\mathbb{F}_3[x]$-module structures on $\mathbb{F}_3^2$, which we know are given by *elementary divisors*, from the Structure Theorem. That is, all the possible structures are parametrized by monic polynomials $d_1 | \ldots | d_t$ where the sum of the degrees is the dimension of the vector space $\mathbb{F}_3^2$, namely 2. Thus, we have a list of irredundant representatives

$$
\begin{cases}
\quad \mathbb{F}_3[x]/\langle Q \rangle & Q \text{ monic quadratic in } \mathbb{F}_3[x] \\[2mm]
\mathbb{F}_3[x]/\langle x - \lambda \rangle \oplus \mathbb{F}_3[x]/\langle x - \lambda \rangle & \lambda \in \mathbb{F}_3^\times
\end{cases}
$$

We *can* write the first case in a so-called rational canonical form, that is, choosing basis $1, x \bmod Q$, so we have two families

$$
\begin{cases}
(1) & \begin{pmatrix} 0 & -b \\ 1 & -a \end{pmatrix} & b \in \mathbb{F}_3, \ a \in \mathbb{F}_3^\times \\[4mm]
(2) & \begin{pmatrix} \lambda & 0 \\ 0 & \lambda \end{pmatrix} & \lambda \mathbb{F}_3^\times
\end{cases}
$$

But the first family can be usefully broken into three subcases, namely, depending upon the reducibility of the quadratic, and whether or not there are repeated roots: there are 3 cases

$$
\begin{aligned}
Q(x) &= & \text{irreducible} \\
Q(x) &= & (x - \lambda)(x - \mu) \quad (\text{with } \lambda \neq \mu) \\
Q(x) &= & (x - \lambda)^2
\end{aligned}
$$

And note that if $\lambda \neq \mu$ then (for a field k)

$$
k[x]/\langle (x - \lambda)(x - \mu) \rangle \approx k[x]/\langle x - \lambda \rangle \oplus k[x]/\langle x - \mu \rangle
$$

Thus, we have

$$
\begin{cases}
(1a) & \begin{pmatrix} 0 & b \\ 1 & a \end{pmatrix} & x^2 + ax + b \text{ irreducible in } \mathbb{F}_3[x] \\[4mm]
(1b) & \begin{pmatrix} \lambda & 0 \\ 0 & \mu \end{pmatrix} & \lambda \neq \mu \text{ both nonzero} & (\text{modulo interchange of } \lambda, \mu) \\[4mm]
(1b) & \begin{pmatrix} \lambda & 1 \\ 0 & \lambda \end{pmatrix} & \lambda \in \mathbb{F}_3^2 \\[4mm]
(2) & \begin{pmatrix} \lambda & 0 \\ 0 & \lambda \end{pmatrix} & \lambda \in \mathbb{F}_3^\times
\end{cases}
$$

One might, further, list the irreducible quadratics in $\mathbb{F}_3[x]$. By counting, we know there are $(3^2 - 3)/2 = 3$ irreducible quadratics, and, thus, the guesses $x^2 - 2$, $x^2 + x + 1$, and $x^2 - x + 1$ (the latter two being cyclotomic, the first using the fact that 2 is not a square mod 3) are all of them.

10.6.6 Example: Determine all conjugacy classes in $GL(3, \mathbb{F}_2)$.

Again, $GL_3(\mathbb{F}_2)$ is the group of *invertible* k-linear endomorphisms of the $\mathbb{F}_2$-vectorspace $\mathbb{F}_2^3$, and conjugacy classes of endomorphisms are in bijection with $\mathbb{F}_2[x]$-module structures on $\mathbb{F}_2^3$, which are given by *elementary divisors*. So all possibilities are parametrized by monic polynomials $d_1 | \ldots | d_t$ where the sum of the degrees is the dimension of the vector space $\mathbb{F}_2^3$, namely 3. Thus, we have a list of irredundant representatives

$$
\begin{cases}
(1) & \mathbb{F}_2[x]/\langle Q \rangle & Q \text{ monic cubic in } \mathbb{F}_2[x] \\[2mm]
(2) & \mathbb{F}_2[x]/\langle x - 1 \rangle \oplus \mathbb{F}_2[x]/\langle (x - 1)^2 \rangle \\[2mm]
(3) & \mathbb{F}_2[x]/\langle x - 1 \rangle \oplus \mathbb{F}_2[x]/\langle x - 1 \rangle \oplus \mathbb{F}_2[x]/\langle x - 1 \rangle
\end{cases}
$$

since the only non-zero element of $\mathbb{F}_2$ is $\lambda = 1$. We *can* write the first case in a so-called rational canonical form, that is, choosing basis $1, x, x^2 \bmod Q$, there are three families

$$
\begin{cases}
(1) & \begin{pmatrix} 0 & 0 & 1 \\ 1 & 0 & -b \\ 0 & 1 & -a \end{pmatrix} \quad x^3 + ax^2 + bx + 1 \text{ in } \mathbb{F}_2[x] \\[18pt]
(2) & \begin{pmatrix} 1 & 0 & 0 \\ 0 & 1 & 0 \\ 0 & 1 & 1 \end{pmatrix} \\[18pt]
(3) & \begin{pmatrix} 1 & 0 & 0 \\ 0 & 1 & 0 \\ 0 & 0 & 1 \end{pmatrix}
\end{cases}
$$

It is useful to look in detail at the possible factorizations in case 1, breaking up the single summand into more summands according to relatively prime factors, giving cases

$$
\begin{cases}
(1a) & \mathbb{F}_2[x]/\langle x^3 + x + 1 \rangle \\[10pt]
(1a') & \mathbb{F}_2[x]/\langle x^3 + x^2 + 1 \rangle \\[10pt]
(1b) & \mathbb{F}_2[x]/\langle (x-1)(x^2 + x + 1) \rangle \\[10pt]
(1c) & \mathbb{F}_2[x]/\langle (x-1)^3 \rangle
\end{cases}
$$

since there are just two irreducible cubics $x^3 + x + 1$ and $x^3 + x^2 + 1$, and a unique irreducible quadratic, $x^2 + x + 1$. (The counting above tells the number, so, after any sort of guessing provides us with the right number of verifiable irreducibles, we can stop.) Thus, the 6 conjugacy classes have irredundant matrix representatives

$$
(1a) \begin{pmatrix} 0 & 0 & 1 \\ 1 & 0 & 1 \\ 0 & 1 & 0 \end{pmatrix} \quad
(1a') \begin{pmatrix} 0 & 0 & 1 \\ 1 & 0 & 0 \\ 0 & 1 & 1 \end{pmatrix} \quad
(1b) \begin{pmatrix} 1 & 0 & 0 \\ 0 & 0 & 1 \\ 0 & 1 & 1 \end{pmatrix} \quad
(1c) \begin{pmatrix} 1 & 0 & 0 \\ 1 & 1 & 0 \\ 0 & 1 & 1 \end{pmatrix}
$$

$$
(2) \begin{pmatrix} 1 & 0 & 0 \\ 0 & 1 & 0 \\ 0 & 1 & 1 \end{pmatrix} \quad
(3) \begin{pmatrix} 1 & 0 & 0 \\ 0 & 1 & 0 \\ 0 & 0 & 1 \end{pmatrix}
$$

10.6.7 Example: Determine all conjugacy classes in $GL(4, \mathbb{F}_2)$.

Again, $GL_4(\mathbb{F}_2)$ is *invertible* k-linear endomorphisms of $\mathbb{F}_2^4$, and conjugacy classes are in bijection with $\mathbb{F}_2[x]$-module structures on $\mathbb{F}_2^4$, given by *elementary divisors*. So all possibilities are parametrized by monic polynomials $d_1 | \ldots | d_t$ where the sum of the degrees is the dimension of the vector space $\mathbb{F}_2^4$, namely 4. Thus, we have a list of irredundant representatives

$$
\begin{cases}
\mathbb{F}_2[x]/\langle Q \rangle & Q \text{ monic quartic} \\[10pt]
\mathbb{F}_2[x]/\langle x-1 \rangle \oplus \mathbb{F}_2[x]/\langle (x-1)Q(x) \rangle & Q \text{ monic quadratic} \\[10pt]
\mathbb{F}_2[x]/\langle x-1 \rangle \oplus \mathbb{F}_2[x]/\langle x-1 \rangle \oplus \mathbb{F}_2[x]/\langle (x-1)^2 \rangle \\[10pt]
\mathbb{F}_2[x]/\langle Q \rangle \oplus \mathbb{F}_2[x]/\langle Q \rangle & Q \text{ monic quadratic} \\[10pt]
\mathbb{F}_2[x]/\langle x-1 \rangle \oplus \mathbb{F}_2[x]/\langle x-1 \rangle \oplus \mathbb{F}_2[x]/\langle x-1 \rangle \oplus \mathbb{F}_2[x]/\langle x-1 \rangle
\end{cases}
$$

162

Modules over PIDs

since the only non-zero element of $\mathbb{F}_2$ is $\lambda = 1$. We could write all cases using rational canonical form, but will not, deferring matrix forms till we've further decomposed the modules. Consider possible factorizations into irreducibles, giving cases

$$(1a) \quad \mathbb{F}_2[x]/\langle x^4 + x + 1 \rangle$$

$$(1a') \quad \mathbb{F}_2[x]/\langle x^4 + x^3 + 1 \rangle$$

$$(1a'') \quad \mathbb{F}_2[x]/\langle x^4 + x^3 + x^2 + x + 1 \rangle$$

$$(1b) \quad \mathbb{F}_2[x]/\langle (x-1)(x^3 + x + 1) \rangle$$

$$(1b') \quad \mathbb{F}_2[x]/\langle (x-1)(x^3 + x^2 + 1) \rangle$$

$$(1c) \quad \mathbb{F}_2[x]/\langle (x-1)^2(x^2 + x + 1) \rangle$$

$$(1d) \quad \mathbb{F}_2[x]/\langle (x^2 + x + 1)^2 \rangle$$

$$(1e) \quad \mathbb{F}_2[x]/\langle (x-1)^4 \rangle$$

$$(2a) \quad \mathbb{F}_2[x]/\langle x-1 \rangle \oplus \mathbb{F}_2[x]/\langle (x-1)(x^2 + x + 1) \rangle$$

$$(2b) \quad \mathbb{F}_2[x]/\langle x-1 \rangle \oplus \mathbb{F}_2[x]/\langle (x-1)^3 \rangle$$

$$(3) \quad \mathbb{F}_2[x]/\langle x-1 \rangle \oplus \mathbb{F}_2[x]/\langle x-1 \rangle \oplus \mathbb{F}_2[x]/\langle (x-1)^2 \rangle$$

$$(4a) \quad \mathbb{F}_2[x]/\langle x^2 + x + 1 \rangle \oplus \mathbb{F}_2[x]/\langle x^2 + x + 1 \rangle$$

$$(4b) \quad \mathbb{F}_2[x]/\langle (x-1)^2 \rangle \oplus \mathbb{F}_2[x]/\langle (x-1)^2 \rangle$$

$$(5) \quad \mathbb{F}_2[x]/\langle x-1 \rangle \oplus \mathbb{F}_2[x]/\langle x-1 \rangle \oplus \mathbb{F}_2[x]/\langle x-1 \rangle \oplus \mathbb{F}_2[x]/\langle x-1 \rangle$$

since there are exactly three irreducible quartics (as indicated), two irreducible cubics, and a single irreducible quadratic. Matrices are, respectively, and unilluminatingly,

$$\begin{bmatrix} 0&0&0&1\\1&0&0&1\\0&1&0&0\\0&0&1&0 \end{bmatrix} \begin{bmatrix} 0&0&0&1\\1&0&0&0\\0&1&0&0\\0&0&1&1 \end{bmatrix} \begin{bmatrix} 0&0&0&1\\1&0&0&1\\0&1&0&1\\0&0&1&1 \end{bmatrix} \begin{bmatrix} 1&0&0&0\\0&0&0&1\\0&1&0&1\\0&0&1&0 \end{bmatrix} \begin{bmatrix} 1&0&0&0\\0&0&0&1\\0&1&0&0\\0&0&1&1 \end{bmatrix}$$

$$\begin{bmatrix} 1&0&0&0\\0&1&0&0\\0&0&0&1\\0&0&1&1 \end{bmatrix} \begin{bmatrix} 0&1&0&0\\1&1&0&0\\0&1&0&1\\0&0&1&1 \end{bmatrix} \begin{bmatrix} 1&0&0&0\\1&1&0&0\\0&1&1&0\\0&0&1&1 \end{bmatrix} \begin{bmatrix} 1&0&0&0\\0&1&0&0\\0&0&0&1\\0&0&1&1 \end{bmatrix}$$

$$\begin{bmatrix} 1&0&0&0\\0&1&0&0\\0&1&1&0\\0&0&1&1 \end{bmatrix} \begin{bmatrix} 1&0&0&0\\0&1&0&0\\0&0&1&0\\0&0&1&1 \end{bmatrix} \begin{bmatrix} 0&1&0&0\\1&1&0&0\\0&0&0&1\\0&0&1&1 \end{bmatrix} \begin{bmatrix} 1&0&0&0\\1&1&0&0\\0&0&1&0\\0&0&1&1 \end{bmatrix} \begin{bmatrix} 1&0&0&0\\0&1&0&0\\0&0&1&0\\0&0&0&1 \end{bmatrix}$$

10.6.8 Example: Tell a p-Sylow subgroup in $GL(3, \mathbb{F}_p)$.

To compute the order of this group in the first place, observe that an automorphism (invertible endomorphism) can take any basis to any other. Thus, letting e_1, e_2, e_3 be the standard basis, for an automorphism g the image ge_1 can be any non-zero vector, of which there are $p^3 - 1$. The image ge_2 can be anything not in the span of ge_1, of which there are $p^3 = p$. The image ge_3 can be anything not in the span of ge_1 and ge_2, of which, because those first two were already linearly independent, there are $p^3 - p^2$. Thus, the order is

$$|GL_3(\mathbb{F}_p)| = (p^3 - 1)(p^3 - p)(p^3 - p^2)$$

The power of p that divides this is p^3. Upon reflection, a person might hit upon considering the subgroup of upper triangular *unipotent* (eigenvalues all 1) matrices

$$\begin{pmatrix} 1 & * & * \\ 0 & 1 & * \\ 0 & 0 & 1 \end{pmatrix}$$

where the super-diagonal entries are all in $\mathbb{F}_p$. Thus, there would be p^3 choices for super-diagonal entries, the right number. By luck, we are done.

10.6.9 Example: Tell a 3-Sylow subgroup in $GL(3, \mathbb{F}_7)$.

As earlier, the order of the group is

$$(7^3 - 1)(7^3 - 7)(7^3 - 7^2) = 2^6 \cdot 3^4 \cdot 7^3 \cdot 19$$

Of course, since $\mathbb{F}_7^\times$ is cyclic, for example, it has a subgroup T of order 3. Thus, one might hit upon the subgroup

$$H = \{ \begin{pmatrix} a & 0 & 0 \\ 0 & b & 0 \\ 0 & 0 & c \end{pmatrix} : a, b, c \in T \}$$

is a subgroup of order 3^3. Missing a factor of 3. But all the permutation matrices (with exactly one non-zero entry in each row, and in each column, and that non-zero entry is 1)

$$\begin{pmatrix} 1 & 0 & 0 \\ 0 & 1 & 0 \\ 0 & 0 & 1 \end{pmatrix} \begin{pmatrix} 1 & 0 & 0 \\ 0 & 0 & 1 \\ 0 & 1 & 0 \end{pmatrix} \begin{pmatrix} 0 & 1 & 0 \\ 1 & 0 & 0 \\ 0 & 0 & 1 \end{pmatrix} \begin{pmatrix} 0 & 0 & 1 \\ 0 & 1 & 0 \\ 1 & 0 & 0 \end{pmatrix} \begin{pmatrix} 0 & 0 & 1 \\ 1 & 0 & 0 \\ 0 & 1 & 0 \end{pmatrix} \begin{pmatrix} 0 & 1 & 0 \\ 0 & 0 & 1 \\ 1 & 0 & 0 \end{pmatrix}$$

These normalize *all* diagonal matrices, and also the subgroup H of diagonal matrices with entries in T. The group of permutation matrices consisting of the identity and the two 3-cycles is order 3, and putting it together with H (as a semi-direct product whose structure is already described for us) gives the order 3^4 subgroup.

10.6.10 Example: Tell a 19-Sylow subgroup in $GL(3, \mathbb{F}_7)$.

Among the Stucture Theorem canonical forms for endomorphisms of $V = \mathbb{F}_7^3$, there are $\mathbb{F}_7[x]$-module structures

$$V \approx \mathbb{F}_7[x]/\langle \text{ irreducible cubic } C \rangle$$

which are *invertible* because of the irreducibility. Let α be the image of x in $\mathbb{F}_7[x]/\langle C \rangle$. Note that $\mathbb{F}_7[\alpha] = \mathbb{F}_7[x]/C$ also has a natural ring structure. Then the action of any $P(x)$ in $k[x]$ on V (via this isomorphism) is, of course,

$$P(x) \cdot Q(\alpha) = P(\alpha) \cdot Q(\alpha) = (P \cdot Q)(x) \bmod C(x)$$

for any $Q(x) \in \mathbb{F}_7[x]$. Since C is irreducible, there are no non-trivial zero divisors in the ring $\mathbb{F}_7[\alpha]$. Indeed, it's a field. Thus, $\mathbb{F}_7[\alpha]^\times$ *injects* to $\mathrm{End}_{\mathbb{F}_7} V$. The point of saying this is that, therefore, if we can find an element of $\mathbb{F}_7[\alpha]^\times$ of order 19 then we have an *endomorphism* of order 19, as well. And it is arguably simpler to hunt around inside $\mathbb{F}_{7^3} = \mathbb{F}_7[\alpha]$ than in groups of matrices.

To compute anything explicitly in $\mathbb{F}_{7^3}$ we need an irreducible cubic. Luckily, $7 = 1 \bmod 3$, so there are many non-cubes mod 7. In particular, there are only two non-zero cubes mod 7, ± 1. Thus, $x^3 - 2$ has no linear factor in $\mathbb{F}_7[x]$, so is irreducible. The *sparseness* (having not so many non-zero coefficients) of this polynomial will be convenient when computing, subsequently.

Now we must find an element of order 19 in $\mathbb{F}_7[x]/\langle x^3 - 2 \rangle$. There seems to be no simple algorithm for choosing such a thing, but there is a reasonable probabilistic approach: since $\mathbb{F}_{7^3}^\times$ is cyclic of order $7^3 - 1 = 19 \cdot 18$, if we pick an element g at random the probability is $(19 - 1)/19$ that its order will be *divisible* by 19. Then, whatever its order is, g^{18} will have order either 19 or 1. That is, if g^{18} is not 1, then it is the desired thing. (Generally, in a cyclic group of order $p \cdot m$ with prime p and p not dividing m, a random element g has probability $(p - 1)/p$ of having order divisible by p, and in any case g^m will be either 1 or will have order p.)

Since elements of the ground field $\mathbb{F}_7^\times$ are all of order 6, these would be bad guesses for the random g. Also, the image of x has cube which is 2, which has order 6, so x itself has order 18, which is not what we want. What to guess next? Uh, maybe $g = x + 1$? We can only try. Compute

$$(x + 1)^{18} = (((x + 1)^3)^2)^3 \bmod x^3 - 2$$

reducing modulo $x^3 - 2$ at intermediate stages to simplify things. So

$$g^3 = x^3 + 3x^2 + 3x + 1 = 3x^2 + 3x + 3 \bmod x^3 - 2 = 3 \cdot (x^2 + x + 1)$$

A minor piece of luck, as far as computational simplicity goes. Then, in $\mathbb{F}_7[x]$,

$$g^6 = 3^2 \cdot (x^2 + x + 1)^2 = 2 \cdot (x^4 + 2x^3 + 3x^2 + 2x + 1) = 2 \cdot (2x + 2 \cdot 2 + 3x^2 + 2x + 1)$$

$$= 2 \cdot (3x^2 + 4x + 5) = 6x^2 + x + 3 \bmod x^3 - 2$$

Finally,

$$g^{18} = (g^6)^3 = (6x^2 + x + 3)^3 \bmod x^3 - 2$$

$$= 6^3 {\cdot} x^6 + (3 {\cdot} 6^2 {\cdot} 1) x^5 + (3 {\cdot} 6^2 {\cdot} 3 + 3 {\cdot} 6 {\cdot} 1^2) x^4 + (6 {\cdot} 6 {\cdot} 1 {\cdot} 3 + 1^3) x^3 + (3 {\cdot} 6 {\cdot} 3^2 + 3 {\cdot} 1^2 {\cdot} 3) x^2 + (3 {\cdot} 1 {\cdot} 3^2) x + 3^3$$

$$= 6x^6 + 3x^5 + 6x^4 + 4x^3 + 3x^2 + 6x + 6 = 6 {\cdot} 4 + 3 {\cdot} 2 {\cdot} x^2 + 6 {\cdot} 2x + 4 {\cdot} 2 + 3x^2 + 6x + 6 = 2x^2 + 4x + 3$$

Thus, if we've not made a computational error, the endomorphism given by multiplication by $2x^2 + 4x + 3$ in $\mathbb{F}_7[x]/\langle x^3 - 2 \rangle$ is of order 19.

To get a matrix, use (rational canonical form) basis $e_1 = 1$, $e_2 = x$, $e_3 = x^2$. Then the matrix of the endomorphism is

$$M = \begin{pmatrix} 3 & 4 & 1 \\ 4 & 3 & 4 \\ 2 & 4 & 3 \end{pmatrix}$$

Pretending to be brave, we check by computing the 19th power of this matrix, modulo 7. Squaring repeatedly, we have (with determinants computed along the way as a sort of parity-check, which in reality did discover a computational error on each step, which was corrected before proceeding)

$$M^2 = \begin{pmatrix} 1 & 0 & 6 \\ 3 & 1 & 0 \\ 0 & 3 & 1 \end{pmatrix} \quad M^4 = \begin{pmatrix} 6 & 3 & 2 \\ 1 & 6 & 3 \\ 5 & 1 & 6 \end{pmatrix} \quad M^8 = \begin{pmatrix} 0 & 4 & 2 \\ 1 & 0 & 4 \\ 2 & 1 & 0 \end{pmatrix} \quad M^{16} = \begin{pmatrix} 6 & 5 & 5 \\ 6 & 6 & 5 \\ 6 & 6 & 6 \end{pmatrix}$$

Then

$$M^{18} = M^2 \cdot M^{16} = \begin{pmatrix} 1 & 0 & 6 \\ 3 & 1 & 0 \\ 0 & 3 & 1 \end{pmatrix} \cdot M^{16} = \begin{pmatrix} 6 & 5 & 5 \\ 6 & 6 & 5 \\ 6 & 6 & 6 \end{pmatrix} = \begin{pmatrix} 0 & 1 & 1 \\ 4 & 0 & 1 \\ 4 & 4 & 0 \end{pmatrix}$$

$$M^{19} = M \cdot M^{18} = \begin{pmatrix} 3 & 4 & 1 \\ 4 & 3 & 4 \\ 2 & 4 & 3 \end{pmatrix} \cdot \begin{pmatrix} 0 & 1 & 1 \\ 4 & 0 & 1 \\ 4 & 4 & 0 \end{pmatrix} = \text{ the identity}$$

Thus, indeed, we have the order 19 element.

Note that, in reality, without some alternative means to verify that we really found an element of order 19, we could easily be suspicious that the numbers were wrong.

Exercises

10.1 Determine all conjugacy classes in $GL_2(\mathbb{F}_5)$.

10.2 Determine all conjugacy classes in $GL_2(\mathbb{F}_4)$.

10.3 Determine all conjugacy classes in $GL_5(\mathbb{F}_2)$.

10.4 Let k be an algebraically closed field. Determine all conjugacy classes in $GL_2(k)$.

10.5 Let k be an algebraically closed field. Determine all conjugacy classes in $GL_3(k)$.

10.6 Find a 31-Sylow subgroup of $GL_3(\mathbb{F}_5)$.

10.7 Find a 2-Sylow subgroup of $GL_2(\mathbb{Q})$.

10.8 Find a 2-Sylow subgroup of $GL_2(\mathbb{Q}(i))$.

10.9 Find a 5-Sylow subgroup of $GL_4(\mathbb{Q})$.

11. Finitely-generated modules

11.1 *Free modules*

The following definition is an example of defining things by *mapping properties*, that is, by the way the object relates to other objects, rather than by internal structure. The first proposition, which says that there is at most one such thing, is typical, as is its proof.

Let R be a commutative ring with 1. Let S be a set. A **free R-module** M on **generators** S is an R-module M and a set map $i : S \longrightarrow M$ such that, for any R-module N and any set map $f : S \longrightarrow N$, there is a unique R-module homomorphism $\tilde{f} : M \longrightarrow N$ such that

$$\tilde{f} \circ i = f : S \longrightarrow N$$

The elements of $i(S)$ in M are an R-**basis** for M.

11.1.1 Proposition: If a free R-module M on generators S exists, it is unique up to unique isomorphism.

Proof: First, we claim that the only R-module homomorphism $F : M \longrightarrow M$ such that $F \circ i = i$ is the identity map. Indeed, by definition,[1] given $i : S \longrightarrow M$ there is a *unique*

---

[1] Letting letting $i : S \longrightarrow M$ take the role of $f : S \longrightarrow N$ in the definition.

$\tilde{i} : M \longrightarrow M$ such that $\tilde{i} \circ i = i$. The identity map on M certainly meets this requirement, so, by uniqueness, $\tilde{i}$ can only be the identity.

Now let M' be another free module on generators S, with $i' : S \longrightarrow M'$ as in the definition. By the defining property of (M, i), there is a unique $\tilde{i}' : M \longrightarrow M'$ such that $\tilde{i}' \circ i = i'$. Similarly, there is a unique $\tilde{i}$ such that $\tilde{i} \circ i' = i$. Thus,

$$i = \tilde{i} \circ i' = \tilde{i} \circ \tilde{i}' \circ i$$

Similarly,

$$i' = \tilde{i}' \circ i = \tilde{i}' \circ \tilde{i} \circ i'$$

From the first remark of this proof, this shows that

$$\tilde{i} \circ \tilde{i}' = \text{ identity map on } M$$

$$\tilde{i}' \circ \tilde{i} = \text{ identity map on } M'$$

So $\tilde{i}'$ and $\tilde{i}$ are mutual inverses. That is, M and M' are isomorphic, and in a fashion that respects the maps i and i'. Further, by uniqueness, there is no *other* map between them that respects i and i', so we have a *unique* isomorphism.   ///

Existence of a free module remains to be demonstrated. We should be relieved that the uniqueness result above assures that any successful construction will invariably yield the same object. Before proving existence, and, thus, before being burdened with irrelevant internal details that arise as artifacts of the construction, we prove the basic facts about free modules.

11.1.2 Proposition: A free R-module M on generators $i : S \longrightarrow M$ is *generated by* $i(S)$, in the sense that the only R-submodule of M containing the image $i(S)$ is M itself.

Proof: Let N be the submodule generated by $i(S)$, that is, the intersection of all submodules of M containing $i(S)$. Consider the quotient M/N, and the map $f : S \longrightarrow M/N$ by $f(s) = 0$ for all $s \in S$. Let $\zeta : M \longrightarrow M/N$ be the 0 map. Certainly $\zeta \circ i = f$. If $M/N \neq 0$, then the quotient map $q : M \longrightarrow M/N$ is not the zero map ζ, and also $q \circ i = f$. But this would contradict the uniqueness in the definition of M.   ////

For a set X of elements of an R-module M, if a relation

$$\sum_{x \in X} r_x \, x = 0$$

with $r_x \in R$ and $x \in M$ (with all but finitely-many coefficients r_x being 0) implies that *all* coefficients r_x are 0, say that the elements of X are **linearly independent** (over R).

11.1.3 Proposition: Let M be a free R-module on generators $i : S \longrightarrow M$. Then any relation (with finitely-many non-zero coefficients $r_s \in R$)

$$\sum_{s \in S} r_s \, i(s) = 0$$

must be trivial, that is, all coefficients r_s are 0. That is, the elements of $i(S)$ are *linearly independent*.

Proof: Suppose $\sum_s r_s\, i(s) = 0$ in the free module M. To show that every coefficient r_s is 0, fix $s_o \in S$ and map $f : S \longrightarrow R$ itself by

$$f(s) = \begin{cases} 0 & (s \neq s_o) \\ 1 & (s = s_o) \end{cases}$$

Let $\tilde{f}$ be the associated R-module homomorphism $\tilde{f} : M \longrightarrow R$. Then

$$0 = \tilde{f}(0) = \tilde{f}(\sum_s r_s\, i(s)) = r_{s_o}$$

This holds for each fixed index s_o, so any such relation is trivial.   ///

11.1.4 Proposition: Let $f : B \longrightarrow C$ be a *surjection* of R-modules, where C is free on generators S with $i : S \longrightarrow C$. Then there is an injection $j : C \longrightarrow B$ such that [2]

$$f \circ j = 1_C \quad\text{and}\quad B = (\ker f) \oplus j(C)$$

11.1.5 Remark: The map $j : C \longrightarrow B$ of this proposition is a **section** of the surjection $f : B \longrightarrow C$.

Proof: Let $\{b_s : s \in S\}$ be any set of elements of B such that $f(b_s) = i(s)$. Invoking the universal property of the free module, given the choice of $\{b_x\}$ there is a unique R-module homomorphism $j : C \longrightarrow B$ such that $(j \circ i)(s) = b_s$. It remains to show that $jC \oplus \ker f = B$. The intersection $jC \cap \ker f$ is trivial, since for $\sum_s r_s\, j(s)$ in the kernel (with all but finitely-many r_s just 0)

$$C \ni 0 = f\left(\sum_s r_s\, j(s)\right) = \sum_s r_s\, i(s)$$

We have seen that any such relation must be trivial, so the intersection $f(C) \cap ker f$ is trivial.

Given $b \in B$, let $f(b) = \sum_s r_s\, i(s)$ (a finite sum), using the fact that the images $i(s)$ generate the free module C. Then

$$f(b - j(f(b))) = f(b - \sum_s r_s b_s) == f(b) - \sum_s r_s f(b_s) = \sum_s r_s i(s) - \sum_s r_s i(s) = 0$$

Thus, $j(C) + \ker f = B$.   ///

We have one more basic result before giving a construction, and before adding any hypotheses on the ring R.

The following result uses an interesting trick, reducing the problem of counting generators for a free module F over a commutative ring R with 1 to counting generators for vector spaces over a field R/M, where M is a maximal proper ideal in R. We see that the number

---

[2] The property which we are about to prove is enjoyed by free modules is the *defining* property of **projective** modules. Thus, in these terms, we are proving that *free modules are projective*.

of generators for a free module over a commutative ring R with unit 1 has a well-defined cardinality, the R-**rank** of the free module.

11.1.6 Theorem: Let F be a free R-module on generators $i : S \longrightarrow F$, where R is a commutative ring with 1. Suppose that F is also a free R-module on generators $j : T \longrightarrow F$. Then $|S| = |T|$.

Proof: Let M be a maximal proper ideal in R, so $k = R/M$ is a field. Let

$$E = M \cdot F = \text{ collection of finite sums of elements } mx, \, m \in M, x \in F$$

and consider the quotient

$$V = F/E$$

with quotient map $q : F \longrightarrow V$. This quotient has a canonical k-module structure

$$(r + M) \cdot (x + M \cdot F) = rx + M \cdot F$$

We claim that V is a *free* k-module on generators $q \circ i : S \longrightarrow V$, that is, is a vector space on those generators. Lagrange's replacement argument shows that the cardinality of the number of generators for a *vector space* over a field is well-defined, so a successful comparison of generators for the original module and this vector space quotient would yield the result.

To show that V is free over k, consider a set map $f : S \longrightarrow W$ where W is a k-vectorspace. The k-vectorspace W has a natural R-module structure compatible with the k-vectorspace structure, given by

$$r \cdot (x + M \cdot F) = rx + M \cdot F$$

Let $\tilde{f} : F \longrightarrow W$ be the unique R-module homomorphism such that $\tilde{f} \circ i = f$. Since $m \cdot w = 0$ for any $m \in M$ and $w \in W$, we have

$$0 = m \cdot f(s) = m \cdot \tilde{f}(i(s)) = \tilde{f}(m \cdot i(s))$$

so

$$\ker \tilde{f} \supset M \cdot F$$

Thus, $\bar{f} : V \longrightarrow W$ defined by

$$\bar{f}(x + M \cdot F) = \tilde{f}(x)$$

is *well*-defined, and $\bar{f} \circ (q \circ i) = f$. This proves the existence part of the defining property of a free module.

For uniqueness, the previous argument can be reversed, as follows. Given $\bar{f} : V \longrightarrow W$ such that $\bar{f} \circ (q \circ i) = f$, let $\tilde{f} = \bar{f} \circ q$. Since there is a unique $\tilde{f} : F \longrightarrow W$ with $\tilde{f} \circ i = f$, there is at most one $\bar{f}$.   ///

Finally, we *construct* free modules, as a proof of existence. [3]

Given a non-empty set S, let M be the set of R-valued functions on S which take value 0 outside a finite subset of S (which may depend upon the function). Map $i : S \longrightarrow M$ by

---

[3] Quite pointedly, the previous results did not use any explicit internal details of what a free module might be, but, rather, only invoked the external mapping properties.

letting $i(s)$ be the function which takes value 1 at $s \in S$ and is 0 otherwise. Add functions value-wise, and let R act on M by value-wise multiplication.

11.1.7 Proposition: The M and i just constructed is a free module on generators S. In particular, given a set map $f : S \longrightarrow N$ for another R-module N, for $m \in M$ define $\tilde{f}(m) \in N$ by[4]

$$\tilde{f}(m) = \sum_{s \in S} m(s) \cdot f(s)$$

Proof: We might check that the explicit expression (with only finitely-many summands non-zero) is an R-module homomorphism: that it respects addition in M is easy. For $r \in R$, we have

$$\tilde{f}(r \cdot m) = \sum_{s \in S} (r \cdot m(s)) \cdot f(s) = r \cdot \sum_{s \in S} m(s) \cdot f(s) = r \cdot \tilde{f}(m)$$

And there should be no *other* R-module homomorphism from M to N such that $\tilde{f} \circ i = f$. Let $F : M \longrightarrow N$ be another one. Since the elements $\{i(s) : s \in S\}$ generate M as an R-module, for an arbitrary collection $\{r_s \in R : s \in S\}$ with all but finitely-many 0,

$$F \left(\sum_{s \in S} r_s \cdot i(s) \right) = \sum_{s \in S} r_s \cdot F(i(s)) = \sum_{s \in S} r_s \cdot f(s) = \tilde{f} \left(\sum_{s \in S} r_s \cdot i(s) \right)$$

so necessarily $F = \tilde{f}$, as desired. ///

11.1.8 Remark: For finite generator sets often one takes

$$S = \{1, 2, \ldots, n\}$$

and then the construction above of the free module on generators S can be identified with the collection R^n of ordered n-tuples of elements of R, as usual.

11.2 *Finitely-generated modules over a domain*

In the sequel, the results will mostly require that R be a domain, or, more stringently, a principal ideal domain. These hypotheses will be carefully noted.

11.2.1 Theorem: Let R be a principal ideal domain. Let M be a free R-module on generators $i : S \longrightarrow M$. Let N be an R-submodule. Then N is a free R-module on at most $|S|$ generators. [5]

Proof: Induction on the cardinality of S. We give the proof for *finite* sets S. First, for $M = R^1 = R$ a free module on a single generator, an R-submodule is an ideal in R. The

---

[4] In this formula, the function m on S is non-zero only at finitely-many $s \in S$, so the sum is finite. And $m(s) \in R$ and $f(s) \in N$, so this expression is a finite sum of R-multiples of elements of N, as required.

[5] The assertion of the theorem is false without some hypotheses on R. For example, even in the case that M has a *single* generator, to know that every submodule needs at most a single generator is exactly to assert that every ideal in R is principal.

hypothesis that R is a PID assures that every ideal in R needs at most one generator. This starts the induction.

Let $M = R^m$, and let $p : R^m \longrightarrow R^{m-1}$ be the map

$$p(r_1, r_2, r_3, \dots, r_m) = (r_2, r_3, \dots, r_m)$$

The image $p(N)$ is free on $\leq m - 1$ generators, by induction. From the previous section, there is always a *section* $j : p(N) \longrightarrow N$ such that $p \circ j = 1_{p(N)}$ and

$$N = \ker p|_N \oplus j(p(N))$$

Since $p \circ j = 1_{p(N)}$, necessarily j is an injection, so is an isomorphism to its image, and $j(p(N))$ is free on $\leq m - 1$ generators. And $\ker p|_N$ is a submodule of R, so is free on at most 1 generator. We would be done if we knew that a direct sum $M_1 \oplus M_2$ of free modules M_1, M_2 on generators $i_1 : S_i \longrightarrow M_1$ and $i_2 : S_2 \longrightarrow M_2$ is a free module on the *disjoint* union $S = S_1 \cup S_2$ of the two sets of generators. We excise that argument to the following proposition.   ///

11.2.2 Proposition: A direct sum [6] $M = M_1 \oplus M_2$ of free modules M_1, M_2 on generators $i_1 : S_i \longrightarrow M_1$ and $i_2 : S_2 \longrightarrow M_2$ is a free module on the *disjoint* union $S = S_1 \cup S_2$ of the two sets of generators. [7]

Proof: Given another module N and a set map $f : S \longrightarrow N$, the restriction f_j of f to S_j gives a unique module homomorphism $\tilde{f}_j : M_j \longrightarrow M$ such that $\tilde{f}_j \circ i_j = f_j$. Then

$$\tilde{f}(m_1, \ m_2) = (f_1 m_1, \ f_2 m_2)$$

is *a* module homomorphism from the direct sum to N with $\tilde{f} \circ i = f$. On the other hand, given any map $g : M \longrightarrow N$ such that $g \circ i = f$, by the uniqueness on the summands M_1 and M_2 inside M, if must be that $g \circ i_j = f_j$ for $j = 1, 2$. Thus, this g is $\tilde{f}$.   ///

For an R-module M, for $m \in M$ the **annihilator** $\mathrm{Ann}_R(m)$ of m in R is

$$\mathrm{Ann}_R(m) = \{r \in R : rm = 0\}$$

It is easy to check that the annihilator is an ideal in R. An element $m \in M$ is a **torsion element** of M if its annihilator is not the 0 ideal. The **torsion submodule** M^{tors} of M is

$$M^{\mathrm{tors}} = \{m \in M : \mathrm{Ann}_R(m) \neq \{0\}\}$$

A module is **torsion free** if its torsion submodule is trivial.

11.2.3 Proposition: For a domain R, the torsion submodule M^{tors} of a given R-module M is an R-submodule of M, and M/M^{tors} is torsion-free.

---

[6] Though we will not use it at this moment, one can give a definition of *direct sum* in the same mapping-theoretic style as we have given for *free module*. That is, the direct sum of a family $\{M_\alpha : \alpha \in A\}$ of modules is a module M and homomorphisms $i_\alpha : M_\alpha \longrightarrow M$ such that, for every family of homomorphisms $f_\alpha : M_\alpha \longrightarrow N$ to another module N, there is a unique $f : M \longrightarrow N$ such that every f_α **factors through** f in the sense that $f_\alpha = f \circ i_\alpha$.

[7] This does not need the assumption that R is a principal ideal domain.

Proof: For torsion elements m, n in M, let x be a non-zero element of $\text{Ann}_R(m)$ and y a non-zero element of $\text{Ann}_{(}n)$. Then $xy \neq 0$, since R is a domain, and

$$(xy)(m + n) = y(xm) + x(yn) = y \cdot 0 + x \cdot 0 = 0$$

And for $r \in R$,

$$x(rm) = r(xm) = r \cdot 0 = 0$$

Thus, the torsion submodule *is* a submodule.

To show that the quotient M/M^{tors} is torsion free, suppose $r \cdot (m + M^{\text{tors}}) \subset M^{\text{tors}}$ for $r \neq 0$. Then $rm \in M^{\text{tors}}$. Thus, there is $s \neq 0$ such that $s(rm) = 0$. Since R is a domain, $rs \neq 0$, so m itself is torsion, so $m + M^{\text{tors}} = M^{\text{tors}}$, which is 0 in the quotient.   ///

An R-module M is **finitely generated** if there are finitely-many $m_1, \ldots, m_n$ such that $\sum_i Rm_i = M$. [8]

11.2.4 Proposition: Let R be a domain. [9]   Given a finitely-generated[10] R-module M, there is a (not necessarily unique) maximal free submodule F, and M/F is a torsion module.

Proof: Let X be a set of generators for M, and let S be a maximal subset of X such that (with inclusion $i : S \longrightarrow M$) the submodule generated by S is free. To be careful, consider why there is such a maximal subset. First, for ϕ not to be maximal means that there is $x_1 \in X$ such that $Rx_1 \subset M$ is free on generator $\{x_1\}$. If $\{x_1\}$ is not maximal with this property, then there is $x_2 \in X$ such that $Rx_1 + Rx_2$ is free on generators $\{x_1, x_2\}$. Since X is finite, there is no issue of infinite ascending unions of free modules. Given $x \in X$ but not in S, by the maximality of S there are coefficients $0 \neq r \in R$ and $r_s \in R$ such that

$$rx + \sum_{s \in S} r_s \cdot i(s) = 0$$

so M/F is torsion.   ///

11.2.5 Theorem: Over a principal ideal domain R a finitely-generated torsion-free module M is free.

Proof: Let X be a finite set of generators of M. From the previous proposition, let S be a maximal subset of X such that the submodule F generated by the inclusion $i : S \longrightarrow M$ is free. Let $x_1, \ldots, x_n$ be the elements of X not in S, and since M/F is torsion, for each x_i there is $0 \neq r_i \in R$ be such that $r_i x_i \in F$. Let $r = \prod_i r_i$. This is a finite product, and is non-zero since R is a domain. Thus, $r \cdot M \subset F$. Since F is free, rM is free on at most

---

[8] This is equivalent to saying that the m_i *generate* M in the sense that the intersection of submodules containing all the m_i is just M itself.

[9] The hypothesis that the ring R is a domain assures that if $r_i x_i = 0$ for $i = 1, 2$ with $0 \neq r_i \in R$ and x_i in an R-module, then not only $(r_1 r_2)(x_1 + x_2) = 0$ but *also* $r_1 r_2 \neq 0$. That is, the notion of *torsion module* has a simple sense over domains R.

[10] The conclusion is false in general without an assumption of finite generation. For example, the $\mathbb{Z}$-module $\mathbb{Q}$ is the ascending union of the free $\mathbb{Z}$-modules $\frac{1}{N} \cdot \mathbb{Z}$, but is itself not free.

$|S|$ generators. Since M is torsion-free, the multiplication by r map $m \longrightarrow rm$ has trivial kernel in M, so $M \approx rM$. That is, M is free.   ///

11.2.6 Corollary: Over a principal ideal domain R a finitely-generated module M is expressible as
$$M \approx M^{\mathrm{tors}} \oplus F$$
where F is a free module and M^{tors} is the torsion submodule of M.

Proof: We saw above that M/M^{tors} is torsion-free, so (being still finitely-generated) is free. The quotient map $M \longrightarrow M/M^{\mathrm{tors}}$ admits a section $\sigma : M/M^{\mathrm{tors}} \longrightarrow M$, and thus
$$M = M^{\mathrm{tors}} \oplus \sigma(M/M^{\mathrm{tors}}) = M^{\mathrm{tors}} \oplus \ \text{free}$$
as desired.   ///

11.2.7 Corollary: Over a principal ideal domain R, a submodule N of a finitely-generated R-module M is finitely-generated.

Proof: Let F be a finitely-generated free module which surjects to M, for example by choosing generators S for M and then forming the free module on S. The inverse image of N in F is a submodule of a free module on finitely-many generators, so (from above) needs at most that many generators. Mapping these generators forward to N proves the finite-generation of N.   ///

11.2.8 Proposition: Let R be a principal ideal domain. Let $e_1, \ldots, e_k$ be elements of a finitely-generated free R-module M which are linearly independent over R, and such that
$$M/(Re_1 + \ldots + Re_k) \quad \text{is torsion-free, hence free}$$
Then this collection can be extended to an R-basis for M.

Proof: Let N be the submodule $N = Re_1 + \ldots + Re_k$ generated by the e_i. The quotient M/N, being finitely-generated and torsion-less, is free. Let $e_{k+1}, \ldots, e_n$ be elements of M whose images in M/N are a basis for M/N. Let $q : M \longrightarrow M/N$ be the quotient map. Then, as above, q has a *section* $\sigma : M/N \longrightarrow M$ which takes $q(e_i)$ to e_i. And, as above,
$$M = \ker q \oplus \sigma(M/N) = N \oplus \sigma(M/N)$$
Since $e_{k+1}, \ldots, e_n$ is a basis for M/N, the collection of *all* $e_1, \ldots, e_n$ is a basis for M.   ///

11.3 *PIDs are UFDs*

We have already observed that *Euclidean* rings are unique factorization domains and are principal ideal domains. The two cases of greatest interest are the ordinary integers $\mathbb{Z}$ and polynomials $k[x]$ in one variable over a field k. But, also, we do have

11.3.1 Theorem: A principal ideal domain is a unique factorization domain.

Before proving this, there are relatively elementary remarks that are of independent interest, and useful in the proof. Before anything else, keep in mind that in a *domain* R (with identity 1), for $x, y \in R$,
$$Rx = Ry \quad \text{if and only if} \quad x = uy \ \text{for some unit} \ u \in R^{\times}$$

Indeed, $x \in Ry$ implies that $x = uy$, while $y \in Rx$ implies $y = vx$ for some v, and then $y = uv \cdot y$ or $(1 - uv)y = 0$. Since R is a domain, either $y = 0$ (in which case this discussion was trivial all along) or $uv = 1$, so u and v are units, as claimed.

Next recall that divisibility $x|y$ is inclusion-reversion for the corresponding ideals, that is

$$Rx \supset Ry \quad \text{if and only if} \quad x|y$$

Indeed, $y = mx$ implies $y \in Rx$, so $Ry \subset Rx$. Conversely, $Ry \subset Rx$ implies $y \in Rx$, so $y = mx$ for some $m \in R$.

Next, given x, y in a PID R, we claim that $g \in R$ such that

$$Rg = Rx + Ry$$

is a greatest common divisor for x and y, in the sense that for any $d \in R$ dividing both x and y, also d divides g (and g itself divides x and y). Indeed, $d|x$ gives $Rx \subset Rd$. Thus, since Rd is closed under addition, any common divisor d of x and y has

$$Rx + Ry \subset Rd$$

Thus, $g \in Rg \subset Rd$, so $g = rd$ for some $r \in R$. And $x \in Rg$ and $y \in Rg$ show that this g does divide both x and y.

Further, note that since a *gcd* $g = \gcd(x, y)$ of two elements x, y in the PID R is a generator for $Rx + Ry$, this *gcd* is expressible as $g = rx + sy$ for some $r, s \in R$.

In particular, a point that starts to address unique factorization is that an *irreducible* element p in a PID R is *prime*, in the sense that $p|ab$ implies $p|a$ or $p|b$. Indeed, the proof is the same as for integers, as follows. If p does *not* divide a, then the irreducibility of p implies that $1 = \gcd(p, a)$, since (by definition of *irreducible*) p has no proper divisors. Let $r, s \in R$ be such that $1 = rp + sa$. Let $ab = tp$. Then

$$b = b \cdot 1 = b \cdot (rp + sa) = br \cdot p + s \cdot ab = p \cdot (br + st)$$

and, thus, b is a multiple of p.

11.3.2 Corollary: *(of proof)* Any ascending chain

$$I_1 \subset I_2 \subset \ldots$$

of ideals in a principal ideal domain is *finite*, in the sense that there is an index i such that

$$I_i = I_{i+1} = I_{i+2} = \ldots$$

That is, a PID is **Noetherian**.

Proof: First, prove the Noetherian property, that any ascending chain of proper ideals

$$I_1 \subset I_2 \subset \ldots$$

in R must be finite. Indeed, the union I is still a proper ideal, since if it contained 1 some I_i would already contain 1, which is not so. Further, $I = Rx$ for some $x \in R$, but x must lie in some I_i, so already $I = I_i$. That is,

$$I_i = I_{i+1} = I_{i+2} = \ldots$$

Let r be a non-unit in R. If r has no proper factorization $r = xy$ (with neither x nor y a unit), then r is irreducible, and we have factored r. Suppose r has no factorization into irreducibles. Then r itself is *not* irreducible, so factors as $r = x_1 y_1$ with neither x_1 nor y_1 a unit. Since r has no factorization into irreducibles, one of x_1 or y_1, say y_1, has no factorization into irreducibles. Thus, $y_1 = x_2 y_2$ with neither x_2 nor y_2 a unit. Continuing, we obtain a chain of inclusions

$$Rr \subset Ry_1 \subset Ry_2 \subset \ldots$$

with all inclusions *strict*. This is impossible, by the Noetherian-ness property just proven. [11] That is, all ring elements have factorizations into irreducibles.

The more serious part of the argument is the *uniqueness* of the factorization, up to changing irreducibles by units, and changing the ordering of the factors. Consider

$$p_1^{e_1} \ldots p_m^{e_m} = (\text{unit}) \cdot q_1^{f_1} \ldots q_n^{f_n}$$

where the p_i and q_j are irreducibles, and the exponents are positive integers. The fact that $p_1 | ab$ implies $p_1 | a$ or $p_1 | b$ (from above) shows that p_1 must differ only by a unit from one of the q_j. Remove this factor from both sides and continue, by induction. ///

11.4 *Structure theorem, again*

The form of the following theorem is superficially stronger than our earlier version, and is more useful.

11.4.1 Theorem: Let R be a principal ideal domain, M a finitely-generated free module over R, and N an R-submodule of M. Then there are elements[12] $d_1 | \ldots | d_t$ of R, uniquely determined up to $\mathbb{R}^\times$, and an R-module basis $m_1, \ldots, m_t$ of M, such that $d_1 e_1, \ldots, d_t e_t$ is an R-basis of N (or $d_i e_i = 0$).

Proof: From above, the quotient M/N has a well-defined torsion submodule T, and $F = (M/N)/T$ is free. Let $q : M \longrightarrow (M/N)/T$ be the quotient map. Let $\sigma : F \longrightarrow M$ be a section of q, such that

$$M = \ker q \oplus \sigma(F)$$

Note that $N \subset \ker q$, and $(\ker q)/N$ is a torsion module. The submodule $\ker q$ of M is canonically defined, though the free complementary submodule [13] $\sigma(F)$ is not. Since $\sigma(F)$ can be described as a sum of a uniquely-determined (from above) number of copies $R/\langle 0 \rangle$, we see that this free submodule in M complementary to $\ker q$ gives the 0 elementary

---

[11] Yes, this proof actually shows that in *any* Noetherian commutative ring with 1 every element has a factorization into irreducibles. This does not accomplish much, however, as the *uniqueness* is far more serious than *existence* of factorization.

[12] Elementary divisors.

[13] Given a submodule A of a module B, a **complementary submodule** A' to A in B is another submodule A' of B such that $B = A \oplus A'$. In general, submodules do not admit complementary submodules. Vector spaces over fields are a marked exception to this failure.

divisors. It remains to treat the finitely-generated torsion module $(\ker q)/N$. Thus, without loss of generality, suppose that M/N is torsion (finitely-generated).

For λ in the set of R-linear *functionals* $\mathrm{Hom}_R(M, R)$ on M, the image $\lambda(M)$ is an ideal in R, as is the image $\lambda(N)$. Let λ be such that $\lambda(N)$ is *maximal* among all ideals occurring as $\lambda(N)$. [14] Let $\lambda(N) = Rx$ for some $x \in R$. We claim that $x \neq 0$. Indeed, express an element $n \in N$ as $n = \sum_i r_i e_i$ for a basis e_i of M with $r_i \in R$, and with respect to this basis define dual functionals $\mu_i \in \mathrm{Hom}_R(M, R)$ by

$$\mu_i(\sum_j s_j\, e_j) = e_i \quad (\text{where } s_j \in R)$$

If $n \neq 0$ then some coefficient r_i is non-zero, and $\mu_i(n) = r_i$. Take $n \in N$ such that $\lambda(n) = x$.

Claim $\mu(n) \in Rx$ for any $\mu \in \mathrm{Hom}_R(M, R)$. Indeed, if not, let $r, s \in R$ such that $r\lambda(n) + s\mu(n)$ is the *gcd* of the two, and $(r\lambda + s\mu)(N)$ is a strictly larger ideal than Rx, contradiction.

Thus, in particular, $\mu_i(n) \in Rx$ for all dual functionals μ_i for a given basis e_i of M. That is, $n = xm$ for some $m \in M$. Then $\lambda(m) = 1$. And

$$M = Rm \oplus \ker \lambda$$

since for any $m' \in M$

$$\lambda(m' - \lambda(m')m) = \lambda(m') - \lambda(m') \cdot 1 = 0$$

Further, for $n' \in N$ we have $\lambda(n') \in Rx$. Let $\lambda(n') = rx$. Then

$$\lambda(n' - r \cdot n) = \lambda(n') - r \cdot \lambda(n) = \lambda(n') - rx = 0$$

That is,
$$N = Rn \oplus \ker \lambda|_N$$

Thus,
$$M/N \approx Rm/Rn \oplus (\ker \lambda)/(\ker \lambda|_N)$$

with $n = xm$. And
$$Rm/Rn = Rm/Rxn \approx R/Rx = R/\langle x \rangle$$

The submodule $\ker \lambda$ is free, being a submodule of a free module over a PID, as is $\ker \lambda|_N$. And the number of generators is reduced by 1 from the number of generators of M. Thus, by induction, we have a basis $m_1, \ldots, m_t$ of M and $x_1, \ldots, x_t$ in R such that $n_i = x_i m_i$ is a basis for N, using functional λ_i whose kernel is $Rm_{i+1} + \ldots + Rm_t$, and $\lambda_i(n_i) = x_i$.

We claim that the above procedure makes $x_i | x_{i+1}$. By construction,

$$n_{i+1} \in \ker \lambda_i \quad \text{and} \quad n_i \in \ker \lambda_{i+1}$$

---

[14] At this point it is not clear that this maximal ideal is unique, but by the end of the proof we will see that it is. The fact that any ascending chain of proper ideals in a PID has a maximal element, that is, that a PID is *Noetherian*, is proven along with the proof that a PID is a unique factorization domain.

Thus, with $r, s \in R$ such that $rx_i + sx_{i+1}$ is the greatest common divisor $g = \gcd(x_i, x_{i+1})$, we have

$$(r\lambda_i + s\lambda_{i+1})(n_i + n_{i+1}) = r \cdot \lambda_i(n_i) + r \cdot \lambda_i(n_{i+1}) + + s \cdot \lambda_{i+1}(n_i) + s \cdot \lambda_{i+1}(n_{i+1})$$

$$= r \cdot x_i + 0 + +0 + s \cdot x_{i+1} = \gcd(x_i, x_{i+1})$$

That is, $Rg \supset Rx_i$ and $Rg \supset Rx_{i+1}$. The maximality property of Rx_i requires that $Rx_i = Rg$. Thus, $Rx_{i+1} \subset Rx_i$, as claimed.

This proves *existence* of a decomposition as indicated. Proof of *uniqueness* is far better treated after introduction of a further idea, namely, *exterior algebra*. Thus, for the moment, we will *not* prove uniqueness, but will defer this until the later point when we treat exterior algebra.

11.5 *Recovering the earlier structure theorem*

The above structure theorem on finitely-generated free modules M over PIDs R and submodules $N \subset M$ gives the structure theorem for finitely-generated modules as a corollary, as follows.

Let F be a finitely-generated R-module with generators [15] $f_1, \ldots, f_n$. Let $S = \{f_1, \ldots, f_n\}$, and let M be the free R-module on generators $i : S \longrightarrow M$. Let

$$q : M \longrightarrow F$$

be the unique R-module homomorphism such that $q(i(f_k)) = f_k$ for each generator f_k. Since $q(M)$ contains all the generators of F, the map q is surjective. [16]

Let $N = \ker q$, so by a basic isomorphism theorem

$$F \approx M/N$$

By the theorem of the last section, M has a basis $m_1, \ldots, m_t$ and there are uniquely determined [17] $r_1 | r_2 | \ldots | r_t \in R$ such that $r_1 m_1, \ldots, r_t m_t$ is a basis for N. Then

$$F \approx M/N \approx (Rm_1/Rr_1m_1) \oplus \ldots \oplus (Rm_t/Rr_tm_t) \approx R/\langle r_1 \rangle \oplus \ldots R/\langle r_t \rangle$$

since

$$Rm_i/Rr_im_i \approx R/\langle r_i \rangle$$

by

$$rm_i + Rr_im_i \longrightarrow r + Rr_i$$

This gives an expression for F of the sort desired.   ///

---

[15] It does not matter whether or not this set is *minimal*, only that it be *finite*.

[16] We will have no further use for the generators f_k of F after having constructed the finitely-generated *free* module M which surjects to F.

[17] Uniquely determined up to units.

11.6 *Submodules of free modules*

Let R be a principal ideal domain. Let A be a well-ordered set, and M a free module on generators e_α for $\alpha \in A$. Let N be a submodule of M.

For $\alpha \in A$, let

$$I_\alpha = \{r \in R : \text{ there exist } r_\beta, \; \beta < \alpha : r \cdot e_\alpha + \sum_{\beta < \alpha} r_\beta \cdot e_\beta \in N\}$$

Since R is a PID, the ideal I_α has a single generator ρ_α (which may be 0). Let $n_\alpha \in N$ be such that

$$n_\alpha = \rho_\alpha \cdot e_\alpha + \sum_{\beta < \alpha} r_\beta \cdot e_\beta$$

for some $r_\beta \in R$. This defines ρ_α and n_α for all $\alpha \in A$ by transfinite induction.

11.6.1 Theorem: N is free on the (non-zero elements among) n_α.

Proof: It is clear that I_α is an ideal in R, so at least one element n_α exists, though it may be 0. For any element $n \in N$ lying in the span of $\{e_\beta : \beta \le \alpha\}$, for some $r \in R$ the difference $n - rn_\alpha$ lies in the span of $\{e_\beta : \beta < \alpha\}$.

We claim that the n_α span N. Suppose not, and let $\alpha \in A$ be the first index such that there is $n \in N$ *not* in that span, with n expressible as $n = \sum_{\beta \le \alpha} r_\beta e_\beta$. Then $r_\alpha = r \cdot \rho_\alpha$ for some $r \in R$, and for suitable coefficients $s_\beta \in R$

$$n - rn_\alpha = \sum_{\beta < \alpha} s_\beta \cdot e_\beta$$

This element must still fail to be in the span of the n_γ's. Since that sum is finite, the supremum of the indices with non-zero coefficient is strictly less than α. This gives a contradiction to the minimality of α, proving that the n_α span N.

Now prove that the (non-zero) n_α's are linearly independent. Indeed, if we have a non-trivial (finite) relation

$$0 = \sum_\beta r_\beta \cdot n_\beta$$

let α be the highest index (among finitely-many) with $r_\alpha \ne 0$ and $n_\alpha \ne 0$. Since n_α is non-zero, it must be that $\rho_\alpha \ne 0$, and then the expression of n_α in terms of the basis $\{e_\gamma\}$ includes e_α with non-zero coefficient (namely, ρ_α). But no n_β with $\beta < \alpha$ needs e_α in its expression, so for suitable $s_\beta \in R$

$$0 = \sum_\beta r_\beta \cdot n_\beta = r_\alpha \rho_\alpha \cdot e_\alpha + \sum_{\beta < \alpha} s_\beta \cdot e_\beta$$

contradicting the linear independence of the e_α's. Thus, we conclude that the n_β's are linearly independent. ///

Exercises

11.1 Find two integer vectors $x = (x_1, x_2)$ and $y = (y_1, y_2)$ such that $\gcd(x_1, x_2) = 1$ and $\gcd(y_1, y_2) = 1$, but $\mathbb{Z}^2/(\mathbb{Z}x + \mathbb{Z}y)$ has non-trivial torsion.

11.2 Show that the $\mathbb{Z}$-module $\mathbb{Q}$ is torsion-free, but is *not* free.

11.3 Let G be the group of positive rational numbers under multiplication. Is G a free $\mathbb{Z}$-module? Torsion-free? Finitely-generated?

11.4 Let G be the quotient group $\mathbb{Q}/\mathbb{Z}$. Is G a free $\mathbb{Z}$-module? Torsion-free? Finitely-generated?

11.5 Let $R = \mathbb{Z}[\sqrt{5}]$, and let $M = R \cdot 2 + R \cdot (1 + \sqrt{5}) \subset \mathbb{Q}(\sqrt{5})$. Show that M is *not* free over R, although it is torsion-free.

11.6 Given an m-by-n matrix M with entries in a PID R, give an existential argument that there are matrices A (n-by-n) and B (m-by-m) with entries in R and with inverses with entries in R, such that AMB is *diagonal*.

11.7 Describe an *algorithm* which, given a 2-by-3 integer matrix M, finds integer matrices A, B (with integer inverses) such that AMB is diagonal.

11.8 Let A be a *torsion* abelian group, meaning that for every $a \in A$ there is $1 \leq n \in Z$ such that $n \cdot a = 0$. Let $A(p)$ be the subgroup of A consisting of elements a such that $p^\ell \cdot a = 0$ for some integer power p^ℓ of a prime p. Show that A is the direct sum of its subgroups $A(p)$ over primes p.

11.9 (*) Let A be a subgroup of $\mathbb{R}^n$ such that in each ball there are finitely-many elements of A. Show that A is a free abelian group on at most n generators.

12. Polynomials over UFDs

The goal here is to give a general result which has as corollary that that rings of polynomials in several variables

$$k[x_1, \ldots, x_n]$$

with coefficients in a field k are *unique factorization domains* in a sense made precise just below. Similarly, polynomial rings in several variables

$$\mathbb{Z}[x_1, \ldots, x_n]$$

with coefficients in $\mathbb{Z}$ form a unique factorization domain. [1]

12.1 *Gauss' lemma*

A **factorization** of an element r into *irreducibles* in an integral domain R is an expression for r of the form

$$r = u \cdot p_1^{e_1} \ldots p_m^{e_m}$$

where u is a unit, p_1 through p_m are *non-associate* [2] irreducible elements, and the e_is are

---

[1] Among other uses, these facts are used to discuss Vandermonde determinants, and in the proof that the *parity* (or *sign*) of a permutation is well-defined.

[2] Recall that two elements x, y of a commutative ring R are *associate* if $x = yu$ for some unit u in R. This terminology is most often applied to prime or irreducible elements.

positive integers. Two factorizations

$$r = u \cdot p_1^{e_1} \ldots p_m^{e_m}$$

$$r = v \cdot q_1^{f_1} \ldots q_n^{f_n}$$

into irreducibles p_i and q_j with units u, v are **equivalent** if $m = n$ and (after possibly renumbering the irreducibles) q_i is *associate* to p_i for all indices i. A domain R is a **unique factorization domain** (UFD) if any two factorizations are equivalent.

12.1.1 Theorem: (*Gauss*) Let R be a unique factorization domain. Then the polynomial ring in one variable $R[x]$ is a unique factorization domain.

12.1.2 Remark: The proof factors $f(x) \in R[x]$ in the larger ring $k[x]$ where k is the *field of fractions* of R (see below), and rearranges constants to get coefficients into R rather than k. Uniqueness of the factorization follows from uniqueness of factorization in R and uniqueness of factorization in $k[x]$.

12.1.3 Corollary: A polynomial ring $k[x_1, \ldots, x_n]$ in a finite number of variables x_1, ..., x_n over a field k is a unique factorization domain. (*Proof by induction.*)   ///

12.1.4 Corollary: A polynomial ring $\mathbb{Z}[x_1, \ldots, x_n]$ in a finite number of variables x_1, ..., x_n over the integers $\mathbb{Z}$ is a unique factorization domain. (*Proof by induction.*)   ////

Before proving the theorem itself, we must verify that unique factorization recovers some naive ideas about divisibility. Recall that for $r, s \in R$ not both 0, an element $g \in R$ dividing both r and s such that any divisor d of both r and s also divides g, is a **greatest common divisor** of r and s, denoted $g = \gcd(r, s)$.

12.1.5 Proposition: Let R be a unique factorization domain. For r, s in R not both 0 there exists $\gcd(r, s)$ unique up to an element of $R^\times$. Factor both r and s into irreducibles

$$r = u \cdot p_1^{e_1} \ldots p_m^{e_m} \qquad s = v \cdot p_1^{f_1} \ldots p_m^{f_n}$$

where u and v are units and the p_i are mutually non-associate irreducibles (allow the exponents to be 0, to use a common set of irreducibles to express both r and s). Then the greatest common divisor has exponents which are the minima of those of r and s

$$\gcd(r, s) = p_1^{\min(e_1, f_1)} \ldots p_m^{\min(e_m, f_m)}$$

Proof: Let

$$g = p_1^{\min(e_1, f_1)} \ldots p_m^{\min(e_m, f_m)}$$

First, g does divide both r and s. On the other hand, let d be any divisor of both r and s. Enlarge the collection of inequivalent irreducibles p_i if necessary such that d can be expressed as

$$d = w \cdot p_1^{h_1} \ldots p_m^{h_m}$$

with unit w and non-negative integer exponents. From $d|r$ there is $D \in R$ such that $dD = r$. Let

$$D = W \cdot p_1^{H_1} \ldots p_m^{H_m}$$

Then

$$wW \cdot p_1^{h_1+H_1} \dots p_m^{h_m+H_m} = d \cdot D = r = u \cdot p_1^{e_1} \dots p_m^{e_m}$$

Unique factorization and non-associateness of the p_i implies that the exponents are the same: for all i

$$h_i + H_i = e_i$$

Thus, $h_i \leq e_i$. The same argument applies with r replaced by s, so $h_i \leq f_i$, and $h_i \leq \min(e_i, f_i)$. Thus, $d|g$. For uniqueness, note that any other greatest common divisor h would have $g|h$, but also $h|r$ and $h|s$. Using the unique (up to units) factorizations, the exponents of the irreducibles in g and h must be the same, so g and h must differ only by a unit. ///

12.1.6 Corollary: Let R be a unique factorization domain. For r and s in R, let $g = \gcd(r, s)$ be the greatest common divisor. Then $\gcd(r/g, s/g) = 1$. ///

12.2 *Fields of fractions*

The **field of fractions** k of an integral domain R is the collection of fractions a/b with $a, b \in R$ and $b \neq 0$ and with the usual rules for addition and multiplication. More precisely, k is the set of ordered pairs (a, b) with $a, b \in R$ and $b \neq 0$, modulo the equivalence relation that

$$(a, b) \sim (c, d)$$

if and only if $ad - bc = 0$. [3] Multiplication and addition are [4]

$$(a, b) \cdot (c, d) = (ac, bd)$$

$$(a, b) + (c, d) = (ad + bc, bd)$$

The map $R \longrightarrow k$ by $r \longrightarrow (r, 1)/ \sim$ is readily verified to be a ring homomorphism. [5] Write a/b rather than $(a, b)/ \sim$. When R is a unique factorization ring, whenever convenient suppose that fractions a/b are *in lowest terms*, meaning that $\gcd(a, b) = 1$.

Extend the notions of divisibility to apply to elements of the fraction field k of R. [6] First, say that $x|y$ for two elements x and y in k if there is $r \in R$ such that $s = rx$. [7] And,

---

[3] This corresponds to the ordinary rule for equality of two fractions.

[4] As usual for fractions.

[5] The assumption that R is a *domain*, is needed to make this work so simply. For commutative rings (with 1) with proper 0-divisors the natural homomorphism $r \longrightarrow (r, 1)$ of the ring to its field of fractions will not be injective. And this construction will later be seen to be a simple extreme example of the more general notion of *localization* of rings.

[6] Of course notions of divisibility in a field itself are trivial, since any non-zero element divides any other. This is *not* what is happening now.

[7] For non-zero r in the domain R, $rx|ry$ if and only if $x|y$. Indeed, if $ry = m \cdot rx$ then by cancellation (using the domain property), $y = m \cdot x$. And $y = m \cdot x$ implies $ry = m \cdot rx$ directly.

for $r_1, \ldots, r_n$ in k, not all 0, a greatest common divisor $\gcd(r_1, \ldots, r_n)$ is an element $g \in k$ such that g divides each r_i and such that if $d \in k$ divides each r_i then $d|g$.

12.2.1 Proposition: In the field of fractions k of a unique factorization domain R (extended) greatest common divisors exist.

Proof: We reduce this to the case that everything is inside R. Given elements $x_i = a_i/b_i$ in k with a_i and b_i all in R, take $0 \neq r \in R$ such that $rx_i \in R$ for all i. Let G be the greatest common divisor of the rx_i, and put $g = G/r$. We claim this g is the greatest common divisor of the x_i. On one hand, from $G|rx_i$ it follows that $g|x_i$. On the other hand, if $d|x_i$ then $rd|rx_i$, so rd divides $G = rg$ and $d|g$.   ///

The **content** $\mathrm{cont}(f)$ of a polynomial f in $k[x]$ is the greatest common divisor [8] of the coefficients of f.

12.2.2 Lemma: *(Gauss)* Let f and g be two polynomials in $k[x]$. Then

$$\mathrm{cont}(fg) = \mathrm{cont}(f) \cdot \mathrm{cont}(g)$$

Proof: From the remark just above for any $c \in k^\times$

$$\mathrm{cont}(c \cdot f) = c \cdot \mathrm{cont}(f)$$

Thus, since

$$\gcd\left(\frac{a}{\gcd(a,b)}, \frac{b}{\gcd(a,b)}\right) = 1$$

without loss of generality $\mathrm{cont}(f) = 1$ and $\mathrm{cont}(g) = 1$. Thus, in particular, both f and g have coefficients in the ring R. Suppose $\mathrm{cont}(fg) \neq 1$. Then there is non-unit irreducible $p \in R$ dividing all the coefficients of fg. Put

$$f(x) = a_0 + a_1 x + a_2 x^2 + \ldots$$

$$g(x) = b_0 + b_1 x + b_2 x^2 + \ldots$$

But p does not divide *all* the coefficients of f, nor *all* those of g. Let i be the smallest integer such that p does not divide a_i, j the largest integer such that p does not divide b_j, and consider the coefficient of x^{i+j} in fg. It is

$$a_0 b_{i+j} + a_1 b_{i+j-1} + \ldots + a_{i-1} b_{j-1} + a_i b_j + a_{i+1} b_{j-1} + \ldots + a_{i+j-1} b_1 + a_{i+j} b_0$$

In summands to the left of $a_i b_j$ the factor a_k with $k < i$ is divisible by p, and in summands to the right of $a_i b_j$ the factor b_k with $k < j$ is divisible by p. This leaves only the summand $a_i b_j$ to consider. Since the whole sum is divisible by p, it follows that $p|a_i b_j$. Since R is a unique factorization domain, either $p|a_i$ or $p|b_j$, contradiction. Thus, it could not have been that p divided all the coefficients of fg.   ///

---

[8] The values of the content function are only well-defined up to units $R^\times$. Thus, Gauss' lemma more properly concerns the *equivalence classes* of irreducibles dividing the respective coefficients.

12.2.3 Corollary: Let f be a polynomial in $R[x]$. If f factors properly in $k[x]$ then f factors properly in $R[x]$. More precisely, if f factors as $f = g \cdot h$ with g and h polynomials in $k[x]$ of positive degree, then there is $c \in k^\times$ such that $cg \in R[x]$ and $h/c \in R[x]$, and

$$f = (cg) \cdot (h/c)$$

is a factorization of f in $R[x]$.

Proof: Since f has coefficients in R, $\mathrm{cont}(f)$ is in R. By replacing f by f/c we may suppose that $\mathrm{cont}(f) = 1$. By Gauss' lemma

$$\mathrm{cont}(g) \cdot \mathrm{cont}(h) = \mathrm{cont}(f) = 1$$

Let $c = \mathrm{cont}(g)$. Then $\mathrm{cont}(h) = 1/c$, and $\mathrm{cont}(g/c) = 1$ and $\mathrm{cont}(c \cdot h) = 1$, so g/c and ch are in $R[x]$, and $(g/c) \cdot (ch) = f$. Thus f is reducible in $R[x]$. ///

12.2.4 Corollary: The irreducibles in $R[x]$ are of two sorts, namely irreducibles in R and polynomials f in $R[x]$ with $\mathrm{cont}(f) = 1$ which are irreducible in $k[x]$.

Proof: If an irreducible p in R factored in $R[x]$ as $p = gh$, then the degrees of g and h would be 0, and g and h would be in R. The irreducibility of p in R would imply that one of g or h would be a unit. Thus, irreducibles in R remain irreducible in $R[x]$.

Suppose p was irreducible in $R[x]$ of positive degree. If $g = \mathrm{cont}(p)$ was a non-unit, then $p = (p/g) \cdot g$ would be a proper factorization of p, contradiction. Thus, $\mathrm{cont}(p) = 1$. The previous corollary shows that p is irreducible in $k[x]$.

Last suppose that f is irreducible in $k[x]$, and has $\mathrm{cont}(f) = 1$. The irreducibility in $k[x]$ implies that if $f = gh$ in $R[x]$ then the degree one of g or h must be 0. Without loss of generality suppose $\deg g = 0$, so $\mathrm{cont}(g) = g$. Since

$$1 = \mathrm{cont}(f) = \mathrm{cont}(g)\mathrm{cont}(h)$$

g is a unit in R, so $f = gh$ is not a proper factorization, and f is irreducible in $R[x]$. ///

Proof: (*of theorem*) We can now combine the corollaries of Gauss' lemma to prove the theorem. Given a polynomial f in $R[x]$, let $c = \mathrm{cont}(f)$, so from above $\mathrm{cont}(f/c) = 1$. The hypothesis that R is a unique factorization domain allows us to factor u into irreducibles in R, and we showed just above that these irreducibles remain irreducible in $R[x]$.

Replace f by $f/\mathrm{cont}(f)$ to assume now that $\mathrm{cont}(f) = 1$. Factor f into irreducibles in $k[x]$ as

$$f = u \cdot p_1^{e_1} \cdots p_m^{e_m}$$

where u is in $k^\times$, the p_is are irreducibles in $k[x]$, and the e_is are positive integers. We can replace each p_i by $p_i/\mathrm{cont}(p_i)$ and replace u by

$$u \cdot \mathrm{cont}(p_1)^{e_1} \cdots \mathrm{cont}(p_m)^{e_m}$$

so then the new p_is are in $R[x]$ and have content 1. Since content is multiplicative, from $\mathrm{cont}(f) = 1$ we find that $\mathrm{cont}(u) = 1$, so u is a unit in R. The previous corollaries demonstrate the irreducibility of the (new) p_is in $R[x]$, so this gives a factorization of f

into irreducibles in $R[x]$. That is, we have an explicit *existence* of a factorization into irreducibles.

Now suppose that we have two factorizations

$$f = u \cdot p_1^{e_1} \cdots p_m^{e_m} = v \cdot q_1^{f_1} \cdots q_n^{f_n}$$

where u, v are in R (and have unique factorizations there) and the p_i and q_j are irreducibles in $R[x]$ of positive degree. From above, all the contents of these irreducibles must be 1. Looking at this factorization in $k[x]$, it must be that $m = n$ and up to renumbering p_i differs from q_i by a constant in $k^\times$, and $e_i = f_i$. Since all these polynomials have content 1, in fact p_i differs from q_i by a unit in R. By equating the contents of both sides, we see that u and v differ by a unit in $R^\times$. Thus, by the unique factorization in R their factorizations into irreducibles in R (and, from above, in $R[x]$) must be essentially the same. Thus, we obtain uniqueness of factorization in $R[x]$.   ///

12.3 *Worked examples*

12.3.1 Example: Let R be a principal ideal domain. Let I be a non-zero prime ideal in R. Show that I is *maximal*.

Suppose that I were strictly contained in an ideal J. Let $I = Rx$ and $J = Ry$, since R is a PID. Then x is a multiple of y, say $x = ry$. That is, $ry \in I$. But y is not in I (that is, not a multiple of p), since otherwise $Ry \subset Rx$. Thus, since I is prime, $r \in I$, say $r = ap$. Then $p = apy$, and (since R is a domain) $1 = ay$. That is, the ideal generated by y contains 1, so is the whole ring R. That is, I is maximal (proper).

12.3.2 Example: Let k be a field. Show that in the polynomial ring $k[x, y]$ in two variables the ideal $I = k[x, y] \cdot x + k[x, y] \cdot y$ is not principal.

Suppose that there were a polynomial $P(x, y)$ such that $x = g(x, y) \cdot P(x, y)$ for some polynomial g and $y = h(x, y) \cdot P(x, y)$ for some polynomial h.

An intuitively appealing thing to say is that since y *does not appear* in the polynomial x, it could not *appear* in $P(x, y)$ or $g(x, y)$. Similarly, since x *does not appear* in the polynomial y, it could not appear in $P(x, y)$ or $h(x, y)$. And, thus, $P(x, y)$ would be in k. It would have to be non-zero to yield x and y as multiples, so would be a unit in $k[x, y]$. Without loss of generality, $P(x, y) = 1$. (Thus, we need to show that I is proper.)

On the other hand, since $P(x, y)$ is supposedly in the ideal I generated by x and y, it is of the form $a(x, y) \cdot x + b(x, y) \cdot y$. Thus, we would have

$$1 = a(x, y) \cdot x + b(x, y) \cdot y$$

Mapping $x \longrightarrow 0$ and $y \longrightarrow 0$ (while mapping k to itself by the identity map, thus sending 1 to $1 \neq 0$), we would obtain

$$1 = 0$$

contradiction. Thus, there is no such $P(x, y)$.

We can be more precise about that admittedly intuitively appealing first part of the argument. That is, let's show that if

$$x = g(x, y) \cdot P(x, y)$$

then the degree of $P(x, y)$ (and of $g(x, y)$) as a polynomial in y (with coefficients in $k[x]$) is 0. Indeed, looking at this equality as an equality in $k(x)[y]$ (where $k(x)$ is the field of rational functions in x with coefficients in k), the fact that degrees *add* in products gives the desired conclusion. Thus,

$$P(x, y) \in k(x) \cap k[x, y] = k[x]$$

Similarly, $P(x, y)$ lies in $k[y]$, so P is in k.

12.3.3 Example: Let k be a field, and let $R = k[x_1, \ldots, x_n]$. Show that the inclusions of ideals

$$Rx_1 \subset Rx_1 + Rx_2 \subset \ldots \subset Rx_1 + \ldots + Rx_n$$

are *strict*, and that all these ideals are *prime*.

One approach, certainly correct in spirit, is to say that *obviously*

$$k[x_1, \ldots, x_n]/Rx_1 + \ldots + Rx_j \approx k[x_{j+1}, \ldots, x_n]$$

The latter ring is a domain (since k is a domain and polynomial rings over domains are domains: proof?) so the ideal was necessarily prime.

But while it is true that certainly $x_1, \ldots, x_j$ go to 0 in the quotient, our intuition uses the explicit construction of polynomials as *expressions* of a certain form. Instead, one might try to give the allegedly trivial and immediate proof that sending $x_1, \ldots, x_j$ to 0 does not somehow cause 1 to get mapped to 0 in k, nor accidentally impose any relations on $x_{j+1}, \ldots, x_n$. A too classical viewpoint does not lend itself to clarifying this. The point is that, given a k-algebra homomorphism $f_o : k \longrightarrow k$, here taken to be the *identity*, and given values 0 for $x_1, \ldots, x_j$ and values $x_{j+1}, \ldots, x_n$ respectively for the other indeterminates, there is a *unique* k-algebra homomorphism $f : k[x_1, \ldots, x_n] \longrightarrow k[x_{j+1}, \ldots, x_n]$ agreeing with f_o on k and sending $x_1, \ldots, x_n$ to their specified targets. Thus, in particular, we *can* guarantee that $1 \in k$ is *not* somehow accidentally mapped to 0, and no relations among the $x_{j+1} \ldots, x_n$ are mysteriously introduced.

12.3.4 Example: Let k be a field. Show that the ideal M generated by $x_1, \ldots, x_n$ in the polynomial ring $R = k[x_1, \ldots, x_n]$ is *maximal* (proper).

We prove the maximality by showing that R/M is a field. The universality of the polynomial algebra implies that, given a k-algebra homomorphism such as the *identity* $f_o : k \longrightarrow k$, and given $\alpha_i \in k$ (take $\alpha_i = 0$ here), there exists a unique k-algebra homomorphism $f : k[x_1, \ldots, x_n] \longrightarrow k$ extending f_o. The kernel of f certainly contains M, since M is generated by the x_i and all the x_i go to 0.

As in the previous exercise, one perhaps should verify that M is *proper*, since otherwise accidentally in the quotient map $R \longrightarrow R/M$ we might *not* have $1 \longrightarrow 1$. If we *do* know that M is a proper ideal, then by the uniqueness of the map f we know that $R \longrightarrow R/M$ is (up to isomorphism) exactly f, so M is maximal proper.

Given a relation

$$1 = \sum_i f_i \cdot x_i$$

with polynomials f_i, using the universal mapping property send all x_i to 0 by a k-algebra homomorphism to k that does send 1 to 1, obtaining $1 = 0$, contradiction.

12.3.5 Remark: One surely is inclined to allege that *obviously* $R/M \approx k$. And, indeed, this quotient is *at most k*, but one should at least acknowledge *concern* that it not be accidentally 0. Making the point that not only can the images of the x_i be chosen, but *also* the k-algebra homomorphism on k, decisively eliminates this possibility.

12.3.6 Example: Show that the maximal ideals in $R = \mathbb{Z}[x]$ are all of the form

$$I = R \cdot p + R \cdot f(x)$$

where p is a prime and $f(x)$ is a monic polynomial which is irreducible modulo p.

Suppose that no non-zero integer n lies in the maximal ideal I in R. Then $\mathbb{Z}$ would inject to the quotient R/I, a field, which then would be of characteristic 0. Then R/I would contain a canonical copy of $\mathbb{Q}$. Let α be the image of x in K. Then $K = \mathbb{Z}[\alpha]$, so certainly $K = \mathbb{Q}[\alpha]$, so α is algebraic over $\mathbb{Q}$, say of degree n. Let $f(x) = a_n x^n + \ldots + a_1 x + a_0$ be a polynomial with rational coefficient such that $f(\alpha) = 0$, and with all denominators multiplied out to make the coefficients *integral*. Then let $\beta = c_n \alpha$: this β is still algebraic over $\mathbb{Q}$, so $\mathbb{Q}[\beta] = \mathbb{Q}(\beta)$, and certainly $\mathbb{Q}(\beta) = \mathbb{Q}(\alpha)$, and $\mathbb{Q}(\alpha) = \mathbb{Q}[\alpha]$. Thus, we still have $K = \mathbb{Q}[\beta]$, but now things have been adjusted so that β satisfies a *monic* equation with coefficients in $\mathbb{Z}$: from

$$0 = f(\alpha) = f(\frac{\beta}{c_n}) = c_n^{1-n}\beta^n + c_{n-1}c_n^{1-n}\beta^{n-1} + \ldots + c_1 c_n^{-1}\beta + c_0$$

we multiply through by c_n^{n-1} to obtain

$$0 = \beta^n + c_{n-1}\beta^{n-1} + c_{n-2}c_n\beta^{n-2} + c_{n-3}c_n^2\beta^{n-3} + \ldots + c_2 c_n^{n-3}\beta^2 + c_1 c_n^{n-2}\beta + c_0 c_n^{n-1}$$

Since $K = \mathbb{Q}[\beta]$ is an n-dimensional Q-vectorspace, we can find rational numbers b_i such that

$$\alpha = b_0 + b_1\beta + b_2\beta^2 + \ldots + b_{n-1}\beta^{n-1}$$

Let N be a large-enough integer such that for every index i we have $b_i \in \frac{1}{N} \cdot \mathbb{Z}$. Note that because we made β satisfy a *monic integer* equation, the set

$$\Lambda = \mathbb{Z} + \mathbb{Z} \cdot \beta + \mathbb{Z} \cdot \beta^2 + \ldots + \mathbb{Z} \cdot \beta^{n-1}$$

is closed under multiplication: β^n is a $\mathbb{Z}$-linear combination of lower powers of β, and so on. Thus, since $\alpha \in N^{-1}\Lambda$, successive powers α^ℓ of α are in $N^{-\ell}\Lambda$. Thus,

$$\mathbb{Z}[\alpha] \subset \bigcup_{\ell \geq 1} N^{-\ell}\Lambda$$

But now let p be a prime not dividing N. We claim that $1/p$ does not lie in $\mathbb{Z}[\alpha]$. Indeed, since $1, \beta, \ldots, \beta^{n-1}$ are linearly independent over $\mathbb{Q}$, there is a *unique* expression for $1/p$ as a $\mathbb{Q}$-linear combination of them, namely the obvious $\frac{1}{p} = \frac{1}{p} \cdot 1$. Thus, $1/p$ is not in $N^{-\ell} \cdot \Lambda$ for any $\ell \in \mathbb{Z}$. This (at last) contradicts the supposition that no non-zero integer lies in a maximal ideal I in $\mathbb{Z}[x]$.

Note that the previous argument uses the infinitude of primes.

Thus, $\mathbb{Z}$ does *not* inject to the field R/I, so R/I has positive characteristic p, and the canonical $\mathbb{Z}$-algebra homomorphism $\mathbb{Z} \longrightarrow R/I$ factors through $\mathbb{Z}/p$. Identifying

$\mathbb{Z}[x]/p \approx (\mathbb{Z}/p)[x]$, and granting (as proven in an earlier homework solution) that for $J \subset I$ we can take a quotient in two stages

$$R/I \approx (R/J)/(\text{image of } J \text{ in } R/I)$$

Thus, the image of I in $(\mathbb{Z}/p)[x]$ is a maximal ideal. The ring $(\mathbb{Z}/p)[x]$ is a PID, since $\mathbb{Z}/p$ is a field, and by now we know that the maximal ideals in such a ring are of the form $\langle f \rangle$ where f is irreducible and of positive degree, and conversely. Let $F \in \mathbb{Z}[x]$ be a polynomial which, when we reduce its coefficients modulo p, becomes f. Then, at last,

$$I = \mathbb{Z}[x] \cdot p + \mathbb{Z}[x] \cdot f(x)$$

as claimed.

12.3.7 Example: Let R be a *PID*, and x, y non-zero elements of R. Let $M = R/\langle x \rangle$ and $N = R/\langle y \rangle$. Determine $\mathrm{Hom}_R(M, N)$.

Any homomorphism $f : M \longrightarrow N$ gives a homomorphism $F : R \longrightarrow N$ by composing with the quotient map $q : R \longrightarrow M$. Since R is a free R-module on one generator 1, a homomorphism $F : R \longrightarrow N$ is completely determined by $F(1)$, and this value can be anything in N. Thus, the homomorphisms from R to N are exactly parametrized by $F(1) \in N$. The remaining issue is to determine which of these maps F *factor through* M, that is, which such F admit $f : M \longrightarrow N$ such that $F = f \circ q$. We could *try* to define (and there is no other choice if it is to succeed)

$$f(r + Rx) = F(r)$$

but this will be well-defined if and only if $\ker F \supset Rx$.

Since $0 = y \cdot F(r) = F(yr)$, the kernel of $F : R \longrightarrow N$ invariably contains Ry, and we need it to contain Rx as well, for F to give a well-defined map $R/Rx \longrightarrow R/Ry$. This is equivalent to

$$\ker F \supset Rx + Ry = R \cdot \gcd(x, y)$$

or

$$F(\gcd(x, y)) = \{0\} \subset R/Ry = N$$

By the R-linearity,

$$R/Ry \ni 0 = F(\gcd(x, y)) = \gcd(x, y) \cdot F(1)$$

Thus, the condition for well-definedness is that

$$F(1) \in R \cdot \frac{y}{\gcd(x, y)} \subset R/Ry$$

Therefore, the desired homomorphisms f are in bijection with

$$F(1) \in R \cdot \frac{y}{\gcd(x, y)}/Ry \subset R/Ry$$

where

$$f(r + Rx) = F(r) = r \cdot F(1)$$

12.3.8 Example: *(A warm-up to Hensel's lemma)* Let p be an odd prime. Fix $a \not\equiv 0 \bmod p$ and suppose $x^2 = a \bmod p$ has a solution x_1. Show that for every positive integer n the congruence $x^2 = a \bmod p^n$ has a solution x_n. (*Hint:* Try $x_{n+1} = x_n + p^n y$ and solve for $y \bmod p$).

Induction, following the hint: Given x_n such that $x_n^2 = a \bmod p^n$, with $n \geq 1$ and $p \neq 2$, show that there will exist y such that $x_{n+1} = x_n + yp^n$ gives $x_{n+1}^2 = a \bmod p^{n+1}$. Indeed, expanding the desired equality, it is equivalent to

$$a = x_{n+1}^2 = x_n^2 + 2x_n p^n y + p^{2n} y^2 \bmod p^{n+1}$$

Since $n \geq 1$, $2n \geq n + 1$, so this is

$$a = x_n^2 + 2x_n p^n y \bmod p^{n+1}$$

Since $a - x_n^2 = k \cdot p^n$ for some integer k, dividing through by p^n gives an equivalent condition

$$k = 2x_n y \bmod p$$

Since $p \neq 2$, and since $x_n^2 = a \neq 0 \bmod p$, $2x_n$ is invertible mod p, so no matter what k is there exists y to meet this requirement, and we're done.

12.3.9 Example: (*Another warm-up to Hensel's lemma*) Let p be a prime not 3. Fix $a \neq 0 \bmod p$ and suppose $x^3 = a \bmod p$ has a solution x_1. Show that for every positive integer n the congruence $x^3 = a \bmod p^n$ has a solution x_n. (*Hint:* Try $x_{n+1} = x_n + p^n y$ and solve for $y \bmod p$.)

Induction, following the hint: Given x_n such that $x_n^3 = a \bmod p^n$, with $n \geq 1$ and $p \neq 3$, show that there will exist y such that $x_{n+1} = x_n + yp^n$ gives $x_{n+1}^3 = a \bmod p^{n+1}$. Indeed, expanding the desired equality, it is equivalent to

$$a = x_{n+1}^3 = x_n^3 + 3x_n^2 p^n y + 3x_n p^{2n} y^2 + p^{3n} y^3 \bmod p^{n+1}$$

Since $n \geq 1$, $3n \geq n + 1$, so this is

$$a = x_n^3 + 3x_n^2 p^n y \bmod p^{n+1}$$

Since $a - x_n^3 = k \cdot p^n$ for some integer k, dividing through by p^n gives an equivalent condition

$$k = 3x_n^2 y \bmod p$$

Since $p \neq 3$, and since $x_n^3 = a \neq 0 \bmod p$, $3x_n^2$ is invertible mod p, so no matter what k is there exists y to meet this requirement, and we're done.

Exercises

12.1 Let k be a field. Show that every non-zero prime ideal in $k[x]$ is maximal.

12.2 Let k be a field. Let x, y, z be indeterminates. Show that the ideal I in $k[x, y, z]$ generated by x, y, z is not principal.

12.3 Let R be a commutative ring with identity that is *not necessarily* an integral domain. Let S be a multiplicative subset of R. The localization $S^{-1}R$ is defined to be the set of pairs (r, s) with $r \in R$ and $s \in S$ modulo the equivalence relation

$$(r, s) \sim (r', s') \iff \text{there is } t \in S \text{ such that} t \cdot (rs' - r's) = 0$$

Show that the natural map $i_S : r \longrightarrow (r, 1)$ is a ring homomorphism, and that $S^{-1}R$ is a ring in which every element of S becomes invertible.

12.4 Indeed, in the situation of the previous exercise, show that every ring homomorphism $\varphi : R \longrightarrow R'$ such that $\varphi(s)$ is invertible in R' for $s \in S$ *factors uniquely through* $S^{-1}R$. That is, there is a unique $f : S^{-1}R \longrightarrow R'$ such that $\varphi = f \circ i_S$ with the natural map i_S.

13. Symmetric groups

13.1 *Cycles, disjoint cycle decompositions*

The **symmetric group** S_n is the group of bijections of $\{1, \ldots, n\}$ to itself, also called **permutations** of n things. A standard notation for the permutation that sends $i \longrightarrow \ell_i$ is

$$\begin{pmatrix} 1 & 2 & 3 & \ldots & n \\ \ell_1 & \ell_2 & \ell_3 & \ldots & \ell_n \end{pmatrix}$$

Under composition of mappings, the permutations of $\{1, \ldots, n\}$ is a *group*.

The **fixed points** of a permutation f are the elements $i \in \{1, 2, \ldots, n\}$ such that $f(i) = i$.

A k-**cycle** is a permutation of the form

$$f(\ell_1) = \ell_2 \quad f(\ell_2) = \ell_3 \quad \ldots \quad f(\ell_{k-1}) = \ell_k \ \text{ and } \ f(\ell_k) = \ell_1$$

for distinct $\ell_1, \ldots, \ell_k$ among $\{1, \ldots, n\}$, and $f(i) = i$ for i not among the ℓ_j. There is standard notation for this cycle:

$$(\ell_1 \ \ell_2 \ \ell_3 \ \ldots \ \ell_k)$$

Note that the same cycle can be written several ways, by cyclically permuting the ℓ_j: for example, it also can be written as

$$(\ell_2 \ \ell_3 \ \ldots \ \ell_k \ \ell_1) \quad \text{or} \quad (\ell_3 \ \ell_4 \ \ldots \ \ell_k \ \ell_1 \ \ell_2)$$

191

Two cycles are **disjoint** when the respective sets of indices *properly moved* are disjoint. That is, cycles $(\ell_1\, \ell_2\, \ell_3\, \ldots\, \ell_k)$ and $(\ell'_1\, \ell'_2\, \ell'_3\, \ldots\, \ell'_{k'})$ are disjoint when the sets $\{\ell_1, \ell_2, \ldots, \ell_k\}$ and $\{\ell'_1, \ell'_2, \ldots, \ell'_{k'}\}$ are disjoint.

13.1.1 Theorem: Every permutation is uniquely expressible as a product of disjoint cycles.

Proof: Given $g \in S_n$, the cyclic subgroup $\langle g \rangle \subset S_n$ generated by g acts on the set $X = \{1, \ldots, n\}$ and decomposes X into disjoint *orbits*

$$O_x = \{g^i x : i \in \mathbb{Z}\}$$

for choices of orbit representatives $x \in X$. For each orbit representative x, let N_x be the order of g when restricted to the orbit $\langle g \rangle \cdot x$, and define a cycle

$$C_x = (x\ gx\ g^2 x\ \ldots\ g^{N_x - 1} x)$$

Since distinct orbits are disjoint, these cycles are disjoint. And, given $y \in X$, choose an orbit representative x such that $y \in \langle g \rangle \cdot x$. Then $g \cdot y = C_x \cdot y$. This proves that g is the product of the cycles C_x over orbit representatives x.   ///

13.2 *Transpositions*

The **(adjacent) transpositions** in the symmetric group S_n are the permutations s_i defined by

$$s_i(j) = \begin{cases} i+1 & \text{(for } j = i) \\ i & \text{(for } j = i+1) \\ j & \text{(otherwise)} \end{cases}$$

That is, s_i is a 2-cycle that interchanges i and $i+1$ and does nothing else.

13.2.1 Theorem: The permutation group S_n on n things $\{1, 2, \ldots, n\}$ is generated by *adjacent transpositions* s_i.

Proof: Induction on n. Given a permutation p of n things, we show that there is a product q of adjacent transpositions such that $(q \circ p)(n) = n$. Then $q \circ p$ can be viewed as a permutation in S_{n-1}, and we do induction on n. We may suppose $p(n) = i < n$, or else we already have $p(n) = n$ and we can do the induction on n.

Do induction on i to get to the situation that $(q \circ p)(n) = n$ for some product q of adjacent transposition. Suppose we have a product q of adjacent transpositions such that $(q \circ p)(n) = i < n$. For example, the empty product q gives $q \circ p = p$. Then $(s_i \circ q \circ p)(n) = i + 1$. By induction on i we're done.   ///

The **length** of an element $g \in S_n$ with respect to the generators $s_1, \ldots, s_{n-1}$ is the smallest integer ℓ such that

$$g = s_{i_1}\, s_{i_2}\, \ldots\, s_{i_{\ell-1}}\, s_{i_\ell}$$

13.3 *Worked examples*

13.3.1 Example: Classify the conjugacy classes in S_n (the *symmetric group* of bijections of $\{1,\ldots,n\}$ to itself).

Given $g \in S_n$, the cyclic subgroup $\langle g \rangle$ generated by g certainly acts on $X = \{1,\ldots,n\}$ and therefore decomposes X into *orbits*

$$O_x = \{g^i x : i \in \mathbb{Z}\}$$

for choices of orbit representatives $x_i \in X$. We claim that the (unordered!) *list of sizes* of the (disjoint!) orbits of g on X uniquely determines the conjugacy class of g, and *vice versa*. (An unordered list that allows the same thing to appear more than once is a **multiset**. It is not simply a *set*!)

To verify this, first suppose that $g = tht^{-1}$. Then $\langle g \rangle$ orbits and $\langle h \rangle$ orbits are related by

$$\langle g \rangle\text{-orbit } O_{tx} \leftrightarrow \langle h \rangle\text{-orbit } O_x$$

Indeed,

$$g^\ell \cdot (tx) = (tht^{-1})^\ell \cdot (tx) = t(h^\ell \cdot x)$$

Thus, if g and h are conjugate, the unordered lists of sizes of their orbits must be the same.

On the other hand, suppose that the unordered lists of sizes of the orbits of g and h are the same. Choose an ordering of orbits of the two such that the cardinalities match up:

$$|O_{x_i}^{(g)}| = |O_{y_i}^{(h)}| \quad (\text{for } i = 1,\ldots,m)$$

where $O_{x_i}^{(g)}$ is the $\langle g \rangle$-orbit containing x_i and $O_{y_i}^{(h)}$ is the $\langle g \rangle$-orbit containing y_i. Fix representatives as indicated for the orbits. Let p be a permutation such that, for each index i, p bijects $O_{x_i}^{(g)}$ to $O_{x_i}^{(g)}$ by

$$p(g^\ell x_i) = h^\ell y_i$$

The only slightly serious point is that this map is well-defined, since there are many exponents ℓ which may give the same element. And, indeed, it is at this point that we use the fact that the two orbits have the same cardinality: we have

$$O_{x_i}^{(g)} \leftrightarrow \langle g \rangle / \langle g \rangle_{x_i} \quad (\text{by } g^\ell \langle g \rangle_{x_i} \leftrightarrow g^\ell x_i)$$

where $\langle g \rangle_{x_i}$ is the isotropy subgroup of x_i. Since $\langle g \rangle$ is cyclic, $\langle g \rangle_{x_i}$ is necessarily $\langle g^N \rangle$ where N is the number of elements in the orbit. The same is true for h, with the same N. That is, $g^\ell x_i$ depends exactly on $\ell \bmod N$, and $h^\ell y_i$ likewise depends exactly on $\ell \bmod N$. Thus, the map p is well-defined.

Then claim that g and h are conjugate. Indeed, given $x \in X$, take $O_{x_i}^{(g)}$ containing $x = g^\ell x_i$ and $O_{y_i}^{(h)}$ containing $px = h^\ell y_i$. The fact that the exponents of g and h are the same is due to the definition of p. Then

$$p(gx) = p(g \cdot g^\ell x_i) = h^{1+\ell}\, y_i = h \cdot h^\ell\, y_i = h \cdot p(g^\ell\, x_i) = h(px)$$

Thus, for all $x \in X$

$$(p \circ g)(x) = (h \circ p)(x)$$

Therefore,

$$p \circ g = h \circ p$$

or

$$pgp^{-1} = h$$

(Yes, there are usually many different choices of p which accomplish this. And we could also have tried to say all this using the more explicit cycle notation, but it's not clear that this would have been a wise choice.)

13.3.2 Example: The **projective linear group** $PGL_n(k)$ is the group $GL_n(k)$ modulo its center k, which is the collection of scalar matrices. Prove that $PGL_2(\mathbb{F}_3)$ is isomorphic to S_4, the group of permutations of 4 things. (*Hint:* Let $PGL_2(\mathbb{F}_3)$ act on **lines** in $\mathbb{F}_3^2$, that is, on one-dimensional $\mathbb{F}_3$-subspaces in $\mathbb{F}_3^2$.)

The group $PGL_2(\mathbb{F}_3)$ acts by permutations on the set X of lines in $\mathbb{F}_3^2$, because $GL_2(\mathbb{F}_3)$ acts on non-zero vectors in $\mathbb{F}_3^2$. The scalar matrices in $GL_2(\mathbb{F}_3)$ certainly stabilize every line (since they act by scalars), so act trivially on the set X.

On the other hand, any non-scalar matrix $\begin{pmatrix} a & b \\ c & d \end{pmatrix}$ acts non-trivially on some line. Indeed, if

$$\begin{pmatrix} a & b \\ c & d \end{pmatrix} \begin{pmatrix} * \\ 0 \end{pmatrix} = \begin{pmatrix} * \\ 0 \end{pmatrix}$$

then $c = 0$. Similarly, if

$$\begin{pmatrix} a & b \\ c & d \end{pmatrix} \begin{pmatrix} 0 \\ * \end{pmatrix} = \begin{pmatrix} 0 \\ * \end{pmatrix}$$

then $b = 0$. And if

$$\begin{pmatrix} a & 0 \\ 0 & d \end{pmatrix} \begin{pmatrix} 1 \\ 1 \end{pmatrix} = \lambda \cdot \begin{pmatrix} 1 \\ 1 \end{pmatrix}$$

for some λ then $a = d$, so the matrix is scalar.

Thus, the map from $GL_2(\mathbb{F}_3)$ to permutations $\text{Aut}_{\text{set}}(X)$ of X has kernel consisting exactly of scalar matrices, so *factors through* (that is, is well defined on) the quotient $PGL_2(\mathbb{F}_3)$, and is *injective* on that quotient. (Since $PGL_2(\mathbb{F}_3)$ is the quotient of $GL_2(\mathbb{F}_3)$ by the kernel of the homomorphism to $\text{Aut}_{\text{set}}(X)$, the kernel of the mapping induced on $PGL_2(\mathbb{F}_3)$ is trivial.)

Computing the order of $PGL_2(\mathbb{F}_3)$ gives

$$|PGL_2(\mathbb{F}_3)| = |GL_2(\mathbb{F}_3)|/|\text{scalar matrices}| = \frac{(3^2-1)(3^2-3)}{3-1} = (3+1)(3^2-3) = 24$$

(The order of $GL_n(\mathbb{F}_q)$ is computed, as usual, by viewing this group as automorphisms of $\mathbb{F}_q^n$.)

This number is the same as the order of S_4, and, thus, an injective homomorphism must be surjective, hence, an isomorphism.

(One might want to verify that the center of $GL_n(\mathbb{F}_q)$ is exactly the scalar matrices, but that's not strictly necessary for this question.)

13.3.3 Example: An automorphism of a group G is **inner** if it is of the form $g \longrightarrow xgx^{-1}$ for fixed $x \in G$. Otherwise it is an **outer automorphism**. Show that every automorphism of the permutation group S_3 on 3 things is *inner*. (*Hint:* Compare the action of S_3 on the set of 2-cycles by conjugation.)

Let G be the group of automorphisms, and X the set of 2-cycles. We note that an automorphism must send order-2 elements to order-2 elements, and that the 2-cycles are exactly the order-2 elements in S_3. Further, since the 2-cycles *generate* S_3, if an automorphism is trivial on all 2-cycles it is the trivial automorphism. Thus, G *injects* to $\mathrm{Aut}_{\mathrm{set}}(X)$, which is permutations of 3 things (since there are three 2-cycles).

On the other hand, the conjugation action of S_3 on itself stabilizes X, and, thus, gives a group homomorphism $f : S_3 \longrightarrow \mathrm{Aut}_{\mathrm{set}}(X)$. The kernel of this homomorphism is trivial: if a non-trivial permutation p conjugates the two-cycle $t = (1\ 2)$ to itself, then

$$(ptp^{-1})(3) = t(3) = 3$$

so $tp^{-1}(3) = p^{-1}(3)$. That is, t fixes the image $p^{-1}(3)$, which therefore is 3. A symmetrical argument shows that $p^{-1}(i) = i$ for all i, so p is trivial. Thus, S_3 injects to permutations of X.

In summary, we have group homomorphisms

$$S_3 \longrightarrow \mathrm{Aut}_{\mathrm{group}}(S_3) \longrightarrow \mathrm{Aut}_{\mathrm{set}}(X)$$

where the map of automorphisms of S_3 to permutations of X is an isomorphism, and the composite map of S_3 to permutations of X is surjective. Thus, the map of S_3 to its own automorphism group is necessarily surjective.

13.3.4 Example: Identify the element of S_n requiring the maximal number of adjacent transpositions to express it, and prove that it is unique.

We claim that the permutation that takes $i \longrightarrow n - i + 1$ is the unique element requiring $n(n-1)/2$ elements, and that this is the maximum number.

For an ordered listing $(t_1, \ldots, t_n)$ of $\{1, \ldots, n\}$, let

$$d_o(t_1, \ldots, t_n) = \text{ number of indices } i < j \text{ such that } t_i > t_j$$

and for a permutation p let

$$d(p) = d_o(p(1), \ldots, p(n))$$

Note that if $t_i < t_j$ for all $i < j$, then the ordering is $(1, \ldots, n)$. Also, given a configuration $(t_1, \ldots, t_n)$ with *some* $t_i > t_j$ for $i < j$, necessarily this inequality holds for some *adjacent* indices (or else the opposite inequality would hold for *all* indices, by transitivity!). Thus, if the ordering is *not* the default $(1, \ldots, n)$, then there is an index i such that $t_i > t_{i+1}$. Then application of the adjacent transposition s_i of $i, i+1$ reduces by exactly 1 the value of the function $d_o()$.

Thus, for a permutation p with $d(p) = \ell$ we can find a product q of exactly ℓ adjacent transpositions such that $q \circ p = 1$. That is, we need *at most* $d(p) = \ell$ adjacent transpositions to express p. (This does not preclude *less efficient* expressions.)

On the other hand, we want to be sure that $d(p) = \ell$ is the *minimum* number of adjacent transpositions needed to express p. Indeed, application of s_i only affects the comparison of $p(i)$ and $p(i+1)$. Thus, it can decrease $d(p)$ by at most 1. That is,

$$d(s_i \circ p) \geq d(p) - 1$$

and possibly $d(s_i \circ p) = d(p)$. This shows that we do need *at least* $d(p)$ adjacent transpositions to express p.

Then the permutation w_o that sends i to $n - i + 1$ has the effect that $w_o(i) > w_o(j)$ for all $i < j$, so it has the maximum possible $d(w_o) = n(n - 1)/2$. For uniqueness, suppose $p(i) > p(j)$ for all $i < j$. Evidently, we must claim that $p = w_o$. And, indeed, the inequalities

$$p(n) < p(n - 1) < p(n - 2) < \ldots < p(2) < p(1)$$

leave no alternative (assigning distinct values in $\{1, \ldots, n\}$) but

$$p(n) = 1 < p(n - 1) = 2 < \ldots < p(2) = n - 1 < p(1) = n$$

(One might want to exercise one's technique by giving a more careful inductive proof of this.)

13.3.5 Example: Let the permutation group S_n on n things act on the polynomial ring $\mathbb{Z}[x_1, \ldots, x_n]$ by $\mathbb{Z}$-algebra homomorphisms defined by $p(x_i) = x_{p(i)}$ for $p \in S_n$. (The universal mapping property of the polynomial ring allows us to define the images of the indeterminates x_i to be whatever we want, and at the same time guarantees that this determines the $\mathbb{Z}$-algebra homomorphism completely.) Verify that this is a group homomorphism

$$S_n \longrightarrow \mathrm{Aut}_{\mathbb{Z}-\mathrm{alg}}(\mathbb{Z}[x_1, \ldots, x_n])$$

Consider

$$D = \prod_{i<j} (x_i - x_j)$$

Show that for any $p \in S_n$

$$p(D) = \sigma(p) \cdot D$$

where $\sigma(p) = \pm 1$. Infer that σ is a (non-trivial) group homomorphism, the **sign** homomorphism on S_n.

Since these polynomial algebras are *free* on the indeterminates, we check that the permutation group *acts* (in the technical sense) on the set of indeterminates. That is, we show associativity and that the identity of the group acts trivially. The latter is clear. For the former, let p, q be two permutations. Then

$$(pq)(x_i) = x_{(pq)(i)}$$

while

$$p(q(x_i)) = p(x_{q(i)} = x_{p(q(i))})$$

Since $p(q(i)) = (pq)(i)$, each $p \in S_n$ gives an automorphism of the ring of polynomials. (The endomorphisms are invertible since the group has inverses, for example.)

Any permutation merely permutes the factors of D, up to sign. Since the group *acts* in the technical sense,

$$(pq)(D) = p(q(D))$$

That is, since the automorphisms given by elements of S_n are $\mathbb{Z}$-linear,

$$\sigma(pq) \cdot D = p(\sigma(q) \cdot D) = \sigma(q)p(D) = \sigma(q) \cdot \sigma(p) \cdot D$$

Thus,

$$\sigma(pq) = \sigma(p) \cdot \sigma(q)$$

which is the homomorphism property of σ. ///

Exercises

13.1 How many distinct k-cycles are there in the symmetric group S_n?

13.2 How many elements of order 35 are there in the symmetric group S_{12}?

13.3 What is the largest order of an element of S_{12}?

13.4 How many elements of order 6 are there in the symmetric group S_{11}?

13.5 Show that the *order* of a permutation is the least common multiple of the lengths of the cycles in a disjoint cycle decomposition of it.

13.6 Let X be the set $\mathbb{Z}/31$, and let $f : X \longrightarrow X$ be the permutation $f(x) = 2 \cdot x$. Decompose this permutation into disjoint cycles.

13.7 Let X be the set $\mathbb{Z}/29$, and let $f : X \longrightarrow X$ be the permutation $f(x) = x^3$. Decompose this permutation into disjoint cycles.

13.8 Show that if a permutation is expressible as a product of an odd number of 2-cycles in *one* way, then *any* expression of it as a product of 2-cycles expresses it as a product of an odd number of 2-cycles.

13.9 Identify the lengths (expressed in terms of *adjacent transpositions*) of all the elements in S_4.

13.10 (*) Count the number of elements of S_n having at least one fixed point.

14. Naive set theory

14.1 *Sets*

Naive definition: A **set** is an *unordered collection* of things (*not* counting multiplicities), its **elements**. Write $x \in S$ or $S \ni x$ for an element x of S. Sets are described either as comma-separated *lists* (whose order is not supposed to be significant)

$$S = \{x_1, x_2, \ldots\}$$

or by a *rule*

$$S = \{x : some\ condition\ on\ x\ is\ met\}$$

The **empty set** is

$$\phi = \{\}$$

14.1.1 Theorem: There is no set S such that $x \in S$ if and only if $x \notin x$.

Proof: Suppose there were such S. Then $S \in S$ if and only if $S \notin S$, contradiction.   ///

Extension Principle *(Leibniz)* Two sets are equal if and only if they have the same elements.

14.1.2 Corollary: There is only one empty set ϕ.   ///

Idea: *Everything is a set.*

A **subset** T of S is a set such that for all elements x of T also x is an element of S. Write $T \subset S$ or $S \supset T$. A subset of S is *proper* if it is neither S itself nor ϕ. The **union** of a set F of sets is

$$\bigcup_{S \in F} S = \{x : x \in S \text{ for some } S \in F\}$$

The **intersection** is

$$\bigcap_{S \in F} S = \{x : x \in S \text{ for all } S \in F\}$$

We make an exception in the case of intersections over F for $F = \phi$, since the defining condition would be vacuous, and (supposedly) *every* set would be an element of that intersection, which is not viable. The union and intersection of a finite number of sets can also be written, respectively, as

$$S_1 \cup \ldots \cup S_n$$

$$S_1 \cap \ldots \cap S_n$$

Proto-definition: The *ordered pair* construct (x, y) with *first* component x and *second* component y should have the property that

$$(x, y) = (z, w) \iff x = z \text{ and } y = w$$

14.1.3 Remark: As sets, taking $(x, y) = \{x, y\}$ fails, since the elements of a set are not ordered. Taking $(x, y) = \{x, \{y\}\}$ fails, since it may be that $x = \{y\}$.

14.1.4 Proposition: We can construct ordered pairs as sets by defining

$$(x, y) = \{\{x\}, \{x, y\}\}$$

Proof: We must prove that $(x, y) = (z, w)$ if and only if the respective components are equal. One direction of the implication is clear. For the other implication, from

$$\{\{x\}, \{x, y\}\} = \{\{z\}, \{z, w\}\}$$

$\{x\}$ is either $\{z\}$ or $\{z, w\}$, and $\{x, y\}$ is either $\{z\}$ or $\{z, w\}$. Treat cases, using the Extension Principle.   ///

For finite n, define recursively **ordered n-tuples** by

$$(x_1, \ldots, x_{n-1}, x_n) = ((x_1, \ldots, x_{n-1}), x_n)$$

14.1.5 Remark: Subsequently we *ignore* the internal details of the construction of ordered pair, and only use its properties. This is a typical ruse.

The **Cartesian product** $X \times Y$ of two sets X and Y is the set of ordered pairs

$$X \times Y = \{(x, y) : x \in X, y \in Y\}$$

A **function** or **map** $f : X \longrightarrow Y$ from X to Y is a subset of $X \times Y$ such that for all $x \in X$ there is a unique y in Y such that $(x,y) \in f$. As usual, this is written $f(x) = y$ or $fx = y$. The **image** $f(X)$ of f is

$$f(X) = \text{ image of } f = \{f(x) : x \in X\}$$

14.1.6 Remark: This definition identifies a function with its graph, rather than by a formula or algorithm by which to *compute* the function.

14.1.7 Definition: A function $f : X \longrightarrow Y$ is **surjective** if for every $y \in Y$ there is $x \in X$ such that $f(x) = y$. It is **injective** if $f(x) = f(x')$ implies $x = x'$. If f is both surjective and injective is it **bijective**.

The **composition** $f \circ g$ of two functions $f : Y \longrightarrow Z$ and $g : X \longrightarrow Y$ is defined by

$$(f \circ g)(x) = f(g(x))$$

A **left inverse** g (if it exists) to a function $f : X \longrightarrow Y$ is a function $g : Y \longrightarrow X$ such that $g \circ f = 1_X$, where 1_X is the **identity function** on X, defined by $1_X(x) = x$ for all $x \in X$. A **right inverse** g (if it exists) to a function $f : X \longrightarrow Y$ is a function $g : Y \longrightarrow X$ such that $f \circ g = 1_Y$

Let F be a set of sets. A **choice function** f on F (if it exists) is any function

$$f : F \longrightarrow \bigcup_{S \in F} S$$

such that

$$f(S) \in S$$

for all S in F. To postulate that at least one choice function exists for any set F of sets is a non-trivial thing, and, roughly, is the **Axiom of Choice**. The collection of all choice functions on F is the **direct product** of the sets, denoted

$$\prod_{S \in F} S$$

Again, to know that this is non-empty (for F infinite) requires something!

K. Godel and P. Cohen proved that the Axiom of Choice is not only not provable from other more mundane axioms for sets, but is *independent* of them, in the sense that it is equally consistent to assume the negation of the Axiom of Choice.

A **relation** R between sets X and Y is a subset of $X \times Y$. A (binary) relation on a set X is a subset of $X \times X$. A relation R on X is
• **reflexive** if $(x,x) \in R$ for all $x \in X$
• **symmetric** if $(x,y) \in R$ implies $(y,x) \in R$ for all $x,y \in X$
• **transitive** if $(x,y) \in R$ and $(y,z) \in R$ implies $(x,z) \in R$ An **equivalence relation** is a relation that enjoys all three of these properties. For an equivalence relation R, the **equivalence class** of x is

$$\text{equivalence class of } x = \{y \in X : (x,y) \in R\}$$

14.2 *Posets, ordinals*

A **partial order** $\leq$ on a set X is a relation R on X, written $x \leq y$ if $(x,y) \in R$, such that
- *(Reflexivity)* $x \leq x$ for all $x \in X$
- If $x \leq y$ and $y \leq x$ then $x = y$
- *(Transitivity)* If $x \leq y$ and $y \leq z$ then $x \leq z$ Then X is a **partially ordered set** or **poset**. We may write $x < y$ if $x \leq y$ and $x \neq y$.

A partial ordering on X is a **total ordering** if for all $x, y \in X$ either $x \leq y$ or $y \leq x$.

A **well ordering** [sic] on a set X is a total ordering on X such that any non-empty subset Y of X has a **minimal element** (also called **least element**). That is, there is an element $y \in Y$ such that for all $z \in Y$ we have $y \leq z$.

14.2.1 Proposition: Let X be a well-ordered set. Let $f : X \longrightarrow X$ be an order-preserving injective map (so $x \leq x'$ implies $f(x) \leq f(x')$). Then

$$f(x) \geq x$$

for all $x \in X$.

Proof: Let Z be the subset of X consisting of elements x such that $f(x) < x$. If Z is non-empty, then it has a least element x. Thus, on one hand, $f(x) < x$. On the other hand, $f(x) \notin Z$, so $f(f(x)) > f(x)$. But, since f preserves order and is injective, $f(x) < x$ implies $f(f(x)) < f(x)$, contradiction. ///

14.2.2 Corollary: The only order-preserving bijection of a well-ordered set X to itself is the identity map. ///

14.2.3 Corollary: There is no order-preserving bijection of a well-ordered set X to a proper **initial segment**
$$X^{<x} = \{y \in X : y < x\}$$
of it for any $x \in X$. ///

14.2.4 Example: The set
$$Z = \{X^{<x} = \{y \in X : y < x\} : x \in X\}$$
of initial segments $X^{<x}$ of a well-ordered set X, with ordering
$$z \leq w \Longleftrightarrow z \subset w$$
has an order-preserving bijection to X by
$$X^{<x} \longleftrightarrow x$$

An **ordinal** is a well-ordered set X such for every element $x \in X$
$$x = X^{<x}$$
That is, x is the set $X^{<x} = \{y \in X : y < x\}$ of its predecessors in X.

14.2.5 Example: The empty set is an ordinal, since the defining condition is met vacuously. Let X be an ordinal that is not the empty set. Then X (being non-empty) has a least element x. Since x is the union of its predecessors, of which there are none, $x = \phi$. So ϕ is the least element of every ordinal.

14.2.6 Example: If X is an ordinal, and $x \in X$, then the **initial segment** below x

$$X^{<x} = \{y \in X : y < x\}$$

is also an ordinal. Indeed, the well-ordering is preserved, and by transitivity the predecessors of y in $X^{<x}$ are exactly the predecessors of y in X, so the defining property of ordinals holds.

14.2.7 Example: If X is an ordinal, then $Y = X \cup \{X\}$, with ordering

$$a \leq b \Longleftrightarrow a \subset b$$

is an ordinal, the **successor** of X. To see this, first note that, for all $y \in Y$ we have $y \subset X$, that is (by definition of the ordering) $y \leq X$. Thus, for $y \in Y$, if $y \neq X$, then $y \subset X$ and (since X is an ordinal) is the set of its predecessors *in* X. And since $y < X$ in Y, X is not among y's predecessors in Y, so y really is the set of its predecessors in Y. And X is the set of its predecessors in Y. ///

Since everything is to be a set, following J. von Neumann, define the initial (finite) ordinals by

$$0 = \phi = \{\}$$

$$1 = \{0\} = \{\phi, \{\phi\}\} = \{\{\}, \{\{\}\}\}$$

$$2 = \{0, 1\} = \{\phi, \{\phi\}, \{\phi, \{\phi\}\}\} = \{\{\}, \{\{\}\}, \{\{\}, \{\{\}\}\}\}$$

$$3 = \{0, 1, 2\} = \{\phi, \{\phi\}, \{\phi, \{\phi\}\}, \{\phi, \{\phi\}, \{\phi, \{\phi\}\}\}\}$$

$$= \{\{\}, \{\{\}\}, \{\{\}, \{\{\}\}\}, \{\{\}, \{\{\}\}, \{\{\}, \{\{\}\}\}\}\}$$

and so on

The set ω of **natural numbers** is [1]

$$\omega = \{0, 1, 2, \ldots\}$$

Define an order $\leq$ on ω by

$$x \leq y \Longleftrightarrow x \subset y$$

It is not at all immediate (with the present definition of the symbols) that ω is an ordinal.

14.2.8 Proposition: If X and Y are ordinals and $Y \subset X$ then there is $x \in X$ such that Y is the initial segment

$$Y = \{y \in X : y < x\} = x$$

Proof: Let Z be the set of elements of X that are not in Y but are below some element of Y. The claim is that Z is empty. If not, let z be the least element in Z. Let $y \in Y$ be such that $z < y$. Since y is the set of its predecessors in X, $x \in y$. But also y is the set of its predecessors in Y, so $x \in y$, contradiction. ///

14.2.9 Theorem: Any two ordinals X, Y are *comparable*, in the sense that either $X = Y$, or X is an initial segment of Y, or Y is an initial segment of X.

---

[1] Shuddering at the casual formation of this ostensibly infinite set is reasonable, since its existence as a set is not formally assured by the existence of the separate finite ordinals.

Proof: The intersection $X \cap Y$ is an ordinal, since for $z \in X \cap Y$

$$\{w \in X \cap Y : w < z\} = \{x \in X : x < z\} \cap \{y \in Y : y < z\} = z \cap z = z$$

Suppose that X is not contained in Y, and Y is not contained in X. From above, $X \cap Y \subset$ is an initial segment

$$X \cap Y = \{z \in X : z < x\} = x$$

in X for some $x \in X$, and also an initial segment

$$X \cap Y = \{w \in Y : z < y\} = y$$

in Y for some $y \in Y$. But then $x = y$, contradiction.   ///

14.2.10 Corollary: Two ordinals admit an order-preserving bijection between them if and only if they are identical, and in that case the only order-preserving bijection is the identity map.

Proof: We already saw that there is at most one order-preserving bijection between two well-ordered sets. Thus, let X and Y be ordinals, and $X \neq Y$. By the theorem, one is an initial segment of the other, so assume without loss of generality that Y is an initial segment

$$Y = \{y \in X : y < x\}$$

for some x in X. Let $f : X \longrightarrow Y$ be an order-preserving bijection. We saw earlier that $f(z) \geq z$ for any well-ordered sets in this situation. But then $f(x) \geq x$, which is impossible.
                                                                                      ///

14.2.11 Corollary: The relation on ordinals defined by $x < y$ if and only if x is an initial segment of y is a total ordering.   ///

14.2.12 Corollary: Given an ordinal x, its successor ordinal $y = x \cup \{x\}$ has the property that $x < y$.   ///

14.2.13 Corollary: There is no largest ordinal.   ///

14.2.14 Theorem: The union of any set of ordinals is an ordinal.

Proof: Let F be a set of ordinals, and

$$E = \bigcup_{X \in F} X$$

is also a set of ordinals. Define a relation $<$ on E by $x < y$ if x is an initial segment in y, that is, is an element of y. The transitivity of $<$ follows (again) from the fact that every element of an ordinal is an ordinal. The comparability of all ordinals (from above) says that this is a *total* ordering. To prove that $<$ is a well-ordering, let D be a non-empty subset of E, and let d be any element of D. If d is least in D, we are done. If d is not least in D, then nevertheless $c \in D$ with $c < d$ are elements of d, since $c < d$ only for c an initial segment of d, that is an element of d. Since d is an ordinal, it is well-ordered, so

$$\{c \in D : c < d\} = D \cap d$$

is well-ordered. Thus, D contains a least element. Finally, we must prove that any element e of E is the set of its predecessors in E. Let X be an element of F such that $e \in X$. Since

X is an ordinal, e is the set of its predecessors d in X. Thus, all such predecessors d are elements of X, so are elements of the union E. Thus,

$$e = \{d \in X : d < e\} \subset \{d \in E : d < e\}$$

On the other hand, for any $d \in E$, the definition of $d < e$ is that d is an initial segment of e, that is, that $d \in e$. In that case, $d \in X$ for *every* ordinal containing e. That is, we have the opposite inclusion

$$e = \{d \in X : d < e\} \supset \{d \in E : d < e\}$$

and e is exactly the set of its predecessors in the union E. ///

14.2.15 Theorem: Every well-ordered set has an order-preserving bijection to exactly one ordinal.

Proof: First, let X be a well-ordered set with each initial segment

$$X^{<x} = \{y \in X : y < x\}$$

for $x \in X$ isomorphic [2] to an ordinal ω_x. We claim that X is isomorphic to an ordinal. From above, since no two distinct ordinals are isomorphic, and since an ordinal admits no non-trivial maps to itself, for each $x \in X$ the ordinal ω_x is uniquely determined and the order-preserving map $f_x : x \longrightarrow \omega_x$ is unique. We view $F : x \longrightarrow \omega_x$ as an ordinal-valued function F on X.

Consider $x < y$ in X. Since x and y are distinct initial segments of X, they are not isomorphic as ordered sets (indeed, there is no order-preserving injection of y to x). Thus, $F(x) = \omega_x$ is not isomorphic to $F(y) = \omega_y$. Thus, since any two ordinals can be compared, either $F(x) = \omega_x$ is an initial segment of $F(y) = \omega_y$ or *vice versa*. Unsurprisingly, if $\omega_y < \omega_x$ then

$$y \approx \omega_y \subset \omega_x \approx x$$

would give an isomorphism of y to a *proper* initial segment x, but (again) this is impossible. Thus, F is an order-preserving bijection of X to a set $\Omega = \{\omega_x = F(x) : x \in X\}$ of ordinals. Since $\Omega = F(X)$ is the image of the well-ordered set X, Ω is well-ordered. To show that Ω is an ordinal, by definition, we must show that for $\omega \in \Omega$ the initial segment

$$\Omega^{<\omega} = \{\omega' \in \Omega : \omega' < \omega\}$$

is equal to ω. Indeed, the hypothesis is exactly this, so Ω is an ordinal, and X is an ordinal (being isomorphic to Ω).

Now we prove the theorem. First we prove that every element of a (non-empty) well-ordered set X is isomorphic to an ordinal. The least element of X is isomorphic to the ordinal ϕ. Given α in X with β isomorphic to an ordinal for all $\beta < \alpha$, then apply the claim to α (in place of X) to conclude that α is isomorphic to an ordinal. And then the claim implies that X is isomorphic to an ordinal. Since two distinct ordinals are not isomorphic, there is exactly one ordinal to which X is isomorphic. ///

The following corollary is sometimes recast as a paradox:

---

[2] As ordered set, of course.

14.2.16 Corollary: *(Burali-Forti)* The collection of all ordinals is not a set.

Proof: Suppose the collection F of all ordinals were a set. Then (by the theorem) the union

$$E = \bigcup_{S \in F} S$$

would be an ordinal. Thus, E would be an element of itself, contradiction.   ////

14.3 *Transfinite induction*

14.3.1 Theorem: Let $P(\alpha)$ be a property that may or may not hold of ordinals α. Suppose that for any ordinal α if $P(\beta)$ for all ordinals $\beta < \alpha$ then $P(\alpha)$ holds. The $P(\alpha)$ holds for *all* ordinals α.

Proof: Let $\omega = \alpha \cup \{\alpha\}$, so ω is an ordinal containing α. Then we can do induction on the *set* ω: prove that $P(\beta)$ holds for all $\beta \in \omega$ (including α). If $P(\gamma)$ *failed* for some γ in ω, then there would be a *least* γ in ω for which it failed. But $P(\delta)$ holds for all $\delta < \gamma$, and the hypothesis assures that $P(\gamma)$ *does* hold, after all. This contradiction shows that $P(\gamma)$ holds for all $\gamma \in \omega$, in particular, for α.   ////

In some situations the induction step, namely, proving that $P(\alpha)$ holds if $P(\beta)$ holds for all $\beta < \alpha$, must be broken into cases, depending on the nature of α.
• The **initial ordinal**, ϕ.
• **Successor ordinals** $\alpha = \beta \cup \{\beta\}$ for some β.
• **Limit ordinals** $\alpha = \bigcup_{\beta < \alpha} \beta$.

14.3.2 Remark: First, contrast the definition of *limit ordinal* with the property enjoyed by *every* ordinal, namely

$$\alpha = \{\beta : \beta \in \alpha\} = \{\beta : \beta < \alpha\}$$

A successor ordinal α is not a limit ordinal, since if $\alpha = \beta \cup \{\beta\}$ then all predecessors of α are subsets of β, and likewise their union, which cannot contain β *as an element*.

14.3.3 Proposition: Every ordinal is either the initial ordinal ϕ, a successor ordinal, or a limit ordinal.

Proof: Suppose α is not ϕ and is not a successor. Let β be the union of the predecessors of α. Since a union of ordinals is an ordinal, β is an ordinal, and $\beta \le \alpha$. If $\beta < \alpha$ then β is among α's predecessors, so is in the union of predecessors, so is the largest among the predecessors of α. The assumption $\beta < \alpha$ gives $\beta \cup \{\beta\} \le \alpha$. It cannot be that $\beta \cup \{\beta\} \le \alpha$ since otherwise $\beta \cup \{\beta\}$ would be a predecessor of α, and thus $\beta \ge \beta \cup \{\beta\}$, which is false. So, then, the successor $\beta \cup \{\beta\}$ of β is α, contradiction to the hypothesis that α is not a successor. Thus, $\beta = \alpha$.   ////

Thus, we can rewrite the first theorem in a manner that refers explicitly to the types of ordinals: to prove a property $P(\alpha)$ holds for all ordinals α:
• Prove $P(\phi)$ holds.
• Prove (for all α) that if $P(\alpha)$ holds then $P(\alpha \cup \{\alpha\})$ holds.
• Prove for every limit ordinal λ that if $P(\alpha)$ holds for all $\alpha < \lambda$ then $P(\lambda)$ holds.

14.4 *Finiteness, infiniteness*

A set S is **Peano finite** if there is some $n \in \omega$ such that there is a bijection of S to

$$n = \{0, 1, 2, \ldots, n-1\}$$

The set is **Peano infinite** if it is *not* Peano finite.

A set S is **Dedekind infinite** if there is an injection from S to a *proper* subset of S. It is Dedekind *finite* if it is *not* Dedekind infinite.

14.4.1 Theorem: (Granting the Axiom of Choice) The two notions of *infinite* are the same.

14.4.2 Remark: To avoid circularity, we should not presume *arithmetic* at this point.

Proof: Let $f : S \longrightarrow S$ be an injection of S to a proper subset of itself. Choose $s_1 \in S$ but not lying in the image $f(S)$. Claim $f(f(S))$ is a proper subset of $f(S)$. Indeed, $f(s_1)$ cannot be in $f(f(S))$, or there would be $t \in f(S)$ such that $f(t) = f(s_1)$, and then by injectivity of f we would have $t = s_1$, contradicting the fact that $s_1 \notin f(S)$. Certainly f restricted to $f(S)$ is still injective.

Thus, $f(f(f(S)))$ is strictly smaller than $f(f(S))$ By induction, we can find $s_1, s_2, \ldots$ such that $s_1 \notin f(S)$, $s_2 \in f(S)$ but $s_2 \notin f(f(S))$, $s_3 \in f(f(S))$ but $s_3 \notin f(f(f(S)))$, etc. In particular, all these s_i are distinct, so we have an injection

$$\{1, 2, 3, \ldots\} \longrightarrow S$$

Thus, Dedekind infinite implies Peano infinite. ///

14.5 *Comparison of infinities*

The **Cantor-Schroeder-Bernstein Theorem** proven here is the key result that allows comparison of infinities. Perhaps it is the first serious theorem in set theory after Cantor's diagonalization argument. Apparently Cantor *conjectured* this result, and it was proven independently by F. Bernstein and E. Schröder in the 1890s. The proof given below is a *natural* proof that one might find after sufficient experimentation and reflection.

It is noteworthy that there is no invocation of the Axiom of Choice, since one can imagine that it would have been needed.

The argument below is not the most succinct possible, but is intended to lend a greater sense of inevitability to the conclusion than would the shortest possible version.

14.5.1 Theorem: (*Cantor-Schroeder-Bernstein*) Let A and B be sets, with injections $f : A \longrightarrow B$ and $g : B \longrightarrow A$. Then there exists a canonical *bijection* $F : A \longrightarrow B$.

Proof: Let

$$A_o = \{a \in A : a \notin g(B)\} \qquad B_o = \{b \in B : b \notin f(A)\}$$

The sets

$$A_{2n} = (g \circ f)^n(A_o) \qquad A_{2n+1} = (g \circ f)^n g(B_o)$$

are disjoint. Let A_∞ be the complement in A to the union $\bigcup_n A_n$. Define F by

$$F(a) = \begin{cases} f(a) & \text{(for } a \in A_n, \ n \in 2\mathbb{Z}) \\ g^{-1}(a) & \text{(for } a \in A_n, \ n \in 1 + 2\mathbb{Z}) \\ f(a) & \text{(for } a \in A_\infty) \end{cases}$$

We must verify that this moderately clever apparent definition really gives a well-defined F, and that F is a bijection. For $n \geq 1$, let

$$B_n = f(A_{n-1})$$

and also let $B_\infty = f(A_\infty)$.

The underlying fact is that $A \cup B$ (disjoint union) is *partitioned* into one-sided or two-sided maximal sequences of elements that map to each other under f and g: we have three patterns. First, one may have

$$a_o \xrightarrow{f} b_1 \xrightarrow{g} a_1 \xrightarrow{f} b_2 \xrightarrow{g} a_2 \longrightarrow \ldots \xrightarrow{f} b_n \xrightarrow{g} a_n \longrightarrow \ldots$$

beginning with $a_o \in A_o$, all $a_i \in A$ and $b_i \in B$. Second, one may have

$$b_o \xrightarrow{g} a_1 \xrightarrow{f} b_1 \xrightarrow{g} a_2 \xrightarrow{f} b_2 \longrightarrow \ldots \xrightarrow{g} a_n \xrightarrow{f} b_n \longrightarrow \ldots$$

with $b_o \in B_o$, and $a_i \in A$ and $b_i \in B$. The third and last possibility is that none of the elements involved is an image of A_o or B_o under any number of iterations of $f \circ g$ or $g \circ f$. Such elements fit into pictures of the form

$$\ldots \xrightarrow{g} a_{-2} \xrightarrow{f} b_{-1} \xrightarrow{g} a_{-1} \xrightarrow{f} b_o \xrightarrow{g} a_o \xrightarrow{f} b_1 \xrightarrow{g} \ldots$$

where $a_i \in A$ and $b_i \in B$. The fundamental point is that any two distinct such sequences of elements are disjoint. And any element certainly lies in such a sequence.

The one-sided sequences of the form

$$a_o \xrightarrow{f} b_1 \xrightarrow{g} a_1 \xrightarrow{f} b_2 \xrightarrow{g} a_2 \longrightarrow \ldots \xrightarrow{f} b_n \xrightarrow{g} a_n \longrightarrow \ldots$$

beginning with $a_o \in A_o$, can be broken up to give part of the definition of F by

$$F : a_o \xrightarrow{f} b_1 \quad F : a_1 \xrightarrow{f} b_2 \ldots$$

The one-sided sequences of the form

$$b_o \xrightarrow{g} a_1 \xrightarrow{f} b_1 \xrightarrow{g} a_2 \xrightarrow{f} b_2 \longrightarrow \ldots \xrightarrow{g} a_n \xrightarrow{f} b_n \longrightarrow \ldots$$

with $b_o \in B_o$, beginning with $b_o \in B_o$, can be broken up to give another part of the definition of F

$$b_o \xrightarrow{g} a_1 \quad b_1 \xrightarrow{g} a_2 \ldots$$

which is to say

$$F : a_1 \xrightarrow{g^{-1}} b_o \quad F : a_2 \xrightarrow{g^{-1}} b_1 \ldots$$

For a double-sided sequence,

$$\ldots \xrightarrow{g} a_{-2} \xrightarrow{f} b_{-1} \xrightarrow{g} a_{-1} \xrightarrow{f} b_o \xrightarrow{g} a_o \xrightarrow{f} b_1 \xrightarrow{g} \ldots$$

there are two equally simple ways to break it up, and we choose

$$F : a_i \xrightarrow{f} b_{i+1}$$

Since the sequences partition $A \cup B$, and since every element of B (and A) appears, F is surely a bijection from A to B. ///

14.6 *Example: transfinite Lagrange replacement*

Let V be a vector space over a field k. Let $E = \{e_\alpha : \alpha \in A\}$ be a set of *linearly independent* elements, and $F = \{f_\beta : \beta \in B\}$ be a *basis* for V.

14.6.1 Theorem: We have an inequality of cardinalities: $|A| \leq |B|$.

Proof: Well order [3] A. We prove by transfinite induction that there is an injection $j : A \longrightarrow B$ such that

$$\{e_\alpha : \alpha \in A\} \cup \{f_\beta : \beta \in B, \ \beta \notin j(A)\}$$

is a basis for V. That is, we can *exchange* (following Lagrange) every element in E for a basis element in F and still have a basis. Thus, since E injects to F we have an inequality of cardinalities.

Fix $\alpha \in A$. Let

$$A^{<\alpha} = \{\gamma \in A : \gamma < \alpha\}$$

For the induction step, suppose that we have an injection

$$j : A^{<\alpha} \longrightarrow B$$

such that

$$\{e_\gamma : \gamma < \alpha\} \cup \{f_\beta : \beta \notin j(A^{<\alpha})\}$$

is a disjoint union, and is still a basis for V. Then, since these elements *span* V, there exist elements a_γ and b_β in the field such that

$$e_\alpha = \sum_{\gamma < \alpha} a_\gamma \cdot e_\gamma + \sum_{\beta \notin j(A^{<\alpha})} b_\beta \cdot f_\beta$$

Since the e's were linearly independent, not all the b_βs can be 0. Pick $\beta \notin j(A^{<\alpha})$ such that $b_\beta \neq 0$, and extend j by defining $j(\alpha) = \beta$.

We must check that

$$\{e_\gamma : \gamma \leq \alpha\} \cup \{f_\beta : \beta \notin j(A^{\leq\alpha})\}$$

is still a basis (and that the union is disjoint). For linear independence, since

$$\{e_\gamma : \gamma < \alpha\} \cup \{f_\delta : \delta \notin j(A^{<\alpha})\}$$

is a basis, any linear relation must properly involve e_α, as

$$e_\alpha = \sum_{\gamma < \alpha} c_\gamma e_\gamma + \sum_{\delta \notin j(A^{\leq\alpha})} d_\delta f_\delta$$

---

[3] To well-order a set is, in effect, an invocation of the Axiom of Choice, and should not be taken lightly, even if it is useful or necessary. See the last section in this chapter.

Replace e_α by its expression

$$e_\alpha = \sum_{\gamma < \alpha} a_\gamma \cdot e_\gamma + \sum_{\delta \notin j(A^{<\alpha})} b_\delta \cdot f_\delta$$

to obtain

$$\sum_{\gamma < \alpha} a_\gamma \cdot e_\gamma + \sum_{\delta \notin j(A^{\leq\alpha})} b_\delta \cdot f_\delta + b_\beta f_\beta = \sum_{\gamma < \alpha} c_\gamma e_\gamma + \sum_{\delta \notin j(A^{\leq\alpha})} d_\delta f_\delta$$

But $b_\beta \neq 0$, f_β occurs only on the left-hand side, and the vectors involved in this sum are a basis, so this is impossible. This proves the linear independence (and disjointness of the union).

To prove the spanning property, use the fact that

$$\{e_\gamma : \gamma < \alpha\} \cup \{f_\beta : \beta \notin j(A^{<\alpha})\}$$

is a basis. That is, given $v \in V$, there are field elements x_γ and y_δ such that

$$v = \sum_{\gamma < \alpha} x_\gamma e_\gamma + + \sum_{\delta \notin j(A^{<\alpha})} y_\delta f_\delta$$

Since $b_\beta \neq 0$ above, we can express f_β in terms of e_α, by

$$f_\beta = b_\beta^{-1} e_\alpha - \sum_{\gamma < \alpha} a_\gamma \cdot e_\gamma + \sum_{\delta \notin j(A^{\leq\alpha})} b_\delta \cdot f_\delta$$

Thus, we can replace f_β by this expression to express v as a linear combination of

$$\{e_\gamma : \gamma \leq \alpha\} \cup \{f_\beta : \beta \notin j(A^{\leq\alpha})\}$$

proving the spanning. By transfinite induction there exists an injection of A to B.   ///

14.6.2 Remark: We could make the invocation of Well-Ordering more explicit: if there were *no* injection $A \longrightarrow B$ as indicated, by Well-Ordering let α be the *first* element in A such that there is no such injection on $A^{<\alpha}$. Then the same discussion yields a contradiction.

We use the *Axiom of Choice* in the guise of the *Well-Ordering Principle*: we *assume* that any set can be *well-ordered*. From the theory of ordinals and well-orderings any well-ordered set is isomorphic (as well-ordered set) to a unique ordinal. From the theory of ordinals, any two ordinals are comparable, in the sense that one is an *initial segment* of the other. Thus, putting these things together, any two sets A, B are comparable in size, in the sense that either A injects to B, or B injects to A.

14.7 *Equivalents of the Axiom of Choice*

There are several statements which are all logically equivalent to each other, and often used to prove *existence* when only existence is required, and no object must be explicitly exhibited. These are **Zorn's Lemma, Hausdorff Maximality Principle, Well-Ordering Principle**, and **Axiom of Choice**. Here we describe these assertions in the context of *naive* set theory, in the style of the discussion above, rather than *formal* or *axiomatic* set theory. [4]

The *Axiom of Choice* or *Zermelo's postulate* asserts that, given a set of sets

$$\{S_i \, : \, i \in I\}$$

with (not necessarily mutually disjoint) non-empty sets S_i (indexed by a *set I*), there exists a set of *choices* s_i, one from each S_i. That is, there *exists* a choice set

$$C \; = \; \{s_i \, : \, i \in I\} \qquad \text{with } s_i \in S_i \text{ for all indices } i \in I$$

This is intuitively obvious for *finite* sets I, but less obviously clear for *infinite* sets of sets. Sometimes this is stated in the form that there is a *choice function* f on the index set I such that $f(i) \in S_i$.

The *Well-ordering Principle* asserts that every set can be well-ordered. More precisely, the assertion is that, given a set S, there is a bijection of S to an *ordinal*.

To state *Zorn's lemma* some preparation is needed. In a poset X, a *chain* is a *totally* ordered subset. An *upper bound* for a totally ordered subset Y of a poset X is an element $b \in X$ (not necessarily in Y) such that $y \le b$ for all $y \in Y$. A *maximal* element $m \in X$ is an element of X such that, for all $x \in X$, $m \le x$ implies $m = x$. Then Zorn's lemma asserts that *every poset in which every chain has an upper bound contains at least one maximal element.*

The *Hausdorf maximality principle* asserts that in any poset, every totally ordered subset is contained in a *maximal* totally ordered subset. Here a *maximal* totally ordered subset is what it sounds like, namely, a totally ordered subset such that any strictly larger subset fails to be totally ordered. A seemingly weaker, but equivalent, form is the assertion that *every poset contains a maximal totally ordered subset.*

We give a representative proof.

Proof: (Axiom of Choice implies the Well-ordering Principle.) Fix a set X. Let c be a choice function on the set of subsets of X. Try to define a function f on ordinals α by transfinite induction, by

$$f(\alpha) \; = \; c\big(X - \{f(\beta) : \text{ordinals } \beta < \alpha\}\big)$$

where for two sets X, A

$$X - A \; = \; \{x : x \in X, \; x \notin A\}$$

This definition *fails* if-and-when

$$X - \{f(\beta) : \text{ordinals } \beta < \alpha\}) \; = \; \phi$$

---

[4] In the late nineteenth and early twentieth centuries, it was unclear whether or not one could expect to prove these assertions from first principles. Further, some mathematicians felt that one or more of these assertions was *obviously* true, while others felt uneasy to varying degrees about invocation of them. In the early 1930's Kurt Gödel proved that the Axiom of Choice is *consistent* (in the Zermelo-Frankel first-order axiomatization) with the other axioms of set theory. In 1963, Paul Cohen proved that the Axiom of Choice was *independent* of the other axioms. In fact, Gödel also proved that the *Continuum Hypothesis* is consistent. This is the hypothesis that there are no cardinals between the countable and the cardinality of the reals. Cohen also proved that the Continuum Hypothesis is independent.

Let us show that each function so defined (as long as we have not run out of elements of X to hit) is *injective*. Indeed, for ordinals $\alpha > \beta$, consider the definition

$$f(\alpha) \;=\; c\big(X - \{f(\gamma) : \text{ordinals } \gamma < \alpha\}\big)$$

The set of values removed from X to choose a value for $f(\alpha)$ includes $f(\beta)$, so necessarily $f(\alpha) \neq f(\beta)$. If at any point

$$X - \{f(\beta) : \text{ordinals } \beta < \alpha\}) \;=\; \phi$$

then f gives a surjection from $\{\beta) : \beta < \alpha\}$ to X, which we have just shown is injective, giving a well-ordering of X. Thus, it suffices to show that it is *impossible* that

$$X - \{f(\beta) : \text{ordinals } \beta < \alpha\}) \;\neq\; \phi$$

for all ordinals α. Indeed, if this were so, then the transfinite induction proceeds uninterrupted, and we have an injective map f from *all* ordinals to X. But the collection of all ordinals is a *class*, not a set, so cannot be injected to any set, contradiction. That is, at some point the transfinite induction fails, and we have the desired well-ordering.   ///

Exercises

14.1 Show that the Well-Ordering Principle implies the Axiom of Choice.

14.2 Show that an arbitrary poset is isomorphic, as a poset, to a set of *sets*, partially ordered by set inclusion.

15. Symmetric polynomials

15.1 *The theorem*

Let S_n be the group of permutations of $\{1, \ldots, n\}$, also called the **symmetric group** on n things.

For indeterminates x_i, let $p \in S_n$ act on $\mathbb{Z}[x_1, \ldots, x_n]$ by

$$p(x_i) = x_{p(i)}$$

A polynomial $f(x_1, \ldots, x_n) \in \mathbb{Z}[x_1, \ldots, x_n]$ is **invariant** under S_n if for all $p \in S_n$

$$f(p(x_1), \ldots, p(x_n)) = f(x_1, \ldots, x_n)$$

The **elementary symmetric polynomials** in $x_1, \ldots, x_n$ are

$$
\begin{aligned}
s_1 &= s_1(x_1, \ldots, x_n) &&= \sum_i x_i \\
s_2 &= s_2(x_1, \ldots, x_n) &&= \sum_{i<j} x_i x_j \\
s_3 &= s_3(x_1, \ldots, x_n) &&= \sum_{i<j<k} x_i x_j x_k \\
s_4 &= s_4(x_1, \ldots, x_n) &&= \sum_{i<j<k<\ell} x_i x_j x_k x_\ell \\
&\quad \cdots \\
s_t &= s_t(x_1, \ldots, x_n) &&= \sum_{i_1<i_2<\ldots<i_t} x_{i_1} x_{i_2} \ldots x_{i_t} \\
&\quad \cdots \\
s_n &= s_n(x_1, \ldots, x_n) &&= x_1 x_2 x_3 \ldots x_n
\end{aligned}
$$

15.1.1 Theorem: A polynomial $f(x_1, \ldots, x_n) \in \mathbb{Z}[x_1, \ldots, x_n]$ is invariant under S_n if and only if it is a polynomial in the *elementary* symmetric functions $s_1, \ldots, s_n$.

15.1.2 Remark: In fact, the proof shows an algorithm which determines the expression for a given S_n-invariant polynomial in terms of the elementary ones.

Proof: Let $f(x_1, \ldots, x_n)$ be S_n-invariant. Let

$$q : \mathbb{Z}[x_1, \ldots, x_{n-1}, x_n] \longrightarrow \mathbb{Z}[x_1, \ldots, x_{n-1}]$$

be the map which kills off x_n, that is

$$q(x_i) = \begin{cases} x_i & (1 \le i < n) \\ 0 & (i = n) \end{cases}$$

If $f(x_1, \ldots, x_n)$ is S_n-invariant, then

$$q(f(x_1, \ldots, x_{n-1}, x_n)) = f(x_1, \ldots, x_{n-1}, 0)$$

is S_{n-1}-invariant, where we take the copy of S_{n-1} inside S_n that fixes n. And note that

$$q(s_i(x_1, \ldots, x_n)) = \begin{cases} s_i(x_1, \ldots, x_{n-1}) & (1 \le i < n) \\ 0 & (i = n) \end{cases}$$

By induction on the number of variables, there is a polynomial P in $n - 1$ variables such that
$$q(f(x_1, \ldots, x_n)) = P(s_1(x_1, \ldots, x_{n-1}), \ldots, s_{n-1}(x_1, \ldots, x_{n-1}))$$

Now use the same polynomial P but with the elementary symmetric functions augmented by insertion of x_n, by

$$g(x_1, \ldots, x_n) = P(s_1(x_1, \ldots, x_n), \ldots, s_{n-1}(x_1, \ldots, x_n))$$

By the way P was chosen,

$$q(f(x_1, \ldots, x_n) - g(x_1, \ldots, x_n)) = 0$$

That is, mapping $x_n \longrightarrow 0$ sends the difference $f - g$ to 0. Using the unique factorization in $\mathbb{Z}[x_1, \ldots, x_n]$, this implies that x_n divides $f - g$. The S_n-invariance of $f - g$ implies that every x_i divides $f - g$. That is, by unique factorization, $s_n(x_1, \ldots, x_n)$ divides $f - g$.

The **total degree** of a monomial $c\,x_1^{e_1} \ldots x_n^{e_n}$ is the sum of the exponents

$$\text{total degree } (c\,x_1^{e_1} \ldots x_n^{e_n}) = e_1 + \ldots + e_n$$

The total degree of a polynomial is the maximum of the total degrees of its monomial summands.

Consider the polynomial

$$\frac{f - g}{s_n} = \frac{f(x_1, \ldots, x_n) - g(x_1, \ldots, x_n)}{s_n(x_1, \ldots, x_n)}$$

It is of lower total degree than the original f. By induction on total degree $(f - g)/s_n$ is expressible in terms of the elementary symmetric polynomials in $x_1, \ldots, x_n$. ///

15.1.3 Remark: The proof also shows that if the total degree of an S_n-invariant polynomial $f(x_1, \ldots, x_{n-1}, x_n)$ in n variables is less than or equal the number of variables, then the expression for $f(x_1, \ldots, x_{n-1}, 0)$ in terms of $s_i(x_1, \ldots, x_{n-1})$ gives the correct formula in terms of $s_i(x_1, \ldots, x_{n-1}, x_n)$.

15.2 *First examples*

15.2.1 Example: Consider

$$f(x_1, \ldots, x_n) = x_1^2 + \ldots + x_n^2$$

The induction on n and the previous remark indicate that the general formula will be found if we find the formula for $n = 2$, since the total degree is 2. Let $q : \mathbb{Z}[x, y] \longrightarrow \mathbb{Z}[x]$ be the $\mathbb{Z}$-algebra map sending $x \longrightarrow x$ and $y \longrightarrow 0$. Then

$$q(x^2 + y^2) = x^2 = s_1(x)^2$$

Then, following the procedure of the proof of the theorem,

$$(x^2 + y^2) - s_1(x, y)^2 = (x^2 + y^2) - (x + y)^2 = -2xy$$

Dividing by $s_2(x, y) = xy$ we obtain -2. (This is visible, anyway.) Thus,

$$x^2 + y^2 = s_1(x, y)^2 - 2s_2(x, y)$$

The induction on the number of variables gives

$$x_1^2 + \ldots + x_n^2 = s_1(x_1, \ldots, x_n)^2 - s_2(x_1, \ldots, x_n)$$

15.2.2 Example: Consider

$$f(x_1, \ldots, x_n) = \sum_i x_i^4$$

Since the total degree is 4, as in the remark just above it suffices to determine the pattern with just 4 variables x_1, x_2, x_3, x_4. Indeed, we start with just 2 variables. Following the procedure indicated in the theorem, letting q be the $\mathbb{Z}$-algebra homomorphism which sends y to 0,

$$q(x^4 + y^4) = x^4 = s_1(x)^4$$

so consider

$$(x^4 + y^4) - s_1(x, y)^4 = -4x^3 y - 6x^2 y^2 - 4xy^3 = -s_1(x, y) \cdot (4x^2 + 6xy + 4y^2)$$

The latter factor of lower total degree is analyzed in the same fashion:

$$q(4x^2 + 6xy + 4y^2) = 4x^2 = 4s_1(x)^2$$

so consider
$$(4x^2 + 6xy + 4y^2) - 4s_1(x,y)^2 = -2xy$$

Going backward,
$$x^4 + y^4 = s_1(x,y)^4 - s_1(x,y) \cdot (4s_1(x,y)^2 - 2s_2(x,y))$$

Passing to three variables,
$$q(x^4 + y^4 + z^4) = x^4 + y^4 = s_1(x,y)^4 - s_1(x,y) \cdot (4s_1(x,y)^2 - 2s_2(x,y))$$

so consider
$$(x^4 + y^4 + z^4) - \left(s_1(x,y,z)^4 - s_1(x,y,z) \cdot (4s_1(x,y,z)^2 - 2s_2(x,y,z))\right)$$

Before expanding this, dreading the 15 terms from the $(x+y+z)^4$, for example, recall that the only terms which will *not* be cancelled are those which involve *all* of x, y, z. Thus, this is

$$-12x^2yz - 12y^2xz - 12z^2xy + (xy+yz+zx) \cdot (4(x+y+z)^2 - 2(xy+yz+zx)) + \text{(irrelevant)}$$

$$= -12x^2yz - 12y^2xz - 12z^2xy + (xy+yz+zx) \cdot (4x^2+4y^2+4z^2+6xy+6yz+6zx) + \text{(irrelevant)}$$

$$= -12x^2yz - 12y^2xz - 12z^2xy + 4xyz^2 + 4yzx^2 + 4zxy^2 + 6xy^2z$$
$$+ 6x^2yz + 6x^2yz + 6xyz^2 + 6xy^2z + 6xyz^2$$
$$= 4xyz(x+y+z) = 4s_3(x,y,z) \cdot s_1(x,y,z)$$

Thus, with 3 variables,
$$x^4 + y^4 + z^4$$
$$= s_1(x,y,z)^4 - s_2(x,y,z) \cdot (4s_1(x,y,z)^2 - 2s_2(x,y,z)) + 4s_3(x,y,z) \cdot s_1(x,y,z)$$

Abbreviating $s_i = s_i(x,y,z,w)$, we anticipate that
$$x^4 + y^4 + z^4 + w^4 - \left(s_1^4 - 4s_1^2 s_2 + 2s_2^2 + 4s_1 s_3\right) = \text{constant} \cdot xyzw$$

We can save a little time by evaluating the constant by taking $x = y = z = w = 1$. In that case
$$\begin{aligned} s_1(1,1,1,1) &= 4 \\ s_2(1,1,1,1) &= 6 \\ s_3(1,1,1,1) &= 4 \end{aligned}$$

and
$$1 + 1 + 1 + 1 - \left(4^4 - 4 \cdot 4^2 \cdot 6 + 2 \cdot 6^2 + 4 \cdot 4 \cdot 4\right) = \text{constant}$$

or
$$\text{constant} = 4 - (256 - 384 + 72 + 64) = -4$$

Thus,
$$x^4 + y^4 + z^4 + w^4 = s_1^4 - 4s_1^2 s_2 + 2s_2^2 + 4s_1 s_3 - 4s_4$$

By the remark above, since the total degree is just 4, this shows that for arbitrary n

$$x_1^4 + \ldots + x_n^4 = s_1^4 - 4s_1^2 s_2 + 2s_2^2 + 4s_1 s_3 - 4s_4$$

15.3 *A variant: discriminants*

Let $x_1, \ldots, x_n$ be indeterminates. Their **discriminant** is

$$D = D(x_1, \ldots, x_n) = \prod_{i<j} (x_i - x_j)$$

Certainly the sign of D depends on the ordering of the indeterminates. But

$$D^2 = \prod_{i \neq j} (x_i - x_j)^2$$

is *symmetric*, that is, is *invariant* under all permutations of the x_i. Therefore, D^2 has an expression in terms of the elementary symmetric functions of the x_i.

15.3.1 Remark: By contrast to the previous low-degree examples, the discriminant (squared) has as high a degree as possible.

15.3.2 Example: With just 2 indeterminates x, y, we have the familiar

$$D^2 = (x-y)^2 = x^2 - 2xy + y^2 = (x+y)^2 - 4xy = s_1^2 - 4s_2$$

Rather than compute the general version in higher-degree cases, let's consider a more accessible variation on the question. Suppose that $\alpha_1, \ldots, \alpha_n$ are roots of an equation

$$X^n + aX + b = 0$$

in a field k, with $a, b \in k$. For simplicity suppose $a \neq 0$ and $b \neq 0$, since otherwise we have even simpler methods to study this equation. Let $f(X) = x^n + aX + b$. The discriminant

$$D(\alpha_1, \ldots, \alpha_n) = \prod_{i<j} (\alpha_i - \alpha_j)$$

vanishes if and only if any two of the α_i coincide. On the other hand, $f(X)$ has a repeated factor in $k[X]$ if and only if $\gcd(f, f') \neq 1$. Because of the sparseness of this polynomial, we can in effect execute the Euclidean algorithm explicitly. Assume that the characteristic of k does not divide $n(n-1)$. Then

$$(X^n + aX + b) - \frac{X}{n} \cdot (nX^{n-1} + a) = a(1 - \frac{1}{n})X + b$$

That is, any repeated factor of $f(X)$ divides $X + \frac{bn}{(n-1)a}$, and the latter linear factor divides $f'(X)$. Continuing, the remainder upon dividing $nX^{n-1} + a$ by the linear factor $X + \frac{bn}{(n-1)a}$ is simply the value of $nX^{n-1} + a$ obtained by evaluating at $\frac{-bn}{(n-1)a}$, namely

$$n \left(\frac{-bn}{(n-1)a} \right)^{n-1} + a = \left(n^n(-1)^{n-1}b^{n-1} + (n-1)^{n-1}a^n \right) \cdot ((n-1)a)^{1-n}$$

Thus, (constraining a to be non-zero)

$$n^n(-1)^{n-1}b^{n-1} + (n-1)^{n-1}a^n = 0$$

if and only if some $\alpha_i - \alpha_j = 0$.

We obviously want to say that with the constraint that all the symmetric functions of the α_i being 0 except the last two, we have computed the discriminant (up to a less interesting constant factor).

A relatively graceful approach would be to show that $R = \mathbb{Z}[x_1,\ldots,x_n]$ admits a *universal* $\mathbb{Z}$-algebra homomorphism $\varphi : R \longrightarrow \Omega$ for some ring Ω that sends the first $n-2$ elementary symmetric functions

$$
\begin{array}{rclcl}
s_1 &=& s_1(x_1,\ldots,x_n) &=& \sum_i x_i \\
s_2 &=& s_2(x_1,\ldots,x_n) &=& \sum_{i<j} x_i x_j \\
s_3 &=& s_3(x_1,\ldots,x_n) &=& \sum_{i<j<k} x_i x_j x_k \\
&\cdots& \\
s_\ell &=& s_\ell(x_1,\ldots,x_n) &=& \sum_{i_1<\ldots<i_\ell} x_{i_1}\cdots x_{i_\ell} \\
&\cdots& \\
s_{n-2} &=& s_{n-2}(x_1,\ldots,x_n) &=& \sum_{i_1<\ldots<i_{n-2}} x_{i_1}\cdots x_{i_{n-2}}
\end{array}
$$

to 0, but imposes no unnecessary further relations on the images

$$ a = (-1)^{n-1}\varphi(s_{n-1}) \quad b = (-1)^n \varphi(s_n) $$

We do not have sufficient apparatus to do this nicely at this moment. [1]   Nevertheless, the computation above does tell us something.

Exercises

15.1 Express $x_1^3 + x_2^3 + \ldots + x_n^3$ in terms of the elementary symmetric polynomials.

15.2 Express $\sum_{i\neq j} x_i x_j^2$ in terms of the elementary symmetric polynomials.

15.3 Let α, β be the roots of a quadratic equation $ax^2 + bx + c = 0$, Show that the *discriminant*, defined to be $(\alpha - \beta)^2$, is $b^2 - 4ac$.

15.4 Consider $f(x) = x^3 + ax + b$ as a polynomial with coefficients in $k(a,b)$ where k is a field not of characteristic 2 or 3. By computing the greatest common divisor of f and f', give a condition for the roots of $f(x) = 0$ to be distinct.

15.5 Express $\sum_{i,j,k \text{ distinct}} x_i x_j x_k^2$ in terms of elementary symmetric polynomials.

---

[1] The key point is that $\mathbb{Z}[x_1,\ldots,x_n]$ is *integral* over $\mathbb{Z}[s_1,s_2,\ldots,s_n]$ in the sense that each x_i is a root of the *monic* equation $X^n - s_1 X^{n-2} + s_2 X^{n-2} - \ldots + (-1)^{n-1}s_{n-1}X + (-1)^n s_n = 0$ It is true that for R an *integral extension* of a ring S, any homomorphism $\varphi_o : S \longrightarrow \Omega$ to an algebraically closed field Ω extends (probably in more than one way) to a homomorphism $\varphi : R \longrightarrow \Omega$. This would give us a justification for our hope that, given $a,b \in \Omega$ we can require that $\varphi_o(s_1) = \varphi_o(s_2) = \ldots = \varphi_o(s_{n-2}) = 0$ while $\varphi_o(s_{n-1}) = (-1)^{n-1}a \quad \varphi_o(s_n) = (-1)^n b$.

16. Eisenstein's criterion

16.1 *Eisenstein's irreducibility criterion*

Let R be a commutative ring with 1, and suppose that R is a *unique factorization domain*. Let k be the field of fractions of R, and consider R as imbedded in k.

16.1.1 Theorem: Let

$$f(x) = x^N + a_{N-1}x^{N-1} + a_{N-2}x^{N-2} + \ldots + a_2 x^2 + a_1 x + a_0$$

be a polynomial in $R[x]$. If p is a prime in R such that p divides every coefficient a_i but p^2 does *not* divide a_0, then $f(x)$ is irreducible in $R[x]$, and is irreducible in $k[x]$.

Proof: Since f has coefficients in R, its *content* (in the sense of Gauss' lemma) is in R. Since it is monic, its content is 1. Thus, by Gauss' lemma, if $f(x) = g(x) \cdot h(x)$ in $k[x]$ we can adjust constants so that the content of both g and h is 1. In particular, we can suppose that both g and h have coefficients in R, and are monic.

Let

$$g(x) = x^m + b_{m-1}x^{m-1} + b_1 x + b_0$$
$$h(x) = x^n + c_{m-1}x^{m-1} + c_1 x + c_0$$

Not both b_0 and c_0 can be divisible by p, since a_0 is not divisible by p^2. Without loss of generality, suppose that $p|b_0$. Suppose that $p|b_i$ for i in the range $0 \le i \le i_1$, and p does *not* divide b_{i_1}. There *is* such an index i_1, since g is monic. Then

$$a_{i_1} = b_{i_1}c_0 + b_{i_1-1}c_1 + \ldots$$

On the right-hand side, since p divides $b_0, \ldots, b_{i_1-1}$, necessarily p divides all summands but possible the first. Since p divides *neither* b_{i_1} nor c_0, and since R is a UFD, p cannot divide $b_{i_1}c_0$, so cannot divide a_{i_1}, contradiction. Thus, after all, f does not factor.   ///

16.2 *Examples*

16.2.1 Example: For a rational prime p, and for any integer $n > 1$, not only does

$$x^n - p = 0$$

not have a *root* in $\mathbb{Q}$, but, in fact, the polynomial $x^n - p$ is *irreducible* in $\mathbb{Q}[x]$.

16.2.2 Example: Let p be a prime number. Consider the pth cyclotomic polynomial

$$\Phi_p(x) = x^{p-1} + x^{p-2} = \ldots + x^2 + x + 1 = \frac{x^p - 1}{x - 1}$$

We claim that $\Phi_p(x)$ is irreducible in $\mathbb{Q}[x]$. Although $\Phi_p(x)$ itself does not directly admit application of Eisenstein's criterion, a minor variant of it does. That is, consider

$$f(x) = \Phi_p(x+1) = \frac{(x+1)^p - 1}{(x+1) - 1} = \frac{x^p + \binom{p}{1}x^{p-1} + \binom{p}{2}x^{p-2} + \ldots + \binom{p}{p-2}x^2 + \binom{p}{p-1}x}{x}$$

$$= x^{p-1} + \binom{p}{1}x^{p-2} + \binom{p}{2}x^{p-3} + \ldots + \binom{p}{p-2}x + \binom{p}{p-1}$$

All the lower coefficients are divisible by p, and the constant coefficient is exactly p, so is not divisible by p^2. Thus, Eisenstein's criterion applies, and f is irreducible. Certainly if $\Phi_p(x) = g(x)h(x)$ then $f(x) = \Phi_p(x+1) = g(x+1)h(x+1)$ gives a factorization of f. Thus, Φ_p has no proper factorization.

16.2.3 Example: Let $f(x) = x^2 + y^2 + z^2$ in $k[x, y, z]$ where k is *not* of characteristic 2. We make identifications like

$$k[x, y, z] = k[y, z][x]$$

via the natural isomorphisms. We want to show that $y^2 + z^2$ is divisible by some prime p in $k[y, z]$, and *not* by p^2. It suffices to show that $y^2 + z^2$ is divisible by some prime p in $k(z)[y]$, and *not* by p^2. Thus, it suffices to show that $y^2 + z^2$ is not a unit, and has no repeated factor, in $k(z)[y]$. Since it is of degree 2, it is certainly not a unit, so has *some* irreducible factor. To test for repeated factors, compute the *gcd* of this polynomial and its derivative, viewed as having coefficients in the field $k(z)$: [1]

$$(y^2 + z^2) - \frac{y}{2}(2y) = z^2 = \text{ non-zero constant}$$

Thus, $y^2 + z^2$ is a square-free non-unit in $k(z)[y]$, so is divisible by some irreducible p in $k[y, z]$ (Gauss' lemma), so Eisenstein's criterion applies to $x^2 + y^2 + z^2$ and p.

---

[1] It is here that the requirement that the characteristic not be 2 is visible.

16.2.4 Example: Let $f(x) = x^2 + y^3 + z^5$ in $k[x, y, z]$ where k is *not* of characteristic dividing 30. We want to show that $y^3 + z^5$ is divisible by some prime p in $k[y, z]$, and *not* by p^2. It suffices to show that $y^3 + z^5$ is divisible by some prime p in $k(z)[y]$, and *not* by p^2. Thus, it suffices to show that $y^2 + z^2$ is not a unit, and has no repeated factor, in $k(z)[y]$. Since it is of degree 2, it is certainly not a unit, so has *some* irreducible factor. To test for repeated factors, compute the *gcd* of this polynomial and its derivative, viewed as having coefficients in the field $k(z)$: [2]

$$(y^2 + z^2) - \frac{y}{2}(2y) = z^2 = \text{ non-zero constant}$$

Thus, $y^2 + z^2$ is a square-free non-unit in $k(z)[y]$, so is divisible by some irreducible p in $k[y, z]$ (Gauss' lemma), so Eisenstein's criterion applies to $x^2 + y^2 + z^2$ and p.

Exercises

16.1 Prove that $x^7 + 48x - 24$ is irreducible in $\mathbb{Q}[x]$.

16.2 Not only does Eisenstein's criterion (with Gauss' lemma) fail to prove that $x^4 + 4$ is irreducible in $\mathbb{Q}[x]$, but, also, this polynomial *does* factor into two irreducible quadratics in $\mathbb{Q}[x]$. Find them.

16.3 Prove that $x^3 + y^3 + z^3$ is irreducible in $k[x, y, z]$ when k is a field not of characteristic 3.

16.4 Prove that $x^2 + y^3 + z^5$ is irreducible in $k[x, y, z]$ even when the underlying field k is of characteristic $2, 3$, or 5.

16.5 Prove that $x^3 + y + y^5$ is irreducible in $\mathbb{C}[x, y]$.

16.6 Prove that $x^n + y^n + 1$ is irreducible in $k[x, y]$ when the characteristic of k does not divide n.

16.7 Let k be a field with characteristic not dividing n. Show that any polynomial $x^n - P(y)$ where $P(y)$ has no repeated factors is irreducible in $k[x, y]$.

---

[2] It is here that the requirement that the characteristic not be 2 is visible.

17. Vandermonde determinants

17.1 *Vandermonde determinants*

A rigorous systematic evaluation of Vandermonde determinants (below) of the following identity uses the fact that a polynomial ring over a UFD is again a UFD. A **Vandermonde matrix** is a square matrix of the form in the theorem.

17.1.1 Theorem:

$$
\det \begin{pmatrix}
1 & 1 & \cdots & 1 \\
x_1 & x_2 & \cdots & x_n \\
x_1^2 & x_2^2 & \cdots & x_n^2 \\
x_1^3 & x_2^3 & \cdots & x_n^3 \\
\vdots & \vdots & & \vdots \\
x_1^{n-1} & x_2^{n-1} & \cdots & x_n^{n-1}
\end{pmatrix} = (-1)^{n(n-1)/2} \cdot \prod_{i<j} (x_i - x_j)
$$

17.1.2 Remark: The most universal version of the assertion uses indeterminates x_i, and proves an identity in

$$\mathbb{Z}[x_1, \ldots, x_n]$$

Proof: First, the idea of the proof. Whatever the determinant may be, it is a polynomial in x_1, ..., x_n. The most universal choice of interpretation of the coefficients is as in $\mathbb{Z}$. If two columns of a matrix are the same, then the determinant is 0. From this we would *want*

223

to conclude that for $i \neq j$ the determinant is *divisible by*[1] $x_i - x_j$ *in the polynomial ring* $\mathbb{Z}[x_1, \ldots, x_n]$. If we can conclude that, then, since these polynomials are pairwise relatively prime, we can conclude that the determinant is divisible by

$$\prod_{i<j} (x_i - x_j)$$

Considerations of *degree* will show that there is no room for further factors, so, up to a constant, this is the determinant.

To make sense of this line of argument, first observe that a determinant is a polynomial function of its entries. Indeed, the formula is

$$\det M = \sum_p \sigma(p)\, M_{1p(1)} M_{2p(2)} \cdots M_{np(n)}$$

where p runs over permutations of n things and $\sigma(p)$ is the *sign* or *parity* of p, that is, $\sigma(p)$ is $+1$ if p is a product of an *even* number of 2-cycles and is -1 if p is the product of an *odd* number of 2-cycles. Thus, for any $\mathbb{Z}$-algebra homomorphism f to a commutative ring R with identity,

$$f : \mathbb{Z}[x_1, \ldots, x_n] \longrightarrow R$$

we have

$$f(\det V) = \det f(V)$$

where by $f(V)$ we mean application of f entry-wise to the matrix V. Thus, if we can prove an identity in $\mathbb{Z}[x_1, \ldots, x_n]$, then we have a corresponding identity in any ring.

Rather than talking about setting x_j equal to x_i, it is safest to try to see divisibility property as directly as possible. Therefore, we do *not* attempt to use the property that the determinant of a matrix with two equal columns is 0. Rather, we use the property[2] that if an element r of a ring R divides every element of a column (or row) of a square matrix, then it divides the determinant. And we are allowed to add any multiple of one column to another without changing the value of the determinant. Subtracting the jth column from the ith column of our Vandermonde matrix (with $i < j$), we have

$$\det V = \det \begin{pmatrix} \cdots & 1-1 & \cdots & 1 & \cdots \\ \cdots & x_i - x_j & \cdots & x_j & \cdots \\ \cdots & x_i^2 - x_j^2 & \cdots & x_j^2 & \cdots \\ \cdots & x_i^3 - x_j^3 & \cdots & x_j^3 & \cdots \\ & \vdots & & \vdots & \\ \cdots & x_i^{n-1} - x_j^{n-1} & \cdots & x_j^{n-1} & \cdots \end{pmatrix}$$

---

[1] If one treats the x_i merely as complex numbers, for example, then one *cannot* conclude that the product of the expressions $x_i - x_j$ with $i < j$ divides the determinant. Attempting to evade this problem by declaring the x_i as somehow *variable* complex numbers is an impulse in the right direction, but is made legitimate only by treating *genuine* indeterminates.

[2] This follows directly from the just-quoted formula for determinants, and also from other descriptions of determinants, but from any viewpoint is still valid for matrices with entries in any commutative ring with identity.

From the identity

$$x^m - y^m = (x - y)(x^{m-1} + x^{m-2}y + \ldots + y^{m-1})$$

it is clear that $x_i - x_j$ divides all entries of the new i^{th} column. Thus, $x_i - x_j$ divides the determinant. This holds for all $i < j$.

Since these polynomials are linear, they are irreducible in $\mathbb{Z}[x_1, \ldots, x_n]$. Generally, the units in a polynomial ring $R[x_1, \ldots, x_n]$ are the units $R^\times$ in R, so the units in $\mathbb{Z}[x_1, \ldots, x_n]$ are just ± 1. Visibly, the various irreducible $x_i - x_j$ are not *associate*, that is, do not merely differ by units. Therefore, their least common multiple is their product. Since $\mathbb{Z}[x_1, \ldots, x_n]$ is a UFD, this product divides the determinant of the Vandermonde matrix.

To finish the computation, we want to argue that the determinant can have no *further* polynomial factors than the ones we've already determined, so up to a constant (which we'll determine) is equal to the latter product. [3] To prove this, we need the notion of **total degree**: the total degree of a monomial $x_1^{m_1} \ldots x_n^{m_n}$ is $m_1 + \ldots + m_n$, and the total degree of a polynomial is the maximum of the total degrees of the monomials occurring in it. We grant for the moment the result of the proposition below, that the total degree of a product is the sum of the total degrees of the factors. The total degree of the product is

$$\sum_{1 \leq i < j \leq n} 1 = \sum_{1 \leq i < n} n - i = \frac{1}{2}n(n-1)$$

To determine the total degree of the determinant, invoke the usual formula for the determinant of a matrix M with entries M_{ij}, namely

$$\det M = \sum_\pi \sigma(\pi) \prod_i M_{i, \pi(i)}$$

where π is summed over permutations of n things, and where $\sigma(\pi)$ is the *sign* of the permutation π. In a Vandermonde matrix all the top row entries have total degree 0, all the second row entries have total degree 1, and so on. Thus, in this permutation-wise sum for a Vandermonde determinant, each summand has total degree

$$0 + 1 + 2 + \ldots + (n-1) = \frac{1}{2}n(n-1)$$

so the total degree of the determinant is the total degree of the product

$$\sum_{1 \leq i < j \leq n} 1 = \sum_{1 \leq i < n} n - i = \frac{1}{2}n(n-1)$$

Thus,

$$\det \begin{pmatrix} 1 & 1 & \ldots & 1 \\ x_1 & x_2 & \ldots & x_n \\ x_1^2 & x_2^2 & \ldots & x_n^2 \\ x_1^3 & x_2^3 & \ldots & x_n^3 \\ \vdots & \vdots & & \vdots \\ x_1^{n-1} & x_2^{n-1} & \ldots & x_n^{n-1} \end{pmatrix} = \text{constant} \cdot \prod_{i<j}(x_i - x_j)$$

---

[3] This is more straightforward than setting up the right viewpoint for the first part of the argument.

Granting this, to determine the constant it suffices to compare a single monomial in both expressions. For example, compare the coefficients of

$$x_1^{n-1} x_2^{n-2} x_3^{n-3} \ldots x_{n-1}^1 x_n^0$$

In the product, the only way x_1^{n-1} appears is by choosing the x_1s in the linear factors $x_1 - x_j$ with $1 < j$. After this, the only way to get x_2^{n-2} is by choosing all the x_2s in the linear factors $x_2 - x_j$ with $2 < j$. Thus, this monomial has coefficient $+1$ in the product.

In the determinant, the only way to obtain this monomial is as the product of entries from lower left to upper right. The indices of these entries are $(n, 1), (n-1, 2), \ldots, (2, n-1), (1, n)$. Thus, the coefficient of this monomial is $(-1)^\ell$ where ℓ is the number of 2-cycles necessary to obtain the permutation p such that

$$p(i) = n + 1 - i$$

Thus, for n even there are $n/2$ two-cycles, and for n odd $(n-1)/2$ two-cycles. For a closed form, as these expressions will appear only as exponents of -1, we only care about values modulo 2. Because of the division by 2, we only care about n modulo 4. Thus, we have values

$$\begin{cases} n/2 & = & 0 \bmod 2 & (\text{for } n = 0 \bmod 4) \\ (n-1)/2 & = & 0 \bmod 2 & (\text{for } n = 1 \bmod 4) \\ n/2 & = & 1 \bmod 2 & (\text{for } n = 3 \bmod 4) \\ (n-1)/2 & = & 1 \bmod 2 & (\text{for } n = 1 \bmod 4) \end{cases}$$

After some experimentation, we find a closed expression

$$n(n-1)/2 \bmod 2$$

Thus, the leading constant is

$$(-1)^{n(n-1)/2}$$

in the expression for the Vandermonde determinant.   ///

Verify the property of total degree:

17.1.3 Lemma: Let $f(x_1, \ldots, x_n)$ and $g(x_1, \ldots, x_n)$ be polynomials in $k[x_1, \ldots, x_n]$ where k is a field. Then the total degree of the product is the sum of the total degrees.

Proof: It is clear that the total degree of the product is less than or equal the sum of the total degrees.

Let $x_1^{e_1} \ldots x_n^{e_n}$ and $x_1^{f_1} \ldots x_n^{f_n}$ be two monomials of highest total degrees $s = e_1 + \ldots + e_n$ and $t = f_1 + \ldots + f_n$ occurring with non-zero coefficients in f and g, respectively. Assume without loss of generality that the exponents e_1 and f_1 of x_1 in the two expressions are the largest among all monomials of total degrees s, t in f and g, respectively. Similarly, assume without loss of generality that the exponents e_2 and f_2 of x_2 in the two expressions are the largest among all monomials of total degrees s, t in f and g, respectively, of degrees e_1 and f_1 in x_1. Continuing similarly, we claim that the coefficient of the monomial

$$M = x_1^{e_1+f_1} \ldots x_n^{e_n+f_n}$$

is simply the product of the coefficients of $x_1^{e_1} \ldots x_n^{e_n}$ and $x_1^{f_1} \ldots x_n^{f_n}$, so non-zero. Let $x_1^{u_1} \ldots x_n^{u_n}$ and $x_1^{v_1} \ldots x_n^{v_n}$ be two other monomials occurring in f and g such that for all indices i we have $u_i + v_i = e_i + f_i$. By the maximality assumption on e_1 and f_1, we have $e_1 \geq u_1$ and $f_1 \geq v_1$, so the only way that the necessary power of x_1 can be achieved is that $e_1 = u_1$ and $f_1 = v_1$. Among exponents with these maximal exponents of x_1, e_2 and f_2 are maximal, so $e_2 \geq u_2$ and $f_2 \geq v_2$, and again it must be that $e_2 = u_2$ and $f_2 = v_2$ to obtain the exponent of x_2. Inductively, $u_i = e_i$ and $v_i = f_i$ for all indices. That is, the only terms in f and g contributing to the coefficient of the monomial M in $f \cdot g$ are monomials $x_1^{e_1} \dots x_n^{e_n}$ and $x_1^{f_1} \dots x_n^{f_n}$. Thus, the coefficient of M is non-zero, and the total degree is as claimed. ///

17.2 *Worked examples*

17.2.1 Example: Show that a *finite* integral domain is necessarily a *field*.

Let R be the integral domain. The integral domain property can be immediately paraphrased as that for $0 \neq x \in R$ the map $y \longrightarrow xy$ has trivial kernel (as R-module map of R to itself, for example). Thus, it is injective. Since R is a finite set, an injective map of it to itself is a bijection. Thus, there is $y \in R$ such that $xy = 1$, proving that x is invertible. ///

17.2.2 Example: Let $P(x) = x^3 + ax + b \in k[x]$. Suppose that $P(x)$ factors into linear polynomials $P(x) = (x - \alpha_1)(x - \alpha_2)(x - \alpha_3)$. Give a polynomial condition on a, b for the α_i to be distinct.

(One might try to do this as a symmetric function computation, but it's a bit tedious.)

If $P(x) = x^3 + ax + b$ has a repeated factor, then it has a common factor with its derivative $P'(x) = 3x^2 + a$.

If the characteristic of the field is 3, then the derivative is the constant a. Thus, if $a \neq 0$, $\gcd(P, P') = a \in k^\times$ is never 0. If $a = 0$, then the derivative is 0, and all the α_i are the same.

Now suppose the characteristic is not 3. In effect applying the Euclidean algorithm to P and P',

$$\left(x^3 + ax + b\right) - \frac{x}{3} \cdot \left(3x^2 + a\right) = ax + b - \frac{x}{3} \cdot a = \frac{2}{3}ax + b$$

If $a = 0$ then the Euclidean algorithm has already terminated, and the condition for distinct roots or factors is $b \neq 0$. Also, possibly surprisingly, at this point we need to consider the possibility that the characteristic is 2. If so, then the remainder is b, so if $b \neq 0$ the roots are always distinct, and if $b = 0$

Now suppose that $a \neq 0$, and that the characteristic is not 2. Then we can divide by $2a$. Continue the algorithm

$$\left(3x^2 + a\right) - \frac{9x}{2a} \cdot \left(\frac{2}{3}ax + b\right) = a + \frac{27b^2}{4a^2}$$

Since $4a^2 \neq 0$, the condition that P have no repeated factor is

$$4a^3 + 27b^2 \neq 0$$

17.2.3 Example: The first three **elementary symmetric functions** in indeterminates $x_1, \ldots, x_n$ are

$$\sigma_1 = \sigma_1(x_1, \ldots, x_n) = x_1 + x_2 + \ldots + x_n = \sum_i x_i$$

$$\sigma_2 = \sigma_2(x_1, \ldots, x_n) = \sum_{i<j} x_i x_j$$

$$\sigma_3 = \sigma_3(x_1, \ldots, x_n) = \sum_{i<j<\ell} x_i x_j x_\ell$$

Express $x_1^3 + x_2^3 + \ldots + x_n^3$ in terms of $\sigma_1, \sigma_2, \sigma_3$.

Execute the algorithm given in the proof of the theorem. Thus, since the degree is 3, if we can derive the right formula for just 3 indeterminates, the same expression in terms of elementary symmetric polynomials will hold generally. Thus, consider $x^3 + y^3 + z^3$. To approach this we first take $y = 0$ and $z = 0$, and consider x^3. This is $s_1(x)^3 = x^3$. Thus, we next consider

$$\left(x^3 + y^3\right) - s_1(x,y)^3 = 3x^2 y + 3xy^2$$

As the algorithm assures, this is divisible by $s_2(x,y) = xy$. Indeed,

$$\left(x^3 + y^3\right) - s_1(x,y)^3 = (3x + 3y)s_2(x,y) = 3s_1(x,y)\,s_2(x,y)$$

Then consider

$$\left(x^3 + y^3 + z^3\right) - \left(s_1(x,y,z)^3 - 3\,s_2(x,y,z)\,s_1(x,y,z)\right) = 3xyz = 3s_3(x,y,z)$$

Thus, again, since the degree is 3, this formula for 3 variables gives the general one:

$$x_1^3 + \ldots + x_n^3 = s_1^3 - 3s_1 s_2 + 3s_3$$

where $s_i = s_i(x_1, \ldots, x_n)$.

17.2.4 Example: Express $\sum_{i \neq j} x_i^2 x_j$ as a polynomial in the elementary symmetric functions of $x_1, \ldots, x_n$.

We could (as in the previous problem) execute the algorithm that proves the theorem asserting that every symmetric (that is, S_n-invariant) polynomial in $x_1, \ldots, x_n$ is a polynomial in the elementary symmetric functions.

But, also, sometimes *ad hoc* manipulations can yield shortcuts, depending on the context. Here,

$$\sum_{i \neq j} x_i^2 x_j = \sum_{i,j} x_i^2 x_j - \sum_{i=j} x_i^2 x_j = \left(\sum_i x_i^2\right)\left(\sum_j x_j\right) - \sum_i x_i^3$$

An easier version of the previous exercise gives

$$\sum_i x_i^2 = s_1^2 - 2s_2$$

and the previous exercise itself gave

$$\sum_i x_i^3 = s_1^3 - 3s_1 s_2 + 3s_3$$

Thus,

$$\sum_{i \neq j} x_i^2 x_j = \left(s_1^2 - 2s_2\right) s_1 - \left(s_1^3 - 3s_1 s_2 + 3s_3\right) = s_1^3 - 2s_1 s_2 - s_1^3 + 3s_1 s_2 - 3s_3 = s_1 s_2 - 3s_3$$

17.2.5 Example: Suppose the characteristic of the field k does not divide n. Let $\ell > 2$. Show that

$$P(x_1, \ldots, x_n) = x_1^n + \ldots + x_\ell^n$$

is irreducible in $k[x_1, \ldots, x_\ell]$.

First, treating the case $\ell = 2$, we claim that $x^n + y^n$ is not a unit and has no repeated factors in $k(y)[x]$. (We take the field of rational functions in y so that the resulting polynomial ring in a single variable is Euclidean, and, thus, so that we understand the behavior of its irreducibles.) Indeed, if we start executing the Euclidean algorithm on $x^n + y^n$ and its derivative nx^{n-1} in x, we have

$$(x^n + y^n) - \frac{x}{n}(nx^{n-1}) = y^n$$

Note that n is invertible in k by the characteristic hypothesis. Since y is invertible (being non-zero) in $k(y)$, this says that the *gcd* of the polynomial in x and its derivative is 1, so there is no repeated factor. And the degree in x is positive, so $x^n + y^n$ has *some* irreducible factor (due to the unique factorization in $k(y)[x]$, or, really, due indirectly to its Noetherian-ness).

Thus, our induction (on n) hypothesis is that $x_2^n + x_3^n + \ldots + x_\ell^n$ is a non-unit in $k[x_2, x_3, \ldots, x_n]$ and has no repeated factors. That is, it is divisible by some irreducible p in $k[x_2, x_3, \ldots, x_n]$. Then in

$$k[x_2, x_3, \ldots, x_n][x_1] \approx k[x_1, x_2, x_3, \ldots, x_n]$$

Eisenstein's criterion applied to $x_1^n + \ldots$ as a polynomial in x_1 with coefficients in $k[x_2, x_3, \ldots, x_n]$ and using the irreducible p yields the irreducibility.

17.2.6 Example: Find the determinant of the **circulant** matrix

$$\begin{pmatrix} x_1 & x_2 & \cdots & x_{n-2} & x_{n-1} & x_n \\ x_n & x_1 & x_2 & \cdots & x_{n-2} & x_{n-1} \\ x_{n-1} & x_n & x_1 & x_2 & \cdots & x_{n-2} \\ \vdots & & & \ddots & & \vdots \\ x_3 & & & & x_1 & x_2 \\ x_2 & x_3 & \cdots & & x_n & x_1 \end{pmatrix}$$

(*Hint:* Let ζ be an nth root of 1. If $x_{i+1} = \zeta \cdot x_i$ for all indices $i < n$, then the $(j+1)^{th}$ row is ζ times the jth, and the determinant is 0.)

Let C_{ij} be the ij^{th} entry of the circulant matrix C. The expression for the determinant

$$\det C = \sum_{p \in S_n} \sigma(p) \, C_{1,p(1)} \cdots C_{n,p(n)}$$

where $\sigma(p)$ is the sign of p shows that the determinant is a polynomial in the entries C_{ij} with integer coefficients. This is the most universal viewpoint that could be taken. However, with some hindsight, some intermediate manipulations suggest or require enlarging the 'constants' to include n^{th} roots of unity ω. Since we do not know that $\mathbb{Z}[\omega]$ is a UFD (and, indeed, it is not, in general), we must adapt. A reasonable adaptation is to work over $\mathbb{Q}(\omega)$. Thus, we will prove an identity in $\mathbb{Q}(\omega)[x_1, \ldots, x_n]$.

Add ω^{i-1} times the ith row to the first row, for $i \geq 2$. The new first row has entries, from left to right,

$$x_1 + \omega x_2 + \omega^2 x_3 + \ldots + \omega^{n-1} x_n$$

$$x_2 + \omega x_3 + \omega^2 x_4 + \ldots + \omega^{n-1} x_{n-1}$$

$$x_3 + \omega x_4 + \omega^2 x_5 + \ldots + \omega^{n-1} x_{n-2}$$

$$x_4 + \omega x_5 + \omega^2 x_6 + \ldots + \omega^{n-1} x_{n-3}$$

$$\ldots$$

$$x_2 + \omega x_3 + \omega^2 x_4 + \ldots + \omega^{n-1} x_1$$

The tth of these is

$$\omega^{-t} \cdot (x_1 + \omega x_2 + \omega^2 x_3 + \ldots + \omega^{n-1} x_n)$$

since $\omega^n = 1$. Thus, in the ring $\mathbb{Q}(\omega)[x_1, \ldots, x_n]$,

$$x_1 + \omega x_2 + \omega^2 x_3 + \ldots + \omega^{n-1} x_n)$$

divides this new top row. Therefore, from the explicit formula, for example, this quantity divides the determinant.

Since the characteristic is 0, the n roots of $x^n - 1 = 0$ are distinct (for example, by the usual computation of *gcd* of $x^n - 1$ with its derivative). Thus, there are n superficially-different linear expressions which divide $\det C$. Since the expressions are linear, they are *irreducible* elements. If we prove that they are *non-associate* (do not differ merely by units), then their product must divide $\det C$. Indeed, viewing these linear expressions in the larger ring

$$\mathbb{Q}(\omega)(x_2, \ldots, x_n)[x_1]$$

we see that they are distinct linear monic polynomials in x_1, so are non-associate.

Thus, for some $c \in \mathbb{Q}(\omega)$,

$$\det C = c \cdot \prod_{1 \leq \ell \leq n} \left(x_1 + \omega^\ell x_2 + \omega^{2\ell} x_3 + \omega^{3\ell} x_4 + \ldots + \omega^{(n-1)\ell} x_n \right)$$

Looking at the coefficient of x_1^n on both sides, we see that $c = 1$.

(One might also observe that the product, when expanded, will have coefficients in $\mathbb{Z}$.)

Exercises

17.1 A k-linear *derivation* D on a commutative k-algebra A, where k is a field, is a k-linear map $D : A \longrightarrow A$ satisfying *Leibniz' identity*

$$D(ab) = (Da) \cdot b + a \cdot (Db)$$

Given a polynomial $P(x)$, show that there is a unique k-linear derivation D on the polynomial ring $k[x]$ sending x to $P(x)$.

17.2 Let A be a commutative k-algebra which is an integral domain, with field of fractions K. Let D be a k-linear derivation on A. Show that there is a unique extension of D to a k-linear derivation on K, and that this extension necessarily satisfies the quotient rule.

17.3 Let $f(x_1, \ldots, x_n)$ be a homogeneous polynomial of *total degree* n, with coefficients in a field k. Let $\partial/\partial x_i$ be partial differentiation with respect to x_i. Prove *Euler's identity*, that

$$\sum_{i=1}^{n} x_i \frac{\partial f}{\partial x_i} = n \cdot f$$

17.4 Let α be algebraic over a field k. Show that any k-linear derivation D on $k(\alpha)$ necessarily gives $D\alpha = 0$.

18. Cyclotomic polynomials II

18.1   Cyclotomic polynomials over $\mathbb{Z}$
18.2   Worked examples

Now that we have Gauss' lemma in hand we can look at cyclotomic polynomials again, not as polynomials with coefficients in various fields, but as *universal* things, having coefficients in $\mathbb{Z}$. [1]   Most of this discussion is simply a rewrite of the earlier discussion with coefficients in fields, especially the case of characteristic 0, paying attention to the invocation of Gauss' lemma. A new point is the fact that the coefficients lie in $\mathbb{Z}$. Also, we note the irreducibility of $\Phi_p(x)$ for prime p in both $\mathbb{Z}[x]$ and $\mathbb{Q}[x]$, via Eisenstein's criterion (and Gauss' lemma, again).

18.1 *Cyclotomic polynomials over $\mathbb{Z}$*

Define

$$\Phi_1(x) = x - 1 \in \mathbb{Z}[x]$$

and for $n > 1$ try to define [2]

$$\Phi_n(x) = \frac{x^n - 1}{\prod_{d|n,\ d<n} \Phi_d(x)}$$

---

[1]   Given any field k, there is a unique $\mathbb{Z}$-algebra homomorphism $\mathbb{Z}[x] \longrightarrow k[x]$ sending x to x and 1 to 1. Thus, *if* we can successfully demonstrate properties of polynomials in $\mathbb{Z}[x]$ *then* these properties descend to any particular $k[x]$. In particular, this may allow us to avoid certain complications regarding the characteristic of the field k.

[2]   It is not immediately clear that the denominator divides the numerator, for example.

We prove inductively that $\Phi_n(x)$ is monic, has integer coefficients, and has constant coefficient ± 1.

First, we claim that $x^n - 1 \in \mathbb{Z}[x]$ has no repeated factors. The greatest common divisor of its coefficients is 1, so by Gauss' lemma any irreducible factors can be taken to be monic polynomials with integer coefficients which are irreducible in $\mathbb{Q}[x]$, not merely in $\mathbb{Z}[x]$. Thus, it suffices to compute the greatest common divisor of $x^n - 1$ and its derivative nx^{n-1} in $\mathbb{Q}[x]$. Since n is invertible in $\mathbb{Q}$,

$$(x^n - 1) - \frac{x}{n} \cdot nx^{n-1} = -1$$

Thus, there are no repeated factors [3] in $x^n - 1$.

Next, note that in $\mathbb{Z}[x]$ we still do have the unlikely-looking

$$\gcd(x^m - 1, x^n - 1) = x^{\gcd(m,n)} - 1$$

Again, the *gcd* of the coefficients of each polynomial is 1, so by Gauss' lemma the *gcd* of the two polynomials can be computed in $\mathbb{Q}[x]$ (and will be a monic polynomial with integer coefficients whose *gcd* is 1). Taking $m \le n$ without loss of generality,

$$(x^n - 1) - x^{n-m}(x^m - 1) = x^{n-m} - 1$$

For $n = qm + r$ with $0 \le r < m$, repeating this procedure q times allows us to reduce n modulo m, finding that the *gcd* of $x^n - 1$ and $x^m - 1$ is the same as the *gcd* of $x^m - 1$ and $x^r - 1$. In effect, this is a single step in the Euclidean algorithm applied to m and n. Thus, by an induction, we obtain the assertion.

Claim that in $\mathbb{Z}[x]$, for $m < n$, $\Phi_m(x)$ and $\Phi_n(x)$ have no common factor. Again, by induction, they have integer coefficients with *gcd* 1, so by Gauss' lemma any common factor has the same nature. Any common factor would be a common factor of $x^n - 1$ and $x^m - 1$, hence, by the previous paragraph, a factor of $x^d - 1$ where $d = \gcd(m, n)$. Since $m \ne n$, d must be a *proper* factor n, and by its recursive definition

$$\Phi_n(x) = \frac{x^n - 1}{\prod_{\delta | n, \, \delta < n} \Phi_\delta(x)} \text{ divides } \frac{x^n - 1}{\prod_{\delta | d} \Phi_\delta(x)} = \text{ divides } \frac{x^n - 1}{x^d - 1}$$

Thus, since $x^n - 1$ has no repeated factors, $\Phi_n(x)$ shares no common factors with $x^d - 1$. Thus, for $m < n$, $\Phi_m(x)$ and $\Phi_n(x)$ have greatest common factor 1.

Therefore, in the attempted definition

$$\Phi_n(x) = \frac{x^n - 1}{\prod_{d | n, \, d < n} \Phi_\delta(x)}$$

by induction the denominators in the right-hand side have no common factor, all divide $x^n - 1$, so their product divides $x^n - 1$, by unique factorization in $\mathbb{Z}[x]$. Thus, the apparent definition of $\Phi_n(x)$ as a polynomial with integer coefficients succeeds. [4]

---

[3] We had noted this earlier, except the conclusion was weaker. Previously, we could only assert that there were no repeated factors in $\mathbb{Q}[x]$, since we knew that the latter ring was Euclidean, hence a PID. One weakness of that viewpoint is that it does not directly tell anything about what might happen over finite fields. Treating *integer* coefficients is the universal.

[4] Proving that the cyclotomic polynomials have integer coefficients is more awkward if one cannot discuss unique factorization in $\mathbb{Z}[x]$.

Also, by induction, from

$$x^n - 1 = \prod_{d|n,\ d \leq n} \Phi_\delta(x)$$

the constant coefficient of $\Phi_n(x)$ is ± 1. And $\Phi_n(x)$ is monic.

Finally, note that for p prime

$$\Phi_p(x+1) = \frac{(x+1)^p - 1}{(x+1) - 1} = x^{p-1} + \binom{p}{1}x^{p-2} + \binom{p}{2}x^{p-3} + \ldots + \binom{p}{p-2}x + \binom{p}{p-1}$$

This has all lower coefficients divisible by p, and the constant coefficient is exactly p, so is not divisible by p^2. Thus, by Eisenstein's criterion, $\Phi_p(x)$ is irreducible in $\mathbb{Z}[x]$. By Gauss' lemma, it is irreducible in $\mathbb{Q}[x]$.

18.2 *Worked examples*

18.2.1 Example: Prove that a *finite division* ring D (a not-necessarily commutative ring with 1 in which any non-zero element has a multiplicative inverse) is commutative. (This is due to *Wedderburn*.) (*Hint:* Check that the center k of D is a field, say of cardinality q. Let $D^\times$ act on D by conjugation, namely $\alpha \cdot \beta = \alpha\beta\alpha^{-1}$, and count orbits, to obtain an equality of the form

$$|D| = q^n = q + \sum_d \frac{q^n - 1}{q^d - 1}$$

where d is summed over some set of integers all strictly smaller than n. Let $\Phi_n(x)$ be the nth cyclotomic polynomial. Show that, on one hand, $\Phi_n(q)$ divides $q^n - q$, but, on the other hand, this is impossible unless $n = 1$. Thus $D = k$.)

First, the *center k* of D is defined to be

$$k = \text{center } D = \{\alpha \in D : \alpha x = x\alpha \quad \text{for all } x \in D\}$$

We claim that k is a field. It is easy to check that k is closed under addition, multiplication, and contains 0 and 1. Since $-\alpha = (-1) \cdot \alpha$, it is closed under taking additive inverses. There is a slight amount of interest in considering closure under taking multiplicative inverses. Let $0 \neq \alpha \in k$, and $x \in D$. Then left-multiply *and* right-multiply $\alpha x = x\alpha$ by α^{-1} to obtain $x\alpha^{-1} = \alpha^{-1}x$. This much proves that k is a division ring. Since its elements commute with every $x \in D$ certainly k is commutative. This proves that k is a field.

The same argument shows that for any $x \in D$ the **centralizer**

$$D_x = \text{centralizer of } x = \{\alpha \in D : \alpha x = x\alpha\}$$

is a division ring, though possibly non-commutative. It certainly contains the center k, so is a k-vectorspace. Noting that $\alpha x = x\alpha$ is equivalent to $\alpha x\alpha^{-1} = x$ for α invertible, we see that $D_x^\times$ is the pointwise fixer of x under the conjugation action.

Thus, the orbit-counting formula gives

$$|D| = |k| + \sum_{\text{non-central orbits } O_x} [D^\times : D_x^\times]$$

where the center k is all singleton orbits and O_x is summed over orbits of non-central elements, choosing representatives x for O_x. This much did not use finiteness of D.

Let $q = |k|$, and $n = \dim_k D$. Suppose $n > 1$. Let $n_x = \dim_k D_x$. Then

$$q^n = q + \sum_{\text{non-central orbits } O_x} \frac{q^n - 1}{q^{n_x} - 1}$$

In all the non-central orbit summands, $n > n_x$. Rearranging,

$$q - 1 = -(q^n - 1) + \sum_{\text{non-central orbits } O_x} \frac{q^n - 1}{q^{n_x} - 1}$$

Let $\Phi_n(x)$ be the nth cyclotomic polynomial, viewed as an element of $\mathbb{Z}[x]$. Then, from the fact that the recursive definition of $\Phi_n(x)$ really does yield a monic polynomial of positive degree with integer coefficients (and so on), and since $n_x < n$ for all non-central orbits, the integer $\Phi_n(q)$ divides the right-hand side, so divides $q - 1$.

We claim that as a complex number $|\Phi_n(q)| > q - 1$ for $n > 1$. Indeed, fix a primitive n^{th} root of unity $\zeta \in \mathbb{C}$. The set of *all* primitive n^{th} roots of unity is $\{\zeta^a\}$ where $1 \le a \le p$ prime to p. Then

$$|\Phi_n(q)|^2 = \prod_{a:\, \gcd(a,n)=1} |q - \zeta^a|^2 = \prod_{a:\, \gcd(a,n)=1} \left[(q - \text{Re}(\zeta^a))^2 + (\text{Im}(\zeta^a))^2 \right]$$

Since $|\zeta| = 1$, the real part is certainly between -1 and $+1$, so $q - \text{Re}(\zeta^a) > q - 1$ unless $\text{Re}(\zeta^a) = 1$, which happens only for $\zeta^a = 1$, which can happen only for $n = 1$. That is, for $n > 1$, the integer $\Phi_n(q)$ is a product of complex numbers each larger than $q - 1$, contradicting the fact that $\Phi_n(q)|(q-1)$. That is, $n = 1$. That is, there are no non-central orbits, and D is commutative.

18.2.2 Example: Let $q = p^n$ be a (positive integer) power of a prime p. Let $F : \mathbb{F}_q \longrightarrow \mathbb{F}_q$ by $F(\alpha) = \alpha^p$ be the Frobenius map over $\mathbb{F}_p$. Let S be a set of elements of $\mathbb{F}_q$ stable under F (that is, F maps S to itself). Show that the polynomial

$$\prod_{\alpha \in S} (x - \alpha)$$

has coefficients in the smaller field $\mathbb{F}_p$.

Since the set S is Frobenius-stable, application of the Frobenius to the polynomial merely permutes the linear factors, thus leaving the polynomial unchanged (since the multiplication of the linear factors is insensitive to ordering.) Thus, the coefficients of the (multiplied-out) polynomial are fixed by the Frobenius. That is, the coefficients are roots of the equation $x^p - x = 0$. On one hand, this polynomial equation has at most p roots in a given field (from unique factorization), and, on the other hand, Fermat's Little Theorem assures that the elements of the field $\mathbb{F}_p$ are roots of that equation. Thus, any element fixed under the Frobenius lies in the field $\mathbb{F}_p$, as asserted.

18.2.3 Example: Let $q = p^n$ be a power of a prime p. Let $F : \mathbb{F}_q \longrightarrow \mathbb{F}_q$ by $F(\alpha) = \alpha^p$ be the Frobenius map over $\mathbb{F}_p$. Show that for every divisor d of n that the fixed points of F^d form the unique subfield $\mathbb{F}_{p^d}$ of $\mathbb{F}_q$ of degree d over the prime field $\mathbb{F}_p$.

This is similar to the previous example, but emphasizing a different part. Fixed points of the dth power F^d of the Frobenius F are exactly the roots of the equation $x^{p^d} - x = 0$ of $x(x^{p^d-1} - 1) = 0$. On one hand, a polynomial has at most as many roots (in a field) as its degree. On the other hand, $\mathbb{F}_{p^d}^\times$ is of order $p^d - 1$, so every element of $\mathbb{F}_{p^d}$ is a root of our equation. There can be no more, so $\mathbb{F}_{p^d}$ is exactly the set of roots.

18.2.4 Example: Let $f(x)$ be a monic polynomial with integer coefficients. Show that f is irreducible in $\mathbb{Q}[x]$ if it is irreducible in $(\mathbb{Z}/p)[x]$ for some p.

First, claim that if $f(x)$ is irreducible in some $(\mathbb{Z}/p)[x]$, then it is irreducible in $\mathbb{Z}[x]$. A factorization $f(x) = g(x) \cdot h(x)$ in $\mathbb{Z}[x]$ maps, under the natural $\mathbb{Z}$-algebra homomorphism to $(\mathbb{Z}/p)[x]$, to the corresponding factorization $f(x) = g(x) \cdot h(x)$ in $(\mathbb{Z}/p)[x]$. (There's little reason to invent a notation for the reduction modulo p of polynomials as long as we are clear what we're doing.) A critical point is that since f is monic both g and h can be taken to be monic also (multiplying by -1 if necessary), since the highest-degree coefficient of a product is simply the product of the highest-degree coefficients of the factors. The irreducibility over $\mathbb{Z}/p$ implies that the degree of one of g and h modulo p is 0. Since they are monic, reduction modulo p does not alter their degrees. Since f is monic, its content is 1, so, by Gauss' lemma, the factorization in $\mathbb{Z}[x]$ is not proper, in the sense that either g or h is just ± 1.

That is, f is irreducible in the ring $\mathbb{Z}[x]$. Again by Gauss' lemma, this implies that f is irreducible in $\mathbb{Q}[x]$.

18.2.5 Example: Let n be a positive integer such that $(\mathbb{Z}/n)^\times$ is *not* cyclic. Show that the nth cyclotomic polynomial $\Phi_n(x)$ factors properly in $\mathbb{F}_p[x]$ for any prime p not dividing n.

(See subsequent text for systematic treatment of the case that p divides n.) Let d be a positive integer such that $p^d - 1 = 0 \bmod n$. Since we know that $\mathbb{F}_{p^d}^\times$ is cyclic, $\Phi_n(x) = 0$ has a root in $\mathbb{F}_{p^d}$ when $p^d - 1 = 0 \bmod n$. For $\Phi_n(x)$ to be irreducible in $\mathbb{F}_p[x]$, it must be that $d = \varphi(n)$ (Euler's totient function φ) is the smallest exponent which achieves this. That is, $\Phi_n(x)$ will be irreducible in $\mathbb{F}_p[x]$ only if $p^{\varphi(n)} = 1 \bmod n$ but no smaller positive exponent achieves this effect. That is, $\Phi_n(x)$ is irreducible in $\mathbb{F}_p[x]$ only if p is of order $\varphi(n)$ in the group $(\mathbb{Z}/n)^\times$. We know that the order of this group is $\varphi(n)$, so any such p would be a generator for the group $(\mathbb{Z}/n)^\times$. That is, the group would be cyclic.

18.2.6 Example: Show that the 15th cyclotomic polynomial $\Phi_{15}(x)$ is irreducible in $\mathbb{Q}[x]$, despite being reducible in $\mathbb{F}_p[x]$ for every prime p.

First, by Sun-Ze

$$(\mathbb{Z}/15)^\times \approx (\mathbb{Z}/3)^\times \times (\mathbb{Z}/5)^\times \approx \mathbb{Z}/2 \oplus \mathbb{Z}/4$$

This is not cyclic (there is no element of order 8, as the maximal order is 4). Thus, by the previous problem, there is no prime p such that $\Phi_{15}(x)$ is irreducible in $\mathbb{F}_p[x]$.

To prove that Φ_{15} is irreducible in $\mathbb{Q}[x]$, it suffices to show that the field extension $\mathbb{Q}(\zeta)$ of $\mathbb{Q}$ generated by any root ζ of $\Phi_{15}(x) = 0$ (in some algebraic closure of $\mathbb{Q}$, if one likes) is of degree equal to the degree of the polynomial Φ_{15}, namely $\varphi(15) = \varphi(3)\varphi(5) = (3-1)(5-1) = 8$. We already know that Φ_3 and Φ_5 are irreducible. And one notes that, given a primitive 15th root of unity ζ, $\eta = \zeta^3$ is a primitive 5th root of unity and $\omega = \zeta^5$ is a primitive third root of unity. And, given a primitive cube root of unity ω and a primitive 5th root of unity η, $\zeta = \omega^2 \cdot \eta^{-3}$ is a primitive 15th root of unity: in fact, if ω and η are produced from ζ, then this formula recovers ζ, since

$$2 \cdot 5 - 3 \cdot 3 = 1$$

Thus,

$$\mathbb{Q}(\zeta) = \mathbb{Q}(\omega)(\eta)$$

By the multiplicativity of degrees in towers of fields

$$[\mathbb{Q}(\zeta) : \mathbb{Q}] = [\mathbb{Q}(\zeta) : \mathbb{Q}(\omega)] \cdot [\mathbb{Q}(\omega) : \mathbb{Q}] = [\mathbb{Q}(\zeta) : \mathbb{Q}(\omega)] \cdot 2 = [\mathbb{Q}(\omega, \eta) : \mathbb{Q}(\omega)] \cdot 2$$

Thus, it would suffice to show that $[\mathbb{Q}(\omega, \eta) : \mathbb{Q}(\omega)] = 4$.

We should not forget that we have shown that $\mathbb{Z}[\omega]$ is Euclidean, hence a PID, hence a UFD. Thus, we are entitled to use Eisenstein's criterion and Gauss' lemma. Thus, it would suffice to prove irreducibility of $\Phi_5(x)$ in $\mathbb{Z}[\omega][x]$. As in the discussion of $\Phi_p(x)$ over $\mathbb{Z}$ with p prime, consider $f(x) = \Phi_5(x+1)$. All its coefficients are divisible by 5, and the constant coefficient is exactly 5 (in particular, not divisible by 5^2). We can apply Eisenstein's criterion and Gauss' lemma if we know, for example, that 5 is a prime in $\mathbb{Z}[\omega]$. (There are other ways to succeed, but this would be simplest.)

To prove that 5 is prime in $\mathbb{Z}[\omega]$, recall the *norm*

$$N(a + b\omega) = (a + b\omega)(a + b\overline{\omega}) = (a + b\omega)(a + b\omega^2) = a^2 - ab + b^2$$

already used in discussing the Euclidean-ness of $\mathbb{Z}[\omega]$. One proves that the norm takes non-negative integer values, is 0 only when evaluated at 0, is *multiplicative* in the sense that $N(\alpha\beta) = N(\alpha)N(\beta)$, and $N(\alpha) = 1$ if and only if α is a unit in $\mathbb{Z}[\omega]$. Thus, if 5 were to factor $5 = \alpha\beta$ in $\mathbb{Z}[\omega]$, then

$$25 = N(5) = N(\alpha) \cdot N(\beta)$$

For a proper factorization, meaning that neither α nor β is a unit, neither $N(\alpha)$ nor $N(\beta)$ can be 1. Thus, both must be 5. However, the equation

$$5 = N(a + b\omega) = a^2 - ab + b^2 = (a - \frac{b}{2})^2 + \frac{3}{4}b^2 = \frac{1}{4}\left((2a - b)^2 + 3b^2\right)$$

has no solution in integers a, b. Indeed, looking at this equation mod 5, since 3 is not a square mod 5 it must be that $b = 0$ mod 5. Then, further, $4a^2 = 0$ mod 5, so $a = 0$ mod 5. That is, 5 divides both a and b. But then 25 divides the norm $N(a + b\omega) = a^2 - ab + b^2$, so it cannot be 5.

Thus, in summary, 5 is prime in $\mathbb{Z}[\omega]$, so we can apply Eisenstein's criterion to $\Phi_5(x + 1)$ to see that it is irreducible in $\mathbb{Z}[\omega][x]$. By Gauss' lemma, it is irreducible in $\mathbb{Q}(\omega)[x]$, so $[\mathbb{Q}(\omega, \eta) : \mathbb{Q}(\omega)] = \varphi(5) = 4$. And this proves that $[\mathbb{Q}(\zeta) : \mathbb{Q}] = 8$, so $\Phi_{15}(x)$ is irreducible over $\mathbb{Q}$.

18.2.7 Example: Let p be a prime. Show that every degree d irreducible in $\mathbb{F}_p[x]$ is a factor of $x^{p^d-1} - 1$. Show that that the $(p^d - 1)^{th}$ cyclotomic polynomial's irreducible factors in $\mathbb{F}_p[x]$ are all of degree d.

Let $f(x)$ be a degree d irreducible in $\mathbb{F}_p[x]$. For a linear factor $x - \alpha$ with α in some field extension of $\mathbb{F}_p$, we know that

$$[\mathbb{F}_p(\alpha) : \mathbb{F}_p] = \text{ degree of minimal poly of } \alpha = \deg f = d$$

Since there is a unique (up to isomorphism) field extension of degree d of $\mathbb{F}_p$, *all* roots of $f(x) = 0$ lie in that field extension $\mathbb{F}_{p^d}$. Since the order of the multiplicative group $\mathbb{F}_{p^d}^\times$ is $p^d - 1$, by Lagrange the order of any non-zero element α of $\mathbb{F}_{p^d}$ is a divisor of $p^d - 1$. That is, α is a root of $x^{p^d-1} - 1 = 0$, so $x - \alpha$ divides $x^{p^d-1} - 1 = 0$. Since f is irreducible, f has no repeated factors, so $f(x) = 0$ has no repeated roots. By unique factorization (these linear factors are mutually distinct irreducibles whose least common multiple is their product), the product of all the $x - \alpha$ divides $x^{p^d-1} - 1$.

For the second part, similarly, look at the linear factors $x - \alpha$ of $\Phi_{p^d-1}(x)$ in a sufficiently large field extension of $\mathbb{F}_p$. Since p does not divide $n = p^d - 1$ there are no repeated factors. The multiplicative group of the field $\mathbb{F}_{p^d}$ is *cyclic*, so contains exactly $\varphi(p^d - 1)$ elements of (maximal possible) order $p^d - 1$, which are roots of $\Phi_{p^d-1}(x) = 0$. The degree of Φ_{p^d-1} is $\varphi(p^d - 1)$, so there are no *other* roots. No proper subfield $\mathbb{F}_{p^e}$ of $\mathbb{F}_{p^d}$ contains *any* elements of order $p^d - 1$, since we know that $e|d$ and the multiplicative group $\mathbb{F}_{p^e}^\times$ is of order $p^e - 1 < p^d - 1$. Thus, any linear factor $x - \alpha$ of $\Phi_{p^d-1}(x)$ has $[\mathbb{F}_p(\alpha) : \mathbb{F}_p] = d$, so the minimal polynomial $f(x)$ of α over $\mathbb{F}_p$ is necessarily of degree d. We claim that f divides Φ_{p^d-1}. Write

$$\Phi_{p^d-1} = q \cdot f + r$$

where q, r are in $\mathbb{F}_p[x]$ and $\deg r < \deg f$. Evaluate both sides to find $r(\alpha) = 0$. Since f was minimal over $\mathbb{F}_p$ for α, necessarily $r = 0$ and f divides the cyclotomic polynomial.

That is, any linear factor of Φ_{p^d-1} (over a field extension) is a factor of a degree d irreducible polynomial in $\mathbb{F}_p[x]$. That is, that cyclotomic polynomial factors into degree d irreducibles in $\mathbb{F}_p[x]$.

18.2.8 Example: Fix a prime p, and let ζ be a primitive p^{th} root of 1 (that is, $\zeta^p = 1$ and no smaller exponent will do). Let

$$V = \det \begin{pmatrix} 1 & 1 & 1 & 1 & \cdots & 1 \\ 1 & \zeta & \zeta^2 & \zeta^3 & \cdots & \zeta^{p-1} \\ 1 & \zeta^2 & (\zeta^2)^2 & (\zeta^2)^3 & \cdots & (\zeta^2)^{p-1} \\ 1 & \zeta^3 & (\zeta^3)^2 & (\zeta^3)^3 & \cdots & (\zeta^3)^{p-1} \\ 1 & \zeta^4 & (\zeta^4)^2 & (\zeta^4)^3 & \cdots & (\zeta^4)^{p-1} \\ \vdots & & & & & \vdots \\ 1 & \zeta^{p-1} & (\zeta^{p-1})^2 & (\zeta^{p-1})^3 & \cdots & (\zeta^{p-1})^{p-1} \end{pmatrix}$$

Compute the rational number V^2.

There are other possibly more natural approaches as well, but the following trick is worth noting. The ij^{th} entry of V is $\zeta^{(i-1)(j-1)}$. Thus, the ij^{th} entry of the square V^2 is

$$\sum_\ell \zeta^{(i-1)(\ell-1)} \cdot \zeta^{(\ell-1)(j-1)} = \sum_\ell \zeta^{(i-1+j-1)(\ell-1)} = \begin{cases} 0 & \text{if } (i-1) + (j-1) \neq 0 \bmod p \\ p & \text{if } (i-1) + (j-1) = 0 \bmod p \end{cases}$$

since

$$\sum_{0 \leq \ell < p} \omega^\ell = 0$$

for any p^{th} root of unity ω other than 1. Thus,

$$V^2 = \begin{pmatrix} p & 0 & 0 & \cdots & 0 & 0 \\ 0 & 0 & 0 & \cdots & 0 & p \\ 0 & 0 & 0 & \cdots & p & 0 \\ & & & \cdot\cdot\cdot & & \\ 0 & 0 & p & \cdots & 0 & 0 \\ 0 & p & 0 & \cdots & 0 & 0 \end{pmatrix}$$

That is, there is a p in the upper left corner, and p's along the anti-diagonal in the lower right $(n-1)$-by-$(n-1)$ block. Thus, granting that the determinant squared is the square of the determinant,

$$(\det V)^2 = \det(V^2) = p^p \cdot (-1)^{(p-1)(p-2)/2}$$

Note that this did not, in fact, depend upon p being prime.

18.2.9 Example: Let $K = \mathbb{Q}(\zeta)$ where ζ is a primitive 15th root of unity. Find 4 fields k strictly between $\mathbb{Q}$ and K.

Let ζ be a primitive 15th root of unity. Then $\omega = \zeta^5$ is a primitive cube root of unity, and $\eta = \zeta^3$ is a primitive fifth root of unity. And $\mathbb{Q}(\zeta) = \mathbb{Q}(\omega)(\eta)$.

Thus, $\mathbb{Q}(\omega)$ is one intermediate field, of degree 2 over $\mathbb{Q}$. And $\mathbb{Q}(\eta)$ is an intermediate field, of degree 4 over $\mathbb{Q}$ (so certainly distinct from $\mathbb{Q}(\omega)$.)

By now we know that $\sqrt{5} \in \mathbb{Q}(\eta)$, so $\mathbb{Q}(\sqrt{5})$ suggests itself as a third intermediate field. But one must be sure that $\mathbb{Q}(\omega) \neq \mathbb{Q}(\sqrt{5})$. We can try a direct computational approach in this simple case: suppose $(a + b\omega)^2 = 5$ with rational a, b. Then

$$5 = a^2 + 2ab\omega + b^2\omega^2 = a^2 + 2ab\omega - b^2 - b^2\omega = (a^2 - b^2) + \omega(2ab - b^2)$$

Thus, $2ab - b^2 = 0$. This requires either $b = 0$ or $2a - b = 0$. Certainly b cannot be 0, or 5 would be the square of a rational number (which we have long ago seen impossible). Try $2a = b$. Then, supposedly,

$$5 = a^2 - 2(2a)^2 = -3a^2$$

which is impossible. Thus, $\mathbb{Q}(\sqrt{5})$ is distinct from $\mathbb{Q}(\omega)$.

We know that $\mathbb{Q}(\omega) = \mathbb{Q}(\sqrt{-3})$. This might suggest

$$\mathbb{Q}(\sqrt{-3} \cdot \sqrt{5}) = \mathbb{Q}(\sqrt{-15})$$

as the fourth intermediate field. We must show that it is distinct from $\mathbb{Q}(\sqrt{-3})$ and $\mathbb{Q}(\sqrt{5})$. If it were equal to either of these, then that field would also contain $\sqrt{5}$ and $\sqrt{-3}$, but we have already checked that (in effect) there is no quadratic field extension of $\mathbb{Q}$ containing both these.

Thus, there are (at least) intermediate fields $\mathbb{Q}(\eta)$, $\mathbb{Q}(\sqrt{-3})$, $\mathbb{Q}(\sqrt{5}$, and $\mathbb{Q}(\sqrt{-15})$.

Exercises

18.1 Find two fields intermediate between $\mathbb{Q}$ and $\mathbb{Q}(\zeta_9)$, where ζ_9 is a primitive 9th root of unity.

18.2 Find two fields intermediate between $\mathbb{Q}$ and $\mathbb{Q}(\zeta_8)$, where ζ_8 is a primitive 8th root of unity.

18.3 Find the smallest exponent ℓ such that the irreducible $x^3 + x + 1$ in $\mathbb{F}_2[x]$ divides $x^{2^\ell} - x$.

18.4 Find the smallest exponent ℓ such that the irreducible $x^3 - x + 1$ in $\mathbb{F}_3[x]$ divides $x^{3^\ell} - x$.

19. Roots of unity

19.1 *Another proof of cyclicness*

Earlier, we gave a more complicated but more elementary proof of the following theorem, using cyclotomic polynomials. There is a cleaner proof using the structure theorem for finite abelian groups, which we give now. [1]   Thus, this result is yet another corollary of the structure theory for finitely-generated free modules over PIDs.

19.1.1 Theorem: Let G be a finite subgroup of the multiplicative group $k^\times$ of a field k. Then G is cyclic.

Proof: By the structure theorem, applied to abelian groups as $\mathbb{Z}$-modules,

$$G \approx \mathbb{Z}/d_1 \oplus \ldots \oplus \mathbb{Z}/d_n$$

where the integers d_i have the property $1 < d_1 | \ldots | d_n$ and no elementary divisor d_i is 0 (since G is finite). All elements of G satisfy the equation

$$x^{d_t} = 1$$

---

[1] The argument using cyclotomic polynomials is wholesome and educational, too, but is much grittier than the present argument.

By unique factorization in $k[x]$, this equation has at most d_t roots in k. Thus, there can be only one direct summand, and G is cyclic.   ///

19.1.2 Remark: Although we will not need to invoke this theorem for our discussion just below of solutions of equations

$$x^n = 1$$

one might take the viewpoint that the traditional pictures of these solutions as points on the unit circle in the complex plane are not at all misleading about more general situations.

19.2 *Roots of unity*

An element ω in any field k with the property that $\omega^n = 1$ for some integer n is a **root of unity**. For positive integer n, if $\omega^n = 1$ and $\omega^t \neq 1$ for positive integers [2] $t < n$, then ω is a **primitive** n^{th} root of unity. [3]

Note that

$$\mu_n = \{\alpha \in k^\times : \alpha^n = 1\}$$

is finite since there are at most n solutions to the degree n equation $x^n = 1$ in any field. This group is known to be *cyclic*, by at least two proofs.

19.2.1 Proposition: Let k be a field and n a positive integer not divisible by the characteristic of the field. An element $\omega \in k^\times$ is a primitive n^{th} root of unity in k if and only if ω is an element of order n in the group μ_n of *all* n^{th} roots of unity in k. If so, then

$$\{\omega^\ell : 1 \leq \ell \leq n, \text{ and } \gcd(\ell, n) = 1\}$$

is a complete (and irredundant) list of all the primitive n^{th} roots of unity in k. A complete and irredundant list of *all* n^{th} roots of unity in k is

$$\{\omega^\ell : 1 \leq \ell \leq n\} = \{\omega^\ell : 0 \leq \ell \leq n - 1\}$$

Proof: To say that ω is a *primitive* n^{th} root of unity is to say that its order in the group $k^\times$ is n. Thus, it generates a cyclic group of order n inside $k^\times$. Certainly any integer power ω^ℓ is in the group μ_n of nth roots of unity, since

$$(\omega^\ell)^n = (\omega^n)^\ell = 1^\ell = 1$$

Since the group generated by ω is inside μ_n and has at least as large cardinality, it is the whole. On the other hand, a generator for μ_n has order n (or else would generate a strictly smaller group). This proves the equivalence of the conditions describing primitive n^{th} roots of unity.

---

[2] If $\omega^n = 1$ then in any case the *smallest* positive integer ℓ such that $\omega^\ell = 1$ is a divisor of n. Indeed, as we have done many times already, write $n = q\ell + r$ with $0 \leq r < |\ell|$, and $1 = \omega^n = \omega^{q\ell+r} = \omega^r$. Thus, since ℓ is least, $r = 0$, and ℓ divides n.

[3] If the characteristic p of the field k divides n, then there are no primitive n^{th} roots of unity in k. Generally, for $n = p^e m$ with p not dividing m, $\Phi_{p^e m}(x) = \Phi_m(x)^{\varphi(p^e)} = \Phi_m(x)^{(p-1)p^{e-1}}$. We'll prove this later.

As in the more general proofs of analogous results for finite cyclic groups, the set of all elements of a cyclic group of order n is the collection of powers $\omega^1, \omega^2, \ldots, \omega^{n-1}, \omega^n$ of any generator ω of the group.

As in the more general proofs of analogous results for cyclic groups, the order of a power ω^ℓ of a generator ω is exactly $n/\gcd(n, \ell)$, since $(\omega^\ell)^t = 1$ if and only if $n|\ell t$. Thus, the set given in the statement of the proposition is a set of primitive n^{th} roots of unity. There are $\varphi(n)$ of them in this set, where φ is Euler's totient-function. ///

19.3 $\mathbb{Q}$ *with roots of unity adjoined*

One of the general uses of *Galois theory* is to understand fields intermediate between a base field k and an algebraic field extension K of k. In the case of *finite fields* we already have simple means to completely understand intermediate fields. Any situation beyond from the finite field case is more complicated. But, to provide further examples, it is possible to consider fields intermediate between $\mathbb{Q}$ and $\mathbb{Q}(\zeta)$ where ζ is a (primitive) n^{th} root of unity.

There are obvious and boring inclusions, since if ζ is a primitive mn^{th} root of unity, then ζ^m is a primitive n^{th} root of unity. That is, we have

$$\mathbb{Q}(\text{primitive } n^{th} \text{ root of unity}) \subset \mathbb{Q}(\text{primitive } mn^{th} \text{ root of unity})$$

In any case, by the *multiplicativity* of field extension degrees in towers, for a primitive n^{th} root of unity ζ, given

$$\mathbb{Q} \subset k \subset \mathbb{Q}(\zeta)$$

we have

$$[\mathbb{Q}(\zeta) : k] \cdot [k : \mathbb{Q}] = [\mathbb{Q}(\zeta) : \mathbb{Q}]$$

In particular, for prime $n = p$, we have already seen that Eisenstein's criterion proves that the pth cyclotomic polynomial $\Phi_p(x)$ is irreducible of degree $\varphi(p) = p - 1$, so

$$[\mathbb{Q}(\zeta) : \mathbb{Q}] = p - 1$$

We will discuss the irreducibility of other cyclotomic polynomials a bit later.

19.3.1 Example: With

$$\zeta_5 = \text{ a primitive fifth root of unity}$$

$$[\mathbb{Q}(\zeta_5) : \mathbb{Q}] = 5 - 1 = 4$$

so any field k intermediate between $\mathbb{Q}(\zeta_5)$ and $\mathbb{Q}$ must be quadratic over $\mathbb{Q}$. In particular, from

$$\zeta_5^4 + \zeta_5^3 + \zeta_5^2 + \zeta_5 + 1 = 0$$

by dividing through by ζ_5^2 we obtain

$$\zeta_5^2 + \zeta_5 + 1 + \zeta_5^{-1} + \zeta_5^{-2} = 0$$

and this can be rearranged to

$$\left(\zeta_5 + \frac{1}{\zeta_5}\right)^2 + \left(\zeta_5 + \frac{1}{\zeta_5}\right) - 1 = 0$$

Letting

$$\xi = \xi_5 = \zeta_5 + \frac{1}{\zeta_5}$$

we have

$$\xi^2 + \xi - 1 = 0$$

so

$$\xi = \frac{-1 \pm \sqrt{1 - 4(-1)}}{2} = \frac{-1 \pm \sqrt{5}}{2}$$

From the standard picture of 5th roots of unity in the complex plane, we have

$$\xi = \zeta_5 + \frac{1}{\zeta_5} = e^{2\pi i/5} + e^{-2\pi i/5} = 2\cos\frac{2\pi}{5} = 2\cos 72^o$$

Therefore,

$$\cos\frac{2\pi}{5} = \frac{-1 + \sqrt{5}}{4}$$

It should be a bit surprising that

$$\mathbb{Q}(\sqrt{5}) \subset \mathbb{Q}(\zeta_5)$$

To prove that there are no *other* intermediate fields will require more work.

19.3.2 Example: With

$$\zeta_7 = \text{ a primitive seventh root of unity}$$

$$[\mathbb{Q}(\zeta_7) : \mathbb{Q}] = 7 - 1 = 6$$

so any field k intermediate between $\mathbb{Q}(\zeta_7)$ and $\mathbb{Q}$ must be *quadratic* or *cubic* over $\mathbb{Q}$. We will find one of each degree. We can use the same front-to-back symmetry of the cyclotomic polynomial that we exploited for a fifth root of 1 in the previous example. In particular, from

$$\zeta_7^6 + \zeta_7^5 + \zeta_7^4 + \zeta_7^3 + \zeta_7^2 + \zeta_7 + 1 = 0$$

by dividing through by ζ_7^3

$$\zeta_7^3 + \zeta_7^2 + \zeta_7 + 1 + \zeta_7^{-1} + \zeta_7^{-2} + \zeta_7^{-3} = 0$$

and thus

$$\left(\zeta_7 + \frac{1}{\zeta_7}\right)^3 + \left(\zeta_7 + \frac{1}{\zeta_7}\right)^2 - 2\left(\zeta_7 + \frac{1}{\zeta_7}\right) - 1 = 0$$

Again letting

$$\xi = \xi_7 = \zeta_7 + \frac{1}{\zeta_7}$$

we have

$$\xi^3 + \xi^2 - 2\xi - 1 = 0$$

and in the complex plane

$$\xi = \zeta_7 + \frac{1}{\zeta_7} = e^{2\pi i/7} + e^{-2\pi i/7} = 2\cos\frac{2\pi}{7}$$

Thus,

$$[\mathbb{Q}(\xi_7) : \mathbb{Q}] = 3$$

We will return to this number in a moment, after we find the intermediate field that is *quadratic* over $\mathbb{Q}$.

Take $n = p$ prime for simplicity. Let's think about the front-to-back symmetry a bit more, to see whether it can suggest something of broader applicability. Again, for any primitive p^{th} root of unity $\zeta = \zeta_p$, and for a relatively prime to p, ζ^a is another primitive p^{th} root of unity. Of course, since $\zeta^p = 1$. ζ^a only depends upon $a \bmod p$. Recalling that $1, \zeta.\zeta^2, \ldots, \zeta^{p-3}, \zeta^{p-2}$ is a $\mathbb{Q}$-basis [4] for $\mathbb{Q}(\zeta)$, we claim that the map

$$\sigma_a : c_0 + c_1\zeta + c_2\zeta^2 + c_3\zeta^3 + \ldots + c_{p-2}\zeta^{p-2} \longrightarrow c_0 + c_1\zeta^a + c_2\zeta^{2a} + c_3\zeta^{3a} + \ldots + c_{p-2}\zeta^{(p-2)a}$$

is a $\mathbb{Q}$-algebra automorphism of $\mathbb{Q}(\zeta)$. That is, σ_a raises each ζ^j to the ath power. Since, again, ζ^j only depends upon $j \bmod p$, all the indicated powers of ζ are primitive p^{th} roots of 1. The $\mathbb{Q}$-linearity of this map is built into its definition, but the multiplicativity is not obvious. Abstracting just slightly, we have

19.3.3 Proposition: Let k be a field, $k(\alpha)$ a finite algebraic extension, where $f(x)$ is the minimal polynomial of α over k. Let $\beta \in k(\alpha)$ be another root [5] of $f(x) = 0$. Then there is a unique field automorphism [6] σ of $k(\alpha)$ over k sending α to β, given by the formula

$$\sigma\left(\sum_{0 \le i < \deg f} c_i\alpha^i\right) = \sum_{0 \le i < \deg f} c_i\beta^i$$

where $c_i \in \mathbb{Q}$.

Proof: Thinking of the universal mapping property of the polynomial ring $k[x]$, let

$$q_\alpha : k[x] \longrightarrow k[\alpha] = k(\alpha)$$

be the unique k-algebra homomorphism sending $x \longrightarrow \alpha$. By definition of the minimal polynomial f of α over k, the kernel of a_α is the principal ideal $\langle f \rangle$ in $k[x]$ generated by f. Let

$$q_\beta : k[x] \longrightarrow k[\alpha] = k(\alpha)$$

be the unique k-algebra homomorphism [7] sending $x \longrightarrow \beta$. Since β satisfies the same monic equation $f(x) = 0$ with f irreducible, the kernel of q_β is also the ideal $\langle f \rangle$. Thus,

---

[4] Yes, the highest index is $p-2$, not $p-1$, and not p. The pth cyclotomic polynomial is of degree $p-1$, and in effect gives a non-trivial linear dependence relation among $1, \zeta, \zeta^2, \ldots, \zeta^{p-2}, \zeta^{p-1}$.

[5] It is critical that the second root lie in the field generated by the first. This issue is a presagement of the idea of *normality* of $k(\alpha)$ over k, meaning that *all* the other roots of the minimal polynomial of α lie in $k(\alpha)$ already. By contrast, for example, the field $\mathbb{Q}(\alpha)$ for any cube root α of 2 does *not* contain any *other* cube roots of 2. Indeed, the ratio of two such would be a primitive cube root of unity lying in $\mathbb{Q}(\alpha)$, which various arguments show is impossible.

[6] This use of the phrase *automorphism over* is standard terminology: a field automorphism $\tau : K \longrightarrow K$ of a field K to itself, with τ fixing every element of a subfield k, is an automorphism of K over k.

[7] Such a homomorphism exists for *any* element β of any k-algebra $k[\alpha]$, whether or not β is related to α.

since
$$\ker q_\beta \supset \ker q_\alpha$$

the map q_β factors through q_α in the sense that there is a unique k-algebra homomorphism
$$\sigma : k(\alpha) \longrightarrow k(\alpha)$$

such that
$$q_\beta = \sigma \circ q_\alpha$$

That is, the obvious attempt at defining σ, by

$$\sigma \left(\sum_{0 \le i < \deg f} c_i \alpha^i \right) = \sum_{0 \le i < \deg f} c_i \beta^i$$

with $c_i \in \mathbb{Q}$ gives a *well-defined* map. [8]   Since

$$\dim_k \sigma(k(\alpha)) = \dim_k q_\beta(k[x]) = \deg f = \dim_k k[\alpha] = \dim_k k(\alpha)$$

the map σ is *bijective*, hence invertible.   ///

19.3.4 Corollary: Let p be prime and ζ a primitive p^{th} root of unity.   The automorphism group $\mathrm{Aut}(\mathbb{Q}(\zeta)/\mathbb{Q})$ is isomorphic to

$$(\mathbb{Z}/p)^\times \approx \mathrm{Aut}(\mathbb{Q}(\zeta)/\mathbb{Q})$$

by the map
$$a \leftrightarrow \sigma_a$$

where
$$\sigma_a(\zeta) = \zeta^a$$

Proof: This uses the irreducibility of $\Phi_p(x)$ in $\mathbb{Q}[x]$. Thus, for all $a \in (\mathbb{Z}/p)^\times$ the power ζ^a is another root of $\Phi_p(x) = 0$, and $\Phi_p(x)$ is the minimal polynomial of both ζ and ζ^a. This gives an injection
$$(\mathbb{Z}/p)^\times \longrightarrow \mathrm{Aut}(\mathbb{Q}(\zeta)/\mathbb{Q})$$

On the other hand, any automorphism σ of $\mathbb{Q}(\zeta)$ over $\mathbb{Q}$ must send ζ to another root of its minimal polynomial, so $\sigma(\zeta) = \zeta^a$ for some $a \in (\mathbb{Z}/p)^\times$, since all primitive p^{th} roots of unity are so expressible. This proves that the map is surjective.   ///

Returning to roots of unity: for a primitive p^{th} root of unity ζ, the map

$$\zeta \longrightarrow \zeta^{-1}$$

maps ζ to another primitive p^{th} root of unity lying in $\mathbb{Q}(\zeta)$, so this map extends to an automorphism
$$\sigma_{-1} : \mathbb{Q}(\zeta) \longrightarrow \mathbb{Q}(\zeta)$$

---

[8] Note that this approach makes the multiplicativity easy, packaging all the issues into the well-definedness, which then itself is a straightforward consequence of the hypothesis that α and β are two roots of the same equation, and that $\beta \in k(\alpha)$.

of $\mathbb{Q}(\zeta)$ over $\mathbb{Q}$. And[9]

$$2\cos\frac{2\pi}{p} = \xi = \zeta + \frac{1}{\zeta} = \zeta + \sigma_{-1}(\zeta)$$

Of course, the identity map on $\mathbb{Q}(\zeta)$ is the automorphism σ_1, and

$$\sigma_{-1}^2 = \sigma_1$$

That is,

$$\{\sigma_1, \sigma_{-1}\}$$

is a *subgroup* of the group of automorphisms of $\mathbb{Q}(\zeta)$ over $\mathbb{Q}$. Indeed, the map

$$a \longrightarrow \sigma_a$$

is a *group homomorphism*

$$(\mathbb{Z}/p)^\times \longrightarrow \mathrm{Aut}(\mathbb{Q}(\zeta)/\mathbb{Q})$$

since

$$\sigma_a\left(\sigma_b(\zeta)\right) = \sigma_a(\zeta^b) = (\sigma_a\zeta)^b$$

since σ_a is a ring homomorphism. Thus, recapitulating a bit,

$$\sigma_a\left(\sigma_b(\zeta)\right) = \sigma_a(\zeta^b) = (\sigma_a\zeta)^b = (\zeta^a)^b = \sigma_{ab}(\zeta)$$

That is, we can take the viewpoint that ξ is formed from ζ by a certain amount of **averaging** or **symmetrizing** over the subgroup $\{\sigma_1, \sigma_{-1}\}$ of automorphisms.

That this symmetrizing or averaging does help isolate elements in smaller subfields of cyclotomic fields $\mathbb{Q}(\zeta)$ is the content of

19.3.5 Proposition: Let G be the group of automorphisms of $\mathbb{Q}(\zeta_p)$ over $\mathbb{Q}$ given by σ_a for $a \in (\mathbb{Z}/p)^\times$. Let $\alpha \in \mathbb{Q}(\zeta_p)$.

$$\alpha \in \mathbb{Q} \quad \text{if and only if} \quad \sigma(\alpha) = \alpha \quad \text{for all } \sigma \in G$$

Proof: Certainly elements of $\mathbb{Q}$ are invariant under all σ_a, by the definition. Let[10]

$$\alpha = \sum_{1 \leq i \leq p-1} c_i \zeta^i$$

with $c_i \in \mathbb{Q}$. The condition $\alpha = \sigma_a(\alpha)$ is

$$\sum_{1 \leq i \leq p-1} c_i \zeta^i = \sum_{1 \leq i \leq p-1} c_i \zeta^{ai}$$

---

[9] Writing an algebraic number in terms of cosine is not quite right, though it is appealing. The problem is that unless we choose an imbedding of $\mathbb{Q}(\zeta)$ into the complex numbers, we cannot really know *which* root of unity we have chosen. Thus, we cannot know which angle's cosine we have. Nevertheless, it is useful to think about this.

[10] It is a minor cleverness to use the $\mathbb{Q}$-basis ζ^i with $1 \leq i \leq p-1$ rather than the $\mathbb{Q}$-basis with $0 \leq i \leq p-2$. The point is that the latter is stable under the automorphisms σ_a, while the former is not.

Since $\zeta^p = 1$, the powers ζ^{ai} only depend upon $ai \bmod p$. The map

$$i \longrightarrow ai \bmod p$$

permutes $\{i : 1 \leq i \leq p-1\}$. Thus, looking at the coefficient of ζ^a as a varies, the equation $\alpha = \sigma_a(\alpha)$ gives

$$c_a = c_1$$

That is, the G-invariance of α requires that α be of the form

$$\alpha = c \cdot (\zeta + \zeta^2 + \ldots + \zeta^{p-1}) = c \cdot (1 + \zeta + \zeta^2 + \ldots + \zeta^{p-1}) - c = -c$$

for $c \in \mathbb{Q}$, using

$$0 = \Phi_p(\zeta) = 1 + \zeta + \zeta^2 + \ldots + \zeta^{p-1}$$

That is, G-invariance implies rationality.   ///

19.3.6 Corollary: Let H be a subgroup of $G = (\mathbb{Z}/p)^\times$, identified with a group of automorphisms of $\mathbb{Q}(\zeta)$ over $\mathbb{Q}$ by $a \longrightarrow \sigma_a$. Let $\alpha \in \mathbb{Q}(\zeta_p)$ be fixed under H. Then

$$[\mathbb{Q}(\alpha) : \mathbb{Q}] \leq [G : H] = \frac{|G|}{|H|}$$

Proof: Since α is H-invariant, the value

$$\sigma_a(\alpha)$$

depends only upon the image of a in G/H, that is, upon the coset $aH \in G/H$. Thus, in

$$f(x) = \prod_{a \in G/H} (x - \sigma_a(\alpha)) \in \mathbb{Q}(\zeta)[x]$$

everything is well-defined. Since it is a ring homomorphism, $\sigma_b \in G$ may be applied to this polynomial factor-wise (acting trivially upon x, of course) merely permuting the $\sigma_a(\alpha)$ among themselves. That is, G fixes this polynomial. On the other hand, multiplying the factors out, this invariance implies that the coefficients of f are G-invariant. By the proposition, the coefficients of f are in $\mathbb{Q}$. Thus, the degree of α over $\mathbb{Q}$ is at most the index $[G : H]$.   ///

19.3.7 Remark: We know that $(\mathbb{Z}/p)^\times$ is *cyclic* of order $p-1$, so we have many explicit subgroups available in any specific numerical example.

19.3.8 Example: Returning to $p = 7$, with $\zeta = \zeta_7$ a primitive 7th root of unity, we want an element of $\mathbb{Q}(\zeta_7)$ of degree 2 over $\mathbb{Q}$. Thus, by the previous two results, we want an element invariant under the (unique[11]) subgroup H of $G = (\mathbb{Z}/7)^\times$ of order 3. Since $2^3 = 1 \bmod 7$, (and $2 \neq 1 \bmod 7$) the automorphism

$$\sigma_2 : \zeta \longrightarrow \zeta^2$$

---

[11] The group $(\mathbb{Z}/7)^\times$ is cyclic, since 7 is prime.

generates the subgroup H of order 3. Thus, consider

$$\alpha = \zeta + \sigma_2(\zeta) + \sigma_2^2(\zeta) = \zeta + \zeta^2 + \zeta^4$$

Note that this α is *not* invariant under σ_3, since

$$\sigma_3(\zeta + \zeta^2 + \zeta^4) = \zeta^3 + \zeta^6 + \zeta^{12} = \zeta^3 + \zeta^5 + \zeta^6$$

That is, $\alpha \notin \mathbb{Q}$. Of course, this is clear from its expression as a linear combination of powers of ζ. Thus, we have not overshot the mark in our attempt to make a field element inside a smaller subfield. The corollary assures that

$$[\mathbb{Q}(\alpha) : \mathbb{Q}] \leq [G : H] = \frac{6}{3} = 2$$

Since $\alpha \notin \mathbb{Q}$, we must have equality. The corollary assures us that

$$f(x) = (x - \alpha)(x - \sigma_3(\alpha))$$

has rational coefficients. Indeed, the linear coefficient is

$$-\left((\zeta + \zeta^2 + \zeta^4) + (\zeta^3 + \zeta^6 + \zeta^{12})\right) = -\left(1 + \zeta + \zeta^2 + \ldots + \zeta^5 + \zeta^6)\right) - 1 = -1$$

since $1 + \zeta + \ldots + \zeta^6 = 0$. The constant coefficient is

$$(\zeta + \zeta^2 + \zeta^4) \cdot (\zeta^3 + \zeta^6 + \zeta^{12})$$

$$= \zeta^{(1+3)} + \zeta^{(1+6)} + \zeta^{(1+12)}\zeta^{(2+3)} + \zeta^{(2+6)} + \zeta^{(2+12)}\zeta^{(4+3)} + \zeta^{(4+6)} + \zeta^{(4+12)}$$

$$= \zeta^4 + 1 + \zeta^6 + \zeta^5 + \zeta + 1 + 1 + \zeta^3 + \zeta^2 = 2$$

Thus, $\alpha = \zeta + \zeta^2 + \zeta^4$ satisfies the quadratic equation

$$x^2 + x + 2 = 0$$

On the other hand, by the quadratic formula we have the roots

$$\alpha = \frac{-1 \pm \sqrt{(-1)^2 - 4 \cdot 2}}{2} = \frac{-1 \pm \sqrt{-7}}{2}$$

That is,

$$\mathbb{Q}(\sqrt{-7}) \subset \mathbb{Q}(\zeta_7)$$

This is not obvious. [12]

19.4 *Solution in radicals, Lagrange resolvents*

As an example, we follow a method of *J.-L. Lagrange* to obtain an expression for

$$\xi = \xi_7 = \zeta_7 + \frac{1}{\zeta_7}$$

---

[12] Not only is this assertion not obvious, but, also, there is the mystery of why it is $\sqrt{-7}$, not $\sqrt{7}$.

in terms of *radicals*, that is, in terms of *roots*. Recall from above that ξ satisfies the cubic equation[13]

$$x^3 + x^2 - 2x - 1 = 0$$

Lagrange's method was to create an expression in terms of the roots of an equation designed to have more accessible symmetries than the original. In this case, let ω be a cube root of unity, not necessarily primitive. For brevity, let $\tau = \sigma_2$. The **Lagrange resolvent** associated to ξ and ω is

$$\lambda = \xi + \omega\tau(\xi) + \omega^2\tau^2(\xi)$$

Since $\sigma_{-1}(\xi) = \xi$, the effect on ξ of σ_a for $a \in G = (\mathbb{Z}/7)^\times$ depends only upon the coset $aH \in G/H$ where $H = \{\pm 1\}$. Convenient representatives for this quotient are $\{1, 2, 4\}$, which themselves form a subgroup.[14] Grant for the moment that we can extend σ_a to an automorphism on $\mathbb{Q}(\xi, \omega)$ over $\mathbb{Q}(\omega)$, which we'll still denote by σ_a.[15] Then the simpler behavior of the Lagrange resolvent λ under the automorphism $\tau = \sigma_2$ is

$$\tau(\lambda) = \tau(\xi + \omega\tau(\xi) + \omega^2\tau^2(\xi)) = \tau(\xi) + \omega\tau^2(\xi) + \omega^2\tau^3(\xi) = \tau(\xi) + \omega\tau^2(\xi) + \omega^2\xi = \omega^{-1} \cdot \lambda$$

since $\tau^3(\xi) = \xi$. Similarly, $\tau^2(\lambda) = \omega^{-2} \cdot \lambda$. Consider

$$f(x) = (x - \lambda)(x - \tau(\lambda))(x - \tau^2(\lambda)) = (x - \lambda)(x - \omega^{-1}\lambda)(x - \omega\lambda)$$

Multiplying this out, since $1 + \omega + \omega^2 = 0$,

$$f(x) = x^3 - \lambda^3$$

And note that, because τ is a ring homomorphism,

$$\tau(\lambda^3) = (\tau(\lambda))^3 = (\omega^{-1}\lambda)^3 = \lambda^3$$

Therefore,[16] $\lambda^3 \in \mathbb{Q}(\omega)$. What is it? Let α, β, γ be the three roots of $x^3 + x^2 - 2x - 1 = 0$.

$$\lambda^3 = \left(\xi + \omega\tau(\xi) + \omega^2\tau^2(\xi)\right)^3 = \left(\alpha + \omega\beta + \omega^2\gamma\right)^3$$

$$= \alpha^3 + \beta^3 + \gamma^3 + 3\omega\alpha^2\beta + 3\omega^2\alpha\beta^2 + 3\omega^2\alpha^2\gamma + 3\omega\alpha\gamma^2 + 3\omega\beta^2\gamma + 3\omega^2\beta\gamma^2 + 6\alpha\beta\gamma$$

---

[13] After some experimentation, one may notice that, upon replacing x by $x + 2$, the polynomial $x^3 + x^2 - 2x - 1$ becomes

$$x^3 + (3 \cdot 2 + 1)x^2 + (3 \cdot 2^2 + 2 \cdot 2 - 2)x + (2^3 + 2^2 - 2 \cdot 2 - 1) = x^3 + 7x^2 - 14x + 7$$

which by Eisenstein's criterion is *irreducible* in $\mathbb{Q}[x]$. Thus, $[\mathbb{Q}(\xi) : \mathbb{Q}] = 3$. This irreducibility is part of a larger pattern involving roots of unity and Eisenstein's criterion.

[14] That there is a set of representatives forming a subgroup ought not be surprising, since a cyclic group of order 6 is isomorphic to $\mathbb{Z}/2 \oplus \mathbb{Z}/3$, by either the structure theorem, or, more simply, by Sun-Ze's theorem.

[15] All of ω, $\zeta = \zeta_7$, and ξ are contained in $\mathbb{Q}(\zeta_{21})$, for a primitive 21th root of unity ζ_{21}. Thus, the *compositum* field $\mathbb{Q}(\xi, \omega)$ can be taken inside $\mathbb{Q}(\zeta_{21})$.

[16] We will return to address this variant of our earlier proposition and corollary about invariant expressions lying in subfields.

$$= \alpha^3 + \beta^3 + \gamma^3 + 3\omega(\alpha^2\beta + \beta^2\gamma + \gamma^2\alpha) + 3\omega^2(\alpha\beta^2 + \beta\gamma^2 + \alpha^2\gamma) + 6\alpha\beta\gamma$$

Since $\omega^2 = -1 - \omega$ this is

$$\alpha^3 + \beta^3 + \gamma^3 + 6\alpha\beta\gamma + 3\omega(\alpha^2\beta + \beta^2\gamma + \gamma^2\alpha) - 3\omega(\alpha\beta^2 + \beta\gamma^2 + \alpha^2\gamma) - 3(\alpha\beta^2 + \beta\gamma^2 + \alpha^2\gamma)$$

In terms of elementary symmetric polynomials,

$$\alpha^3 + \beta^3 + \gamma^3 = s_1^3 - 3s_1 s_2 + 3s_3$$

Thus,

$$\lambda^3 = s_1^3 - 3s_1 s_2 + 9s_3 + 3\omega(\alpha^2\beta + \beta^2\gamma + \gamma^2\alpha) - 3\omega(\alpha\beta^2 + \beta\gamma^2 + \alpha^2\gamma) - 3(\alpha\beta^2 + \beta\gamma^2 + \alpha^2\gamma)$$

Note that neither of the two expressions

$$\alpha^2\beta + \beta^2\gamma + \gamma^2\alpha \qquad \alpha\beta^2 + \beta\gamma^2 + \alpha^2\gamma$$

is invariant under *all* permutations of α, β, γ, but only under powers of the *cycle*

$$\alpha \longrightarrow \beta \longrightarrow \gamma \longrightarrow \alpha$$

Thus, we cannot expect to use the symmetric polynomial algorithm to express the two parenthesized items in terms of elementary symmetric polynomials. A more specific technique is necessary.

Writing α, β, γ in terms of the 7th root of unity ζ gives

$$\alpha\beta^2 + \beta\gamma^2 + \gamma\alpha^2 = (\zeta + \zeta^6)(\zeta^2 + \zeta^5)^2 + (\zeta^2 + \zeta^5)(\zeta^4 + \zeta^3)^2 + (\zeta^4 + \zeta^3)(\zeta + \zeta^6)^2$$

$$= (\zeta + \zeta^6)(\zeta^4 + 2 + \zeta^3) + (\zeta^2 + \zeta^5)(\zeta + 2 + \zeta^6) + (\zeta^4 + \zeta^3)(\zeta^2 + 2\zeta^5)$$

$$= (\zeta + \zeta^6)(\zeta^4 + 2 + \zeta^3) + (\zeta^2 + \zeta^5)(\zeta + 2 + \zeta^6) + (\zeta^4 + \zeta^3)(\zeta^2 + 2 + \zeta^5)$$

$$= 4(\zeta + \zeta^2 + \zeta^3 + \zeta^4 + \zeta^5 + \zeta^6)$$

$$= -4$$

since [17] $\Phi_7(\zeta) = 0$. This is one part of the second parenthesized expression. The other is superficially very similar, but in fact has different details:

$$\alpha^2\beta + \beta^2\gamma + \gamma^2\alpha = (\zeta + \zeta^6)^2(\zeta^2 + \zeta^5) + (\zeta^2 + \zeta^5)^2(\zeta^4 + \zeta^3) + (\zeta^4 + \zeta^3)^2(\zeta + \zeta^6)$$

$$= (\zeta^2 + 2 + \zeta^5)(\zeta^2 + \zeta^5) + (\zeta^4 + 2 + \zeta^3)(\zeta^4 + \zeta^3) + (\zeta + 2 + \zeta^6)(\zeta + \zeta^6)$$

$$= 6 + 3(\zeta + \zeta^2 + \zeta^3 + \zeta^4 + \zeta^5 + \zeta^6) = 3$$

From the equation $x^3 + x^2 - 2x - 1 = 0$ we have

$$s_1 = -1 \qquad s_2 = -2 \qquad s_3 = 1$$

----

[17] Anticipating that ζ must not appear in the final outcome, we could have managed some slightly clever economies in this computation. However, the element of redundancy here is a useful check on the accuracy of the computation.

Putting this together, we have

$$\lambda^3 = s_1^3 - 3s_1s_2 + 9s_3 + 3\omega \cdot 3 - 3\omega \cdot (-4) - 3(-4)$$

$$= (-1)^3 - 3(-1)(-2) + 9(1) + 3\omega \cdot 3 - 3\omega \cdot (-4) - 3(-4)$$

$$= -1 - 6 + 9 + +9\omega + 12\omega + 12 = 14 + 21\omega$$

That is,

$$\lambda = \sqrt[3]{14 + 21\omega}$$

or, in terms of ξ

$$\xi + \omega\tau(\xi) + \omega^2\tau^2(\xi) = \sqrt[3]{14 + 21\omega}$$

Now we will obtain a system of three linear equations which we can solve for ξ.

The same computation works for ω^2 in place of ω, since ω^2 is another primitive cube root of 1. The computation is much easier when ω is replaced by 1, since

$$(\alpha + 1 \cdot \beta + 1^2 \cdot \gamma)^3$$

is already $s_1^3 = -1$ Thus, fixing a primitive cube root ω of 1, we have

$$\begin{cases} \xi + \tau(\xi) + \tau^2(\xi) & = & -1 \\ \xi + \omega\tau(\xi) + \omega^2\tau^2(\xi) & = & \sqrt[3]{14 + 21\omega} \\ \xi + \omega^2\tau(\xi) + \omega\tau^2(\xi) & = & \sqrt[3]{14 + 21\omega^2} \end{cases}$$

Solving for ξ gives

$$\xi = \frac{-1 + \sqrt[3]{14 + 21\omega} + \sqrt[3]{14 + 21\omega^2}}{3}$$

Despite appearances, we know that ξ can in some sense be expressed without reference to the cube root of unity ω, since

$$\xi^3 + \xi^2 - 2\xi - 1 = 0$$

and this equations has rational coefficients. The apparent entanglement of a cube root of unity is an artifact of our demand to express ξ in terms of root-taking.

19.4.1 Remark: There still remains the issue of being sure that the automorphisms σ_a of $\mathbb{Q}(\zeta)$ over $\mathbb{Q}$ (with ζ a primitive 7th root of unity) can be extended to automorphisms of $\mathbb{Q}(\zeta_7, \omega)$ over $\mathbb{Q}(\omega)$. As noted above, for a primitive 21th root of unity η, we have

$$\zeta = \eta^3 \qquad \omega = \eta^7$$

so all the discussion above can take place inside $\mathbb{Q}(\eta)$.

We can take advantage of the fact discussed earlier that $\mathbb{Z}[\omega]$ is Euclidean, hence a PID.[18] Note that 7 is no longer prime in $\mathbb{Z}[\omega]$, since

$$7 = (2 - \omega)(2 - \omega^2) = (2 - \omega)(3 + \omega)$$

Let

$$N(a + b\omega) = (a + b\omega)(a + b\omega^2)$$

---

[18] We will eventually give a systematic proof that all cyclotomic polynomials are irreducible in $\mathbb{Q}[x]$.

be the *norm* discussed earlier. It is a multiplicative map $\mathbb{Z}[\omega] \longrightarrow \mathbb{Z}$, $N(a+b\omega) = 0$ only for $a + b\omega = 0$, and $N(a+b\omega) = 1$ if and only if $a + b\omega$ is a unit in $\mathbb{Z}[\omega]$. One computes directly

$$N(a+b\omega) = a^2 - ab + b^2$$

Then both $2-\omega$ and $3+\omega$ *are* prime in $\mathbb{Z}[\omega]$, since their norms are 7. They are not associate, however, since the hypothesis $3 + \omega = \mu \cdot (2 - \omega)$ gives

$$5 = (3+\omega) + (2-\omega) = (1+\mu)(2-\omega)$$

and then taking norms gives

$$25 = 7 \cdot N(1+\mu)$$

which is impossible. Thus, 7 is not a unit, and is square-free in $\mathbb{Z}[\omega]$.

In particular, we can still apply Eisenstein's criterion and Gauss' lemma to see that $\Phi_7(x)$ is irreducible in $\mathbb{Q}(\omega)[x]$. In particular,

$$[\mathbb{Q}(\zeta_7,\omega) : \mathbb{Q}(\omega)] = 6$$

And this allows an argument parallel to the earlier one for $\mathrm{Aut}(\mathbb{Q}(\zeta_7)/\mathbb{Q})$ to show that

$$(\mathbb{Z}/7)^\times \approx \mathrm{Aut}(\mathbb{Q}(\zeta_7,\omega)/\mathbb{Q}(\omega))$$

by

$$a \longrightarrow \tau_a$$

where

$$\tau_a(\zeta_7) = \zeta_7^a$$

Then the automorphisms σ_a of $\mathbb{Q}(\zeta)$ over $\mathbb{Q})$ are simply the *restrictions* of τ_a to $\mathbb{Q}(\zeta)$.

19.4.2 Remark: If we look for zeros of the cubic $f(x) = x^3 + x^2 - 2x - 1$ in the real numbers $\mathbb{R}$, then we find three real roots. Indeed,

$$\begin{cases} f(2) & = & 7 \\ f(1) & = & -1 \\ f(-1) & = & 1 \\ f(-2) & = & -1 \end{cases}$$

Thus, by the intermediate value theorem there is a root in the interval $[1,2]$, a second root in the interval $[-1,1]$, and a third root in the interval $[-2,-1]$. All the roots are real. Nevertheless, the expression for the roots in terms of radicals involves primitive cube roots of unity, none of which is real. [19]

[19] Beginning in the Italian renaissance, it was observed that the formula for *real* roots to cubic equations involved complex numbers. This was troubling, both because complex numbers were certainly not widely accepted at that time, and because it seemed jarring that natural expressions for real numbers should necessitate complex numbers.

19.5 *Quadratic fields, quadratic reciprocity*

This discussion will do two things: show that all field extensions $\mathbb{Q}(\sqrt{D})$ lie inside fields $\mathbb{Q}(\zeta_n)$ obtained by adjoining primitive n^{th} roots of unity [20] to $\mathbb{Q}$, and prove *quadratic reciprocity*. [21]

Let p be an odd prime and x an integer. The **quadratic symbol** is defined to be

$$\left(\frac{a}{p}\right)_2 = \begin{cases} 0 & \text{(for } a = 0 \bmod p) \\ 1 & \text{(for } a \text{ a non-zero square mod } p) \\ -1 & \text{(for } a \text{ a non-square mod } p) \end{cases}$$

One part of quadratic reciprocity is an easy consequence of the cyclicness of $(\mathbb{Z}/p)^\times$ for p prime, and amounts to a restatement of earlier results:

19.5.1 Proposition: For p an odd prime

$$\left(\frac{-1}{p}\right)_2 = (-1)^{(p-1)/2} = \begin{cases} 1 & \text{(for } p = 1 \bmod 4) \\ -1 & \text{(for } p = 3 \bmod 4) \end{cases}$$

Proof: If -1 is a square mod p, then a square root of it has order 4 in $(\mathbb{Z}p)^\times$, which is of order $p - 1$. Thus, by Lagrange, $4|(p - 1)$. This half of the argument does not need the cyclicness. On the other hand, suppose $4|(p - 1)$. Since $(\mathbb{Z}/p)^\times$ is cyclic, there are exactly two elements α, β of order 4 in $(\mathbb{Z}/p)^\times$, and exactly one element -1 of order 2. Thus, the squares of α and β must be -1, and -1 has two square roots.   ///

Refining the previous proposition, as a corollary of the cyclicness of $(\mathbb{Z}/p)^\times$, we have **Euler's criterion**:

19.5.2 Proposition: *(Euler)* Let p be an odd prime. For an integer a

$$\left(\frac{a}{p}\right)_2 = a^{(p-1)/2} \bmod p$$

Proof: If $p|a$, this equality certainly holds. For $a \neq 0 \bmod p$ certainly $a^{(p-1)/2} = \pm 1 \bmod p$, since

$$\left(a^{(p-1)/2}\right)^2 = a^{p-1} = 1 \bmod p$$

and the only square roots of 1 in $\mathbb{Z}/p$ are ± 1. If $a = b^2 \bmod p$ is a non-zero square mod p, then

$$a^{(p-1)/2} = (b^2)^{(p-1)/2} = b^{p-1} = 1 \bmod p$$

---

[20] The fact that every *quadratic* extension of $\mathbb{Q}$ is contained in a field generated by roots of unity is a very special case of the *Kronecker-Weber* theorem, which asserts that any *galois extension* of $\mathbb{Q}$ with *abelian* galois group lies inside a field generated over $\mathbb{Q}$ by roots of unity.

[21] Though Gauss was the first to give a proof of quadratic reciprocity, it had been conjectured by Lagrange some time before, and much empirical evidence supported the conclusion.

This was the easy half. For the harder half we need the cyclicness. Let g be a generator for $(\mathbb{Z}/p)^\times$. Let $a \in (\mathbb{Z}/p)^\times$, and write $a = g^t$. If a is not a square mod p, then t must be odd, say $t = 2s + 1$. Then

$$a^{(p-1)/2} = g^{t(p-1)/2} = g^{(2s+1)(p-1)/2} = g^{s(p-1)} \cdot g^{(p-1)/2} = g^{(p-1)/2} = -1$$

since g is of order $p - 1$, and since -1 is the unique element of order 2. ///

19.5.3 Corollary: The quadratic symbol has the multiplicative property

$$\left(\frac{ab}{p}\right)_2 = \left(\frac{a}{p}\right)_2 \cdot \left(\frac{b}{p}\right)_2$$

Proof: This follows from the expression for the quadratic symbol in the previous proposition. ///

A more interesting special case [22] is

19.5.4 Theorem: For p an odd prime, we have the formula [23]

$$\left(\frac{2}{p}\right)_2 = (-1)^{(p^2-1)/8} = \begin{cases} 1 & (\text{for } p = 1 \bmod 8) \\ -1 & (\text{for } p = 3 \bmod 8) \\ -1 & (\text{for } p = 5 \bmod 8) \\ 1 & (\text{for } p = 7 \bmod 8) \end{cases}$$

Proof: Let i denote a square root of -1, and we work in the ring $\mathbb{Z}[i]$. Since the binomial coefficients $\binom{p}{k}$ are divisible by p for $0 < k < p$, in $\mathbb{Z}[i]$

$$(1 + i)^p = 1^p + i^p = 1 + i^p$$

Also, $1 + i$ is roughly a square root of 2, or at least of 2 times a unit in $\mathbb{Z}[i]$, namely

$$(1 + i)^2 = 1 + 2i - 1 = 2i$$

Then, using Euler's criterion, in $\mathbb{Z}[i]$ modulo the ideal generated by p

$$\left(\frac{2}{p}\right)_2 = 2^{(p-1)/2} = (2i)^{(p-1)/2} \cdot i^{-(p-1)/2}$$

$$= \left((1 + i)^2\right)^{(p-1)/2} \cdot i^{-(p-1)/2} = (1 + i)^{p-1} \cdot i^{-(p-1)/2} \bmod p$$

Multiply both sides by $1 + i$ to obtain, modulo p,

$$(1 + i) \cdot \left(\frac{2}{p}\right)_2 = (1 + i)^p \cdot i^{-(p-1)/2} = (1 + i^p) \cdot i^{-(p-1)/2} \bmod p$$

---

[22] Sometimes called a *supplementary* law of quadratic reciprocity.

[23] The expression of the value of the quadratic symbol as a power of -1 is just an interpolation of the values. That is, the expression $(p^2 - 1)/8$ does not present itself naturally in the argument.

The right-hand side depends only on p modulo 8, and the four cases given in the statement of the theorem can be computed directly.   ///

The main part of quadratic reciprocity needs somewhat more preparation. Let p and q be distinct odd primes. Let $\zeta = \zeta_q$ be a primitive q^{th} root of unity. The **quadratic Gauss sum** mod q is

$$g = \sum_{b \bmod q} \zeta_q^b \cdot \left(\frac{b}{q}\right)_2$$

19.5.5 Proposition: Let q be an odd prime, ζ_q a primitive q^{th} root of unity. Then

$$g^2 = \left(\sum_{b \bmod q} \zeta_q^b \cdot \left(\frac{b}{q}\right)_2\right)^2 = \left(\frac{-1}{q}\right)_2 \cdot q$$

That is, either $\sqrt{q}$ or $\sqrt{-q}$ is in $\mathbb{Q}(\zeta_q)$, depending upon whether q is 1 or 3 modulo 4.

Proof: Compute

$$g^2 = \sum_{a,b \bmod q} \zeta_q^{a+b} \cdot \left(\frac{ab}{q}\right)_2$$

from the multiplicativity of the quadratic symbol. And we may restrict the sum to a, b not $0 \bmod q$. Then

$$g^2 = \sum_{a,b \bmod q} \zeta_q^{a+ab} \cdot \left(\frac{a^2 b}{q}\right)_2$$

by replacing b by $ab \bmod q$. Since $a \neq 0 \bmod q$ this is a bijection of $\mathbb{Z}/q$ to itself. Then

$$g^2 = \sum_{a \neq 0, b \neq 0} \zeta_q^{a+ab} \cdot \left(\frac{a^2}{q}\right)_2 \left(\frac{b}{q}\right)_2 = \sum_{a \neq 0, b \neq 0} \zeta_q^{a(1+b)} \cdot \left(\frac{b}{q}\right)_2$$

For fixed b, if $1 + b \neq 0 \bmod q$ then we can replace $a(1 + b)$ by a, since $1 + b$ is invertible mod q. With $1 + b \neq 0 \bmod q$, the inner sum over a is

$$\sum_{a \neq 0 \bmod q} \zeta_q^a = \left(\sum_{a \bmod q} \zeta_q^a\right) - 1 = 0 - 1 = -1$$

When $1 + b = 0 \bmod q$, the sum over a is $q - 1$. Thus, the whole is

$$g^2 = \sum_{b = -1 \bmod q} (q-1) \cdot \left(\frac{b}{q}\right)_2 - \sum_{b \neq 0, -1 \bmod q} \left(\frac{b}{q}\right)_2 = (q-1) \cdot \left(\frac{-1}{q}\right)_2 - \sum_{b \bmod q} \left(\frac{b}{q}\right)_2 + \left(\frac{-1}{q}\right)_2$$

Let c be a non-square mod q. Then $b \longrightarrow bc$ is a bijection of $\mathbb{Z}/q$ to itself, and so

$$\sum_{b \bmod q} \left(\frac{b}{q}\right)_2 = \sum_{b \bmod q} \left(\frac{bc}{q}\right)_2 = \left(\frac{c}{q}\right)_2 \cdot \sum_{b \bmod q} \left(\frac{b}{q}\right)_2 = -\sum_{b \bmod q} \left(\frac{b}{q}\right)_2$$

Since $A = -A$ implies $A = 0$ for integers A, we have

$$\sum_{b \bmod q} \left(\frac{b}{q}\right)_2 = 0$$

Then we have

$$g^2 = (q-1) \cdot \left(\frac{-1}{q}\right)_2 - \sum_{b \bmod q} \left(\frac{b}{q}\right)_2 + \left(\frac{-1}{q}\right)_2 = q \cdot \left(\frac{-1}{q}\right)_2$$

as claimed. ///

Now we can prove

19.5.6 Theorem: *(Quadratic Reciprocity)* Let p and q be distinct odd primes. Then

$$\left(\frac{p}{q}\right)_2 = (-1)^{(p-1)(q-1)/4} \cdot \left(\frac{q}{p}\right)_2$$

Proof: Using Euler's criterion and the previous proposition, modulo p in the ring $\mathbb{Z}[\zeta_q]$,

$$\left(\frac{q}{p}\right)_2 = q^{(p-1)/2} = \left(g^2 \left(\frac{-1}{q}\right)_2\right)^{(p-1)/2}$$

$$= g^{p-1} \left(\frac{-1}{q}\right)_2^{(p-1)/2} = g^{p-1} \left((-1)^{(q-1)/2}\right)^{(p-1)/2}$$

Multiply through by the Gauss sum g, to obtain

$$g \cdot \left(\frac{q}{p}\right)_2 = g^p \cdot (-1)^{(p-1)(q-1)/4} \bmod p$$

Since p divides the middle binomial coefficients, and since p is odd (so $(b/q)_2^p = (b/q)_2$ for all b),

$$g^p = \left(\sum_{b \bmod q} \zeta_q^b \cdot \left(\frac{b}{q}\right)_2\right)^p = \sum_{b \bmod q} \zeta_q^{bp} \cdot \left(\frac{b}{q}\right)_2 \bmod p$$

Since p is invertible modulo q, we can replace b by $bp^{-1} \bmod q$ to obtain

$$g^p = \sum_{b \bmod q} \zeta_q^b \cdot \left(\frac{bp^{-1}}{q}\right)_2 = \left(\frac{p^{-1}}{q}\right)_2 \cdot \sum_{b \bmod q} \zeta_q^b \cdot \left(\frac{b}{q}\right)_2 = \left(\frac{p}{q}\right)_2 \cdot g \bmod p$$

Putting this together,

$$g \cdot \left(\frac{q}{p}\right)_2 = \left(\frac{p}{q}\right)_2 \cdot g \cdot (-1)^{(p-1)(q-1)/4} \bmod p$$

We obviously want to cancel the factor of g, but we must be sure that it is invertible in $\mathbb{Z}[\zeta_q]$ modulo p. Indeed, since

$$g^2 = q \cdot \left(\frac{-1}{q}\right)_2$$

we could *multiply* both sides by g to obtain

$$q \left(\frac{-1}{q}\right)_2 \cdot \left(\frac{q}{p}\right)_2 \cdot q \left(\frac{-1}{q}\right)_2 = \left(\frac{p}{q}\right)_2 \cdot q \left(\frac{-1}{q}\right)_2 \cdot (-1)^{(p-1)(q-1)/4} \bmod p$$

Since $\pm q$ is invertible mod p, we cancel the $q(-1/q)_2$ to obtain

$$\left(\frac{q}{p}\right)_2 = \left(\frac{p}{q}\right)_2 \cdot (-1)^{(p-1)(q-1)/4} \bmod p$$

Both sides are ± 1 and $p > 2$, so we have an *equality* of integers

$$\left(\frac{q}{p}\right)_2 = \left(\frac{p}{q}\right)_2 \cdot (-1)^{(p-1)(q-1)/4}$$

which is the assertion of quadratic reciprocity.   ///

19.6 *Worked examples*

19.6.1 Example: Let ζ be a primitive n^{th} root of unity in a field of characteristic 0. Let M be the n-by-n matrix with ij^{th} entry ζ^{ij}. Find the multiplicative inverse of M.

Some experimentation (and an exercise from the previous week) might eventually suggest consideration of the matrix A having ij^{th} entry $\frac{1}{n}\zeta^{-ij}$. Then the ij^{th} entry of MA is

$$(MA)_{ij} = \frac{1}{n}\sum_k \zeta^{ik-kj} = \frac{1}{n}\sum_k \zeta^{(i-j)k}$$

As an example of a *cancellation principle* we claim that

$$\sum_k \zeta^{(i-j)k} = \begin{cases} 0 & (\text{for } i-j \neq 0) \\ n & (\text{for } i-j = 0) \end{cases}$$

The second assertion is clear, since we'd be summing n 1's in that case. For $i-j \neq 0$, we can change variables in the indexing, replacing k by $k+1$ mod n, since ζ^a is well-defined for $a \in \mathbb{Z}/n$. Thus,

$$\sum_k \zeta^{(i-j)k} = \sum_k \zeta^{(i-j)(k+1)} = \zeta^{i-j}\sum_k \zeta^{(i-j)k}$$

Subtracting,

$$(1 - \zeta^{i-j})\sum_k \zeta^{(i-j)k} = 0$$

For $i-j \neq 0$, the leading factor is non-zero, so the sum must be zero, as claimed.   ///

19.6.2 Example: Let $\mu = \alpha\beta^2 + \beta\gamma^2 + \gamma\alpha^2$ and $\nu = \alpha^2\beta + \beta^2\gamma + \gamma^2\alpha$. Show that these are the two roots of a quadratic equation with coefficients in $\mathbb{Z}[s_1, s_2, s_3]$ where the s_i are the elementary symmetric polynomials in α, β, γ.

Consider the quadratic polynomial

$$(x - \mu)(x - \nu) = x^2 - (\mu + \nu)x + \mu\nu$$

We will be done if we can show that $\mu + \nu$ and $\mu\nu$ are symmetric polynomials as indicated. The sum is

$$\mu + \nu = \alpha\beta^2 + \beta\gamma^2 + \gamma\alpha^2 + \alpha^2\beta + \beta^2\gamma + \gamma^2\alpha$$

$$= (\alpha + \beta + \gamma)(\alpha\beta + \beta\gamma + \gamma\alpha) - 3\alpha\beta\gamma = s_1 s_2 - 3s_3$$

This expression is plausibly obtainable by a few trial-and-error guesses, and examples nearly identical to this were done earlier. The product, being of higher degree, is more daunting.

$$\mu\nu = (\alpha\beta^2 + \beta\gamma^2 + \gamma\alpha^2)(\alpha^2\beta + \beta^2\gamma + \gamma^2\alpha)$$

$$= \alpha^3 + \alpha\beta^4 + \alpha^2\beta^2\gamma^2 + \alpha^2\beta^2\gamma^2 + \beta^3\gamma^3 + \alpha\beta\gamma^4 + \alpha^4\beta\gamma + \alpha^2\beta^2\gamma^2 + \alpha^3\gamma^3$$

Following the symmetric polynomial algorithm, at $\gamma = 0$ this is $\alpha^3\beta^3 = s_2(\alpha, \beta)^3$, so we consider

$$\frac{\mu\nu - s_2^3}{s_3} = \alpha^3 + \beta^3 + \gamma^3 - 3s_3 - 3(\mu + \nu)$$

where we are lucky that the last 6 terms were $\mu + \nu$. We have earlier found the expression for the sum of cubes, and we have expressed $\mu + \nu$, so

$$\frac{\mu\nu - s_2^3}{s_3} = (s_1^3 - 3s_1 s_2 + 3s_3) - 3s_3 - 3(s_1 s_2 - 3s_3) = s_1^3 - 6s_1 s_2 + 9s_3$$

and, thus,

$$\mu\nu = s_2^3 + s_1^3 s_3 - 6s_1 s_2 s_3 + 9s_3^2$$

Putting this together, μ and ν are the two roots of

$$x^2 - (s_1 s_2 - 3s_3)x + (s_2^3 + s_1^3 s_3 - 6s_1 s_2 s_3 + 9s_3^2) = 0$$

(One might also speculate on the relationship of μ and ν to solution of the general cubic equation.) ///

19.6.3 Example: The 5th cyclotomic polynomial $\Phi_5(x)$ factors into two irreducible quadratic factors over $\mathbb{Q}(\sqrt{5})$. Find the two irreducible factors.

We have shown that $\sqrt{5}$ occurs inside $\mathbb{Q}(\zeta)$, where ζ is a primitive fifth root of unity. Indeed, the discussion of Gauss sums in the proof of quadratic reciprocity gives us the convenient

$$\zeta - \zeta^2 - \zeta^3 + \zeta^4 = \sqrt{5}$$

We also know that $[\mathbb{Q}(\sqrt{5}) : \mathbb{Q}] = 2$, since $x^2 - 5$ is irreducible in $\mathbb{Q}[x]$ (Eisenstein and Gauss). And $[\mathbb{Q}(\zeta) : \mathbb{Q}] = 4$ since $\Phi_5(x)$ is irreducible in $\mathbb{Q}[x]$ of degree $5 - 1 = 4$ (again by Eisenstein and Gauss). Thus, by multiplicativity of degrees in towers of fields, $[\mathbb{Q}(\zeta) : \mathbb{Q}(\sqrt{5})] = 2$.

Thus, since none of the 4 primitive fifth roots of 1 lies in $\mathbb{Q}(\sqrt{5})$, each is necessarily quadratic over $\mathbb{Q}(\sqrt{5})$, so has minimal polynomial over $\mathbb{Q}(\sqrt{5})$ which is quadratic, in contrast to the minimal polynomial $\Phi_5(x)$ over $\mathbb{Q}$. Thus, the 4 primitive fifth roots break up into two (disjoint) bunches of 2, grouped by being the 2 roots of the same quadratic over $\mathbb{Q}(\sqrt{5})$. That is, the fifth cyclotomic polynomial factors as the product of those two minimal polynomials (which are necessarily irreducible over $\mathbb{Q}(\sqrt{5})$).

In fact, we have a trick to determine the two quadratic polynomials. Since

$$\zeta^4 + \zeta^3 + \zeta^2 + \zeta + 1 = 0$$

divide through by ζ^2 to obtain

$$\zeta^2 + \zeta + 1 + \zeta^{-1} + \zeta^{-2} = 0$$

Thus, regrouping,

$$\left(\zeta + \frac{1}{\zeta}\right)^2 + \left(\zeta + \frac{1}{\zeta}\right)^2 - 1 = 0$$

Thus, $\xi = \zeta + \zeta^{-1}$ satisfies the equation

$$x^2 + x - 1 = 0$$

and $\xi = (-1 \pm \sqrt{5})/2$. Then, from

$$\zeta + \frac{1}{\zeta} = (-1 \pm \sqrt{5})/2$$

multiply through by ζ and rearrange to

$$\zeta^2 - \frac{-1 \pm \sqrt{5}}{2}\zeta + 1 = 0$$

Thus,

$$x^4 + x^3 + x^2 + x + 1 = \left(x^2 - \frac{-1 + \sqrt{5}}{2}x + 1\right)\left(x^2 - \frac{-1 - \sqrt{5}}{2}x + 1\right)$$

Alternatively, to see what can be done similarly in more general situations, we recall that $\mathbb{Q}(\sqrt{5})$ is the subfield of $\mathbb{Q}(\zeta)$ fixed pointwise by the automorphism $\zeta \longrightarrow \zeta^{-1}$. Thus, the 4 primitive fifth roots of unity should be paired up into the orbits of this automorphism. Thus, the two (irreducible in $\mathbb{Q}(\sqrt{5})[x]$) quadratics are

$$(x - \zeta)(x - \zeta^{-1}) = x^2 - (\zeta + \zeta^{-1})x + 1$$

$$(x - \zeta^2)(x - \zeta^{-2}) = x^2 - (\zeta^2 + \zeta^{-2})x + 1$$

Again, without imbedding things into the complex numbers, etc., there is no canonical one of the two square roots of 5, so the $\pm\sqrt{5}$ just means that whichever one we pick first the other one is its negative. Similarly, there is no distinguished one among the 4 primitive fifth roots unless we imbed them into the complex numbers. There is no need to do this. Rather, specify one ζ, and specify a $\sqrt{5}$ by

$$\zeta + \zeta^{-1} = \frac{-1 + \sqrt{5}}{2}$$

Then necessarily

$$\zeta^2 + \zeta^{-2} = \frac{-1 - \sqrt{5}}{2}$$

And we find the same two quadratic equations again. Since they are necessarily the minimal polynomials of ζ and of ζ^2 over $\mathbb{Q}(\sqrt{5})$ (by the degree considerations) they are irreducible in $\mathbb{Q}(\sqrt{5})[x]$.   ///

19.6.4 Example: The 7th cyclotomic polynomial $\Phi_7(x)$ factors into two irreducible cubic factors over $\mathbb{Q}(\sqrt{-7})$. Find the two irreducible factors.

Let ζ be a primitive 7th root of unity. Let $H = \langle \tau \rangle$ be the order 3 subgroup of the automorphism group $G \approx (\mathbb{Z}/7)^\times$ of $\mathbb{Q}(\zeta)$ over $\mathbb{Q}$, where $\tau = \sigma_2$ is the automorphism

$\tau(\zeta) = \zeta^2$, which has order 3. We have seen that $\mathbb{Q}(\sqrt{-7})$ is the subfield fixed pointwise by H. In particular, $\alpha = \zeta + \zeta^2 + \zeta^4$ should be at most quadratic over $\mathbb{Q}$. Recapitulating the earlier discussion, α is a zero of the quadratic polynomial

$$(x - (\zeta + \zeta^2 + \zeta^4))(x - (\zeta^3 + \zeta^6 + \zeta^5))$$

which will have coefficients in $\mathbb{Q}$, since we have arranged that the coefficients are G-invariant. Multiplying out and simplifying, this is

$$x^2 + x + 2$$

with zeros $(-1 \pm \sqrt{-7})/2$.

The coefficients of the polynomial

$$(x - \zeta)(x - \tau(\zeta))(x - \tau^2(\zeta)) = (x - \zeta)(x - \zeta^2)(x - \zeta^4)$$

will be H-invariant and therefore will lie in $\mathbb{Q}(\sqrt{-7})$. In parallel, taking the primitive 7th root of unity ζ^3 which is not in the H-orbit of ζ, the cubic

$$(x - \zeta^3)(x - \tau(\zeta^3))(x - \tau^2(\zeta^3)) = (x - \zeta^3)(x - \zeta^6)(x - \zeta^5)$$

will also have coefficients in $\mathbb{Q}(\sqrt{-7})$. It is no coincidence that the exponents of ζ occuring in the two cubics are disjoint and exhaust the list $1, 2, 3, 4, 5, 6$.

Multiplying out the first cubic, it is

$$(x - \zeta)(x - \zeta^2)(x - \zeta^4) = x^3 - (\zeta + \zeta^2 + \zeta^4)x^2 + (\zeta^3 + \zeta^5 + \zeta^6)x - 1$$

$$= x^3 - \left(\frac{-1 + \sqrt{-7}}{2}\right)x^2 + \left(\frac{-1 - \sqrt{-7}}{2}\right)x - 1$$

for a choice of ordering of the square roots. (Necessarily!) the other cubic has the roles of the two square roots reversed, so is

$$(x - \zeta^3)(x - \zeta^6)(x - \zeta^2) = x^3 - (\zeta^3 + \zeta^5 + \zeta^6)x + (\zeta + \zeta^2 + \zeta^4)x - 1$$

$$= x^3 - \left(\frac{-1 - \sqrt{-7}}{2}\right)x^2 + \left(\frac{-1 + \sqrt{-7}}{2}\right)x - 1$$

Since the minimal polynomials of primitive 7th roots of unity are of degree 3 over $\mathbb{Q}(\sqrt{-7})$ (by multiplicativity of degrees in towers), these cubics are irreducible over $\mathbb{Q}(\sqrt{-7})$. Their product is $\Phi_7(x)$, since the set of all 6 roots is all the primitive 7th roots of unity, and there is no overlap between the two sets of roots. ///

19.6.5 Example: Let ζ be a primitive 13th root of unity in an algebraic closure of $\mathbb{Q}$. Find an element α in $\mathbb{Q}(\zeta)$ which satisfies an irreducible cubic with rational coefficients. Find an element β in $\mathbb{Q}(\zeta)$ which satisfies an irreducible quartic with rational coefficients. Determine the cubic and the quartic explicitly.

Again use the fact that the automorphism group G of $\mathbb{Q}(\zeta)$ over $\mathbb{Q}$ is isomorphic to $(\mathbb{Z}/13)^\times$ by $a \longrightarrow \sigma_a$ where $\sigma_a(\zeta) = \zeta^a$. The unique subgroup A of order 4 is generated by $\mu = \sigma_5$.

From above, an element $\alpha \in \mathbb{Q}(\zeta)$ fixed by A is of degree at most $|G|/|A| = 12/4 = 3$ over $\mathbb{Q}$. Thus, try symmetrizing/averaging ζ itself over the subgroup A by

$$\alpha = \zeta + \mu(\zeta) + \mu^2(\zeta) + \mu^3(\zeta) = \zeta + \zeta^5 + \zeta^{12} + \zeta^8$$

The unique subgroup B of order 3 in G is generated by $\nu = \sigma_3$. Thus, necessarily the coefficients of

$$(x - \alpha)(x - \nu(\alpha))(x - \nu^2(\alpha))$$

are in $\mathbb{Q}$. Also, one can see directly (because the ζ^i with $1 \leq i \leq 12$ are linearly independent over $\mathbb{Q}$) that the images $\alpha, \nu(\alpha), \nu^2(\alpha)$ are distinct, assuring that the cubic is irreducible over $\mathbb{Q}$.

To multiply out the cubic and determine the coefficients as rational numbers it is wise to be as economical as possible in the computation. Since we know *a priori* that the coefficients are rational, we need not drag along all the powers of ζ which appear, since there will necessarily be cancellation. Precisely, we compute in terms of the $\mathbb{Q}$-basis

$$1, \zeta, \zeta^2, \ldots, \zeta^{10}, \zeta^{11}$$

Given ζ^n appearing in a sum, reduce the exponent n modulo 13. If the result is 0, add 1 to the sum. If the result is 12, add -1 to the sum, since

$$\zeta^{12} = -(1 + \zeta + \zeta^2 + \ldots + \zeta^{11})$$

expresses ζ^{12} in terms of our basis. If the reduction mod 13 is anything else, drop that term (since we know it will cancel). And we can go through the monomial summand in lexicographic order. Using this bookkeeping strategy, the cubic is

$$\left(x - (\zeta + \zeta^5 + \zeta^{12} + \zeta^8)\right)\left(x - (\zeta^3 + \zeta^2 + \zeta^{10} + \zeta^{11})\right)\left(x - (\zeta^9 + \zeta^6 + \zeta^4 + \zeta^7)\right)$$

$$= x^3 - (-1)x^2 + (-4)x - (-1) = x^3 + x^2 - 4x + 1$$

Yes, there are $3 \cdot 4^2$ terms to sum for the coefficient of x, and 4^3 for the constant term. Most give a contribution of 0 in our bookkeeping system, so the workload is not completely unreasonable. (A numerical computation offers a different sort of check.) Note that Eisenstein's criterion (and Gauss' lemma) gives another proof of the irreducibility, by replacing x by $x + 4$ to obtain

$$x^3 + 13x^2 + 52x + 65$$

and noting that the prime 13 fits into the Eisenstein criterion here. This is yet another check on the computation.

For the quartic, reverse the roles of μ and ν above, so put

$$\beta = \zeta + \nu(\zeta) + \nu^2(\zeta) = \zeta + \zeta^3 + \zeta^9$$

and compute the coefficients of the quartic polynomial

$$(x - \beta)(x - \mu(\beta))(x - \mu^2(\beta))(x - \mu^3(\beta))$$

$$= \left(x - (\zeta + \zeta^3 + \zeta^9)\right)\left(x - (\zeta^5 + \zeta^2 + \zeta^6)\right)\left(x - (\zeta^{12} + \zeta^{10} + \zeta^4)\right)\left(x - (\zeta^8 + \zeta^{11} + \zeta^7)\right)$$

Use the same bookkeeping approach as earlier, to allow a running tally for each coefficient. The sum of the 4 triples is -1. For the other terms some writing-out seems necessary. For example, to compute the constant coefficient, we have the product

$$(\zeta + \zeta^3 + \zeta^9)(\zeta^5 + \zeta^2 + \zeta^6)(\zeta^{12} + \zeta^{10} + \zeta^4)(\zeta^8 + \zeta^{11} + \zeta^7)$$

which would seem to involve 81 summands. We can lighten the burden by writing only the exponents which appear, rather than recopying zetas. Further, multiply the first two factors and the third and fourth, leaving a multiplication of two 9-term factors (again, retaining only the exponents)

$$(6 \quad 3 \quad 7 \quad 8 \quad 5 \quad 9 \quad 1 \quad 11 \quad 2)(7 \quad 10 \quad 6 \quad 5 \quad 8 \quad 4 \quad 12 \quad 2 \quad 11)$$

As remarked above, a combination of an exponent from the first list of nine with an exponent from the second list will give a non-zero contribution only if the sum (reduced modulo 13) is either 0 or 12, contributing 1 or -1 respectively. For each element of the first list, we can keep a running tally of the contributions from each of the 9 elements from the second list. Thus, grouping by the elements of the first list, the contributions are, respectively,

$$(1-1) + (1) + (1-1) + (1-1) + (-1+1) + (1) + (1-1) + (1)(-1+1) = 3$$

The third symmetric function is a sum of 4 terms, which we group into two, writing in the same style

$$(1 \quad 3 \quad 9 \quad 5 \quad 2 \quad 6)(7 \quad 10 \quad 6 \quad 5 \quad 8 \quad 4 \quad 12 \quad 2 \quad 11)$$

$$+(6 \quad 3 \quad 7 \quad 8 \quad 5 \quad 9 \quad 1 \quad 11 \quad 2)(12 \quad 10 \quad 4 \quad 8 \quad 11 \quad 7)$$

In each of these two products, for each item in the lists of 9, we tally the contributions of the 6 items in the other list, obtaining,

$$(0+0-1+0+1+1+1+0+0) + (1+1+0-1+0+1+0+0+0) = 4$$

The computation of the second elementary symmetric function is, similarly, the sum

$$(1 \quad 3 \quad 9)(5 \quad 2 \quad 6 \quad 12 \quad 10 \quad 4 \quad 8 \quad 11 \quad 7)$$

$$+(5 \quad 2 \quad 6)(12 \quad 10 \quad 4 \quad 8 \quad 11 \quad 7) + (12 \quad 10 \quad 4)(8 \quad 11 \quad 7)$$

Grouping the contributions for each element in the lists $1, 3, 9$ and $5, 2, 6$ and $12, 10, 4$, this gives

$$[(1-1) + (1) + (1)] + [(1-1) + (-1+1) + (1)] + [0+0+(-1)] = 2$$

Thus, in summary, we have

$$x^4 + x^3 + 2x^2 - 4x + 3$$

Again, replacing x by $x+3$ gives

$$x^4 + 13x^3 + 65x^2 + 143x + 117$$

All the lower coefficients are divisible by 13, but not by 13^2, so Eisenstein proves irreducibility. This again gives a sort of verification of the correctness of the numerical computation. ///

19.6.6 Example: Let $f(x) = x^8 + x^6 + x^4 + x^2 + 1$. Show that f factors into two irreducible quartics in $\mathbb{Q}[x]$. Show that

$$x^8 + 5x^6 + 25x^4 + 125x^2 + 625$$

also factors into two irreducible quartics in $\mathbb{Q}[x]$.

The first assertion can be verified by an elementary trick, namely

$$x^8 + x^6 + x^4 + x^2 + 1 = \frac{x^{10} - 1}{x^2 - 1} = \frac{\Phi_1(x)\Phi_2(x)\Phi_5(x)\Phi_{10}(x)}{\Phi_1(x)\Phi_2(x)}$$

$$= \Phi_5(x)\Phi_{10}(x) = (x^4 + x^3 + x^2 + x + 1)(x^4 - x^3 + x^2 - x + 1)$$

But we do learn something from this, namely that the factorization of the first octic into linear factors naturally has the eight linear factors occurring in two bunches of four, namely the primitive 5th roots of unity and the primitive 10th roots of unity. Let ζ be a primitive 5th root of unity. Then $-\zeta$ is a primitive 10th. Thus, the 8 zeros of the *second* polynomial will be $\sqrt{5}$ times primitive 5th and 10th roots of unity. The question is how to group them together in two bunches of four so as to obtain rational coefficients of the resulting two quartics.

The automorphism group G of $\mathbb{Q}(\zeta)$ over $\mathbb{Q}$ is isomorphic to $(\mathbb{Z}/10)^\times$, which is generated by $\tau(\zeta) = \zeta^3$. That is, taking a product of linear factors whose zeros range over an orbit of ζ under the automorphism group G,

$$x^4 + x^3 + x^2 + x + 1 = (x - \zeta)(x - \zeta^3)(x - \zeta^9)(x - \zeta^7)$$

has coefficients in $\mathbb{Q}$ and is the minimal polynomial for ζ over $\mathbb{Q}$. Similarly looking at the orbit of $-\zeta$ under the automorphism group G, we see that

$$x^4 - x^3 + x^2 - x + 1 = (x + \zeta)(x + \zeta^3)(x + \zeta^9)(x + \zeta^7)$$

has coefficients in $\mathbb{Q}$ and is the minimal polynomial for $-\zeta$ over $\mathbb{Q}$.

The discussion of Gauss sums in the proof of quadratic reciprocity gives us the convenient

$$\zeta - \zeta^2 - \zeta^3 + \zeta^4 = \sqrt{5}$$

Note that this expression allows us to see what effect the automorphism $\sigma_a(\zeta) = \zeta^a$ has on $\sqrt{5}$

$$\sigma_a(\sqrt{5}) = \sigma_a(\zeta - \zeta^2 - \zeta^3 + \zeta^4) = \begin{cases} \sqrt{5} & \text{(for } a = 1, 9) \\ -\sqrt{5} & \text{(for } a = 3, 7) \end{cases}$$

Thus, the orbit of $\sqrt{5}\zeta$ under G is

$$\sqrt{5}\zeta, \quad \tau(\sqrt{5}\zeta) = -\sqrt{5}\zeta^3, \quad \tau^2(\sqrt{5}\zeta) = \sqrt{5}\zeta^4, \quad \tau^3(\sqrt{5}\zeta) = -\sqrt{5}\zeta^2$$

giving quartic polynomial

$$(x - \sqrt{5}\zeta)(x + \sqrt{5}\zeta^3)(x - \sqrt{5}\zeta^4)(x + \sqrt{5}\zeta^2)$$

$$= x^4 - \sqrt{5}(\zeta - \zeta^2 - \zeta^3 + \zeta^4)x^3 + 5(-\zeta^4 + 1 - \zeta^3 - \zeta^2 + 1 - \zeta)x^2 - 5\sqrt{5}(\zeta^4 - \zeta^2 + \zeta - \zeta^3)x + 25$$

$$= x^4 - 5x^3 + 15x^2 - 25x + 25$$

We might anticipate what happens with the other bunch of four zeros, but we can also compute directly (confirming the suspicion). The orbit of $-\sqrt{5}\zeta$ under G is

$$-\sqrt{5}\zeta,\ \tau(-\sqrt{5}\zeta)=\sqrt{5}\zeta^3,\ \tau^2(-\sqrt{5}\zeta)=-\sqrt{5}\zeta^4,\ \tau^3(-\sqrt{5}\zeta)=\sqrt{5}\zeta^2$$

giving quartic polynomial

$$(x+\sqrt{5}\zeta)(x-\sqrt{5}\zeta^3)(x+\sqrt{5}\zeta^4)(x-\sqrt{5}\zeta^2)$$

$$=x^4+\sqrt{5}(\zeta-\zeta^2-\zeta^3+\zeta^4)x^3+5(-\zeta^4+1-\zeta^3-\zeta^2+1-\zeta)x^2+5\sqrt{5}(\zeta^4-\zeta^2+\zeta-\zeta^3)x+25$$

$$=x^4+5x^3+15x^2+25x+25$$

Thus, we expect that

$$x^8+5x^6+25x^4+125x^2+625=(x^4-5x^3+15x^2-25x+25)\cdot(x^4+5x^3+15x^2+25x+25)$$

Because of the sign flips in the odd-degree terms in the quartics, the octic is also

$$x^8+5x^6+25x^4+125x^2+625=(x^4+15x^2+25)^2-25(x^3+5x)^2$$

(This factorization of an altered product of two cyclotomic polynomials is an *Aurifeuille-LeLasseur* factorization, after two amateur mathematicians who studied them, brought to wider attention by E. Lucas in the late 19th century.)

19.6.7 Example: Let p be a prime not dividing m. Show that in $\mathbb{F}_p[x]$

$$\Phi_{mp}(x)=\Phi_m(x)^{p-1}$$

From the recursive definition,

$$\Phi_{pm}(x)=\frac{x^{pm}-1}{\prod_{d|m}\Phi_{p^e d}(x)\cdot\prod_{d|m,\ d<m}\Phi_{pd}(x)}$$

In characteristic p, the numerator is $(x^m-1)^p$. The first product factor in the denominator is x^m-1. Thus, the whole is

$$\Phi_{pm}(x)=\frac{(x^m-1)^p}{(x^m-1)\cdot\prod_{d|m,\ d<m}\Phi_{pd}(x)}$$

By induction on $d<m$, in the last product in the denominator has factors

$$\Phi_{pd}(x)=\Phi_d(x)^{p-1}$$

Cancelling,

$$\Phi_{pm}(x)=\frac{(x^m-1)^p}{(x^m-1)\cdot\prod_{d|m,\ d<m}\Phi_d(x)^{p-1}}=\frac{(x^m-1)^{p-1}}{\prod_{d|m,\ d<m}\Phi_d(x)^{p-1}}$$

$$=\left(\frac{x^m-1}{\prod_{d|m,\ d<m}\Phi_d(x)}\right)^{p-1}$$

which gives $\Phi_m(x)^{p-1}$ as claimed, by the recursive definition. ///

Exercises

19.1   Find two fields intermediate between $\mathbb{Q}$ and $\mathbb{Q}(\zeta_{11})$, where ζ_{11} is a primitive 11th root of unity.

19.2   The 5th cyclotomic polynomial factors into two irreducibles in $\mathbb{F}_{19}[x]$. Find these two irreducibles.

19.3   The 8th cyclotomic polynomial factors into two irreducibles in $\mathbb{F}_7[x]$. Find these two irreducibles.

19.4   The 8th cyclotomic polynomial factors into two irreducible quadratics in $\mathbb{Q}(\sqrt{2})[x]$. Find these two irreducibles.

20. Cyclotomic III

The main goal is to prove that all cyclotomic polynomials $\Phi_n(x)$ are irreducible in $\mathbb{Q}[x]$, and to see what happens to $\Phi_n(x)$ over $\mathbb{F}_p$ when $p|n$.

The irreducibility over $\mathbb{Q}$ allows us to conclude that the automorphism group of $\mathbb{Q}(\zeta_n)$ over $\mathbb{Q}$ (with ζ_n a primitive n^{th} root of unity) is

$$\text{Aut}(\mathbb{Q}(\zeta_n)/\mathbb{Q}) \approx (\mathbb{Z}/n)^\times$$

by the map

$$(\zeta_n \longrightarrow \zeta_n^a) \longleftarrow a$$

The case of prime-power cyclotomic polynomials in $\mathbb{Q}[x]$ needs only Eisenstein's criterion, but the case of general n seems to admit no comparably simple argument. The proof given here uses ideas already in hand, but also an unexpected trick. We will give a different, less elementary, but possibly more natural argument later using p-adic numbers and Dirichlet's theorem on primes in an arithmetic progression.

20.1 *Prime-power cyclotomic polynomials over* $\mathbb{Q}$

The proof of the following is just a slight generalization of the prime-order case.

20.1.1 Proposition: For p prime and for $1 \le e \in \mathbb{Z}$ the prime-power p^e-th cyclotomic polynomial $\Phi_{p^e}(x)$ is irreducible in $\mathbb{Q}[x]$.

Proof: Not unexpectedly, we use Eisenstein's criterion to prove that $\Phi_{p^e}(x)$ is irreducible in $\mathbb{Z}[x]$, and the invoke Gauss' lemma to be sure that it is irreducible in $\mathbb{Q}[x]$. Specifically, let

$$f(x) = \Phi_{p^e}(x+1)$$

If $e = 1$, we are in the familiar prime-order case. Since p divides binomial coefficients $\binom{p}{i}$ for $0 < i < p$

$$\Phi_p(x+1) = \frac{(x+1)^p - 1}{(x+1) - 1} = x^{p-1} + \binom{p}{1}x^{p-2} + \ldots + \binom{p}{2}x + \binom{p}{1}$$

reaching the usual conclusion directly in this case.

Now consider $e > 1$. Let

$$f(x) = \Phi_{p^e}(x+1)$$

Recall that

$$\Phi_{p^e}(x) = \Phi_p(x^{p^{e-1}}) = \frac{x^{p^e} - 1}{x^{p^{e-1}} - 1}$$

First, we check that p divides all but the highest-degree coefficient of $f(x)$. To do so, map everything to $\mathbb{F}_p[x]$, by reducing coefficients modulo p. For $e \geq 1$

$$(x+1)^{p^{e-1}} = x^{p^{e-1}} + 1 \bmod p$$

Therefore, in $\mathbb{F}_p[x]$

$$f(x) = \Phi_p((x+1)^{p^{e-1}}) = \frac{(x+1)^{p^e} - 1}{(x+1)^{p^{e-1}} - 1}$$

$$= ((x+1)^{p^{e-1}})^{p-1} + ((x+1)^{p^{e-1}})^{p-2} + \ldots + ((x+1)^{p^{e-1}}) + 1$$

$$= (x^{p^{e-1}}+1)^{p-1} + (x^{p^{e-1}}+1)^{p-2} + \ldots + (x^{p^{e-1}}+1) + 1$$

$$= \frac{(x^{p^{e-1}}+1)^p - 1}{(x^{p^{e-1}}+1) - 1} = \frac{x^{p^{e-1}}+1-1}{x^{p^{e-1}}} = \frac{x^{p^e}}{x^{p^{e-1}}} = x^{p^{e-1}(p-1)}$$

in $\mathbb{F}_p[x]$. Thus, all the lower coefficients are divisible by p. [1]   To determine the constant coefficient of $f(x)$, again use

$$\Phi_{p^e}(x) = \Phi_p(x^{p^{e-1}})$$

to compute

$$\text{constant coefficient of } f = f(0) = \Phi_{p^e}(1) = \Phi_p(1^{p^{e-1}}) = \Phi_p(1) = p$$

as in the prime-order case. Thus, p^2 does not divide the constant coefficient of f. Then apply Eisenstein's criterion and Gauss' lemma to obtain the irreducibility.   ///

20.1.2 Corollary: Let ζ be a primitive p^e-th root of unity. The automorphism group $\text{Aut}(\mathbb{Q}(\zeta)/\mathbb{Q})$ is isomorphic

$$(\mathbb{Z}/p^e)^\times \approx \text{Aut}(\mathbb{Q}(\zeta)/\mathbb{Q})$$

by

$$a \longrightarrow \sigma_a$$

where

$$\sigma_a(\zeta) = \zeta^a$$

---

[1] Note that this argument in $\mathbb{F}_p[x]$ by itself cannot prove that p^2 does not divide the constant coefficient, since we are computing only in $\mathbb{F}_p[x]$.

Proof: This follows from the irreducibility of $\Phi_{p^e}(x)$ in $\mathbb{Q}[x]$ and the fact that all primitive p^e-th roots of unity are expressible as ζ^a with a in $(\mathbb{Z}/p^e)^\times$. More precisely, we saw earlier that for any other root β of $f(x) = 0$ in $\mathbb{Q}(\alpha)$ with f the minimal polynomial of α over $\mathbb{Q}$, there is an automorphism of $\mathbb{Q}(\alpha)$ sending α to β. Thus, for any a relatively prime to p there is an automorphism which sends $\zeta \longrightarrow \zeta^a$. On the other hand, any automorphism must send ζ to a root of $\Phi_{p^e}(x) = 0$, and these are all of the form ζ^a. Thus, we have an isomorphism. ///

20.2 *Irreducibility of cyclotomic polynomials over* $\mathbb{Q}$

Now consider general n, and ζ a primitive n^{th} root of unity. We prove irreducibility over $\mathbb{Q}$ of the nth cyclotomic polynomial, and the very useful corollary that the automorphism group of $\mathbb{Q}(\zeta_n)$ over $\mathbb{Q}$ (for a primitive n^{th} root of unity ζ_n) is isomorphic to $(\mathbb{Z}/n)^\times$.

20.2.1 Theorem: The nth cyclotomic polynomial $\Phi_n(x)$ is irreducible in $\mathbb{Q}[x]$.

Proof: Suppose that $\Phi_n(x) = f(x)g(x)$ in $\mathbb{Q}[x]$ with f of positive degree. Via Gauss' lemma we can suppose that both f and g are monic and are in $\mathbb{Z}[x]$. Let $x - \zeta$ be a linear factor of $f(x)$ in $k[x]$ for an extension field k of $\mathbb{Q}$. We wish to show that $x - \zeta^a$ is also a linear factor of f for every $a \in (\mathbb{Z}/n)^\times$, and thus that

$$\deg f = \varphi(n) = \deg \Phi_n$$

concluding that $f = \Phi_n$.

Since each $a \in (\mathbb{Z}/n)^\times$ is a product of primes p not dividing n, it suffices to show that $x - \zeta^p$ is a linear factor of $f(x)$ for all primes p not dividing n. If not, then $x - \zeta^p$ is necessarily a linear factor of $g(x)$, by unique factorization in $k[x]$. That is, ζ is a root of $g(x^p) = 0$ in k, so $x - \zeta$ divides $g(x^p)$ in $k[x]$.

Thus, in $\mathbb{Q}[x]$ the *gcd* of $f(x)$ and $g(x^p)$ is not 1: otherwise, there would be $r(x), s(x) \in Q[x]$ such that

$$1 = r(x) \cdot f(x) + s(x) \cdot g(x^p)$$

Mapping $\mathbb{Q}[x]$ to k by $x \longrightarrow \zeta$ would give the impossible

$$1 = r(\zeta) \cdot 0 + s(\zeta) \cdot 0 = 0$$

Thus, $d(x) = \gcd(f(x), g(x^p))$ in $\mathbb{Q}[x]$ is of positive degree. Let $a(x)$ and $b(x)$ be in $\mathbb{Q}[x]$ such that

$$f(x) = a(x) \cdot d(x) \qquad g(x^p) = b(x) \cdot d(x)$$

We can certainly take d to be in $\mathbb{Z}[x]$ and have content 1. By Gauss' lemma, $a(x)$ and $b(x)$ are in $\mathbb{Z}[x]$ and have content 1. In fact, adjusting by at most ± 1, we can take $a(x)$, $b(x)$, and $d(x)$ all to be monic.

Map everything to $\mathbb{F}_p[x]$. There $g(x^p) = g(x)^p$, so

$$\begin{cases} f(x) & = & a(x) \cdot d(x) \\ g(x)^p & = & g(x^p) & = & b \cdot d \end{cases}$$

Let $\delta(x) \in \mathbb{F}_p[x]$ be an irreducible dividing $d(x)$ in $\mathbb{F}_p[x]$. Then since $\delta(x)$ divides $g(x)^p$ in $\mathbb{F}_p[x]$ it divides $g(x)$. Also $\delta(x)$ divides $f(x)$ in $\mathbb{F}_p[x]$, so $\delta(x)^2$ apparently divides

$\Phi_n(x) = f(x) \cdot g(x)$ in $\mathbb{F}_p[x]$. But p does not divide n, so $\Phi_n(x)$ has no repeated factor in $\mathbb{F}_p[x]$, contradiction. Thus, it could not have been that $\Phi_n(x)$ factored properly in $\mathbb{Q}[x]$.

///

20.2.2 Corollary: Let ζ be a primitive n-th root of unity. The automorphism group $\mathrm{Aut}(\mathbb{Q}(\zeta)/\mathbb{Q})$ is isomorphic

$$(\mathbb{Z}/n)^\times \approx \mathrm{Aut}(\mathbb{Q}(\zeta)/\mathbb{Q})$$

by

$$a \longrightarrow \sigma_a$$

where

$$\sigma_a(\zeta) = \zeta^a$$

Proof: This follows from the irreducibility of $\Phi_n(x)$ in $\mathbb{Q}[x]$ and the fact that all primitive n-th roots of unity are expressible as ζ^a with a in $(\mathbb{Z}/n)^\times$. More precisely, we saw earlier that for any other root β of $f(x) = 0$ in $\mathbb{Q}(\alpha)$ with f the minimal polynomial of α over $\mathbb{Q}$, there is an automorphism of $\mathbb{Q}(\alpha)$ sending α to β. Thus, for any a relatively prime to n there is an automorphism which sends $\zeta \longrightarrow \zeta^a$. On the other hand, any automorphism must send ζ to a root of $\Phi_n(x) = 0$, and these are all of the form ζ^a, because of the nature of cyclic groups. Thus, we have an isomorphism.

///

20.3 *Factoring* $\Phi_n(x)$ *in* $\mathbb{F}_p[x]$ *with* $p|n$

It turns out that a sensible proof of the following can be given using only the inductive definition of $\Phi_n(x)$ in $\mathbb{Z}[x]$.

20.3.1 Theorem: For a prime p, integer m not divisible by p, and integer $e \geq 1$, in $\mathbb{F}_p[x]$ the $p^e m^{th}$ cyclotomic polynomial $\Phi_{p^e m}(x)$ is

$$\Phi_{p^e m}(x) = \Phi_m(x)^{\varphi(p^e)} = \Phi_m(x)^{(p-1)(p^{e-1})}$$

where φ is Euler's totient function.

Proof: From the recursive definition, for $1 \leq e \in \mathbb{Z}$,

$$\Phi_{p^e m}(x) = \frac{x^{p^e m} - 1}{\prod_{d|m,\, 0 \leq \varepsilon < e} \Phi_{p^\varepsilon d}(x) \cdot \prod_{d|m,\, d < m} \Phi_{p^e d}(x)}$$

In characteristic p, the numerator is $(x^m - 1)^{p^e}$. The first product factor in the denominator is $x^{p^{e-1} m} - 1$, which in characteristic p is $(x^m - 1)^{p^{e-1}}$. Thus, the whole is

$$\Phi_{p^e m}(x) = \frac{(x^m - 1)^{p^e}}{(x^m - 1)^{p^{e-1}} \cdot \prod_{d|m,\, d < m} \Phi_{p^e d}(x)}$$

By induction on $d < m$, in the last product in the denominator has factors

$$\Phi_{p^e d}(x) = \Phi_d(x)^{\varphi(p^e)}$$

Cancelling,

$$\Phi_{p^e m}(x) = \frac{(x^m - 1)^{p^e}}{(x^m - 1)^{p^{e-1}} \cdot \prod_{d|m,\, d<m} \Phi_d(x)^{\varphi(p^e)}}$$

$$= \frac{(x^m - 1)^{(p-1)p^{e-1}}}{\prod_{d|m,\, d<m} \Phi_d(x)^{\varphi(p^e)}} = \left(\frac{x^m - 1}{\prod_{d|m,\, d<m} \Phi_d(x)} \right)^{\varphi(p^e)}$$

which gives $\Phi_m(x)^{\varphi(p^e)}$ as claimed, by the recursive definition. ///

20.4 Worked examples

20.4.1 Example: Prove that a prime p such that $p = 1 \bmod 3$ factors *properly* as $p = ab$ in $\mathbb{Z}[\omega]$, where ω is a primitive cube root of unity. (*Hint:* If p were prime in $\mathbb{Z}[\omega]$, then $\mathbb{Z}[\omega]/p$ would be a integral domain.)

The hypothesis on p implies that $(\mathbb{Z}/p)^\times$ has order divisible by 3, so there is a primitive third root of unity ζ in $\mathbb{Z}/p$. That is, the third cyclotomic polynomial $x^2 + x + 1$ factors mod p. Recall the isomorphisms

$$\mathbb{Z}[\omega]/p \approx (\mathbb{Z}[x]/(x^2 + x + 1))/p \approx (\mathbb{Z}/p)[x]/(x^2 + x + 1)$$

Since $x^2 + x + 1$ factors mod p, the right-most quotient is *not* an integral domain. Recall that a commutative ring modulo an ideal is an integral domain if and only if the ideal is prime. Thus, looking at the left-most quotient, the ideal generated by p in $\mathbb{Z}[\omega]$ is not prime. Since we have seen that $\mathbb{Z}[\omega]$ is Euclidean, hence a PID, the element p must factor properly. ///

20.4.2 Example: Prove that a prime p such that $p = 2 \bmod 5$ generates a prime ideal in the ring $\mathbb{Z}[\zeta]$, where ζ is a primitive fifth root of unity.

The hypothesis on p implies that $\mathbb{F}_{p^n}^\times$ has order divisible by 5 only for n divisible by 4. Thus, the fifth cyclotomic polynomial Φ_5 is irreducible modulo p: (If it had a linear factor then $\mathbb{F}_p^\times$ would contain a primitive fifth root of unity, so have order divisible by 5. If it had a quadratic factor then $\mathbb{F}_{p^2}^\times$ would contain a primitive fifth root of unity, so have order divisible by 5.) Recall the isomorphisms

$$\mathbb{Z}[\zeta]/p \approx (\mathbb{Z}[x]/\Phi_5)/p \approx (\mathbb{Z}/p)[x]/(\Phi_5)$$

Since Φ_5 is irreducible mod p, the right-most quotient is an integral domain. As recalled in the previous example, a commutative ring modulo an ideal is an integral domain if and only if the ideal is prime. Thus, looking at the left-most quotient, the ideal generated by p in $\mathbb{Z}[\zeta]$ is prime. ///

20.4.3 Example: Find the monic irreducible polynomial with rational coefficients which has as zero

$$\alpha = \sqrt{3} + \sqrt{5}$$

In this simple example, we can take a rather *ad hoc* approach to find a polynomial with α as 0. Namely,

$$\alpha^2 = 3 + 2\sqrt{3}\sqrt{5} + 5 = 8 + 2\sqrt{15}$$

Then
$$(\alpha^2 - 8)^2 = 4 \cdot 15 = 60$$

Thus,
$$\alpha^4 - 16\alpha^2 + 4 = 0$$

But this approach leaves the question of the irreducibility of this polynomial over $\mathbb{Q}$.

By Eisenstein, $x^2 - 3$ and $x^2 - 5$ are irreducible in $\mathbb{Q}[x]$, so the fields generated over $\mathbb{Q}$ by the indicated square roots are of degree 2 over $\mathbb{Q}$. Since (inside a fixed algebraic closure of $\mathbb{Q}$) $[\mathbb{Q}(\sqrt{3}, \sqrt{5}) : \mathbb{Q}] \leq [\mathbb{Q}(\sqrt{3}) : \mathbb{Q}] \cdot [\mathbb{Q}(\sqrt{3}) : \mathbb{Q}]$,

$$[\mathbb{Q}(\sqrt{3}, \sqrt{5}) : \mathbb{Q}] \leq 4$$

It is natural to claim that we have equality. To prove equality, one approach is to show that there is no $\sqrt{5}$ in $\mathbb{Q}(\sqrt{3})$: supposed that $(a + b\sqrt{3})^2 = 5$ with $a, b \in \mathbb{Q}$. Then

$$(a^2 - 3b^2) + 2ab\sqrt{3} = 5 = 5 + 0 \cdot \sqrt{3}$$

Since $\sqrt{3}$ and 1 are linearly independent over $\mathbb{Q}$ (this is what the field degree assertions are), this requires that either $a = 0$ or $b = 0$. In the latter case, we would have $a^2 = 5$. In the former, $3b^2 = 5$. In either case, Eisenstein's criterion (or just unique factorization in $\mathbb{Z}$) shows that the corresponding polynomials $x^2 - 5$ and $3x^2 - 5$ are irreducible, so this is impossible.

To prove that the quartic of which $\alpha = \sqrt{3} + \sqrt{5}$ is a root is irreducible, it suffices to show that α generates $\mathbb{Q}(\sqrt{3}, \sqrt{5})$. Certainly

$$\frac{\alpha^2 - 8}{2} = \sqrt{15}$$

(If we were in characteristic 2 then we could not divide by 2. But, also, in that case $3 = 5$.) Then
$$(\frac{\alpha^2 - 8}{2}) \cdot \alpha = \sqrt{15} \cdot \alpha = 3\sqrt{5} + 5\sqrt{3}$$

The system of two linear equations
$$\begin{aligned} \sqrt{3} + \sqrt{5} &= \alpha \\ 5\sqrt{3} + 3\sqrt{5} &= (\tfrac{\alpha^2-8}{2}) \cdot \alpha \end{aligned}$$

can be solved for $\sqrt{3}$ and $\sqrt{5}$. Thus, α generates the quartic field extension, so has a quartic minimal polynomial, which must be the monic polynomial we found.   ///

A more extravagant proof (which generalizes in an attractive manner) that

$$[\mathbb{Q}(\sqrt{3}, \sqrt{5}) : \mathbb{Q}] = 4$$

uses cyclotomic fields and (proto-Galois theoretic) facts we already have at hand about them. Let ζ_n be a primitive n^{th} root of unity. We use the fact that

$$\text{Aut}(\mathbb{Q}(\zeta_n)/\mathbb{Q}) \approx (\mathbb{Z}/n)^\times$$

by
$$(\sigma_a : \zeta_n \longrightarrow \zeta_n^a) \longleftarrow a$$

Letting $n = 4pq$ with distinct odd primes p, q, by Sun-Ze's theorem

$$\mathbb{Z}/n \approx \mathbb{Z}/4 \oplus \mathbb{Z}/p \oplus \mathbb{Z}/q$$

Thus, given an automorphism τ_1 of $\mathbb{Q}(\zeta_p)$ over $\mathbb{Q}$, an automorphism τ_2 of $\mathbb{Q}(\zeta_q)$ over $\mathbb{Q}$, and an automorphism τ_3 of $\mathbb{Q}(i)$ over $\mathbb{Q}$, there is an automorphism σ of $\mathbb{Q}(\zeta_{4pq})$ over $\mathbb{Q}$ which restricts to τ_1 on $\mathbb{Q}(\zeta_p)$, to τ_2 on $\mathbb{Q}(\zeta_2)$, and to τ_3 on $\mathbb{Q}(i)$. Also,

$$\sqrt{p \cdot \left(\frac{-1}{p}\right)_2} \in \mathbb{Q}(\text{primitive } p^{th} \text{ root of unity})$$

In particular, letting ζ_p be a primitive p^{th} root of unity, the Gauss sum expression

$$\sqrt{p \cdot \left(\frac{-1}{p}\right)_2} = \sum_{b \bmod p} \left(\frac{b}{p}\right)_2 \cdot \zeta_p^b$$

shows (as observed earlier) that

$$\sigma_a\left(\sqrt{p \cdot \left(\frac{-1}{p}\right)_2}\right) = \left(\frac{a}{p}\right)_2 \cdot \sqrt{p \cdot \left(\frac{-1}{p}\right)_2}$$

The signs under the radicals can be removed by removing a factor of i, if necessary. Thus, we can choose $a \in (\mathbb{Z}/4pq)^\times$ with $a = 1 \bmod 4$ to assure that $\sigma_a(i) = i$, and

$$\begin{cases} \sigma_a(\sqrt{p}) &= -\sqrt{p} \\ \sigma_a(\sqrt{q}) &= \sqrt{q} \end{cases}$$

That is, a is any non-zero square modulo q and is a non-square modulo p. That is, σ_a is an automorphism of $\mathbb{Q}(\zeta_{4pq})$ which properly moves $\sqrt{p}$ but does not move $\sqrt{q}$. Thus, σ_a is trivial on $\mathbb{Q}(\sqrt{q})$, so this field cannot contain $\sqrt{p}$. Thus, the degree $[\mathbb{Q}(\sqrt{p}, \sqrt{q}) : \mathbb{Q}] > 2$. But also this degree is at most 4, and is divisible by $[\mathbb{Q}(\sqrt{q}) : \mathbb{Q}] = 2$. Thus, the degree is 4, as desired. ///

20.4.4 Example: Find the monic irreducible polynomial with rational coefficients which has as zero

$$\alpha = \sqrt{3} + \sqrt[3]{5}$$

Eisenstein's criterion shows that $x^2 - 3$ and $x^3 - 5$ are irreducible in $\mathbb{Q}[x]$, so the separate field degrees are as expected: $[\mathbb{Q}\sqrt{3}) : \mathbb{Q}] = 2$, and $[\mathbb{Q}(\sqrt[3]{5}) : \mathbb{Q}] = 3$. This case is somewhat simpler than the case of two square roots, since the degree $[\mathbb{Q}(\sqrt{3}, \sqrt[3]{5}) : \mathbb{Q}]$ of any compositum is divisible by both $2 = [\mathbb{Q}(\sqrt{3})]$ and $3 = [\mathbb{Q}(\sqrt[3]{5}) : \mathbb{Q}] = 3$, so is divisible by $6 = \text{lcm}(2, 3)$. On the other hand, it is at most the product $6 = 2 \cdot 3$ of the two degrees, so is exactly 6.

To find a sextic over $\mathbb{Q}$ satisfied by α, we should be slightly more clever. Note that immediately

$$(\alpha - \sqrt{3})^3 = 5$$

which is

$$\alpha^3 - 3\sqrt{3}\alpha^2 + 3 \cdot 3\alpha - 3\sqrt{3} = 5$$

Moving all the square roots to one side,

$$\alpha^3 + 9\alpha - 5 = \sqrt{3} \cdot 3 \cdot (\alpha^2 + 1)$$

and then square again to obtain

$$\alpha^6 + 81\alpha^2 + 25 + 18\alpha^4 - 10\alpha^3 - 90\alpha = 27(\alpha^4 + 2\alpha^2 + 1)$$

Rearranging gives
$$\alpha^6 - 9\alpha^4 - 10\alpha^3 + 27\alpha^2 - 90\alpha - 2 = 0$$

Thus, since α is of degree 6 over $\mathbb{Q}$, the polynomial

$$x^6 - 9x^4 - 10x^3 + 27x^2 - 90x - 2$$

of which α is a zero is irreducible.   ///

20.4.5 Example: Find the monic irreducible polynomial with rational coefficients which has as zero
$$\alpha = \frac{1 + \sqrt[3]{10} + \sqrt[3]{10}^2}{3}$$

First, by Eisenstein's criterion $x^3 - 10$ is irreducible over $\mathbb{Q}$, so $\sqrt[3]{10}$ generates a cubic extension of $\mathbb{Q}$, and thus 1, $\sqrt[3]{10}$, and $\sqrt[3]{10}^2$ are linearly independent over $\mathbb{Q}$. Thus, α is not in $\mathbb{Q}$. Since it lies inside a cubic field extension of $\mathbb{Q}$, it satisfies a monic cubic equation with rational coefficients. The issue, then, is to find the cubic.

First we take advantage of the special nature of the situation. A little more generally, let $\beta^3 = A$ with $A \neq 1$. We note that

$$\beta^2 + \beta + 1 = \frac{\beta^3 - 1}{\beta - 1} = \frac{A - 1}{\beta - 1}$$

From $\beta^3 - A = 0$, using $\beta = (b\eta - 1) + 1$, we have

$$(\beta - 1)^3 + 3(\beta - 1)^2 + 3(\beta - 1)^2 - (A - 1) = 0$$

Dividing through by $(\beta - 1)^3$ gives

$$1 + 3(\frac{1}{\beta - 1}) + 3(\frac{1}{\beta - 1})^2 - \frac{A - 1}{(\beta - 1)^3} = 0$$

Multiplying through by $-(A-1)^2$ and reversing the order of the terms gives

$$(\frac{A-1}{\beta-1})^3 - 3(\frac{A-1}{\beta-1})^2 - 3(A-1)(\frac{A-1}{\beta-1}) - (A-1)^2 = 0$$

That is, $1 + \sqrt[3]{A} + \sqrt[3]{A}^2$ is a root of

$$x^3 - 3x^2 - 3(A-1)x - (A-1)^2 = 0$$

Then $(1 + \sqrt[3]{A} + \sqrt[3]{A}^2)/3$ is a root of

$$x^3 - x^2 - (\frac{A-1}{3})x - \frac{(A-1)^2}{27} = 0$$

When $(A-1)^2$ is divisible by 27 we have a nice simplification, as with $A = 10$, in which case the cubic is

$$x^3 - x^2 - 3x - 3 = 0$$

which has *integral* coefficients. ///

20.4.6 Remark: The fact that the coefficients are integral despite the apparent denominator of α is entirely parallel to the fact that $\frac{-1\pm\sqrt{D}}{2}$ satisfies the quadratic equation

$$x^2 - x + \frac{1 - D}{4} = 0$$

which has *integral coefficients* if $D = 1 \bmod 4$.

There is a more systematic approach to finding minimal polynomials that will work in more general circumstances, which we can also illustrate in this example. Again let $\beta = \sqrt[3]{A}$ where A is not a cube in the base field k. Then, again, we know that $1 + \beta + \beta^2$ is not in the ground field k, so, since it lies in a cubic field extension, has minimal polynomial over k which is an irreducible (monic) *cubic*, say $x^3 + ax^2 + bx + c$. We can determine a, b, c systematically, as follows. Substitute $1 + \beta + \beta^2$ for x and require

$$(1 + \beta + \beta^2)^3 + a(1 + \beta + \beta^2)^2 + b(1 + \beta + \beta^2) + c = 0$$

Multiply out, obtaining

$$(\beta^6 + \beta^3 + 1 + 3\beta^5 + 3\beta^4 + 3\beta^2 + 3\beta^4 + 3\beta^2 + 3\beta + 6\beta^3)$$

$$+ a(\beta^4 + \beta^2 + 1 + 2\beta^3 + 2\beta^2 + 2\beta) + b(\beta^2 + \beta + 1) + c$$

$$= 0$$

Use the fact that $\beta^3 = A$ (if β satified a more complicated cubic this would be messier, but still succeed) to obtain

$$(3A + 6 + 3a + b)\beta^2 + (6A + 3 + (A + 2)a + +b)\beta$$

$$+ (A^2 + 7A + 1 + (2A + 1)a + b + c) = 0$$

Again, $1, \beta, \beta^2$ are linearly independent over the ground field k, so this condition is equivalent to the system

$$\begin{cases} 3a & + \ b & & = & -(3A + 6) \\ (A + 2)a & + \ b & & = & -(6A + 3) \\ (2A + 1)a & + \ b & + \ c & = & -(A^2 + 7A + 1) \end{cases}$$

From the first two equations $a = -3$, and then $b = -3(A - 1)$, and from the last $c = -(A - 1)^2$, exactly as earlier. ///

20.4.7 Remark: This last approach is only palatable if there's no other recourse.

20.4.8 Example: Let p be a prime number, and $a \in \mathbb{F}_p^\times$. Prove that $x^p - x + a$ is irreducible in $\mathbb{F}_p[x]$. (*Hint*: Verify that if α is a root of $x^p - x + a = 0$, then so is $\alpha + 1$.)

Comment: It might have been even more helpful to recommend to look at the effect of Frobenious $b \longrightarrow b^p$, but the hint as given reveals an interesting fact in its own right, and takes us part of the way to understanding the situation.

If α is a root in an algebraic closure, then

$$(\alpha + 1)^p - (\alpha + 1) + a = \alpha^p + 1 - \alpha - 1 + a = 0$$

so $\alpha + 1$ is another root. Thus, the roots of this equation are exactly

$$\alpha, \; \alpha + 1, \; \alpha + 2, \; \ldots, \; \alpha + (p - 1)$$

which are distinct. (The polynomial is of degree p, so there are no more than p zeros.)

Similarly, but even more to the point is that the Frobenius automorphism F has the effect

$$F(\alpha) = \alpha^p = (\alpha^p - \alpha + a) + \alpha - a = \alpha - a$$

Let A be a subset of this set of zeros. We have shown that a polynomial

$$\prod_{\beta \in A} (x - \beta)$$

has coefficients in $\mathbb{F}_p$ if and only if A is stable under the action of the Frobenius. Since $a \neq 0$, the smallest F-stable subset of A is necessarily the whole, since the values

$$F^\ell(\alpha) = \alpha - \ell \cdot a$$

are distinct for $\ell = 0, 1, \ldots, p - 1$. By unique factorization, any factor of $x^p - x + 1$ is a product of linear factors $x - F^\ell(\alpha)$, and we have shown that a product of such factors has coefficients in $\mathbb{F}_p$ only if *all* these factors are included. That is, $x^p - x + a$ is irreducible in $\mathbb{F}_p[x]$.   ///

20.4.9 Example: Let $k = \mathbb{F}_p(t)$ be the field of rational expressions in an indeterminate t with coefficients in $\mathbb{F}_p$. Show that the polynomial $X^p - t \in k[X]$ is irreducible in $k[X]$, but has properly repeated factors over an algebraic closure of k.

That polynomial meets Eisenstein's criterion in $\mathbb{F}_p[t][X]$, since t is a prime element in the UFD $\mathbb{F}_p[t]$, so (via Gauss' lemma) $X^p - t$ is irreducible in $\mathbb{F}_p(t)[X]$. Let α be any root of $X^p - t = 0$. Then, because the inner binomial coefficients $\binom{p}{i}$ are divisible by p,

$$(X - \alpha)^p = X^p - \alpha^p = X^p - t$$

That is, over an algebraic closure of $\mathbb{F}_p(t)$, the polynomial $X^p - t$ is a linear polynomial raised to the pth power.

20.4.10 Example: Let x be an indeterminate over $\mathbb{C}$. For a, b, c, d in $\mathbb{C}$ with $ad - bc \neq 0$, let

$$\sigma(x) = \sigma_{a,b,c,d}(x) = \frac{ax + b}{cx + d}$$

and define

$$\sigma\left(\frac{P(x)}{Q(x)}\right) = \frac{P(\sigma(x))}{Q(\sigma(x))}$$

for P and Q polynomials. Show that σ gives a field automorphism of the field of rational functions $\mathbb{C}(x)$ over $\mathbb{C}$.

The argument uses no properties of the complex numbers, so we discuss an arbitrary field k instead of $\mathbb{C}$.

Since the polynomial algebra $k[x]$ is the free k-algebra on one generator, by definition for any k-algebra A and chosen element $a \in A$, there is a unique k-algebra map $k[x] \longrightarrow A$ such that $x \longrightarrow a$. And, second, for any *injective* k-algebra map f of $k[x]$ to a *domain* R the field of fractions $k(x)$ of $k[x]$ has an associated map $\tilde{f}$ to the field of fractions of R, by

$$\tilde{f}(P/Q) = f(P)/f(Q)$$

where P and Q are polynomials.

In the case at hand, any choice $\sigma(x) = g(x)/h(x)$ in $k(x)$ (with polynomials g, h with h not the 0 polynomial) gives a unique k-algebra homomorphism $k[x] \longrightarrow k(x)$, by

$$\sigma(P(x)) = P(\sigma(x)) = P(\frac{g(x)}{h(x)})$$

To know that we have an extension to the field of fractions $k(x)$ of $k[x]$, we must check that the kernel of the map $k[x] \longrightarrow k(x)$ is non-zero. That is, we must verify for a positive-degree polynomial (assume without loss of generality that $a_n \neq 0$)

$$P(x) = a_n x^n + \ldots + a_o$$

that

$$0 \neq \sigma(P(x)) \in k(x)$$

Again,

$$\sigma(P(x)) = P(\sigma(x)) = P(\frac{g(x)}{h(x)}) = a_n(\frac{g}{h})^n + \ldots + a_o$$

$$= h^{-n} \cdot (a_n g^n + a_{n-1} g^{n-1} h + \ldots + a_1 g h^{n-1} + a_o h^n)$$

We could have assumed without loss of generality that g and h are relatively prime in $k[x]$. If the degree of g is positive, let $p(x)$ be an irreducible factor of $g(x)$. Then an equality

$$0 = a_n g^n + a_{n-1} g^{n-1} h + \ldots + a_1 g h^{n-1} + a_o h^n$$

would imply that $p|h$, contradiction. But if $\deg h > 0$ we reach a nearly identical contradiction. That is, a field map $k(x) \longrightarrow k(x)$ can send x to any element of $k(x)$ not lying in k. Thus, certainly, for $ad - bc \neq 0$, $(ax + b)/(cx + d)$ is not in k, and is a legitimate field map image of x.

To prove surjectivity of $\sigma(x) = (ax+b)/(cx+d)$, we find an inverse τ, specifically such that $\sigma \circ \tau = 1$. It may not be surprising that

$$\tau : x \longrightarrow \frac{dx - b}{-cx + a}$$

is such an inverse:

$$(\sigma \circ \tau)(x) = \frac{a(\frac{dx-b}{-cx+a}) + b}{c(\frac{dx-b}{-cx+a}) + d} = \frac{a(dx - b) + b(-cx + a)}{c(dx - b) + d(-cx + a)}$$

$$= \frac{(ad - bc)x - ab + ba}{cdx - cb - dcx + ad} = \frac{(ad - bc)x}{ad - bc} = x$$

That is, the given field maps are surjective. All field maps that do not map all elements to 0 are injective, so these maps are field automorphisms of $k(x)$.

20.4.11 Example: In the situation of the previous exercise, show that *every* automorphism of $\mathbb{C}(x)$ over $\mathbb{C}$ is of this form.

We did also show in the previous example that for g and h polynomials, not both constant, h not 0,

$$\sigma(x) = \frac{g(x)}{h(x)}$$

determines a field map $k(x) \longrightarrow k(x)$. If it were surjective, then there would be coefficients a_i and b_j in k such that x is expressible as

$$x = \frac{a_m \sigma(x)^m + \ldots + a_0}{b_n \sigma(x)^n + \ldots + b_0}$$

with $a_m \neq 0$ and $b_n \neq 0$. Let $\sigma(x) = p/q$ where p and q are relatively prime polynomials. Then

$$x \cdot q^{-n}(b_n p^n + b_{n-1} p^{n-1} q + \ldots + b_0 q^n) = q^{-m}(a_m p^m + a_{m-1} p^{m-1} q + \ldots + a_0 q^m)$$

or

$$x \cdot q^m(b_n p^n + b_{n-1} p^{n-1} q + \ldots + b_0 q^n) = q^n(a_m p^m + a_{m-1} p^{m-1} q + \ldots + a_0 q^m)$$

Collecting the only two terms lacking an explicit factor of p, we find that

$$(b_0 x - a_0) \cdot q^{m+n}$$

is visibly a multiple of p. Since p and q are relatively prime and $k[x]$ is a UFD, necessarily p divides $b_0 x - a_0$. Since degrees add in products, the degree of p is at most 1.

One argument to prove that $\deg q \leq 1$ is to observe that if p/q generates all of a field then so does its inverse q/p. Thus, by the previous paragraph's argument which showed that $\deg p \leq 1$, we have $\deg q \leq 1$.

For another argument concerning the denominator: a more direct computation approach does illustrate something useful about polynomial algebra: For $m > n$, we would have a polynomial equation

$$x \cdot q^{m-n}(b_n p^n + b_{n-1} p^{n-1} q + \ldots + b_0 q^n) = a_m p^m + a_{m-1} p^{m-1} q + \ldots + a_0 q^m$$

The only term not visibly divisible by q is $a_m p^m$, so apparently q divides $a_m p^m$. Since p, q are relatively prime, this would imply that $\deg q = 0$. Similarly, for $m < n$, the polynomial equation

$$x \cdot (b_n p^n + b_{n-1} p^{n-1} q + \ldots + b_0 q^n) = q^{n-m}(a_m p^m + a_{m-1} p^{m-1} q + \ldots + a_0 q^m)$$

implies that q divides $x \cdot b_n p^n$, and the coprimality of p, q implies that $\deg q \leq 1$. If $m = n$, then the polynomial equation

$$x \cdot (b_n p^n + b_{n-1} p^{n-1} q + \ldots + b_0 q^n) = a_m p^m + a_{m-1} p^{m-1} q + \ldots + a_0 q^m$$

implies that q divides (keeping in mind that $m = n$)

$$x \cdot b_n p^n - a_m p^m = (x b_n - a_n) \cdot p^n$$

The coprimality of p, q implies that q divides $x b_n - a_n$, so $\deg q \leq 1$ again in this case.

Thus, if $\sigma(x) = p/q$ gives a surjection of $k(x)$ to itself, the maximum of the degrees of p and q is 1. ///

20.4.12 Example: Let s and t be indeterminates over $\mathbb{F}_p$, and let $\mathbb{F}_p(s^{1/p}, t^{1/p})$ be the field extension of the rational function field $\mathbb{F}_p(s, t)$ obtained by adjoining roots of $X^p - s = 0$ and of $X^p - t = 0$. Show that there are infinitely-many (distinct) fields intermediate between $\mathbb{F}_p(s, t)$ and $\mathbb{F}_p(s^{1/p}, t^{1/p})$.

By Eisenstein's criterion in $k[s, t][X]$ we see that both $X^p - s$ and $X^p - t$ are irreducible in $k(s, t)[X]$, so $s^{1/p}$ and $t^{1/p}$ each generates a degree p extension of $k(s, t)$. We show that $[k(s^{1/p}, t^{1/p}) : k(s, t)] = p^2$. By Eisenstein's criterion in $\mathbb{F}_p(t)[s][X]$ the polynomial $X^p - s$ is irreducible, since the prime s in $\mathbb{F}_p(t)[s]$, but not its square, divides all but the highest term. And then $X^p - t$ is irreducible in $k(s^{1/p})[t][X]$ since the prime t in $k(s^{1/p}(s))[t]$ divides all the lower coefficients and its square does not divide the constant term.

Observe that for any polynomial $f(s, t)$, because the characteristic is p,

$$(s^{1/p} + f(s, t)t^{1/p})^p = s + f(s, t)^p t$$

For example, for any positive integer n

$$(s^{1/p} + s^n t^{1/p})^p = s + s^{np} t$$

Again, by Eisenstein's criterion in $\mathbb{F}_p(t)[s][X]$ the polynomial

$$X^p - (s + s^{np} t)$$

is irreducible, since the prime s in $\mathbb{F}_p(t)[s]$, but not its square, divides all but the highest term. Thus, the pth root of any $s + s^{np}t$ generates a degree p extension of $\mathbb{F}_p(s, t)$.

We claim that for distinct positive integers m, n

$$\mathbb{F}_p(s, t, (s + s^{mp}t)^{1/p}) \neq \mathbb{F}_p(s, t, (s + s^{np}t)^{1/p})$$

To prove this, we will show that any subfield of $\mathbb{F}_p(s^{1/p}, t^{1/p})$ which contains both $(s + s^{mp}t)^{1/p}$ and $(s + s^{np}t)^{1/p}$ is the whole field $\mathbb{F}_p(s^{1/p}, t^{1/p})$, which is of degree p^2 (rather than p). Indeed,

$$(s + s^{mp}t)^{1/p} - (s + s^{np}t)^{1/p} = s^{1/p} + s^m t^{1/p} - (s^{1/p} + s^n t^{1/p}) = (s^m - s^n)t^{1/p}$$

Since $m \neq n$ we can divide by $s^m - s^n$ to obtain $t^{1/p}$. Then we can surely express $s^{1/t}$ as well. Thus, for $m \neq n$, the field obtained by adjoining the two different p^{th} roots is of degree p^2 over $\mathbb{F}_p(s, t)$, so the two degree p extensions cannot be identical (or the whole degree would be just p). ///

20.4.13 Remark: From a foundational viewpoint, the above discussion is a bit glib about the interaction of s and t, and the interaction of $s^{1/n}$ and t. Though this is not the main point at the moment, detection of *implied relations* among *variables* can become serious. At present, the idea is that there are *no* relations between s and t, so relations between $s^{1/n}$ and t will not pop up. This *can* be made more precise in preparation for coping with more complicated situations later.

20.4.14 Example: Determine the degree of $\mathbb{Q}(\sqrt{65 + 56i})$ over $\mathbb{Q}$, where $i = \sqrt{-1}$.

We show that $65 + 56i$ is not a square in $\mathbb{Q}(i)$. We use the *norm*

$$N(\alpha) = \alpha \cdot \alpha^\sigma$$

from $\mathbb{Q}(i)$ to $\mathbb{Q}$, where as usual $(a + bi)^\sigma = a - bi$ for rational a, b. Since $-i$ is the other zero of the minimal polynomial $x^2 + 1$ of i over $\mathbb{Q}$, the map σ is a field automorphism of $\mathbb{Q}(i)$ over $\mathbb{Q}$. (Indeed, we showed earlier that there exists a $\mathbb{Q}$-linear field automorphism of $\mathbb{Q}(i)$ taking i to $-i$.) Since σ is a field automorphism, N is *multiplicative*, in the sense that

$$N(\alpha\beta) = N(\alpha) \cdot N(\beta)$$

Thus, if $\alpha = \beta^2$, we would have

$$N(\alpha) = N(\beta^2) = N(\beta)^2$$

and the latter is a square in $\mathbb{Q}$. Thus, if $\alpha = 65 + 56i$ were a square, then

$$N(65 + 56i) = 65^2 + 56^2 = 7361$$

would be a square. One could factor this into primes in $\mathbb{Z}$ to see that it is not a square, or hope that it is not a square modulo some relatively small prime. Indeed, modulo 11 it is 2, which is not a square modulo 11 (by brute force, or by Euler's criterion (using the cyclicness of $(\mathbb{Z}/11)^\times$) $2^{(11-1)/2} = -1 \bmod 11$, or by recalling the part of Quadratic Reciprocity that asserts that 2 is a square mod p only for $p = \pm 1 \bmod 8$).

20.4.15 Example: Fix an algebraically closed field k. Find a simple condition on $w \in k$ such that the equation $z^5 + 5zw + 4w^2 = 0$ has no repeated roots z in k.

Use some form of the Euclidean algorithm to compute the greatest common divisor in $k(w)[z]$ of $f(z) = z^5 + 5zw + 4w^2$ and its (partial?) derivative (with respect to z, not w). If the characteristic of k is 5, then we are in trouble, since the derivative (in z) vanishes identically, and therefore it is impossible to avoid repeated roots. So suppose the characteristic is not 5. Similarly, if the characteristic is 2, there will always be repeated roots, since the polynomial becomes $z(z^4 + w)$. So suppose the characteristic is not 2.

$$(z^5 + 5zw + 4w^2) - \tfrac{z}{5} \cdot (5z^4 + 5w) = 4zw + 4w^2$$
$$(z^4 + w) - \tfrac{1}{4w}(z^3 - z^2 w + zw^2 - w^3) \cdot (4zw + 4w^2) = w - w^4$$

where we also assumed that $w \neq 0$ to be able to divide. The expression $w - w^4$ is in the ground field $k(w)$ for the polynomial ring $k(w)[z]$, so if it is non-zero the polynomial and its derivative (in z) have no common factor. We know that this implies that the polynomial has no repeated factors. Thus, in characteristic not 5 or 2, for $w(1 - w^3) \neq 0$ we are assured that there are no repeated factors.

20.4.16 Remark: The algebraic closedness of k did not play a role, but may have helped avoid various needless worries.

20.4.17 Example: Fix a field k and an indeterminate t. Fix a positive integer $n > 1$ and let $t^{1/n}$ be an nth root of t in an algebraic closure of the field of rational functions $k(t)$. Show that $k[t^{1/n}]$ is isomorphic to a polynomial ring in one variable.

(There are many legitimate approaches to this question...)

We show that $k[t^{1/n}]$ is a free k-algebra on one generator $t^{1/n}$. That is, given a k-algebra A, a k-algebra homomorphism $f : k \longrightarrow A$, and an element $a \in A$, we must show that there is a unique k-algebra homomorphism $F : k[t^{1/n}] \longrightarrow A$ extending $f : k \longrightarrow A$ and such that $F(t^{1/n}) = a$.

Let $k[x]$ be a polynomial ring in one variable, and let $f : k[x] \longrightarrow k[t^{1/n}]$ be the (surjective) k-algebra homomorphism taking x to $t^{1/n}$. If we can show that the kernel of f is trivial, then f is an isomorphism and we are done.

Since $k[t]$ is a free k-algebra on one generator, it is infinite-dimensional as a k-vectorspace. Thus, $k[t^{1/n}]$ is infinite-dimensional as a k-vectorspace. Since $f : k[x] \longrightarrow k[t^{1/n}]$ is surjective, its image $k[x]/(\ker f) \approx f(k[x])$ is infinite-dimensional as a k-vectorspace.

Because $k[x]$ is a principal ideal domain, for an ideal I, either a quotient $k[x]/I$ is finite-dimensional as a k-vector space, or else $I = \{0\}$. There are no (possibly complicated) intermediate possibilities. Since $k[x]/(\ker f)$ is infinite-dimensional, $\ker f = \{0\}$. That is, $f : k[x] \longrightarrow k[t^{1/n}]$ is an isomorphism. ///

20.4.18 Remark: The vague and mildly philosophical point here was to see why an nth root of an *indeterminate* is still such a thing. It is certainly possible to use different language to give structurally similar arguments, but it seems to me that the above argument captures the points that occur in any version. For example, use of the notion of field elements *transcendental* over some ground field does suggest a good intuition, but still requires attention to similar details.

20.4.19 Example: Fix a field k and an indeterminate t. Let $s = P(t)$ for a monic polynomial P in $k[x]$ of positive degree. Find the monic irreducible polynomial $f(x)$ in $k(s)[x]$ such that $f(t) = 0$.

Perhaps this yields to direct computation, but we will do something a bit more conceptual.

Certainly s is a root of the equation $P(x) - s = 0$. It would suffice to prove that $P(x) - s$ is irreducible in $k(s)[x]$. Since P is monic and has coefficients in k, the coefficients of $P(x) - s$ are in the subring $k[s]$ of $k(s)$, and their gcd is 1. In other words, as a polynomial in x, $P(x) - s$ has *content* 1. Thus, from Gauss' lemma, $P(x) - s$ is irreducible in $k(s)[x]$ if and only if it is irreducible in $k[s][x] \approx k[x][s]$. As a polynomial in s (with coefficients in $k[x]$), $P(x) - s$ has content 1, since the coefficient of s is -1. Thus, $P(x) - s$ is irreducible in $k[x][s]$ if and only if it is irreducible in $k(x)[s]$. In the latter ring it is simply a linear polynomial in s, so is irreducible.

20.4.20 Remark: The main trick here is to manage to interchange the roles of x and s, and then use the fact that $P(x) - s$ is much simpler as a polynomial in s than as a polynomial in x.

20.4.21 Remark: The notion of irreducibility in $k[s][x] \approx k[x][s]$ does not depend upon how we view these polynomials. Indeed, irreducibility of $r \in R$ is equivalent to the irreducibility of $f(r)$ in S for any ring isomorphism $f : R \longrightarrow S$.

20.4.22 Remark: This approach generalizes as follows. Let $s = P(t)/Q(t)$ with relatively prime polynomials P, Q (and $Q \neq 0$). Certainly t is a zero of the polynomial $Q(x)s - P(s)$, and we claim that this is a (not necessarily monic) polynomial over $k(x)$ of minimal degree of which t is a 0. To do this we show that $Q(x)s - P(x)$ is irreducible in $k(s)[x]$. First, we claim that its content (as a polynomial in x with coefficients in $k[s]$) is 1. Let $P(x) = \sum_i a_i x^i$ and $Q(x) = \sum_j b_j x^j$, where $a_i, b_j \in k$ and we allow some of them to be 0. Then

$$Q(x)s - P(x) = \sum_i (b_i t - a_i) x^i$$

The content of this polynomial is the *gcd* of the linear polynomials $b_i t - a_i$. If this *gcd* were 1, then all these linear polynomials would be scalar multiples of one another (or 0). But that would imply that P, Q are scalar multiples of one another, which is impossible since they are relatively prime. So (via Gauss' lemma) the content is 1, and the irreducibility of $Q(x)s - P(x)$ in $k(s)[x]$ is equivalent to irreducibility in $k[s][x] \approx k[x][s]$. Now we verify that the content of the polynomial in t (with coefficient in $k[x]$) $Q(x)s - P(x)$ is 1. The content is the *gcd* of the coefficients, which is the *gcd* of P, Q, which is 1 by assumption. Thus, $Q(x)s - P(x)$ is irreducible in $k[x][s]$ if and only if it is irreducible in $k(x)[s]$. In the latter, it is a polynomial of degree at most 1, with non-zero top coefficients, so in fact linear. Thus, it is irreducible in $k(x)[s]$. We conclude that $Q(x)s - P(x)$ was irreducible in $k(s)[x]$.

Further, this approach shows that $f(x) = Q(x) - sP(x)$ is indeed a polynomial of minimal degree, over $k(x)$, of which t is a zero. Thus,

$$[k(t) : k(s)] = \max(\deg P, \deg Q)$$

Further, this proves a much sharper fact than that automorphisms of $k(t)$ only map $t \longrightarrow (at + b)/(ct + d)$, since any rational expression with higher-degree numerator or denominator generates a strictly smaller field, with the degree down being the maximum of the degrees.

20.4.23 Example: Let $p_1, p_2, \ldots$ be any ordered list of the prime numbers. Prove that $\sqrt{p_1}$ is *not* in the field

$$\mathbb{Q}(\sqrt{p_2}, \sqrt{p_3}, \ldots)$$

generated by the square roots of all the *other* primes.

First, observe that any rational expression for $\sqrt{p_1}$ in terms of the other square roots can only involve finitely many of them, so what truly must be proven is that $\sqrt{p_1}$ is not in the field

$$\mathbb{Q}(\sqrt{p_2}, \sqrt{p_3}, \ldots, \sqrt{p_N})$$

generated by any finite collection of square roots of *other* primes.

Probably an induction based on direct computation can succeed, but this is not the most interesting or informative. Instead:

Let ζ_n be a primitive n^{th} root of unity. Recall that for an odd prime p

$$\sqrt{p \cdot \left(\frac{-1}{p}\right)_2} \in \mathbb{Q}(\zeta_p)$$

Certainly $i = \sqrt{-1} \in \mathbb{Q}(\zeta_4)$. Thus, letting $n = 4p_1 p_2 \ldots p_N$, all the $\sqrt{p_1}, \ldots \sqrt{p_N}$ are in $K = \mathbb{Q}(\zeta_n)$. From the Gauss sum expression for these square roots, the automorphism

$\sigma_a(\zeta_n) = \zeta_n^a$ of $\mathbb{Q}(\zeta_n)$ has the effect

$$\sigma_a\sqrt{p_i \cdot \left(\frac{-1}{p_i}\right)_2} = \left(\frac{a}{p_i}\right)_2 \cdot \sqrt{p_i \cdot \left(\frac{-1}{p_i}\right)_2}$$

Thus, for $a = 1 \bmod 4$, we have $\sigma_a(i) = i$, and

$$\sigma_a(\sqrt{p_i}) = \left(\frac{a}{p_i}\right)_2 \cdot \sqrt{p_i}$$

Since $(\mathbb{Z}/p_i)^\times$ is cyclic, there *are* non-squares modulo p_i. In particular, let b be a non-square mod p_1. if we have a such that

$$\begin{cases} a & = & 1 \bmod 4 \\ a & = & b \bmod p_1 \\ a & = & 1 \bmod p_2 \\ & \vdots & \\ a & = & 1 \bmod p_N \end{cases}$$

then σ_a fixes $\sqrt{p_2}, \ldots, \sqrt{p_N}$, so when restricted to $K = \mathbb{Q}(\sqrt{p_2}, \ldots, \sqrt{p_N})$ is trivial. But by design $\sigma_a(\sqrt{p_1}) = -\sqrt{p_1}$, so this square root cannot lie in K. ///

20.4.24 Example: Let $p_1, \ldots, p_n$ be distinct prime numbers. Prove that

$$\mathbb{Q}(\sqrt{p_1}, \ldots, \sqrt{p_N}) = \mathbb{Q}(\sqrt{p_1} + \ldots + \sqrt{p_N})$$

Since the degree of a compositum KL of two field extensions K, L of a field k has degree *at most* $[K : k] \cdot [L : k]$ over k,

$$[\mathbb{Q}(\sqrt{p_1}, \ldots, \sqrt{p_N}) : \mathbb{Q}] \leq 2^N$$

since $[\mathbb{Q}(\sqrt{p_i}) : \mathbb{Q}] = 2$, which itself follows from the irreducibility of $x^2 - p_i$ from Eisenstein's criterion. The previous example shows that the bound 2^N is the actual degree, by multiplicativity of degrees in towers.

Again, a direct computation might succeed here, but might not be the most illuminating way to proceed. Instead, we continue as in the previous solution. Let

$$\alpha = \sqrt{p_1} + \ldots + \sqrt{p_n}$$

Without determining the minimal polynomial f of α over $\mathbb{Q}$ directly, we note that any automorphism τ of $\mathbb{Q}(\zeta_n)$ over $\mathbb{Q}$ can *only* send α f to other zeros of f, since

$$f(\tau\alpha) = \tau(f(\alpha)) = \tau(0) = 0$$

where the first equality follows exactly because the coefficients of f are fixed by τ. Thus, if we show that α has at least 2^N distinct images under automorphisms of $\mathbb{Q}(\zeta_n)$ over $\mathbb{Q}$, then the degree of f is at least 2^N. (It is at most 2^N since α does lie in that field extension, which has degree 2^N, from above.)

As in the previous exercise, for each index i among $1, \ldots, N$ we can find a_i such that

$$\sigma_{a_i}(\sqrt{p_j}) = \begin{cases} +\sqrt{p_j} & \text{for } j \neq i \\ -\sqrt{p_j} & \text{for } j = i \end{cases}$$

Thus, among the images of α are

$$\pm\sqrt{p_1} \pm \sqrt{p_2} \pm \ldots \pm \sqrt{p_N}$$

with all 2^N sign choices. These elements are all distinct, since any equality would imply, for some non-empty subset $\{i_1, \ldots, i_\ell\}$ of $\{1, \ldots, N\}$, a relation

$$\sqrt{p_{i_1}} + \ldots + \sqrt{p_{i_\ell}} = 0$$

which is precluded by the previous problem (since no one of these square roots lies in the field generated by the others). Thus, there are at least 2^N images of α, so α is of degree at least over 2^N, so is of degree exactly that. By multiplicativity of degrees in towers, it must be that α generates all of $\mathbb{Q}(\sqrt{p_1}, \ldots, \sqrt{p_N})$.   ///

20.4.25 Example: Let $\alpha = xy^2 + yz^2 + zx^2$, $\beta = x^2y + y^2z + z^2x$ and let s_1, s_2, s_3 be the elementary symmetric polynomials in x, y, z. Describe the relation between the quadratic equation satisfied by α and β over the field $\mathbb{Q}(s_1, s_2, s_3)$ and the quantity

$$\Delta^2 = (x-y)^2(y-z)^2(z-x)^2$$

Letting the quadratic equation be $ax^2 + bx + c$ with $a = 1$, the usual $b^2 - 4ac$ will turn out to be this Δ^2. (Thus, there is perhaps some inconsistency in whether these are *discriminants* or their squares.) The interesting question is how to best be sure that this is so. As usual, *in principle* a direct computation would work, but it is more interesting to give a less computational argument.

Let

$$\delta = b^2 - 4ac = (-\alpha - \beta)^2 - 4 \cdot 1 \cdot \alpha\beta = (\alpha - \beta)^2$$

The fact that this δ is the *square* of something is probably unexpected, unless one has anticipated what happens in the sequel. Perhaps the least obvious point is that, if any two of x, y, z are identical, then $\alpha = \beta$. For example, if $x = y$, then

$$\alpha = xy^2 + yz^2 + zx^2 = x^3 + xz^2 + zx^2$$

and

$$\beta = x^2y + y^2z + z^2x = x^3 + x^2z + z^2x = \alpha$$

The symmetrical arguments show that $x - y$, $x - z$, and $y - z$ all divide $\alpha - \beta$, in the (UFD, by Gauss) polynomial ring $\mathbb{Q}[x, y, z]$. The UFD property implies that the product $(x - y)(x - z)(y - z)$ divides $\alpha - \beta$. Since $\delta = (\alpha - \beta)^2$, and since Δ is the *square* of that product of three linear factors, up to a constant they are equal.

To determine the constant, we need only look at a single monomial. For example, the x^4y^2 term in $(\alpha - \beta)^2$ can be determined with $z = 0$, in which case

$$(\alpha - \beta)^2 = (xy^2 - x^2y)^2 = 1 \cdot x^4y^2 + \text{other}$$

Similarly, in Δ^2, the coefficient of x^4y^2 can be determined with $z = 0$, in which case

$$\Delta^2 = (x - y)^2(x)^2(y)^2 = x^4y^2 + \text{other}$$

That is, the coefficient is 1 in both cases, so, finally, we have $\delta = \Delta^2$, as claimed.   ///

20.4.26 Example: Let t be an integer. If the image of t in $\mathbb{Z}/p$ is a square for every prime p, is t necessarily a square?

Yes, but we need not only Quadratic Reciprocity but also Dirichlet's theorem on primes in arithmetic progressions to see this. Dirichlet's theorem, which has no intelligible *purely algebraic* proof, asserts that for a positive integer N and integer a with $\gcd(a, N) = 1$, there are infinitely many primes p with $p = a \bmod N$.

Factor t into prime powers $t = \varepsilon p_1^{m_1} \ldots p_n^{m_n}$ where $\varepsilon = \pm 1$, the p_i are primes, and the m_i are positive integers. Since t is not a square either $\varepsilon = -1$ or some exponent m_i is *odd*.

If $\varepsilon = -1$, take q to be a prime different from all the p_i and $q = 3 \bmod 4$. The latter condition assures (from the cyclicness of $(\mathbb{Z}/q)^\times$) that -1 is not a square mod q, and the first condition assures that t is not 0 modulo q. We will arrange further congruence conditions on q to guarantee that each p_i is a (non-zero) *square* modulo q. For each p_i, if $p_i = 1 \bmod 4$ let $b_i = 1$, and if $p_i = 3 \bmod 4$ let b_i be a non-square mod p_i. Require of q that $q = 7 \bmod 8$ and $q = b_i \bmod p_i$ for odd p_i. (The case of $p_i = 2$ is handled by $q = 7 \bmod 8$, which assures that 2 is a square mod q, by Quadratic Reciprocity.) Sun-Ze's theorem assures us that these conditions can be met simultaneously, by *integer* q. Then by the main part of Quadratic Reciprocity, for $p_i > 2$,

$$\left(\frac{p_i}{q} \right)_2 = (-1)^{(p_i-1)(q-1)} \cdot \left(\frac{q}{p_i} \right)_2$$

$$= \begin{cases} (-1) \cdot \left(\frac{q}{p_i} \right)_2 & (\text{for } p_i = 3 \bmod 4) \\ (+1) \cdot \left(\frac{q}{p_i} \right)_2 & (\text{for } p_i = 1 \bmod 4) \end{cases} = 1 \qquad (\text{in either case})$$

That is, all the p_i are squares modulo q, but $\varepsilon = -1$ is not, so t is a non-square modulo q, since Dirichlet's theorem promises that there are infinitely many (hence, at least one) primes q meeting these congruence conditions.

For $\varepsilon = +1$, there must be some odd m_i, say m_1. We want to devise congruence conditions on primes q such that all p_i with $i \geq 2$ are squares modulo q but p_1 is *not* a square mod q. Since we do not need to make $q = 3 \bmod 4$ (as was needed in the previous case), we can take $q = 1 \bmod 4$, and thus have somewhat simpler conditions. If $p_1 = 2$, require that $q = 5 \bmod 8$, while if $p_1 > 2$ then fix a non-square $b \bmod p_1$ and let $q = b \bmod p_1$. For $i \geq 2$ take $q = 1 \bmod p_i$ for odd p_i, and $q = 5 \bmod 8$ for $p_i = 2$. Again, Sun-Ze assures us that these congruence conditions are equivalent to a single one, and Dirichlet's theorem assures that there are *primes* which meet the condition. Again, Quadratic Reciprocity gives, for $p_i > 2$,

$$\left(\frac{p_i}{q} \right)_2 = (-1)^{(p_i-1)(q-1)} \cdot \left(\frac{q}{p_i} \right)_2 = \left(\frac{q}{p_i} \right)_2 = \begin{cases} -1 & (\text{for } i = 1) \\ +1 & (\text{for } i \geq 2) \end{cases}$$

The case of $p_i = 2$ was dealt with separately. Thus, the product t is the product of a *single* non-square mod q and a bunch of squares modulo q, so is a non-square mod q.

20.4.27 Remark: And in addition to everything else, it is worth noting that for the 4 choices of odd q modulo 8, we achieve all 4 of the different effects

$$\left(\frac{-1}{q}\right)_2 = \pm 1 \qquad \left(\frac{2}{q}\right)_2 = \pm 1$$

20.4.28 Example: Find the irreducible factors of $x^5 - 4$ in $\mathbb{Q}[x]$. In $\mathbb{Q}(\zeta)[x]$ with a primitive fifth root of unity ζ.

First, by Eisenstein's criterion, $x^5 - 2$ is irreducible over $\mathbb{Q}$, so the fifth root of 2 generates a quintic extension of $\mathbb{Q}$. Certainly a fifth root of 4 lies in such an extension, so must be either rational or generate the quintic extension, by multiplicativity of field extension degrees in towers. Since $4 = 2^2$ is not a fifth power in $\mathbb{Q}$, the fifth root of 4 generates a quintic extension, and its minimal polynomial over $\mathbb{Q}$ is necessarily quintic. The given polynomial is at worst a multiple of the minimal one, and has the right degree, so is *it*. That is, $x^5 - 4$ is irreducible in $\mathbb{Q}[x]$. (*Comment:* I had overlooked this trick when I thought the problem up, thinking, instead, that one would be forced to think more in the style of the *Kummer* ideas indicated below.)

Yes, it is true that irreducibility over the larger field would imply irreducibility over the smaller, but it might be difficult to see directly that 4 is not a fifth power in $\mathbb{Q}(\zeta)$. For example, we do not know anything about the behavior of the ring $\mathbb{Z}[\zeta]$, such as whether it is a UFD or not, so we cannot readily attempt to invoke Eisenstein. Thus, our *first* method to prove irreducibility over $\mathbb{Q}(\zeta)$ uses the irreducibility over $\mathbb{Q}$.

Instead, observe that the field extension obtained by adjoining ζ is quartic over $\mathbb{Q}$, while that obtained by adjoining a fifth root β of 4 is quintic. Any field K containing both would have degree divisible by both degrees (by multiplicativity of field extension degrees in towers), and at most the product, so in this case exactly 20. As a consequence, β has *quintic* minimal polynomial over $\mathbb{Q}(\zeta)$, since $[K : \mathbb{Q}(\zeta)] = 5$ (again by multiplicativity of degrees in towers). That is, the given quintic must be that minimal polynomial, so is irreducible.   ///

Another approach to prove irreducibility of $x^5 - 4$ in $\mathbb{Q}[x]$ is to prove that it is irreducible modulo some prime p. To have some elements of $\mathbb{Z}/p$ not be 5th powers we need $p = 1 \bmod 5$ (by the cyclicness of $(\mathbb{Z}/p)^\times$), and the smallest candidate is $p = 11$. First, 4 is not a fifth power in $\mathbb{Z}/11$, since the only fifth powers are ± 1 (again using the cyclicness to make this observation easy). In fact, $2^5 = 32 = -1 \bmod 11$, so we can infer that 2 is a generator for the order 11 cyclic group $(\mathbb{Z}/11)^\times$. Then if $4 = \alpha^5$ for some $\alpha \in \mathbb{F}_{11^2}$, also $\alpha^{11^2 - 1} = 1$ and $4^5 = 1 \bmod 11$ yield

$$1 = \alpha^{11^2 - 1} = (\alpha^5)^{24} = 4^{24} = 4^4 = 5^2 = 2 \bmod 11$$

which is false. Thus, $x^5 - 4$ can have no linear or quadratic factor in $\mathbb{Q}[x]$, so is irreducible in $\mathbb{Q}[x]$. (*Comment:* And I had overlooked *this* trick, too, when I thought the problem up.)

Yet another approach, which illustrates more what happens in Kummer theory, is to grant ourselves just that a is not a 5th power in $\mathbb{Q}(\zeta)$, and prove irreducibility of $x^5 - a$. That a is not a 5th power in $\mathbb{Q}(\zeta)$ can be proven without understanding much about the ring $\mathbb{Z}[\zeta]$ (if we are slightly lucky) by taking *norms* from $\mathbb{Q}(\zeta)$ to $\mathbb{Q}$, in the sense of writing

$$N(\beta) = \prod_{\tau \in \mathrm{Aut}(\mathbb{Q}(\zeta)/\mathbb{Q})} \tau(\beta)$$

In fact, we know that $\text{Aut}(\mathbb{Q}(\zeta)/\mathbb{Q}) \approx (\mathbb{Z}/5)^\times$, generated (for example) by $\sigma_2(\zeta) = \zeta^2$. We compute directly that N takes values in $\mathbb{Q}$: for lightness of notation let $\tau = \sigma_2$, and then

$$\tau(N\beta) = \tau\left(\beta \cdot \tau\beta \cdot \tau^2\beta \cdot \tau^3\beta\right) = \tau\beta \cdot \tau^2\beta \cdot \tau^3\beta \cdot \tau^4\beta$$

$$= \beta \cdot \tau\beta \cdot \tau^2\beta \cdot \tau^3\beta = N(\beta)$$

since $\tau^4 = 1$, by rearranging. Since we are inside a cyclotomic field, we already know the (proto-Galois theory) fact that invariance under all automorphisms means the thing lies inside $\mathbb{Q}$, as claimed. And since τ is an automorphism, the norm N is multiplicative (as usual). Thus, if $\beta = \gamma^5$ is a fifth power, then

$$N(\beta) = N(\gamma^5) = N(\gamma)^5$$

is a fifth power of a rational number. The norm of $\beta = 4$ is easy to compute, namely

$$N(4) = 4 \cdot 4 \cdot 4 \cdot 4 = 2^8$$

which is not a fifth power in $\mathbb{Q}$ (by unique factorization). So, without knowing much about the ring $\mathbb{Z}[\zeta]$, we do know that 4 does not become a fifth power there.

Let α be a fifth root of 4. Then, in fact, the complete list of fifth roots of 4 is $\alpha, \zeta\alpha, \zeta^2\alpha, \zeta^3\alpha, \zeta^4\alpha$. If $x^5 - 4$ factored properly in $\mathbb{Q}(\zeta)[x]$, then it would have a linear or quadratic factor. There can be no linear factor, because (as we just showed) there is no fifth root of 4 in $\mathbb{Q}(\zeta)$. If there were a proper *quadratic* factor it would have to be of the form (with $i \neq j$ mod 5)

$$(x - \zeta^i\alpha)(x - \zeta^j\alpha) = x^2 - (\zeta^i + \zeta^j)\alpha x + \zeta^{i+j}\alpha^2$$

Since $\alpha \notin \mathbb{Q}(\zeta)$, this would require that $\zeta^i + \zeta^j = 0$, or $\zeta^{i-j} = -1$, which does not happen. Thus, we have irreducibility.

20.4.29 Remark: This last problem is a precursor to *Kummer theory*. As with cyclotomic extensions of fields, extensions by n^{th} roots have the simplicity that we have an explicit and simple form for *all* the roots in terms of a given one. This is not typical.

Exercises

20.1 Prove that a prime p such that $p = 3$ mod 7 generates a prime ideal in $\mathbb{Z}[\zeta]$ where ζ is a primitive 7th root of unity.

20.2 Let $P(y)$ be an irreducible polynomial in $k[x]$. Let n be an integer not divisible by the characteristic of the field k. Show that $x^n - P(y)$ is irreducible in $k[x, y]$.

20.3 Let x be an indeterminate over a field k. Show that there is a field automorphism of $k(x)$ sending x to $c \cdot x$ for any non-zero element c of k.

20.4 Let x be an indeterminate over a field k of characteristic p, a prime. Show that there are only finitely-many fields between $k(x)$ and $k(x^{1/p})$.

20.5 Let k be an algebraically closed field of characteristic 0. Find a polynomial condition on $a \in k$ such that $z^5 - z + a = 0$ has distinct roots.

21. Primes in arithmetic progressions

Dirichlet's theorem is a strengthening of Euclid's theorem that there are infinitely many primes p. Dirichlet's theorem allows us to add the condition that $p = a \bmod N$ for fixed a invertible modulo fixed N, and still be assured that there are infinitely-many primes meeting this condition.

The most intelligible proof of this result uses a bit of analysis, in addition to some interesting algebraic ideas. The analytic idea already arose with Euler's proof of the infinitude of primes, which we give below. New algebraic ideas due to Dirichlet allowed him to isolate primes in different congruence classes modulo N.

In particular, this issue is an opportunity to introduce the **dual group**, or **group of characters**, of a finite abelian group. This idea was one impetus to the development of a more abstract notion of *group*, and also of *group representations* studied by Schur and Frobenious.

21.1 *Euler's theorem and the zeta function*

To illustrate how to use special functions of the form

$$Z(s) = \sum_{n=1}^{\infty} \frac{a_n}{n^s}$$

called **Dirichlet series** to prove things about primes, we first give Euler's proof of the infinitude of primes. [1]

The simplest Dirichlet series is the **Euler-Riemann** zeta function[2]

$$\zeta(s) = \sum_{n=1}^{\infty} \frac{1}{n^s}$$

This converges absolutely and (uniformly in compacta) for real $s > 1$. For *real $s > 1$*

$$\frac{1}{s-1} = \int_1^{\infty} \frac{dx}{x^s} = 1 + \frac{1}{s-1} \le \zeta(s) \le 1 + \int_1^{\infty} \frac{dx}{x^s} = 1 + \frac{1}{s-1}$$

This proves that

$$\lim_{s \longrightarrow 1^+} \zeta(s) = +\infty$$

The relevance of this to a study of primes is the **Euler product expansion**[3]

$$\zeta(s) = \sum_{n=1}^{\infty} \frac{1}{n^s} = \prod_{p \text{ prime}} \frac{1}{1 - \frac{1}{p^s}}$$

To prove that this holds, observe that

$$\sum_{n=1}^{\infty} \frac{1}{n^s} = \prod_{p \text{ prime}} \left(1 + \frac{1}{p^s} + \frac{1}{p^{2s}} + \frac{1}{p^{3s}} + \dots \right)$$

by unique factorization into primes. [4]   Summing the indicated geometric series gives

$$\zeta(s) = \prod_{p \text{ prime}} \frac{1}{1 - \frac{1}{p^s}}$$

Since sums are more intuitive than products, take a logarithm

$$\log \zeta(s) = \sum_p -\log(1 - \frac{1}{p^s}) = \sum_p \left(\frac{1}{p^s} + \frac{1}{2p^{2s}} + \frac{1}{3p^{3s}} + \dots \right)$$

by the usual expansion (for $|x| < 1$)

$$-\log(1 - x) = x + \frac{x^2}{2} + \frac{x^3}{3} + \dots$$

---

[1] Again, the 2000 year old elementary proof of the infinitude of primes, ascribed to *Euclid* perhaps because his texts survived, proceeds as follows. Suppose there were only finitely many primes altogether, $p_1, \dots, p_n$. Then $N = 1 + p_1 \dots p_n$ cannot be divisible by any p_i in the list, yet has *some* prime divisor, contradiction. This viewpoint does not give much indication about how to make the argument more *quantitative*. Use of $\zeta(s)$ seems to be the way.

[2] Studied by many other people before and since.

[3] Valid only for $s > 1$.

[4] Manipulation of this infinite product of infinite sums is not completely trivial to justify.

Taking a derivative in s gives

$$-\frac{\zeta'(s)}{\zeta(s)} = \sum_{p \text{ prime}, \, m \geq 1} \frac{\log p}{p^{ms}}$$

Note that, for each fixed $p > 1$,

$$\sum_{m \geq 1} \frac{\log p}{p^{ms}} = \frac{(\log p) \, p^{-s}}{1 - p^{-s}}$$

converges absolutely for real $s > 0$.

Euler's argument for the infinitude of primes is that, if there were only finitely-many primes, then the right-hand side of

$$-\frac{\zeta'(s)}{\zeta(s)} = \sum_{p \text{ prime}, \, m \geq 1} \frac{\log p}{p^{ms}}$$

would converge for real $s > 0$. However, we saw that $\zeta(s) \longrightarrow +\infty$ as s approaches 1 from the right. Thus, $\log \zeta(s) \longrightarrow +\infty$, and $\frac{d}{ds}(\log \zeta(s)) = \zeta'(s)/\zeta(s) \longrightarrow -\infty$ as $s \longrightarrow 1^+$. This contradicts the convergence of the sum over (supposedly finitely-many) primes. Thus, there must be infinitely many primes. ///

21.2 *Dirichlet's theorem*

In addition to Euler's observation (above) that the analytic behavior [5] of $\zeta(s)$ at $s = 1$ implied the existence of infinitely-many primes, Dirichlet found an algebraic device to focus attention on single congruence classes modulo N.

21.2.1 Theorem: *(Dirichlet)* Fix an integer $N > 1$ and an integer a such that $\gcd(a, N) = 1$. Then there are infinitely many primes p with

$$p = a \bmod N$$

21.2.2 Remark: If $\gcd(a, N) > 1$, then there is at most one prime p meeting the condition $p = a \bmod n$, since any such p would be divisible by the *gcd*. Thus, the necessity of the *gcd* condition is obvious. It is noteworthy that beyond this *obvious* condition there is nothing further needed.

21.2.3 Remark: For $a = 1$, there is a simple purely algebraic argument using cyclotomic polynomials. For general a the most intelligible argument involves a little analysis.

---

[5] Euler's proof uses only very crude properties of $\zeta(s)$, and only of $\zeta(s)$ as a function of a *real*, rather than *complex*, variable. Given the status of complex number and complex analysis in Euler's time, this is not surprising. It is slightly more surprising that Dirichlet's original argument also was a real-variable argument, since by that time, a hundred years later, complex analysis was well-established. Still, until Riemann's memoir of 1858 there was little reason to believe that the behavior of $\zeta(s)$ off the real line was of any interest.

Proof: A **Dirichlet character** modulo N is a group homomorphism

$$\chi : (\mathbb{Z}/N)^{\times} \longrightarrow \mathbb{C}^{\times}$$

extended by 0 to all of $\mathbb{Z}/n$, that is, by defining $\chi(a) = 0$ if a is not invertible modulo N. This extension-by-zero then allows us to compose χ with the reduction-mod-N map $\mathbb{Z} \longrightarrow \mathbb{Z}/N$ and also consider χ as a function on $\mathbb{Z}$. Even when extended by 0 the function χ is still *multiplicative* in the sense that

$$\chi(mn) = \chi(m) \cdot \chi(n)$$

where or not one of the values is 0. The **trivial** character χ_o modulo N is the character which takes only the value 1 (and 0).

The standard **cancellation trick** is that

$$\sum_{a \bmod N} \chi(a) = \begin{cases} \varphi(N) & \text{(for } \chi = \chi_o) \\ 0 & \text{(otherwise)} \end{cases}$$

where φ is Euler's totient function. The proof of this is easy, by changing variables, as follows. For $\chi = \chi_o$, all the values for a invertible mod N are 1, and the others are 0, yielding the indicated sum. For $\chi \neq \chi_o$, there is an *invertible* b mod N such that $\chi(b) \neq 1$ (and is not 0, either, since b is invertible). Then the map $a \longrightarrow a \cdot b$ is a *bijection* of $\mathbb{Z}/N$ to itself, so

$$\sum_{a \bmod N} \chi(a) = \sum_{a \bmod N} \chi(a \cdot b) = \sum_{a \bmod N} \chi(a) \cdot \chi(b) = \chi(b) \cdot \sum_{a \bmod N} \chi(a)$$

That is,

$$(1 - \chi(b)) \cdot \sum_{a \bmod N} \chi(a) = 0$$

Since $\chi(b) \neq 1$, it must be that $1 - \chi(b) \neq 0$, so the sum is 0, as claimed.

Dirichlet's *dual* trick is to sum over characters χ mod N evaluated at fixed a in $(\mathbb{Z}/N)^{\times}$. We claim that

$$\sum_{\chi} \chi(a) = \begin{cases} \varphi(N) & \text{(for } a = 1 \bmod N) \\ 0 & \text{(otherwise)} \end{cases}$$

We will prove this in the next section.

Granting that, we have also, for b invertible modulo N,

$$\sum_{\chi} \chi(a)\chi(b)^{-1} = \sum_{\chi} \chi(ab^{-1}) = \begin{cases} \varphi(N) & \text{(for } a = b \bmod N) \\ 0 & \text{(otherwise)} \end{cases}$$

Given a Dirichlet character χ modulo N, the corresponding **Dirichlet L-function** is

$$L(s, \chi) = \sum_{n \geq 1} \frac{\chi(n)}{n^s}$$

Since we have the multiplicative property $\chi(mn) = \chi(m)\chi(n)$, each such L-function has an **Euler product** expansion

$$L(s, \chi) = \prod_{p \text{ prime}, \, p \nmid N} \frac{1}{1 - \chi(p)\, p^{-s}}$$

This follows as it did for $\zeta(s)$, by

$$L(s,\chi) = \sum_{n \text{ with } \gcd(n,N)=1} \frac{\chi(n)}{n^s}$$

$$= \prod_{p \text{ prime}, \, p \nmid N} \left(1 + \chi(p)p^{-s} + \chi(p)^2 p^{-2s} + \ldots\right) = \prod_{p \text{ prime}, \, p \nmid N} \frac{1}{1 - \chi(p)\,p^{-s}}$$

by summing geometric series. Taking a logarithmic derivative (as with zeta) gives

$$-\frac{L'(s,\chi)}{L(s,\chi)} = \sum_{p \nmid N \text{ prime}, \, m \geq 1} \frac{\log p}{\chi(p)^m \, p^{ms}} = \sum_{p \nmid N \text{ prime}} \frac{\log p}{\chi(p) \, p^s} + \sum_{p \nmid N \text{ prime}, \, m \geq 2} \frac{\log p}{\chi(p)^m \, p^{ms}}$$

The second sum on the right will turn out to be subordinate to the first, so we aim our attention at the first sum, where $m = 1$.

To pick out the primes p with $p = a \bmod N$, use the sum-over-χ trick to obtain

$$\sum_{\chi \bmod N} \chi(a) \cdot \frac{\log p}{\chi(p) \, p^s} = \begin{cases} \varphi(N) \cdot (\log p)\, p^{-s} & (\text{for } p = a \bmod N) \\ 0 & (\text{otherwise}) \end{cases}$$

Thus,

$$-\sum_{\chi \bmod N} \chi(a)\frac{L'(s,\chi)}{L(s,\chi)} = \sum_{\chi \bmod N} \chi(a) \sum_{p \nmid N \text{ prime}, \, m \geq 1} \frac{\log p}{\chi(p)^m \, p^{ms}}$$

$$= \sum_{p = a \bmod N} \frac{\varphi(N) \log p}{p^s} + \sum_{\chi \bmod N} \chi(a) \sum_{p \nmid N \text{ prime}, \, m \geq 2} \frac{\log p}{\chi(p)^m \, p^{ms}}$$

We do not care about whether cancellation does or does not occur in the second sum. All that we care is that it is absolutely convergent for $\mathrm{Re}(s) > \frac{1}{2}$. To see this we do *not* need any subtle information about primes, but, rather, dominate the sum over primes by the corresponding sum over integers ≥ 2. Namely,

$$\left| \sum_{p \nmid N \text{ prime}, \, m \geq 2} \frac{\log p}{\chi(p)^m \, p^{ms}} \right| \leq \sum_{n \geq 2, \, m \geq 2} \frac{\log n}{n^{m\sigma}} = \sum_{n \geq 2} \frac{(\log n)/n^{2\sigma}}{1 - n^{-\sigma}} \leq \frac{1}{1 - 2^{-\sigma}} \sum_{n \geq 2} \frac{\log n}{n^{2\sigma}}$$

where $\sigma = \mathrm{Re}(s)$. This converges for $\mathrm{Re}(s) > \frac{1}{2}$.

That is, for $s \longrightarrow 1^+$,

$$-\sum_{\chi \bmod N} \chi(a)\frac{L'(s,\chi)}{L(s,\chi)} = \varphi(N) \sum_{p = a \bmod N} \frac{\log p}{p^s} + (\text{something continuous at } s = 1)$$

We have isolated primes $p = a \bmod N$. Thus, as Dirichlet saw, to prove the infinitude of primes $p = a \bmod N$ it would suffice to show that the left-hand side of the last inequality blows up at $s = 1$. In particular, for the **trivial** character $\chi_o \bmod N$, with values

$$\chi(b) == \begin{cases} 1 & (\text{for } \gcd(b, N) = 1) \\ 0 & (\text{for } \gcd(b, N) > 1) \end{cases}$$

the associated L-function is barely different from the zeta function, namely

$$L(s, \chi_o) = \zeta(s) \cdot \prod_{p|N} \left(1 - \frac{1}{p^s}\right)$$

Since none of those finitely-many factors for primes dividing N is 0 at $s = 1$, $L(s, \chi_o)$ still blows up at $s = 1$.

By contrast, we will show below that for **non-trivial** character χ mod N, $\lim_{s \to 1+} L(s, \chi)$ is *finite*, and

$$\lim_{s \to 1+} L(s, \chi) \neq 0$$

Thus, for non-trivial character, the logarithmic derivative is finite and non-zero at $s = 1$. Putting this all together, we will have

$$\lim_{s \to 1+} - \sum_{\chi \bmod N} \chi(a) \frac{L'(s, \chi)}{L(s, \chi)} = +\infty$$

Then necessarily

$$\lim_{s \to 1+} \varphi(N) \sum_{p = a \bmod N} \frac{\log p}{p^s} = +\infty$$

and there must be infinitely many primes $p = a$ mod N.   ///

21.2.4 Remark: The non-vanishing of the non-trivial L-functions at 1, which we prove a bit further belo, is a crucial technical point.

21.3 *Dual groups of abelian groups*

Before worrying about the non-vanishing of L-functions at $s = 1$ for non-trivial characters χ, we explain Dirichlet's innovation, the use of group characters to isolate primes in a specified congruence class modulo N.

These ideas were the predecessors of the group theory work of Frobenious and Schur 50 years later, and one of the ancestors of *representation theory* of groups.

The **dual group** or **group of characters** $\widehat{G}$ of a finite abelian group G is by definition

$$\widehat{G} = \{\text{group homomorphisms } \chi : G \longrightarrow \mathbb{C}^\times\}$$

This $\widehat{G}$ is itself an abelian group under the operation on characters defined for $g \in G$ by

$$(\chi_1 \cdot \chi_2)(g) = \chi_1(g) \cdot \chi_2(g)$$

21.3.1 Proposition: Let G be a cyclic group of order n with specified generator g_1. Then $\widehat{G}$ is isomorphic to the group of complex n^{th} roots of unity, by

$$(g_1 \longrightarrow \zeta) \longleftarrow \zeta$$

That is, an nth root of unity ζ gives the character χ such that

$$\chi(g_1^\ell) = \zeta^\ell$$

In particular, $\widehat{G}$ is cyclic of order n.

Proof: First, the value of a character χ on g_1 determines all values of χ, since g_1 is a generator for G. And since $g_1^n = e$,

$$\chi(g_1)^n = \chi(g_1^n) = \chi(e) = 1$$

it follows that the only possible values of $\chi(g_1)$ are n^{th} roots of unity. At the same time, for an nth root of unity ζ the formula

$$\chi(g_1^\ell) = \zeta^\ell$$

does give a *well-defined* function on G, since the ambiguity on the right-hand side is by changing ℓ by multiples of n, but g_1^ℓ does only depend upon $\ell \bmod n$. Since the formula gives a well-defined function, it gives a homomorphism, hence, a character. ///

21.3.2 Proposition: Let $G = A \oplus B$ be a direct sum of finite abelian groups. Then there is a *natural* isomorphism of the dual groups

$$\widehat{G} \approx \widehat{A} \oplus \widehat{B}$$

by

$$((a \oplus b) \longrightarrow \chi_1(a) \cdot \chi_2(b)) \quad \longleftarrow \quad \chi_1 \oplus \chi_2$$

Proof: The indicated map is certainly an injective homomorphism of abelian groups. To prove surjectivity, let χ be an arbitrary element of $\widehat{G}$. Then for $a \in A$ and $b \in B$

$$\chi_1(a) = \chi(a \oplus 0) \quad \chi_2(a) = \chi(0 \oplus b)$$

gives a pair of characters χ_1 and χ_2 in $\widehat{A}$ and $\widehat{B}$. Unsurprisingly, $\chi_1 \oplus \chi_2$ maps to the given χ, proving surjectivity. ///

21.3.3 Corollary: Invoking the Structure Theorem for finite abelian groups, write a finite abelian group G as

$$G \approx \mathbb{Z}/d_1 \oplus \ldots \mathbb{Z}/d_t$$

for some elementary divisors d_i. [6] Then

$$\widehat{G} \approx \widehat{\mathbb{Z}/d_1} \oplus \ldots \widehat{\mathbb{Z}/d_t} \approx \mathbb{Z}/d_1 \oplus \ldots \mathbb{Z}/d_t \approx G$$

In particular,

$$|\widehat{G}| = |G|$$

Proof: The leftmost of the three isomorphisms is the assertion of the previous proposition. The middle isomorphism is the sum of isomorphisms of the form (for $d \neq 0$ and integer)

$$\widehat{\mathbb{Z}/d} \approx \mathbb{Z}/d$$

proven just above in the guise of cyclic groups. ///

---

[6] We do not need to know that $d_1 | \ldots | d_t$ for present purposes.

21.3.4 Proposition: Let G be a finite abelian group. For $g \neq e$ in G, there is a character $\chi \in \widehat{G}$ such that $\chi(g) \neq 1$. [7]

Proof: Again expressing G as a sum of cyclic groups

$$G \approx \mathbb{Z}/d_1 \oplus \ldots \mathbb{Z}/d_t$$

given $g \neq e$ in G, there is some index i such that the projection g_i of g to the ith summand $\mathbb{Z}/d_i$ is non-zero. If we can find a character on $\mathbb{Z}/d_i$ which gives value $\neq 1$ on g_i, then we are done. And, indeed, sending a generator of $\mathbb{Z}/d_i$ to a *primitive* d_i^{th} root of unity sends every non-zero element of $\mathbb{Z}/d_i$ to a complex number other than 1.   ///

21.3.5 Corollary: *(Dual version of cancellation trick)* For g in a finite abelian group,

$$\sum_{\chi \in \widehat{G}} \chi(g) = \begin{cases} |G| & \text{(for } g = e) \\ 0 & \text{(otherwise)} \end{cases}$$

Proof: If $g = e$, then the sum counts the characters in $\widehat{G}$. From just above,

$$|\widehat{G}| = |G|$$

On the other hand, given $g \neq e$ in G, by the previous proposition let χ_1 be in $\widehat{G}$ such that $\chi_1(g) \neq 1$. The map on $\widehat{G}$

$$\chi \longrightarrow \chi_1 \cdot \chi$$

is a bijection of $\widehat{G}$ to itself, so

$$\sum_{\chi \in \widehat{G}} \chi(g) = \sum_{\chi \in \widehat{G}} (\chi \cdot \chi_1)(g) = \chi_1(g) \cdot \sum_{\chi \in \widehat{G}} \chi(g)$$

which gives

$$(1 - \chi_1(g)) \cdot \sum_{\chi \in \widehat{G}} \chi(g) = 0$$

Since $1 - \chi_1(g) \neq 0$, it must be that the sum is 0.   ///

21.4 *Non-vanishing on* $\mathrm{Re}(s) = 1$

Dirichlet's argument for the infinitude of primes $p = a \bmod N$ (for $\gcd(a, N) = 1$) requires that $L(1, \chi) \neq 0$ for all $\chi \bmod N$. We prove this now, granting that these functions have meromorphic extensions to some neighborhood of $s = 1$. We also need to know that for the trivial character $\chi_o \bmod N$ the L-function $L(s, \chi_o)$ has a *simple* pole at $s = 1$. These analytical facts are proven in the next section.

21.4.1 Theorem: For a Dirichlet character $\chi \bmod N$ other than the trivial character $\chi_o \bmod N$,

$$L(1, \chi) \neq 0$$

---

[7] This idea that characters can distinguish group elements from each other is just the tip of an iceberg.

Proof: To prove that the *L*-functions $L(s, \chi)$ do not vanish at $s = 1$, and in fact do not vanish on the whole line[8] $\text{Re}(s) = 1$, any direct argument involves a trick similar to what we do here. [9]

For χ whose square is not the trivial character χ_o modulo N, the standard trick is to consider

$$\lambda(s) = L(s, \chi_o)^3 \cdot L(s, \chi)^4 \cdot L(s, \chi^2)$$

Then, letting $\sigma = \text{Re}(s)$, from the Euler product expressions for the *L*-functions noted earlier, in the region of convergence,

$$|\lambda(s)| = \left| \exp\left(\sum_{m,p} \frac{3 + 4\chi(p^m) + \chi^2(p^m)}{mp^{ms}} \right) \right| = \exp\left| \sum_{m,p} \frac{3 + 4\cos\theta_{m,p} + \cos 2\theta_{m,p}}{mp^{m\sigma}} \right|$$

where for each m and p we let

$$\theta_{m,p} = (\text{the argument of } \chi(p^m)) \in \mathbb{R}$$

The trick[10] is that for *any* real θ

$$3 + 4\cos\theta + \cos 2\theta = 3 + 4\cos\theta + 2\cos^2\theta - 1 = 2 + 4\cos\theta + 2\cos^2\theta = 2(1 + \cos\theta)^2 \geq 0$$

Therefore, all the terms inside the large sum being exponentiated are non-negative, and, [11]

$$|\lambda(s)| \geq e^0 = 1$$

In particular, if $L(1, \chi) = 0$ were to be 0, then, since $L(s, \chi_o)$ has a simple pole at $s = 1$ and since $L(s, \chi^2)$ does *not* have a pole (since $\chi^2 \neq \chi_o$), the multiplicity ≥ 4 of the 0 in the product of *L*-functions would overwhelm the three-fold pole, and $\lambda(1) = 0$. This would contradict the inequality just obtained.

For $\chi^2 = \chi_o$, instead consider

$$\lambda(s) = L(s, \chi) \cdot L(s, \chi_o) = \exp\left(\sum_{p,m} \frac{1 + \chi(p^m)}{mp^{ms}} \right)$$

---

[8] Non-vanishing of $\zeta(s)$ on the whole line $\text{Re}(s) = 1$ yields the Prime Number Theorem: let $\pi(x)$ be the number of primes less than x. Then $\pi(x) \sim x/\ln x$, meaning that the limit of the ratio of the two sides as $x \longrightarrow \infty$ is 1. This was first proven in 1896, separately, by Hadamard and de la Vallée Poussin. The same sort of argument also gives an analogous *asymptotic* statement about primes in each congruence class modulo N, namely that $\pi_{a,N}(x) \sim x/[\varphi(N) \cdot \ln x]$, where $\gcd(a, N) = 1$ and φ is Euler's totient function.

[9] A more natural (and dignified) but considerably more demanding argument for non-vanishing would entail following the Maaß-Selberg discussion of the spectral decomposition of $SL(2, \mathbb{Z}) \backslash SL(2, \mathbb{R})$.

[10] Presumably found after considerable fooling around.

[11] Miraculously...

If $L(1,\chi) = 0$, then this would cancel the simple pole of $L(s,\chi_o)$ at 1, giving a non-zero finite value at $s = 1$. The series inside the exponentiation is a Dirichlet series with non-negative coefficients, and for real s

$$\sum_{p,m} \frac{1+\chi(p^m)}{mp^{ms}} \geq \sum_{p,\,m\text{ even}} \frac{1+1}{mp^{ms}} = \sum_{p,m} \frac{1+1}{2mp^{2ms}} = \sum_{p,m} \frac{1}{mp^{2ms}} = \log\zeta(2s)$$

Since $\zeta(2s)$ has a simple pole at $s = \frac{1}{2}$ the series

$$\log\left(L(s,\chi)\cdot L(s,\chi_o)\right) = \sum_{p,m} \frac{1+\chi(p^m)}{mp^{ms}} \geq \log\zeta(2s)$$

necessarily blows up as $s \longrightarrow \frac{1}{2}^+$. But by **Landau's Lemma** (in the next section), a Dirichlet series with non-negative coefficients cannot blow up as $s \longrightarrow s_o$ along the real line unless the function represented by the series fails to be holomorphic at s_o. Since the function given by $\lambda(s)$ is holomorphic at $s = 1/2$, this gives a contradiction to the supposition that $\lambda(s)$ is holomorphic at $s = 1$ (which had allowed this discussion at $s = 1/2$). That is, $L(1,\chi) \neq 0$. ///

21.5 *Analytic continuations*

Dirichlet's original argument did not emphasize holomorphic functions, but by now we know that discussion of vanishing and blowing-up of functions is most clearly and simply accomplished if the functions are meromorphic when viewed as functions of a complex variable.

For the purposes of Dirichlet's theorem, it suffices to meromorphically continue [12] the L-functions to $\text{Re}(s) > 0$. [13]

21.5.1 Theorem: The Dirichlet L-functions

$$L(s,\chi) = \sum_n \frac{\chi(n)}{n^s} = \prod_p \frac{1}{1-\chi(p)\,p^{-s}}$$

have meromorphic continuations to $\text{Re}(s) > 0$. For χ non-trivial, $L(s,\chi)$ is *holomorphic* on that half-plane. For χ trivial, $L(s,\chi_o)$ has a *simple* pole at $s = 1$ and is holomorphic otherwise.

---

[12] An extension of a holomorphic function to a larger region, on which it may have some poles, is called a **meromorphic continuation**. There is *no* general methodology for proving that functions have meromorphic continuations, due in part to the fact that, generically, functions *do not* have continuations beyond some natural region where they're defined by a convergent series or integral. Indeed, to be able to prove a meromorphic continuation result for a given function is tantamount to proving that it has some deeper significance.

[13] Already prior to Riemann's 1858 paper, it was known that the Euler-Riemann zeta function and all the L-functions we need here did indeed have meromorphic continuations to the whole complex plane, have no poles unless the character χ is trivial, and have functional equations similar to that of zeta, namely that $\pi^{-s/2}\Gamma(s/2)\zeta(s)$ is invariant under $s \longrightarrow 1-s$.

Proof: First, to treat the trivial character χ_o mod N, recall, as already observed, that the corresponding L-function differs in an elementary way from $\zeta(s)$, namely

$$L(s, \chi_o) = \zeta(s) \cdot \prod_{p|N} \left(1 - \frac{1}{p^s}\right)$$

Thus, we analytically continue $\zeta(s)$ instead of $L(s, \chi_o)$. To analytically continue $\zeta(s)$ to $\text{Re}(s) > 0$ observe that the sum for $\zeta(s)$ is fairly well approximated by a more elementary function

$$\zeta(s) - \frac{1}{s-1} = \sum_{n=1}^{\infty} \frac{1}{n^s} - \int_1^{\infty} \frac{dx}{x^s} = \sum_{n=1}^{\infty} \left[\frac{1}{n^s} - \frac{\left(\frac{1}{n^{s-1}} - \frac{1}{(n+1)^{s-1}}\right)}{1-s}\right]$$

Since

$$\frac{\left(\frac{1}{n^{s-1}} - \frac{1}{(n+1)^{s-1}}\right)}{1-s} = \frac{1}{n^s} + O(\frac{1}{n^{s+1}})$$

with a uniform O-term, we obtain

$$\zeta(s) - \frac{1}{s-1} = \sum_n O(\frac{1}{n^{s+1}}) = \text{holomorphic for } \text{Re}(s) > 0$$

The obvious analytic continuation of $1/(s-1)$ allows analytic continuation of $\zeta(s)$.

A relatively elementary analytic continuation argument for *non-trivial* characters uses **partial summation**. That is, let $\{a_n\}$ and $\{b_n\}$ be sequences of complex numbers such that the partial sums $A_n = \sum_{i=1}^n a_i$ are *bounded*, and $b_n \longrightarrow 0$. Then it is useful to rearrange (taking $A_0 = 0$ for notational convenience)

$$\sum_{n=1}^{\infty} a_n b_n = \sum_{n=1}^{\infty} (A_n - A_{n-1})b_n = \sum_{n=0}^{\infty} A_n b_n - \sum_{n=0}^{\infty} A_n b_{n+1} = \sum_{n=0}^{\infty} A_n (b_n - b_{n+1})$$

Taking $a_n = \chi(n)$ and $b_n = 1/n^s$ gives

$$L(s, \chi) = \sum_{n=0}^{\infty} \left(\sum_{i=1}^n \chi(n)\right) (\frac{1}{n^s} - \frac{1}{(n+1)^s})$$

The difference $1/n^s - 1/(n+1)^s$ is s/n^{s+1} up to higher-order terms, so this expression gives a holomorphic function for $\text{Re}(s) > 0$. ///

21.6 *Dirichlet series with positive coefficients*

Now we prove Landau's result on Dirichlet series with positive coefficients. (More precisely, the coefficients are *non-negative*.)

21.6.1 Theorem: *(Landau)* Let

$$f(s) = \sum_{n=1}^{\infty} \frac{a_n}{n^s}$$

be a Dirichlet series with real coefficients $a_n \geq 0$. Suppose that the series defining $f(s)$ converges for $\text{Re}(s) \geq \sigma_o$. Suppose further that the function f extends to a function holomorphic in a neighborhood of $s = \sigma_o$. Then, in fact, the series defining $f(s)$ converges for $\text{Re}(s) > \sigma_o - \varepsilon$ for some $\varepsilon > 0$.

Proof: First, by replacing s by $s - \sigma_o$ we lighten the notation by reducing to the case that $\sigma_o = 0$. Since the function $f(s)$ given by the series is holomorphic on $\mathrm{Re}(s) > 0$ and on a neighborhood of 0, there is $\varepsilon > 0$ such that $f(s)$ is holomorphic on $|s - 1| < 1 + 2\varepsilon$, and the power series for the function converges nicely on this open disk. Differentiating the original series termwise, we evaluate the derivatives of $f(s)$ at $s = 1$ as

$$f^{(i)}(1) = \sum_n \frac{(-\log n)^i \, a_n}{n} = (-1)^i \sum_n \frac{(\log n)^i \, a_n}{n}$$

and Cauchy's formulas yield, for $|s - 1| < 1 + 2\varepsilon$,

$$f(s) = \sum_{i \geq 0} \frac{f^{(i)}(1)}{i!} (s - 1)^i$$

In particular, for $s = -\varepsilon$, we are assured of the convergence to $f(-\varepsilon)$ of

$$f(-\varepsilon) = \sum_{i \geq 0} \frac{f^{(i)}(1)}{i!} (-\varepsilon - 1)^i$$

Note that $(-1)^i f^{(i)}(1)$ is a positive Dirichlet series, so we move the powers of -1 a little to obtain

$$f(-\varepsilon) = \sum_{i \geq 0} \frac{(-1)^i f^{(i)}(1)}{i!} (\varepsilon + 1)^i$$

The series

$$(-1)^i f^{(i)}(1) = \sum_n (\log n)^i \frac{a_n}{n}$$

has positive terms, so the double series (convergent, with positive terms)

$$f(-\varepsilon) = \sum_{n,i} \frac{a_n (\log n)^i}{i!} (1 + \varepsilon)^i \frac{1}{n}$$

can be rearranged to

$$f(-\varepsilon) = \sum_n \frac{a_n}{n} \left(\sum_i \frac{(\log n)^i (1 + \varepsilon)^i}{i!} \right) = \sum_n \frac{a_n}{n} n^{(1+\varepsilon)} = \sum_n \frac{a_n}{n^{-\varepsilon}}$$

That is, the latter series converges (absolutely).   ///

22. Galois theory

The main result here is that inside nice [1] finite-degree field extensions L of k, the intermediate fields K are in (inclusion-reversing) bijection with subgroups H of the **Galois group**

$$G = \mathrm{Gal}(L/k) = \mathrm{Aut}(L/k)$$

of *automorphisms* of L over k, by

$$\text{subgroup } H \leftrightarrow \text{ subfield } K \text{ fixed by } H$$

This is depicted as

$$G \left\{ \begin{array}{c} L \\ | \\ K \\ | \\ k \end{array} \right\} H$$

For K the fixed field of subgroup H there is the equality

$$[L:K] = |H|$$

---

[1] Namely **Galois** field extensions, which are by definition both **separable** and **normal**, defined momentarily.

303

Further, if H is a *normal* subgroup of G, then

$$\mathrm{Gal}(K/k) \approx G/H$$

In the course of proving these things we also elaborate upon the situations in which these ideas apply.

Galois' original motivation for this study was solution of equations in radicals (roots), but by now that classical problem is of much less importance than the general structure revealed by these results.

Also, notice that much of our earlier discussion of finite fields, cyclotomic polynomials, and roots of unity amounted to explicit examples of the statements here. In fact, there are few computationally accessible examples beyond those we have already discussed.

This whole discussion is more technical than the previous examples, but this is not surprising, considering the scope.

22.1 *Field extensions, imbeddings, automorphisms*

A more flexible viewpoint on field extensions, imbeddings, and automorphisms will be useful in what follows. Some of this is review.

A **field extension** K of a given field k is a field which is a k-algebra. That is, in addition to being a field, K is a k-module, *and* with the commutativity property

$$\xi(\alpha \cdot \eta) \; = \; \alpha \cdot (\xi \eta) \qquad \text{(for } \alpha \in k \text{ and } \xi, \eta \in K\text{)}$$

and with the *unital* property

$$1_k \cdot \xi \; = \; \xi \qquad \text{(for all } \xi \in K\text{)}$$

Note that the unital-ness promises that the map

$$\alpha \longrightarrow \alpha \cdot 1_K$$

gives an isomorphism of k to a subfield of K. Thus, when convenient, we may identify k with a subfield of K. However, it would be inconvenient if we did not have the flexibility to treat field extensions of k as k-algebras in this sense.

A field K is *algebraically closed* if, for every polynomial $f(x) \in K[x]$ of positive degree n, the equation $f(x) = 0$ has n roots in K.

An *algebraic closure* $\overline{k}$ of a field k is an algebraically closed field extension $\overline{k}$ of k such that every element of $\overline{k}$ is *algebraic* over k. We proved that algebraic closures exist, and are essentially unique, in the sense that, for two algebraic closures K_1 and K_2 of k, there is a field isomorphism $\sigma : K_1 \longrightarrow K_2$ which is the identity map on k.

It is immediate from the definition that an algebraic closure $\overline{k}$ of a field k is an algebraic closure of any intermediate field $k \subset K \subset \overline{k}$.

As a matter of traditional terminology, when K and L are field extensions of k and $\varphi : K \longrightarrow L$ is a k-algebra map, we may also say that $\varphi : K \longrightarrow L$ is a field map *over k*.

Next, for an irreducible $f(x) \in k[x]$, and for β a root of $f(x) = 0$ in a field extension K of k, there is a k-algebra map $k(\alpha) \longrightarrow K$ such that $\sigma\alpha = \beta$. To prove this, first note that, by the universal property of $k[x]$, there is a unique k-algebra homomorphism $k[x] \longrightarrow \overline{k}$ sending x to β. The kernel is the ideal generated by the minimal polynomial of β, which is $f(x)$. That is, this map factors through $k[x]/f$, which isomorphic to $k(\alpha)$.

In particular, an algebraic extension $k(\alpha)$ of k can be imbedded by a k-algebra map into an algebraic closure $\overline{k}$ of k in at least one way. In fact, for each root β of $f(x) = 0$ in $\overline{k}$, there is a k-algebra homomorphism $k(\alpha) \longrightarrow \overline{k}$ sending α to β. Conversely, any β in $\overline{k}$ which is the image $\sigma\alpha$ of α under a k-algebra homomorphism σ must be a root of $f(x) = 0$: compute

$$f(\beta) = f(\sigma\alpha) = \sigma\Big(f(\alpha)\Big) = \sigma(0) = 0$$

As a corollary of this last discussion, we see that any k-algebra automorphism of $k(\alpha)$ must send α to another root of its minimal polynomial over k, of which there are at most

$$\deg f = [k(\alpha) : k]$$

An induction based on the previous observations will show that any *finite* (hence, algebraic) field extension K of k admits at least one k-algebra homomorphism $\sigma : K \longrightarrow \overline{k}$ to a given algebraic closure $\overline{k}$ of k. [2] This is proven as follows. Using the finiteness of K over k, there are finitely-many elements $\alpha_1, \ldots, \alpha_n$ in K such that

$$K = k(\alpha_1, \alpha_2, \ldots, \alpha_n) = k(\alpha_1)(\alpha_2) \ldots (\alpha_n)$$

Then there is a k-algebra imbedding of $k(\alpha_1)$ into $\overline{k}$. Classically, one would say that we will *identify* $k(\alpha_1)$ with its image by this map. It is better to say that we give K a $\sigma(k(\alpha_1))$-algebra structure by

$$\sigma(\beta) \cdot \xi = \beta \cdot \xi \qquad \text{(for } \beta \in k(\alpha_1) \text{ and } \xi \in K)$$

Since $\overline{k}$ is an algebraic closure of $\sigma k(\alpha_1)$, the same principle shows that there is a $k(\alpha_1)$-linear field homomorphism of $k(\alpha_1, \alpha_2)$ to $\overline{k}$. Continuing inductively, we obtain a field homomorphism of K to $\overline{k}$.

Then a similar argument proves that, given a finite extension L of a finite extension K of k, any k-algebra imbedding of K into an algebraic closure $\overline{k}$ *extends* to an imbedding of L into $\overline{k}$. To see this, let $\sigma : K \longrightarrow \overline{k}$ be given. View the finite field extension L of K as a finite field extension of σK as indicated above. Since $\overline{k}$ is also an algebraic closure of K, there is at least one K-algebra imbedding of L into $\overline{k}$.

22.2 Separable field extensions

The notion of *separability* of a field extension has several useful equivalent formulations. We will rarely be interested in non-separable field extensions. Happily, in characteristic

---

[2] In fact, *any* algebraic extension K of k imbeds into an algebraic closure of k, but the proof requires some equivalent of the Axiom of Choice, such as Well-Ordering, or Zorn's Lemma.

0, *all* extensions are separable (see below). Also, even in positive characteristic, all finite extensions of *finite* fields are separable. That is, for our purposes, non-separable field extensions are a pathology that we can avoid. Indeed, the results of this section can be viewed as proving that we can avoid non-separable extensions by very mild precautions.

A finite (hence, algebraic) field extension K of k is **separable** if the number of (nonzero) field maps

$$\sigma : K \longrightarrow \overline{k}$$

of K to an algebraic closure $\overline{k}$ of k is equal to the degree $[K : k]$.

22.2.1 Proposition: Let $k(\alpha)$ be a field extension of k with α a zero of an irreducible monic polynomial f in $k[x]$. Then $k(\alpha)$ is separable over k if and only if f has no repeated factors. [3]

Proof: As noted much earlier, the only possible images of α in $\overline{k}$ are zeros of the irreducible polynomial $f(x)$ of α over k, since

$$f(\sigma\alpha) = \sigma\big(f(\alpha)\big) = \sigma(0) = 0$$

because σ is a field homomorphism fixing the field k, in which the coefficients of f lie. We have already seen that

$$[k(\alpha) : k] = \deg f$$

regardless of separability. Thus, there are *at most* $[k(\alpha) : k]$ imbeddings of $k(\alpha)$ into $\overline{k}$. If the roots are *not* distinct, then there are strictly fewer than $[k(\alpha) : k]$ imbeddings into $\overline{k}$.

We recall the earlier argument that every root β of $f(x) = 0$ in $\overline{k}$ can be hit by some imbedding of $k(\alpha)$. Because the polynomial ring $k[x]$ is the free k-algebra on one generator, there is a homomorphism

$$\sigma_o : k[x] \longrightarrow \overline{k}$$

sending x to β and the identity on k. The kernel is the ideal generated by the minimal polynomial of β over k, which is f. Thus, this homomorphism factors through the quotient $k(\alpha) \approx k[x]/f$.   ///

22.2.2 Example: The simplest example of a *non*-separable extension is $\mathbb{F}_p(t^{1/p})$ over $\mathbb{F}_p(t)$, where $\mathbb{F}_p$ is the field with p elements and t is an indeterminate. The minimal polynomial for $\alpha = t^{1/p}$ is

$$x^p - t = (x - t^{1/p})^p$$

It is reasonable to view this as an avoidable pathology.

Now we give an iterated version of the first proposition:

22.2.3 Proposition: A finite field extension $k(\alpha_1, \alpha_2, \ldots, \alpha_n)$ of k is separable if and only each intermediate extension

$$k(\alpha_1, \alpha_2, \ldots, \alpha_i) \,/\, k(\alpha_1, \alpha_2, \ldots, \alpha_{i-1})$$

is separable.

---

[3] In many examples one easily tests for repeated factors by computing the *gcd* of f and its derivative.

Proof: The notation means to consider the large field as obtained by repeatedly adjoining single elements:

$$k(\alpha_1, \alpha_2, \ldots, \alpha_n) = k(\alpha_1)(\alpha_2) \ldots (\alpha_n)$$

Let

$$[k(\alpha_1, \ldots, \alpha_i) : k(\alpha_1, \ldots, \alpha_{i-1})] = d_i$$

Since degrees multiply in towers,

$$[K : k] = d_1 \cdot_2 \cdot \ldots \cdot d_n$$

An imbedding of K into an algebraic closure $\bar{k}$ of k can be given by first imbedding $k(\alpha_1)$, then extending this to an imbedding of $k(\alpha_1, \alpha_2)$, and so on, noting that an algebraic closure of k is an algebraic closure of any of these finite extensions. There are at most d_1, d_2, ..., d_n such imbeddings at the respective stages, with equality achieved if and only if the intermediate extension is separable. ///

Now a version which de-emphasizes elements:

22.2.4 Proposition: If K is a finite separable extension of k and L is a finite separable extension of K, then L is a finite separable extension of k.

Proof: By the finiteness, we can write

$$K = k(\alpha_1, \ldots, \alpha_m)$$

By the separability assumption on K/k, by the previous proposition, each intermediate extension

$$k(\alpha_1, \ldots, \alpha_i) \,/\, k(\alpha_1, \ldots, \alpha_{i-1})$$

is separable. Further, write

$$L = k(\alpha_1, \ldots, \alpha_m, \alpha_{n+1}, \ldots, \alpha_{m+n})$$

The separability hypothesis on K/L and the previous proposition imply that all the further intermediate extensions are separable. Then apply the previous proposition in the opposite order to see that L/k is separable. ///

22.2.5 Proposition: If K and L are finite separable extensions of k inside a fixed algebraic closure $\bar{k}$ of k, then their compositum KL (inside $\bar{k}$) is a finite separable extension of k.

Proof: By the previous proposition, it suffices to prove that the compositum KL is separable over K. Let $L = k(\beta_1, \ldots, \beta_n)$. By the second proposition, the separability of L/k implies that all the intermediate extensions

$$k(\beta_1, \ldots, \beta_i) \,/\, k(\beta_1, \ldots, \beta_{i-1})$$

are separable. Thus, the minimal polynomial f_i of β_i over $k(\beta_1, \ldots, \beta_{i-1})$ has no repeated factors. Since the minimal polynomial of β_i over $K(\beta_1, \ldots, \beta_{i-1})$ is a factor of f_i, it has no repeated factors. Going back in the other direction again, this means that

$$K(\beta_1, \ldots, \beta_i) \,/\, K(\beta_1, \ldots, \beta_{i-1})$$

is separable, for every i. Then L/K is separable. ///

22.3 *Primitive elements*

The following finiteness result is stronger than one might suspect, and gives further evidence that finite separable extensions are well-behaved.

22.3.1 Proposition: Let K be a finite field extension of k. There is a single generator α such that $K = k(\alpha)$ if and only if there are only finitely-many fields between k and K.

Proof: First, suppose that $K = k(\alpha)$. Let E be an intermediate field, and let $g(x) \in E[x]$ be the minimal polynomial of α over E. Adjoining the *coefficients* of g to k gives a field F between k and E. Since g is irreducible in $E[x]$, it is certainly irreducible in the smaller $F[x]$. Since $K = k(\alpha) = F(\alpha)$, the degree of K over E is equal to its degree over F. By the multiplicativity of degrees in towers, $E = F$. That is, E is uniquely determined by the monic polynomial g. Since g divides f, and since there are only finitely-many monic divisors of f, there are only finitely-many possible intermediate fields.

Conversely, assume that there are only finitely-many fields between k and K. For k *finite*, the intermediate fields are k vector subspaces of the finite-dimensional k vector space K, so there are only finitely-many. Now consider *infinite* k. It suffices to show that for any two algebraic elements α, β over k, there is a single γ such that $k(\alpha, \beta) = k(\gamma)$. Indeed, let $\gamma = \alpha + t\beta$ with $t \in k$ to be determined. Since there are finitely-many intermediate fields and k is infinite, there are $t_1 \neq t_2$ such that

$$k(\alpha + t_1\beta) = k(\alpha + t_2\beta)$$

Call this intermediate field E. Then

$$(t_2 - t_1)\beta = (\alpha + t_1\beta) - (\alpha + t_2\beta) \in E$$

We can divide by $t_1 - t_2 \in k^\times$, so $\beta \in E$, and then $\alpha \in E$. Thus, the singly-generated E is equal to $k(\alpha, \beta)$. The finite-dimensional extension K is certainly finitely generated, so an induction proves that K is singly generated over k.   ///

22.3.2 Remark: There can be infinitely-many fields between a base field and a finite extension, as the example of the degree p^2 extension $\mathbb{F}_p(s^{1/p}, t^{1/p})$ of $\mathbb{F}_p(s, t)$ with independent indeterminates s, t showed earlier.

In classical terminology, a single element α of K such that $K = k(\alpha)$ is called a **primitive elements** for K over k.

22.3.3 Corollary: Let K be a finite separable extension of k. Then there are finitely-many fields between K and k, and K can be generated by a single element over k.

Proof: The issue is to show that a separable extension with two generators can be generated by a single element. Let $E = k(\alpha, \beta)$, with α, β separable over k. Let X be the set of distinct imbeddings of E into $\overline{k}$ over k, and put

$$f(x) = \Pi_{\sigma \neq \tau, \text{ in } X}(\sigma\alpha + x \cdot \sigma\beta - \tau\alpha - x\tau\beta)$$

This f is not the 0 polynomial, so there is $t \in \overline{k}$ such that $f(t) \neq 0$. Then the $\sigma(\alpha + t\beta)$ are n distinct field elements. Thus, $k(\alpha + t\beta)$ has degree at least n over k. On the other hand, this n is the degree of $k(\alpha, \beta)$ over k, so $k(\alpha, \beta) = k(\alpha + t\beta)$.   ///

22.4 *Normal field extensions*

In contrast to separability, the condition that a finite field extension K of k be *normal* is *not* typical. [4] There are several different useful characterizations of the property.

A finite field extension K of k is **normal** if all k-algebra homomorphisms of K into a fixed algebraic closure $\overline{k}$ of k have the same *image*.

22.4.1 Remark: Thus, in discussions of a normal extensions, it is not surprising that an algebraic closure can serve a useful auxiliary role.

22.4.2 Example: To illustrate that normal extensions are arguably atypical, note that the field extension $\mathbb{Q}(\sqrt[3]{2})$ of $\mathbb{Q}$ is *not* normal, since one imbedding into a copy of $\overline{\mathbb{Q}}$ inside $\mathbb{C}$ sends the cube root to a *real* number, while two others send it to *complex* (non-real) numbers.

22.4.3 Example: All cyclotomic extensions of $\mathbb{Q}$ are normal. Indeed, let ζ be a primitive n^{th} root of unity. We have already seen that *every* primitive n^{th} root of unity is of the form ζ^k where k is relatively prime to n. Since any mapping of $\mathbb{Q}(\zeta)$ to an algebraic closure of $\mathbb{Q}$ sends ζ to a primitive n^{th} root of unity, the image is unavoidably the same.

22.4.4 Remark: Note that the key feature of roots of unity used in the last example was that by adjoining *one* root of an equation to a base field we include *all*. This motivates:

22.4.5 Proposition: Let $f(x)$ be the minimal polynomial of a generator α of a finite field extension $k(\alpha)$ of k. The extension $k(\alpha)/k$ is normal if and only if every root β of $f(x) = 0$ lies in $k(\alpha)$, if and only if $f(x)$ factors into linear factors in $k(\alpha)[x]$.

Proof: The equivalence of the last two conditions is elementary. As we have seen several times by now, the k-algebra imbeddings $\sigma : k(\alpha) \longrightarrow \overline{k}$ are in bijection with the roots of $f(x) = 0$ in $\overline{k}$, with each root getting hit by a unique imbedding. If $k(\alpha)/k$ is normal, then $k(\sigma(\alpha)) = k(\tau(\alpha))$ for any two roots $\sigma(\alpha)$ and $\tau(\alpha)$ in $\overline{k}$. That is, any one of these images of $k(\alpha)$ contains every root of $f(x) = 0$ in $\overline{k}$. Since $k(\alpha) \approx k(\sigma(\alpha))$ for any such imbedding σ, the same conclusion applies to $k(\alpha)$.

On the other hand, suppose that $f(x)$ factors into linear factors in $k(\alpha)[x]$. Then it certainly factors into linear factors in $k(\sigma(\alpha))[x]$, for every $\sigma : k(\alpha) \longrightarrow \overline{k}$. That is, any $k(\sigma(\alpha))$ contains all the roots of $f(x) = 0$ in $\overline{k}$. That is,

$$k(\sigma(\alpha)) = k(\tau(\alpha))$$

for any two such imbeddings, which is to say that the two images are the same. ///

22.4.6 Proposition: If L is a finite normal field extension of k, and $k \subset K \subset L$, then L is normal over K.

Proof: An algebraic closure $\overline{k}$ of k is also an algebraic closure of any image $\sigma(K)$ of K in $\overline{k}$, since K is algebraic over k. The collection of imbeddings of L into $\overline{k}$ that extend $\sigma : K \longrightarrow \sigma(K)$ is a subset of the collection of *all* k-algebra imbeddings of L to $\overline{k}$. Thus,

---

[4] Thus, this terminology is potentially misleading, in essentially the same manner as the terminology *normal subgroups*.

the fact that all the latter images are the same implies that all the former images are the same. ///

22.4.7 Remark: In the situation of the last proposition, it is certainly *not* the case that K/k is normal. This is in sharp contrast to the analogous discussion regarding separability.

22.4.8 Proposition: A finite field extension K of k is normal if and only if, for every irreducible polynomial f in $k[x]$, if $f(x)$ has *one* linear factor in $K[x]$, then it factors completely into linear factors in $K[x]$.

Proof: First, suppose that K is normal over k, sitting inside an algebraic closure $\overline{k}$ of k. Let $\alpha \in K$ be a root in $\overline{k}$ of an irreducible polynomial $f(x) \in k[x]$. As recalled at the beginning of this chapter, given a root β of $f(x) = 0$ in $\overline{k}$, there is a k-algebra homomorphism $\sigma : k(\alpha) \longrightarrow K$ sending α to β. As recalled above, σ extends to a k-algebra homomorphism $\sigma : K \longrightarrow \overline{k}$. By the assumption of normality, $\sigma K = K$. Thus, $\beta \in K$.

For the converse, first let $K = k(\alpha)$. Let $f(x)$ be the minimal polynomial of α over k. By assumption, all the other roots of $f(x) = 0$ in $\overline{k}$ are in K. As recalled above, any k-algebra map of $k(\alpha) \longrightarrow \overline{k}$ must send α to some such root β, any k-algebra map of $k(\alpha)$ to $\overline{k}$ sends $k(\alpha)$ to K. Since $k(\alpha)$ is a finite-dimensional k-vectorspace, the injectivity of a field map implies surjectivity. That is, every k-algebra image of K in $\overline{k}$ is the original copy of K in $\overline{k}$.

For the general case of the converse, let $K = k(\alpha_1, \ldots, \alpha_n)$. Let $f_i(x)$ be the minimal polynomial of α_i over $k(\alpha_1, \ldots, \alpha_{i-1})$. Do induction on i. By hypothesis, and by the previous proposition, all the other roots of $f_i(x) = 0$ lie in K. Since any $k(\alpha_1, \ldots, \alpha_{i-1})$-linear map must send α_i to one of these roots, every image of $k(\alpha_1, \ldots, \alpha_i)$ is inside K. The induction implies that every k-algebra image of K in $\overline{k}$ is inside K. The finite-dimensionality implies that the image must be *equal* to K. ///

The idea of the latter proof can be re-used to prove a slightly different result:

22.4.9 Proposition: Let f be a not-necessarily irreducible polynomial in $k[x]$. Let $\overline{k}$ be a fixed algebraic closure of k. Any finite field extension K of k obtained as

$$K = k(\text{all roots of } f(x) = 0 \text{ in } \overline{k})$$

is *normal* over k.

Proof: First, suppose K is obtained from k by adjoining all the roots $\alpha_1, \ldots, \alpha_n$ of an irreducible $f(x)$ in $k[x]$. Certainly $K = k(\alpha_1, \ldots, \alpha_n)$. Let $f_i(x)$ be the minimal polynomial of α_i over $k(\alpha_1, \ldots, \alpha_{i-1})$. Any k-algebra homomorphism $k(\alpha_1) \longrightarrow \overline{k}$ must send α_1 to some $\alpha_i \in K$, so any such image of $k(\alpha_1)$ is inside K. Do induction on i. Since $f_i(x)$ is a factor of $f(x)$, all the other roots of $f_i(x) = 0$ lie in K. Since any $k(\alpha_1, \ldots, \alpha_{i-1})$-linear map must send α_i to one of these roots, every image of $k(\alpha_1, \ldots, \alpha_i)$ is inside K. By induction, every k-algebra image of K in $\overline{k}$ is inside K. The finite-dimensionality implies that the image must be *equal* to K. Thus, K is normal over k. ///

Given a (not necessarily irreducible) polynomial f in $k[x]$, a **splitting field** for f over k is a field extension K obtained by adjoining to k all the zeros of f (in some algebraic closure $\overline{k}$ of k). Thus, the assertion of the previous proposition is that *splitting fields are normal*.

The same general idea of proof gives one more sort of result, that moves in a slightly new conceptual direction:

22.4.10 Proposition: Let K be a normal field extensions of k. Let $f(x)$ be an irreducible in $k[x]$. Let α, β be two roots of $f(x) = 0$ in K. Then there is a k-algebra automorphism $\sigma : K \longrightarrow K$ such that

$$\sigma(\alpha) = \beta$$

Proof: Let $\bar{k}$ be an algebraic closure of k, and take $K \subset \bar{k}$ without loss of generality. By now we know that there is a k-algebra map $k(\alpha) \longrightarrow \bar{k}$ sending α to β, and that this map extends to a k-algebra homomorphism $K \longrightarrow \bar{k}$. By the normality of K over k, every image of K in $\bar{k}$ is K. Thus, the extended map is an automorphism of K over k.   ///

22.4.11 Remark: For K normal over k and L normal over K, it is not necessarily the case that L is normal over k. For example, $\mathbb{Q}(\sqrt{2})$ is normal over $\mathbb{Q}$, and $\mathbb{Q}(\sqrt{1 + \sqrt{2}})$ is normal over $\mathbb{Q}(\sqrt{2})$, but $\mathbb{Q}(\sqrt{1 + \sqrt{2}})$ is *not* normal over $\mathbb{Q}$.

22.5 *The main theorem*

A finite field extension K of k is **Galois** if it is both separable and normal over k. Let K be a finite Galois field extension of k. The **Galois group** of K over k is the automorphism group

$$G = \operatorname{Gal}(K/k) = \operatorname{Aut}(K/k)$$

The **Galois group** of a polynomial f in $k[x]$ over k is the Galois group of the *splitting field* of f over k.

22.5.1 Theorem: Let L be a finite Galois extension of k. The intermediate fields K between k and L are in *inclusion-reversing* bijection with subgroups H of the Galois group $G = \operatorname{Gal}(L/k)$ by

$$\text{subgroup } H \leftrightarrow \text{ subfield } K \text{ fixed by } H$$

For K the fixed field of subgroup H there is the equality

$$[L : K] = |H|$$

Further, H is a *normal* subgroup of G if and only if its fixed field K is Galois over k. If so, then

$$\operatorname{Gal}(K/k) \approx G/H$$

The standard picture for this is

$$G \left\{ \begin{matrix} L \\ | \\ K \\ | \\ k \end{matrix} \right\} H$$

22.5.2 Remark: The bijection between subgroups and intermediate fields is *inclusion-reversing*.

Proof: This proof is complicated. The first part goes from intermediate fields to subgroups of the Galois group. The second part goes from subgroups to intermediate fields. Then a few odds and ends are cleaned up.

First, we prove that the pointwise-fixed field

$$L^G = \{\alpha \in L : g \cdot \alpha = \alpha \text{ for all } g \in G\}$$

is just k itself, as opposed to being anything larger. Suppose that $\alpha \in L$ but $\alpha \notin k$. Let $f(x)$ be the minimal polynomial of α over k. Since L is separable over k, α is separable over k, so there is a root $\beta \neq \alpha$ of $f(x)$ in $\overline{k}$. Since L is normal over k, in fact $\beta \in L$. The last proposition of the previous section shows that there is an automorphism of L sending α to β. Thus, $\alpha \notin k$. This proves that the pointwise-fixed field of L^G is k.

Upon reflection, this argument proves that for an intermediate field K between k and L, the pointwise-fixed field of $\mathrm{Gal}(L/K)$ is K itself. In symbols, $K = L^{\mathrm{Gal}(L/K)}$.

Next, we show that the map $K \longrightarrow \mathrm{Gal}(L/K)$ of intermediate fields to subgroups of the Galois group is *injective*. For an intermediate field K, L/K is Galois. We just proved that K is the fixed field of $\mathrm{Gal}(L/K)$ inside L. Likewise, for another intermediate field $K' \neq K$, the pointwise-fixed field of $\mathrm{Gal}(L/K')$ in L is K'. Thus, these two subgroups must be distinct.

Next, show that, for two intermediate fields K, K' between L and k, with $H = \mathrm{Gal}(L/K)$ and $H' = \mathrm{Gal}(L/K')$, the Galois group of L over the compositum KK' is

$$H \cap H' = \mathrm{Gal}(L/KK')$$

Indeed, every element of $H \cap H'$ leaves KK' fixed pointwise. On the other hand, every element of $\mathrm{Gal}(L/k)$ leaving KK' fixed pointwise certainly fixes both K and K'.

Next, with the notation of the previous pragraph, we claim that the pointwise-fixed field of the smallest subgroup A of G containing both H and H' is $K \cap K'$. Indeed, this fixed field must lie inside the fixed field of H, which is K, and must lie inside the fixed field of H', which is K'. Thus, the fixed field of A is contained in $K \cap K'$. On the other hand, every element of $K \cap K'$ *is* fixed by H and by H', so is fixed by the subgroup of $\mathrm{Gal}(L/k)$ generated by them.

Keeping this notation, next we claim that $K \subset K'$ if and only if $H \supset H'$. Indeed, $g \in \mathrm{Gal}(L/k)$ leaving K' fixed certainly leaves K fixed, so $g \in H$. This is one direction of the equivalence. On the other hand, when $H \supset H'$, certainly the fixed field of the larger group H is contained in the fixed field of the smaller.

Now, following Artin, we go from subgroups of the Galois group to intermediate fields. That is, we prove that every subgroup of a Galois group is the Galois group of the top field L over an intermediate field. Let E be an arbitrary field, and B a group of field automorphisms of E, with $|B| = n$. Let $K = E^B$ be the pointwise-fixed field of B inside E. Then E/K is Galois, with Galois group B. To see this, let $\alpha \in E$ and let $b_1, \ldots, b_n \in B$ be a *maximal* collection of elements of B such that the $b_i\alpha$ are *distinct*. Certainly α is a root of the polynomial

$$f(x) = \Pi_{i=1}^n (x - b_i\alpha)$$

For any $b \in B$ the list $bb_1\alpha, \ldots, bb_n\alpha$ must be merely a permutation of the original list $b_1\alpha, \ldots, b_n$, or else the maximality is contradicted. Thus, the polynomial f^b obtained by letting $b \in B$ act on the coefficients of f is just f itself. That is, the coefficients lie in the pointwise-fixed field $K = E^B$. By construction, the roots of f are distinct. This shows that every element of E is separable of degree at most n over K, and the minimal polynomial over K of every $\alpha \in E$ splits completely in E. Thus, E is separable and normal over K, hence, Galois. By the theorem of the primitive element, $E = K(\alpha)$ for some α in E, and $[E : K] \leq n$ since the degree of the minimal polynomial of α over K is *at most n*. On the other hand, we saw that the number of automorphisms of $E = K(\alpha)$ over $K(\alpha)$ is *at most* the degree of the extension. Thus, B is the whole Galois group.

Incidentally, this last discussion proves that the order of the Galois group is equal to the degree of the field extension.

Finally, for an intermediate field K between k and L, as shown earlier, the top field L is certainly separable over K, and is also normal over K. Thus, L is Galois over K. The last paragraph does also show that $\mathrm{Gal}(L/K)$ is the subgroup of $\mathrm{Gal}(L/k)$ pointwise-fixing K.

Finally, we must prove that an intermediate field K between k and L is *normal* over the bottom field k if and only if its pointwise-fixer subgroup N in $G = \mathrm{Gal}(L/k)$ is a normal subgroup of G. First, for K normal over k, any element of G stabilizes K, giving a group homomorphism $G \longrightarrow \mathrm{Gal}(K/k)$. The kernel is a normal subgroup of G, and by definition is the subgroup of G fixing K. On the other hand, if K is *not* normal over k, then there is an imbedding σ of K to $\bar{k}$ whose image is *not* K itself. Early on, we saw that such an imbedding extends to L, and, since L is normal over k, the image of L is L. Thus, this map gives an element σ of the Galois group $\mathrm{Gal}(L/k)$. We have $\sigma K \neq K$. Yet it is immediate that $\mathrm{Gal}(L/K)$ and $\mathrm{Gal}(L/\sigma K)$ are conjugate by σ. By now we know that these pointwise-fixer groups are unequal, so neither one is normal in $\mathrm{Gal}(L/k)$.

This finishes the proof of the main theorem of Galois theory. ///

22.6 Conjugates, trace, norm

Let K/k be a finite Galois field extension with Galois group. For $\alpha \in K$, the **(Galois) conjugates** of α over k are the images $\sigma\alpha$ for $\sigma \in G$.

The **(Galois) trace** from K to k is the map

$$\mathrm{trace}_{K/k}\,\alpha \;=\; \mathrm{tr}_{K/k} \;=\; \sum_{\sigma \in G} \sigma\alpha \qquad (\text{for } \alpha \in K)$$

The **(Galois) norm** from K to k is the map

$$\mathrm{norm}_{K/k}\,\alpha \;=\; N_{K/k} \;=\; \Pi_{\sigma \in G}\sigma\alpha \qquad (\text{for } \alpha \in K)$$

Of course, usually an element α in K has a non-trivial *isotropy subgroup* in G, so there may be fewer *distinct* conjugates of α than conjugates altogether. For that matter, sometimes *conjugates* insinuates that one is to take distinct conjugates.

When K is the splitting field over k of the minimal polynomial $f(x) \in k[x]$ for α separable algebraic over k,

$$f(x) \;=\; \Pi_{\sigma \in G}(x - \sigma\alpha) \;=\; x^n - \mathrm{tr}_{K/k}\alpha\, x^{n-1} + \ldots + (-1)^n \cdot N_{K/k}\alpha$$

where $n = [K : k]$. The other symmetric polynomials in α do not have names as common as the trace and norm.

22.7 *Basic examples*

A Galois extension is called **cyclic** if its Galois group is cyclic. Generally, any adjective that can be applied to a *group* can be applied to a *Galois field extension* if its Galois group has that property.

22.7.1 Example: Let $k = \mathbb{F}_q$ be a **finite** field $\mathbb{F}_q$ with q elements. Although the result was not couched as Galois theory, we have already seen (essentially) that every extension K of k is *cyclic*, generated by the Frobenius automorphism

$$\text{Frob}_q : \alpha \longrightarrow \alpha^q$$

Thus, without citing the main theorem of Galois theory, we already knew that

$$[K : k] = |\text{Gal}(K/k)|$$

22.7.2 Example: Let $k = \mathbb{Q}$, and ζ a primitive n^{th} root of unity. Again, the result was not portrayed as Galois theory, but we already saw (essentially) that $K = \mathbb{Q}(\zeta)$ is an *abelian* Galois extension, with

$$\text{Gal}(\mathbb{Q}(\zeta)/\mathbb{Q}) \approx (\mathbb{Z}/n)^{\times}$$

by

$$(\zeta \longrightarrow \zeta^a) \longleftarrow a$$

22.7.3 Example: *(Kummer extensions)* Fix a prime p, let $k = \mathbb{Q}(\zeta)$ where ζ is a primitive p^{th} root of unity, and take a to be *not* a pth power in $k^{\times}$. Let K be the splitting field over $\mathbb{Q}(\zeta)$ of $x^p - a$. Then $K = k(\alpha)$ for any p^{th} root α of a,

$$[K : \mathbb{Q}(\zeta)] = p$$

and the Galois group is

$$\text{Gal}(K/\mathbb{Q}(\zeta)) \approx \mathbb{Z}/p \quad \text{(with addition)}$$

by

$$(\alpha \longrightarrow \zeta^{\ell} \cdot \alpha) \longleftarrow \ell$$

(Proof of this is left as an exercise.)

22.7.4 Example: Fix a prime p, let $k = \mathbb{Q}$ and take a to be *not* a pth power in $\mathbb{Q}^{\times}$. Let K be the splitting field over $\mathbb{Q}(\zeta)$ of $x^p - a$. Then $K = k(\alpha, \zeta)$ for any p^{th} root α of a, and for ζ a primitive p^{th} root of unity. We have

$$[K : \mathbb{Q}] = p(p - 1)$$

and the Galois group is a semi-direct product

$$\text{Gal}(K/\mathbb{Q}) \approx \mathbb{Z}/p \times_f (\mathbb{Z}/p)^{\times}$$

(Proof of this is left as an exercise at the end of this section.)

22.7.5 Example: Let $t_1, \ldots, t_n$ be independent indeterminates over a field E. Let $K = E(t_1, \ldots, t_n)$ be the field of fractions of the polynomial ring $E[t_1, \ldots, t_n]$. Let the permutation group $G = S_n$ on n things act on K by permutations of the t_i, namely, for a permutation π, let

$$\sigma_\pi(t_i) = t_{\pi(i)}$$

We prove below that the fixed field in K of G is the field

$$k = E(s_1, \ldots, s_n)$$

generated by the elementary symmetric polynomials s_i in the t_i.

22.7.6 Remark: The content of the last example is that **generic** polynomials of degree n have Galois groups S_n, even though various *particular* polynomials may have much smaller Galois groups.

22.8 *Worked examples*

22.8.1 Example: Show that $\mathbb{Q}(\sqrt{2})$ is normal over $\mathbb{Q}$.

We must show that all imbeddings $\sigma : \mathbb{Q}(\sqrt{2}) \longrightarrow \overline{\mathbb{Q}}$ to an algebraic closure of $\mathbb{Q}$ have the same image. Since (by Eisenstein and Gauss) $x^2 - 2$ is irreducible in $\mathbb{Q}[x]$, it is the minimal polynomial for any square root of 2 in any field extension of $\mathbb{Q}$. We know that (non-zero) field maps $\mathbb{Q}(\alpha) \longrightarrow \overline{\mathbb{Q}}$ over $\mathbb{Q}$ can only send roots of an irreducible $f(x) \in \mathbb{Q}[x]$ to roots of the same irreducible in $\overline{\mathbb{Q}}$. Let β be a square root of 2 in $\overline{\mathbb{Q}}$. Then $-\beta$ is another, and is the *only* other square root of 2, since the irreducible is of degree 2. Thus, $\sigma(\sqrt{2}) = \pm\beta$. Whichever sign occurs, the image of the whole $\mathbb{Q}(\sqrt{2})$ is the same. ///

22.8.2 Example: Show that $\mathbb{Q}(\sqrt[3]{5})$ is not normal over $\mathbb{Q}$.

By Eisenstein and Gauss, $x^3 - 5$ is irreducible in $\mathbb{Q}[x]$, so $[\mathbb{Q}(\sqrt[3]{5}) : \mathbb{Q}] = 3$. Let α be one cube root of 5 in an algebraic closure $\overline{\mathbb{Q}}$ of $\mathbb{Q}$. Also, observe that $x^3 - 5$ has no repeated factors, since its derivative is $3x^2$, and the *gcd* is readily computed to be 1. Let β be *another* cube root of 5. Then $(\alpha/beta)^3 = 1$ and $\alpha/beta \neq 1$, so that ratio is a primitive cube root of unity ω, whose minimal polynomial over $\mathbb{Q}$ we know to be $x^2 + x + 1$ (which is indeed irreducible, by Eisenstein and Gauss). Thus, the cubic field extension $\mathbb{Q}(\alpha)$ over $\mathbb{Q}$ cannot contain β, since otherwise it would have a quadratic subfield $\mathbb{Q}(\omega)$, contradicting the multiplicativity of degrees in towers.

Since

$$\mathbb{Q}(\alpha) \approx \mathbb{Q}[x]/\langle x^3 - 5 \rangle \approx \mathbb{Q}(\beta)$$

we can map a copy of $\mathbb{Q}(\sqrt[3]{5})$ to either $\mathbb{Q}(\alpha)$ or $\mathbb{Q}(\beta)$, sending $\sqrt[3]{5}$ to either α or β. But inside $\overline{\mathbb{Q}}$ the two fields $\mathbb{Q}(\alpha)$ and $\mathbb{Q}(\beta)$ are distinct sets. That is, $\mathbb{Q}(\sqrt[3]{5})$ is not normal. ///

22.8.3 Example: Find all fields intermediate between $\mathbb{Q}$ and $\mathbb{Q}(\zeta_{13})$ where ζ_{13} is a primitive 13th root of unity.

We already know that the Galois group G of the extension is isomorphic to $(\mathbb{Z}/13)^\times$ by

$$a \longrightarrow (\sigma_a : \zeta \longrightarrow \zeta^a)$$

and that group is cyclic. Thus, the subgroups are in bijection with the divisors of the order, 12, namely 1,2,3,4,6,12. By the main theorem of Galois theory, the intermediate fields are in bijection with the *proper* subgroups, which will be the fixed fields of the subgroups of orders $2, 3, 4, 6$. We have already identified the quadratic-over-$\mathbb{Q}$ subfield of any cyclotomic field $\mathbb{Q}(\zeta_p)$ with a primitive p^{th} root of unity ζ_p with p *prime*, via Gauss sums, as $\mathbb{Q}(\sqrt{\pm p})$ with the sign being the quadratic symbol $(-1/p)_2$. Thus, here, the subgroup fixed by the subgroup of order 6 is quadratic over $\mathbb{Q}$, and is $\mathbb{Q}(\sqrt{13})$.

We claim that the subfield fixed by $\zeta \longrightarrow \zeta^{\pm 1}$ is $\mathbb{Q}(\xi)$, where $\xi = \zeta + \zeta^{-1}$ is obtained by averaging ζ over that group of automorphisms. First, ξ is not 0, since those two powers of ζ are linearly independent over $\mathbb{Q}$. Second, to show that ξ is not accidentally invariant under any *larger* group of automorphisms, observe that

$$\sigma_a(\xi) = \zeta^a + \zeta^{-a} = \zeta^a + \zeta^{13-a}$$

Since $\zeta^1, \zeta^2, \ldots, \zeta^{11}, \zeta^{12}$ are a $\mathbb{Q}$-basis for $\mathbb{Q}(\zeta)$, an equality $\sigma_a(\xi) = \xi$ is

$$\zeta^a + \zeta^{13-a} = \sigma_a(\xi) = \xi = \zeta + \zeta^{12}$$

which by the linear independence implies $a = \pm 1$. This proves that this ξ generates the sextic-over-$\mathbb{Q}$ subextension.

To give a second description of ξ by telling the irreducible in $\mathbb{Q}[x]$ of which it is a zero, divide through the equation satisfied by ζ by ζ^6 to obtain

$$\zeta^6 + \zeta^5 + \ldots + \zeta + 1 + \zeta^{-1} + \ldots + \zeta^{-6} = 0$$

Thus,

$$\xi^6 + \xi^5 + (1 - \binom{6}{1})\xi^4 + (1 - \binom{5}{1})\xi^3 + (1 - \binom{6}{2}) + 5 \cdot \binom{4}{1})\xi^2$$

$$+ (1 - \binom{5}{2} + 4 \cdot \binom{3}{1})\xi + (1 - \binom{6}{3} + 5 \cdot \binom{4}{2} - 6\binom{2}{1})$$

$$= \xi^6 + \xi^5 - 5\xi^4 - 4\xi^3 + 6\xi^2 + 3\xi - 1 = 0$$

To describe ξ as a root of this sextic is an alternative to describing it as $\xi = \zeta + \zeta^{-1}$. Since we already know that ξ is of degree 6 over $\mathbb{Q}$, this sextic is necessarily irreducible.

The quartic-over-$\mathbb{Q}$ intermediate field is fixed by the (unique) order 3 subgroup $\{1, \sigma_3, \sigma_9\}$ of automorphisms. Thus, we form the average

$$\alpha = \zeta + \zeta^3 + \zeta^9$$

and claim that α generates that quartic extension. Indeed, if σ_a were to fix α, then

$$\zeta^2 + \zeta^{3a} + \zeta^{9a} = \sigma_a(\alpha) = \alpha = \zeta + \zeta^3 + \zeta^9$$

By the linear independence of $\zeta^2, \zeta^2, \ldots, \zeta^{12}$, this is possible only for a among $1, 3, 9$ modulo 13. This verifies that this α exactly generates the quartic extension.

To determine the quartic irreducible of which α is a root, we may be a little clever. Namely, we first find the irreducible *quadratic* over $\mathbb{Q}(\sqrt{13})$ of which α is a root. From Galois theory, the non-trivial automorphism of $\mathbb{Q}(\alpha)$ over $\mathbb{Q}(\sqrt{13})$ is (the restriction of) σ_4, since 4 is of order 6 in $(\mathbb{Z}/13)^\times$. Thus, the irreducible of α over $\mathbb{Q}(\sqrt{13})$ is

$$(x - \alpha)(x - \sigma_4\alpha)$$

in

$$\alpha + \sigma_4\alpha = \zeta + \zeta^3 + \zeta^9 + \zeta^4 + \zeta^{12} + \zeta^{10} \in \mathbb{Q}(\sqrt{13})$$

the exponents appearing are exactly the non-zero squares modulo 13, so

$$\alpha + \sigma_4\alpha = \sum_{\ell:\,\left(\frac{\ell}{13}\right)_2 = 1} \zeta^\ell = \frac{1}{2} \cdot \left(\sum_{1 \le \ell \le 12} \left(\frac{\ell}{13}\right)_2 \zeta^\ell + \sum_{1 \le \ell \le 12} \zeta^\ell \right) = \frac{\sqrt{13} - 1}{2}$$

from discussion of Gauss sums. And

$$\alpha \cdot \sigma_4\alpha = 3 + \zeta^5 + \zeta^{11} + \zeta^7 + \zeta^2 + \zeta^8 + \zeta^6 \in \mathbb{Q}(\sqrt{13})$$

The exponents are exactly the non-squares modulo 13, so this is

$$3 - \frac{1}{2} \cdot \left(\sum_{1 \le \ell \le 12} \left(\frac{\ell}{13}\right)_2 \zeta^\ell - \sum_{1 \le \ell \le 12} \zeta^\ell \right) = 3 - \frac{\sqrt{13} + 1}{2} = \frac{-\sqrt{13} + 5}{2}$$

Thus, the quadratic over $\mathbb{Q}(\sqrt{13})$ is

$$x^2 - \frac{\sqrt{13} - 1}{2} x + \frac{-\sqrt{13} + 5}{2}$$

It is interesting that the discriminant of this quadratic is

$$\sqrt{13} \cdot \frac{3 - \sqrt{13}}{2}$$

and that (taking the *norm*)

$$\frac{3 - \sqrt{13}}{2} \cdot \frac{3 + \sqrt{13}}{2} = -1$$

To obtain the quartic over $\mathbb{Q}$, multiply this by the same expression with $\sqrt{13}$ replaced by its negative, to obtain

$$\left(x^2 + \frac{x}{2} + \frac{5}{2}\right)^2 - 13\left(\frac{x}{2} + \frac{1}{2}\right)^2 = x^4 + \frac{x^2}{4} + \frac{25}{4} + x^3 + 5x^2 + \frac{5x}{2} - \frac{13x^2}{4} - \frac{13x}{2} - \frac{13}{4}$$

$$= x^4 + x^3 + 2x^2 - 4x + 3$$

Finally, to find the cubic-over-$\mathbb{Q}$ subfield fixed by the subgroup $\{1, \sigma_5, \sigma-1, \sigma_8\}$ of the Galois group, first consider the expression

$$\beta = \zeta + \zeta^5 + \zeta^{12} + \zeta^8$$

obtained by averaging ζ by the action of this subgroup. This is not zero since those powers of ζ are linearly independent over $\mathbb{Q}$. And if

$$\zeta^a + \zeta^{5a} + \zeta^{12a} + \zeta^{8a} = \sigma_a(\beta) = \beta = \zeta + \zeta^5 + \zeta^{12} + \zeta^8$$

the the linear independence implies that a is among $1, 5, 12, 8$ mod 13. Thus, β is not accidentally invariant under a larger group.

Of course we might want a second description of β by telling the irreducible cubic it satisfies. This was done by brute force earlier, but can also be done in other fashions to illustrate other points. For example, we know *a priori* that it *does* satisfy a cubic.

The linear coefficient is easy to determine, as it is the negative of

$$\beta + \sigma_2(\beta) + \sigma_2^2(\beta) = (\zeta + \zeta^5 + \zeta^{12} + \zeta^8) + (\zeta^2 + \zeta^{10} + \zeta^{11} + \zeta^3) + (\zeta^4 + \zeta^7 + \zeta^9 + \zeta^6) = -1$$

since the powers of ζ are ζ^i with i running from 1 to 12. Thus, the cubic is of the form $x^3 + x^2 + ax + b$ for some a, b in $\mathbb{Q}$.

We know that $\beta = \zeta + \zeta^5 + \zeta^{12} + \zeta^8$ is a zero of this equation, and from

$$\beta^3 + \beta^2 + a\beta + b = 0$$

we can determine a and b. Expanding β^3 and β^2, we have

$$(\zeta^3 + \zeta^2 + \zeta^{10} + \zeta^{11}$$

$$+ 3(\zeta^7 + \zeta^4 + \zeta + \zeta^{12} + \zeta^{10} + \zeta^4 + \zeta^9 + \zeta^3 + \zeta^5 + |zeta^8 + \zeta^6 + \zeta^2)$$

$$+ 6(\zeta^5 + \zeta + \zeta^8 + \zeta^{12})$$

$$+ (\zeta^2 + \zeta^{10} + \zeta^{11} + \zeta^3 + 2(\zeta^6 + 1 + \zeta^9 + \zeta^4 + 1 + \zeta^7))$$

$$+ a \cdot (\zeta + \zeta^5 + \zeta^{12} + \zeta^8) + b = 0$$

Keeping in mind that
$$\zeta^{12} = -(1 + \zeta + \zeta^2 + \ldots + \zeta^{10} + \zeta^{11})$$

using the linear independence of $1, \zeta, \zeta^2, \ldots, \zeta^{10}, \zeta^{11}$ by looking at the coefficients of 1, ζ, ζ^2, ζ^3, ... we obtain relations, respectively,

$$\begin{aligned} -3 - 6 + 2 \cdot 2 - a + b &= 0 \\ 0 &= 0 \\ 1 - 6 + 1 - a &= 0 \\ 1 - 6 + 1 - a &= 0 \\ \ldots \end{aligned}$$

From this, $a = -4$ and $b = 1$, so
$$x^3 + x^2 - 4x + 1$$

is the cubic of which $\beta = \zeta + \zeta^5 + \zeta^{12} + \zeta^8$ is a zero.   ///

22.8.4 Remark: It is surprising that the product of β and its two conjugates is -1.

22.8.5 Example: Find all fields intermediate between $\mathbb{Q}$ and a splitting field of $x^3 - x + 1$ over $\mathbb{Q}$.

First, we check the irreducibility in $\mathbb{Q}[x]$. By Gauss this is irreducible in $\mathbb{Q}[x]$ if and only if so in $\mathbb{Z}[x]$. For irreducibility in the latter it suffices to have irreducibility in $(\mathbb{Z}/p)[x]$, for example for $\mathbb{Z}/3$, as suggested by the exponent. Indeed, an earlier example showed that for prime p and $a \neq 0 \mod p$ the polynomial $x^p - x + a$ is irreducible modulo p. So $x^3 - x + 1$ is irreducible mod 3, so irreducible in $\mathbb{Z}[x]$, so irreducible in $\mathbb{Q}[x]$.

Even though we'll see shortly that in characteristic 0 irreducible polynomials always have distinct zeros, we briefly note why: if $f = g^2 h$ over an extension field, then $\deg \gcd(f, f') > 0$, where as usual f' is the derivative of f. If $f' \neq 0$, then the *gcd* has degree at most $\deg f' = \deg f - 1$, and is in $\mathbb{Q}[x]$, contradicting the irreducibility of f. And the derivative can be identically 0 if the characteristic is 0.

Thus, any of the three distinct zeros α, β, γ of $x^3 - x + 1$ generates a cubic extension of $\mathbb{Q}$.

Now things revolve around the discriminant

$$\Delta = (\alpha - \beta)^2 (\beta - \gamma)^2 (\gamma - \alpha)^2 = -27 \cdot 1^3 - 4 \cdot (-1)^3 = -27 + 4 = -23$$

from the computations that show that the discriminant of $x^3 + bx + c$ is $-27c^2 - 4b^3$. From its explicit form, if two (or all) the roots of a cubic are adjoined to the groundfield $\mathbb{Q}$, then the square root of the discriminant also lies in that (splitting) field. Since -23 is *not* a square of a rational number, the field $\mathbb{Q}(\sqrt{-23})$ is a subfield of the splitting field.

Since the splitting field K is normal (and in characteristic 0 inevitably separable), it is Galois over $\mathbb{Q}$. Any automorphism σ of K over $\mathbb{Q}$ must permute the 3 roots among themselves, since

$$\sigma(\alpha)^3 - \sigma(\alpha) + 1 = \sigma(\alpha^3 - \alpha + 1) = \sigma(0) = 0$$

Thus, the Galois group is a *subgroup* of the permutation group S_3 on 3 things. Further, the Galois group is *transitive* in its action on the roots, so cannot be merely of order 1 or 2. That is, the Galois group is either cyclic of order 3 or is the full permutation group S_3. Since the splitting field has a quadratic subfield, via the main theorem of Galois theory we know that the order of the Galois group is *even*, so is the full S_3.

By the main theorem of Galois theory, the intermediate fields are in inclusion-reversing bijection with the proper subgroups of S_3. Since the discriminant is not a square, the 3 subfields obtained by adjoining the different roots of the cubic are distinct (since otherwise the square root of the discriminant would be there), so these must give the subfields corresponding to the 3 subgroups of S_3 of order 2. The field $\mathbb{Q}(\sqrt{-23})$ must correspond to the single remaining subgroup of order 3 containing the 3-cycles. There are no other subgroups of S_3 (by Lagrange and Sylow, or even by direct observation), so there are no other intermediate fields. ///

22.8.6 Example: Find all fields intermediate between $\mathbb{Q}$ and $\mathbb{Q}(\zeta_{21})$ where ζ_{21} is a primitive 21st root of unity.

We have already shown that the Galois group G is isomorphic to

$$(\mathbb{Z}/21)^\times \approx (\mathbb{Z}/7)^\times \times (\mathbb{Z}/3)^\times \approx \mathbb{Z}/6 \oplus \mathbb{Z}/2 \approx \mathbb{Z}/3 \oplus \mathbb{Z}/2 \oplus \mathbb{Z}/2$$

(isomorphisms via Sun-Ze's theorem), using the fact that $(\mathbb{Z}/p)^\times$ for p prime is *cyclic*.

Invoking the main theorem of Galois theory, to determine all intermediate fields (as fixed fields of subgroups) we should determine all subgroups of $\mathbb{Z}/3 \oplus \mathbb{Z}/2 \oplus \mathbb{Z}/2$. To understand the collection of all subgroups, proceed as follows. First, a subgroup H either contains an element of order 3 or not, so H either contains that copy of $\mathbb{Z}/3$ or not. Second, $\mathbb{Z}/2 \oplus \mathbb{Z}/2$ is a two-dimensional vector space over $\mathbb{F}_2$, so its proper subgroups correspond to one-dimensional subspaces, which correspond to non-zero vectors (since the scalars are just $\{0, 1\}$), of which there are exactly 3. Thus, combining these cases, the complete list of *proper* subgroups of G is

$$
\begin{aligned}
H_1 &= \mathbb{Z}/3 \oplus 0 \oplus 0 \\
H_2 &= \mathbb{Z}/3 \oplus \mathbb{Z}/2 \oplus 0 \\
H_3 &= \mathbb{Z}/3 \oplus 0 \oplus \mathbb{Z}/2 \\
H_4 &= \mathbb{Z}/3 \oplus \mathbb{Z}/2 \cdot (1, 1) \\
H_5 &= \mathbb{Z}/3 \oplus \mathbb{Z}/2 \oplus \mathbb{Z}/2 \\
H_6 &= 0 \oplus \mathbb{Z}/2 \oplus 0 \\
H_7 &= 0 \oplus 0 \oplus \mathbb{Z}/2 \\
H_8 &= 0 \oplus \mathbb{Z}/2 \cdot (1, 1) \\
H_9 &= 0 \oplus \mathbb{Z}/2 \oplus \mathbb{Z}/2
\end{aligned}
$$

At worst by trial and error, the cyclic subgroup of order 3 in $(\mathbb{Z}/21)^\times$ is $\{1, 4, 16\}$, and the $\mathbb{Z}/2 \oplus \mathbb{Z}/2$ subgroup is $\{1, 8, 13, -1\}$.

An auxiliary point which is useful and makes things conceptually clearer is to verify that in $\mathbb{Q}(\zeta_n)$, where $n = p_1 \ldots p_t$ is a product of *distinct* primes p_i, and ζ_n is a primitive n^{th} root of unity, the powers

$$
\{\zeta^t : 1 \le t < n, \ \text{with} \ \gcd(t, n) = 1\}
$$

is (as you might be hoping[5]) a $\mathbb{Q}$-basis for $\mathbb{Q}(\zeta_n)$.

Prove this by induction. Let ζ_m be a primitive m^{th} root of unity for any m. The assertion holds for n prime, since for p prime

$$
\frac{x^p - 1}{x - 1}
$$

is the minimal polynomial for a primitive p^{th} root of unity. Suppose the assertion is true for n, and let p be a prime not dividing n. By now we know that the np^{th} cyclotomic polynomial is irreducible over $\mathbb{Q}$, so the degree of $\mathbb{Q}(\zeta_{np})$ over $\mathbb{Q}$ is (with Euler's totient function φ)

$$
[\mathbb{Q}(\zeta_{np})\mathbb{Q}] = \varphi(np) = \varphi(n) \cdot \varphi(p) = [\mathbb{Q}(\zeta_n)\mathbb{Q}] \cdot [\mathbb{Q}(\zeta_p)\mathbb{Q}]
$$

since p and n are relatively prime. Let a, b be integers such that $1 = an + bp$. Also note that $\zeta = \zeta_n \cdot \zeta_p$ is a primitive np^{th} root of unity. Thus, in the explicit form of Sun-Ze's theorem, given $i \bmod p$ and $j \bmod n$ we have

$$
an \cdot i + bp \cdot j = \begin{cases} i & \bmod p \\ j \bmod n \end{cases}
$$

Suppose that there were a linear dependence relation

$$
0 = \sum_i c_\ell \, \zeta_{np}^\ell
$$

with $c_i \in \mathbb{Q}$ and with ℓ summed over $1 \le \ell < np$ with $\gcd(\ell, np) = 1$. Let $i = \ell \bmod p$ and $j = \ell \bmod n$. Then

$$
\zeta_{np}^{ani+bpj} = \zeta_n^j \cdot \zeta_p^i
$$

---

[5] For $n = 4$ and $n = 9$ the assertion is definitely false, for example.

and

$$0 = \sum_{i=1}^{p} \zeta_p^i \left(\sum_j c_{ani+bpj} \, \zeta_n^j \right)$$

where j is summed over $1 \le j < n$ with $\gcd(j,n) = 1$. Such a relation would imply that $\zeta_p, \ldots, \zeta_p^{p-1}$ would be linearly dependent over $\mathbb{Q}(\zeta_n)$. But the minimal polynomial of ζ_p over this larger field is the same as it is over $\mathbb{Q}$ (because the degree of $\mathbb{Q}(\zeta_n, \zeta_p)$ over $\mathbb{Q}(\zeta_n)$ is still $p-1$), so this implies that all the coefficients are 0. ///

22.8.7 Example: Find all fields intermediate between $\mathbb{Q}$ and $\mathbb{Q}(\zeta_{27})$ where ζ_{27} is a primitive 27th root of unity.

We know that the Galois group G is isomorphic to $(\mathbb{Z}/27)^\times$, which we also know is *cyclic*, of order $(3-1)3^{3-1} = 18$, since 27 is a power of an odd prime (namely, 3). The subgroups of a cyclic group are in bijection with the divisors of the order, so we have subgroups precisely of orders $1, 2, 3, 6, 9, 18$. The proper ones have orders $2, 3, 6, 9$. We can verify that $g = 2$ is a generator for the cyclic group $(\mathbb{Z}/27)^\times$, and the subgroups of a cyclic group are readily expressed in terms of powers of this generator. Thus, letting $\zeta = \zeta_{27}$, indexing the alphas by the order of the subgroup fixing them,

$$
\begin{aligned}
\alpha_2 &= \zeta + \zeta^{-1} \\
\alpha_3 &= \zeta + \zeta^{2^6} + \zeta^{2^{12}} \\
\alpha_6 &= \zeta + \zeta^{2^3} + \zeta^{2^6} + \zeta^{2^9} + \zeta^{2^{12}} + \zeta^{2^{15}} \\
\alpha_9 &= \zeta + \zeta^{2^2} + \zeta^{2^4} + \zeta^{2^6} + \zeta^{2^8} + \zeta^{2^{10}} \zeta^{2^{12}} + \zeta^{2^{14}} + \zeta^{2^{16}}
\end{aligned}
$$

But there are some useful alternative descriptions, some of which are clearer. Since ζ_{27}^3 is a primitive 9th root of unity ζ_9, which is of degree $\varphi(9) = 6$ over $\mathbb{Q}$, this identifies the degree 6 extension generated by α_3 ($3 \cdot 6 = 18$) more prettily. Similarly, ζ_{27}^9 is a primitive cube root of unity ζ_3, and $\mathbb{Q}(\zeta_3) = \mathbb{Q}(\sqrt{-3})$ from earlier examples. This is the quadratic subfield also generated by α_9. And from

$$0 = \frac{\zeta_9^9 - 1}{\zeta_9^3 - 1} = \zeta_9^6 + \zeta_9^3 + 1$$

we use our usual trick

$$\zeta_9^3 + 1 + \zeta_9^{-3} = 0$$

and then

$$(\zeta_9 + \zeta_9^{-1})^3 - 3(\zeta_9 + \zeta_9^{-1}) - 1 = 0$$

so a root of

$$x^3 - 3x - 1 = 0$$

generates the degree 3 field over $\mathbb{Q}$ also generated by α_6. ///

22.8.8 Example: Find all fields intermediate between $\mathbb{Q}$ and $\mathbb{Q}(\sqrt{2}, \sqrt{3}, \sqrt{5})$.

Let $K = \mathbb{Q}(\sqrt{2}, \sqrt{3}, \sqrt{5})$. Before invoking the main theorem of Galois theory, note that it really is true that $[K : \mathbb{Q}] = 2^3$, as a special case of a more general example we did earlier, with an arbitrary list of primes.

To count the proper subgroups of the Galois group $G \approx \mathbb{Z}/2 \oplus \mathbb{Z}/2 \oplus \mathbb{Z}/2$, it is useful to understand the Galois group as a 3-dimensional vector space over $\mathbb{F}_2$. Thus, the proper subgroups are the one-dimensional subspace and the two-dimensional subspaces, as vector spaces.

There are $2^3 - 1$ non-zero vectors, and since the field is $\mathbb{F}_2$, this is the number of subgroups of order 2. Invoking the main theorem of Galois theory, these are in bijection with the intermediate fields which are of degree 4 over $\mathbb{Q}$. We can easily think of several quartic fields over $\mathbb{Q}$, namely $\mathbb{Q}(\sqrt{2}, \sqrt{3})$, $\mathbb{Q}(\sqrt{2}, \sqrt{5})$, $\mathbb{Q}(\sqrt{3}, \sqrt{5})$, $\mathbb{Q}(\sqrt{6}, \sqrt{5})$, $\mathbb{Q}(\sqrt{10}, \sqrt{3})$, $\mathbb{Q}(\sqrt{2}, \sqrt{15})$, and the least obvious $\mathbb{Q}(\sqrt{6}, \sqrt{15})$. The argument that no two of these are the same is achieved most efficiently by use of the automorphisms σ, τ, ρ of the whole field which have the effects

$$\begin{array}{ccc} \sigma(\sqrt{2}) = -\sqrt{2} & \sigma(\sqrt{3}) = \sqrt{3} & \sigma(\sqrt{5}) = \sqrt{5} \\ \tau(\sqrt{2}) = \sqrt{2} & \tau(\sqrt{3}) = -\sqrt{3} & \tau(\sqrt{5}) = \sqrt{5} \\ \rho(\sqrt{2}) = \sqrt{2} & \rho(\sqrt{3}) = \sqrt{3} & \rho(\sqrt{5}) = -\sqrt{5} \end{array}$$

which are restrictions of automorphisms of the form $\zeta \longrightarrow \zeta^a$ of the cyclotomic field containing all these quadratic extensions, for example $\mathbb{Q}(\zeta_{120})$ where ζ_{120} is a primitive 120th root of unity.

To count the subgroups of order $4 = 2^2$, we might be a little clever and realize that the two-dimensional $\mathbb{F}_2$-vectorsubspaces are exactly the kernels of non-zero linear maps $\mathbb{F}_2^3 \longrightarrow \mathbb{F}_2$. Thus, these are in bijection with the non-zero vectors in the $\mathbb{F}_2$-linear dual to $\mathbb{F}_2^3$, which is again 3-dimensional. Thus, the number of two-dimensional subspaces is again $2^3 - 1$.

Or, we can count these two-dimensional subspaces by counting ordered pairs of two linearly independent vectors (namely $(2^3-1)(2^3-2) = 42$) and dividing by the number of changes of bases possible in a two-dimensional space. The latter number is the cardinality of $GL(2, \mathbb{F}_2)$, which is $(2^2 - 1)(2^2 - 2) = 6$. The quotient is 7 (unsurprisingly).

We can easily write down several quadratic extensions of $\mathbb{Q}$ inside the whole field, namely $\mathbb{Q}(\sqrt{2})$, $\mathbb{Q}(\sqrt{3})$, $\mathbb{Q}(\sqrt{5})$, $\mathbb{Q}(\sqrt{6})$, $\mathbb{Q}(\sqrt{10})$, $\mathbb{Q}(\sqrt{15})$, $\mathbb{Q}(\sqrt{30})$. That these are distinct can be shown, for example, by observing that the effects of the automorphisms σ, τ, ρ differ.   ///

22.8.9 Example: Let a, b, c be independent indeterminates over a field k. Let z be a zero of the cubic

$$x^3 + ax^2 + bx + c$$

in some algebraic closure of $K = k(a, b, c)$. What is the degree $[K(z) : K]$? What is the degree of the splitting field of that cubic over K?

First, we prove that $f(x) = x^3 + ax^2 + bx + c$ is irreducible in $k(a, b, c)[x]$. As a polynomial in x with coefficients in the ring $k(a, b)[c]$, it is monic and has *content* 1, so its irreducibility in $k(a, b, c)[x]$ is equivalent to its irreducibility in $k(a, b)[c][x] \approx k(a, b)[x][c]$. As a polynomial in c it is monic and linear, hence irreducible. This proves the irreducibility in $k(a, b, c)[x]$. Generally, $[K(z) : K]$ is equal to the degree of the minimal polynomial of z over K. Since f is irreducible it *is* the minimal polynomial of z over K, so $[K(z) : K] = 3$.

To understand the degree of the *splitting field*, let the three roots of $x^3 + ax^2 + bx + c = 0$ be z, u, v. Then (the discriminant)

$$\Delta = (z - u)^2 (u - v)^2 (v - z)^2$$

certainly lies in the splitting field, and is a *square* in the splitting field. But if Δ is *not* a square in the ground field K, then the splitting field contains the quadratic field $K(\sqrt{\Delta})$, which is of degree 2 over K. Since $\gcd(2, 3) = 1$, this implies that the splitting field is of degree at least 6 over K. But $f(x)/(x-z)$ is of degree 2, so the degree of the splitting field cannot be *more* than 6, so it is *exactly* 6 if the discriminant is *not* a square in the ground field K.

Now we use the fact that the a, b, c are indeterminates. Gauss' lemma assures us that a polynomial A in a, b, c is a square in $k(a, b, c)$ if and only it is a square in $k[a, b, c]$, since the reducibilities of $x^2 - A$ in the two rings are equivalent. Further, if A is square in $k[a, b, c]$ then it is a square in any homomorphic image of $k[a, b, c]$. If the characteristic of k is not 2, map $a \longrightarrow 0$, $c \longrightarrow 0$, so that $f(x)$ becomes $x^3 + bx$. The zeros of this are 0 and $\pm\sqrt{b}$, so the discriminant is

$$\Delta = (0 - \sqrt{b})^2 (0 + \sqrt{b})^2 (-\sqrt{b} - \sqrt{b})^2 = b \cdot b \cdot 4b = 4b^3 = (2b)^2 \cdot b$$

The indeterminate b is not a square. (For example, $x^2 - b$ is irreducible by Gauss, using Eisenstein's criterion.) That is, because this image is not a square, we know that the genuine discriminant is not a square in $k(a, b, c)$ without computing it.

Thus, the degree of the splitting field is always 6, for characteristic not 2.

For characteristic of k equal to 2, things work differently, since the cubic expression $(z - u)(u - v)(v - z)$ is already invariant under any group of permutations of the three roots. But, also, in characteristic 2, separable quadratic extensions are not all obtained via square roots, but, rather, by adjoining zeros of *Artin-Schreier* polynomials $x^2 - x + a$. ...

///

22.8.10 Example: Let $x_1, \ldots, x_n$ be independent indeterminates over a field k, with elementary symmetric polynomials $s_1, \ldots, s_n$. Prove that the Galois group of $k(x_1, \ldots, x_n)$ over $k(s_1, \ldots, s_n)$ is the symmetric group S_n on n things.

Since $k[x_1, \ldots, x_n]$ is the free (commutative) k-algebra on those n generators, for a given permutation p we can certainly map $x_i \longrightarrow x_{p(i)}$. Then, since this has trivial kernel, we can extend it to a map on the fraction field $k(x_1, \ldots, x_n)$. So the permutation group S_n on n things does act by automorphisms of $k(x_1, \ldots, x_n)$. Certainly such permutations of the indeterminates leaves $k[s_1, \ldots, s_n]$ pointwise fixed, so certainly leaves the fraction field $k(s_1, \ldots, s_n)$ pointwise fixed.

Each x_i is a zero of

$$f(X) = X^n - s_1 X^{n-1} + s_2 X^{n-2} - \ldots + (-1)^n s_n$$

so certainly $k(x_1, \ldots, x_n)$ is *finite* over $k(s_1, \ldots, s_n)$. Indeed, $k(x_1, \ldots, x_n)$ is a splitting field of $f(X)$ over $k(s_1, \ldots, s_n)$, since no smaller field could contain $x_1, \ldots, x_n$ (with or without $s_1, \ldots, s_n$). So the extension is *normal* over $k(s_1, \ldots, s_n)$. Since the x_i are mutually independent indeterminates, certainly no two are equal, so $f(X)$ is separable, and the splitting field is separable over $k(s_1, \ldots, s_n)$. That is, the extension is Galois.

The degree of $k(x_1, \ldots, x_n)$ over $k(s_1, \ldots, s_n)$ is *at most* $n!$, since x_1 is a zero of $f(X)$, x_2 is a zero of the polynomial $f(X)/(X - x_1)$ in $k(x_1)[X]$, x_3 is a zero of the polynomial $f(X)/(X - x_1)(X - x_2)$ in $k(x_1, x_2)[X]$, and so on. Since the Galois group contains S_n, the degree is *at least* $n!$ (the order of S_n). Thus, the degree is exactly $n!$ and the Galois group is exactly S_n.

Incidentally, this proves that $f(X) \in k(s_1, \ldots, s_n)[X]$ is irreducible, as follows. Note first that the degree of the splitting field of *any* polynomial $g(X)$ of degree d is at most $d!$, proven best by induction: given one root α_1, in $k(\alpha_1)[X]$ the polynomial $g(X)/(X-\alpha_1)$ has splitting field of degree at most $(d-1)!$, and with that number achieved *only* if $g(X)/(X-\alpha_1)$ is *irreducible* in $k(\alpha_1)[X]$. And $[k(\alpha_1):k] \le d$, with the maximum achieved if and only if $g(X)$ is irreducible in $k[X]$. Thus, by induction, the maximum possible degree of the splitting field of a degree d polynomial is $d!$, and for this to occur it is *necessary* that the polynomial be irreducible.

Thus, in the case at hand, if $f(X)$ were *not* irreducible, its splitting field could not be of degree $n!$ over $k(s_1, \ldots, s_n)$, contradiction.   ///

22.8.11 Example: Let K/k be a finite separable extension, $\overline{k}$ an algebraic closure of k, and $\sigma_1, \ldots, \sigma_n$ distinct field homomorphisms of K to $\overline{k}$. These σ are *linearly independent* over $\overline{k}$, in the following sense. If $\alpha_1, \ldots, \alpha_n \in \overline{k}$ are such that for all $\beta \in K$

$$\alpha_1\,\sigma_1(\beta) + \ldots + \alpha_n\,\sigma_n(\beta) \;=\; 0$$

then all α_i are 0.

Renumbering if necessary, let

$$\alpha_1\,\sigma_1(\beta) + \ldots + \alpha_n\,\sigma_n(\beta) \;=\; 0$$

be the *shortest* such relation with all α_i nonzero. Let $\gamma \in K^\times$ be a *primitive element* for K/k, that is, $K = k(\gamma)$. Then all the $\sigma_i(\gamma)$ are distinct. Replacing β by $\gamma \cdot \beta$ in the displayed relation and dividing by $\sigma_1(\gamma)$ gives another relation

$$\alpha_1\,\sigma_1(\beta) + \frac{\alpha_2 \cdot \sigma_2(\gamma)}{\sigma_1(\gamma)}\,\sigma(\beta) + \ldots + \frac{\alpha_n\,\sigma_n(\gamma)}{\sigma_1(\gamma)}\sigma_n(\beta) \;=\; 0$$

Since the ratios $\chi_i(\gamma)/\chi_1(\gamma)$ are not 1 for $i > 1$, subtraction of this relation from the first relation gives a shorter relation, contradiction.   ///

22.8.12 Example: Let K be a finite separable extension of a field k. Show that the Galois trace tr $: K \longrightarrow k$ is not the 0 map.

Let $\sigma_1, \ldots, \sigma_n$ be the distinct field homomorphisms of K into a chosen algebraic closure $\overline{k}$ of k. The trace is

$$\mathrm{tr}\,(\beta) \;=\; \sigma_1(\beta) + \ldots + \sigma_n(\beta) \;=\; 1 \cdot \sigma_1(\beta) + \ldots + 1 \cdot \sigma_n(\beta)$$

The previous example shows that this linear combination of the imbeddings σ_i is not the 0 map.   ///

22.8.13 Example: Let K/k be a finite separable extension. Show that the *trace pairing*

$$\langle,\rangle \;:\; K \times K \longrightarrow k$$

defined by

$$\langle \alpha, \beta \rangle \;=\; \mathrm{tr}\,_{K/k}(\alpha \cdot beta)$$

is *non-degenerate*.

That is, we must prove that, for any non-zero $\alpha \in K$, there is $\beta \in K$ such that $\mathrm{tr}\,(\alpha\beta) \ne 0$. The previous example shows that the trace of a primitive element γ is non-zero. Thus, given $\alpha \ne 0$, let $\beta = \gamma/\alpha$.   ///

Exercises

22.8.1 Example: Show that any quadratic extension of $\mathbb{Q}$ is normal over $\mathbb{Q}$.

22.8.2 Example: Take an integer d which is not a cube or a rational number. Show that $\mathbb{Q}(\sqrt[3]{d})$ is *not* normal over $\mathbb{Q}$.

22.8.3 Example: Find all fields intermediate between $\mathbb{Q}$ and $\mathbb{Q}(\zeta_{11})$ where ζ_{11} is a primitive 13th root of unity.

22.8.4 Example: Find all fields intermediate between $\mathbb{Q}$ and $\mathbb{Q}(\zeta_8)$ where ζ_8 is a primitive 27th root of unity.

22.8.5 Example: Find all fields intermediate between $\mathbb{Q}$ and $\mathbb{Q}(\sqrt{3}, \sqrt{5}, \sqrt{7})$.

22.6 What is the Galois group of $x^3 - x - 1$ over $\mathbb{Q}$?

22.7 What is the Galois group of $x^3 - 2$ over $\mathbb{Q}$?

22.8 What is the Galois group of $x^3 - x - 1$ over $\mathbb{Q}(\sqrt{23})$?

22.9 What is the Galois group of $x^4 - 5$ over $\mathbb{Q}$, over $\mathbb{Q}(\sqrt{5}$, over $\sqrt{-5}$, over $\mathbb{Q}(i)$, and over $\mathbb{Q}(\sqrt{2})$?

22.10 Let K/k be a finite separable extension. Show that for every intermediate field $k \subset E \subset K$, the extensions E/k and K/E are separable.

22.11 Show that $\mathbb{Q}(\sqrt{2})$ is normal over $\mathbb{Q}$, and $\mathbb{Q}(\sqrt{1+\sqrt{2}})$ is normal over $\mathbb{Q}(\sqrt{2})$, but $\mathbb{Q}(\sqrt{1+\sqrt{2}})$ is *not* normal over $\mathbb{Q}$.

22.12 Find all subfields of the splitting field over $\mathbb{Q}$ of $x^4 + 2$.

22.13 Let k be a field. Let $\alpha_1, \ldots, \alpha_n$ be distinct elements of $k^\times$. Suppose that $c_1, \ldots, c_n$ in k are such that for all positive integers ℓ

$$\sum_i c_i \alpha_i^\ell = 0$$

Show that all the c_i are 0.

22.14 Let K be a finite normal field extension of a field k. Let P be a monic irreducible in $k[x]$. Let Q and R be two monic irreducible factors of P in $K[x]$. Show that there is $\sigma \in \text{Aut}(K/k)$ such that $Q^\sigma = R$ (with σ acting on the coefficients).

22.15 Show that every finite algebraic extension of a finite field is normal and separable, hence Galois.

22.16 Show that any cyclotomic field (that is, an extension of $\mathbb{Q}$ obtained by adjoining a root of unity) is normal and separable, hence Galois.

22.17 Fix a prime p. Let k be a field *not* of characteristic p, containing a primitive p^{th} root of unity ζ. Let $a \in k$ *not* be a pth power of any element of k, and let α be a pth root of α. Prove that the *Kummer extension* $K = k(\alpha)$ is normal and separable, hence Galois.

Prove that the Galois group is cyclic of order p, given by automorphisms

$$\alpha \longrightarrow \zeta^\ell \cdot \alpha \qquad \text{(for } 0 \le \ell < p)$$

22.18 Let $t_1, \ldots, t_n$ be independent indeterminates over a field E. Let $K = E(t_1, \ldots, t_n)$ be the field of fractions of the polynomial ring $E[t_1, \ldots, t_n]$. Let

$$k = E(s_1, \ldots, s_n)$$

be the subfield generated by the elementary symmetric polynomials s_i in the t_i. Prove that the extension K/k is normal and separable, hence Galois. (Then, from our earlier discussion, its Galois group is the permutation group on n things.)

22.19 Show that the Galois trace $\sigma : \mathbb{F}_{q^n} \longrightarrow \mathbb{F}_q$ is

$$\sigma(\alpha) = \alpha + \alpha^q + \alpha^{q^2} + \ldots + \alpha^{q^{n-1}}$$

22.20 Show that the Galois norm $\nu : \mathbb{F}_{q^n} \longrightarrow \mathbb{F}_q$ is

$$\nu(\alpha) = \alpha^{\frac{q^n-1}{q-1}}$$

22.21 Let k be a finite field, and K a finite extension. Show that trace and norm maps $K \longrightarrow k$ are surjective.

22.22 Let k be a finite field with q elements. Fix a positive integer n. Determine the order of the largest cyclic subgroup in $GL(n, k)$.

22.23 Let m and n be coprime. Let ζ be a primitive m^{th} root of unity. Show that the cyclotomic polynomial $\varphi_n(x)$ is irreducible in $\mathbb{Q}(\zeta)[x]$.

22.24 (*Artin*) Let $\overline{\mathbb{Q}}$ be a fixed algebraic closure of $\mathbb{Q}$. Let k be a maximal subfield of $\overline{\mathbb{Q}}$ not containing $\sqrt{2}$. Show that every finite extension of k is cyclic.

22.25 (*Artin*) Let σ be an automorphism of $\overline{\mathbb{Q}}$ over $\mathbb{Q}$, with fixed field k. Show that every finite extension of k is cyclic.

23. Solving equations by radicals

Around 1800, Ruffini sketched a proof, completed by Abel, that the general quintic equation is *not* solvable in radicals, by contrast to cubics and quartics whose solutions by radicals were found in the Italian renaissance, not to mention quadratic equations, understood in antiquity. Ruffini's proof required classifying the possible forms of radicals. By contrast, Galois' systematic development of the idea of *automorphism group* replaced the study of the expressions themselves with the study of their movements.

Galois theory solves some classical problems. Ruler-and-compass constructions, in coordinates, can *only* express quantities in repeated quadratic extensions of the field generated by given points, but nothing else. Thus, *trisection of angles* by ruler and compass is impossible for general-position angles, since the general trisection requires a cube root.

The examples and exercises continue with other themes.

23.1 *Galois' criterion*

We will not prove all the results in this section, for several reasons. First, solution of equations in radicals is no longer a critical or useful issue, being mostly of historical interest. Second, in general it is non-trivial to verify (or disprove) Galois' condition for solvability in radicals. Finally, to understand that Galois' condition is *intrinsic* requires the Jordan-Hölder theorem on *composition series* of groups (stated below). While its statement is clear, the proof of this result is technical, difficult to understand, and not re-used elsewhere here.

23.1.1 Theorem: Let G be the Galois group of the splitting field K of an irreducible polynomial f over k. If G has a sequence of subgroups

$$\{1\} \subset G_1 \subset G_2 \subset \ldots \subset G_m = G$$

such that G_i is *normal* in G_{i+1} and G_{i+1}/G_i is *cyclic* for every index i, then a root of $f(x) = 0$ can be expressed in terms of radicals. Conversely, if roots of f can be expressed in terms of radicals, then the Galois group G has such a chain of subgroups.

Proof: (*Sketch*) On one hand, adjunction of n roots is cyclic of degree n if the primitive n^{th} roots of unity are in the base field. If the nth roots of unity are *not* in the base field, we can *adjoin* them by taking a field extension obtainable by successive root-taking of orders strictly less than n. Thus, root-taking amounts to successive cyclic extensions, which altogether gives a *solvable* extension. On the other hand, a solvable extension is given by successive cyclic extensions. After n^{th} roots of unity are adjoined (which requires successive cyclic extensions of degrees less than n), one can prove that any cyclic extension is obtained by adjoining roots of $x^n - a$ for some a in the base. This fact is most usefully proven by looking at *Lagrange resolvents*.   ///

23.1.2 Theorem: The *general* n^{th} degree polynomial equation is not solvable in terms of radicals for $n > 4$.

Proof: The meaning of *general* is that the Galois group is the largest possible, namely the symmetric group S_n on n things. Then we invoke the theorem to see that we must prove that S_n is *not solvable* for $n > 4$. In fact, the normal subgroup A_n of S_n is *simple* for $n > 4$ (see just below), in the sense that it has no proper normal subgroups (and is not cyclic). In particular, A_n has no chain of subgroups normal in each other with cyclic quotients. This *almost* finishes the proof. What is missing is verifying the plausible claim that the simplicity of A_n means that no *other* possible chain of subgroups inside S_n can exist *with* cyclic quotients. We address this just below.   ///

A group is **simple** if it has not proper normal subgroups (and maybe is not a cyclic group of prime order, and is not the trivial group). A group G with a chain of subgroups G_i, each normal in the next, with the quotients *cyclic*, is a **solvable** group, because of the conclusion of this theorem.

23.1.3 Proposition: For $n \geq 5$ the alternating group A_n on n things is *simple*.

Proof: (*Sketch*) The trick is that for $n \geq 5$ the group A_n is generated by 3-cycles. Keeping track of 3-cycles, one can prove that the commutator subgroup of A_n, generated by expressions $xyx^{-1}y^{-1}$, for $x, y \in A_n$, is A_n itself. This yields the simplicity of A_n.   ///

23.1.4 Remark: A similar discussion addresses the question of **constructibility by ruler and compass**. One can prove that a point is *constructible* by ruler and compass if and only if its coordinates lie in a field extension of $\mathbb{Q}$ obtained by successive *quadratic* field extensions. Thus, for example, a regular n-gon can be constructed by ruler and compass exactly when $(\mathbb{Z}/n)^\times$ is a two-group. This happens exactly when n is of the form

$$n = 2^m \cdot p_1 \ldots p_\ell$$

where each p_i is a *Fermat prime*, that is, is a prime of the form $p = 2^{2^t} + 1$. Gauss constructed a regular 17-gon. The next Fermat prime is 257. Sometime in the early 19th century someone *did* literally construct a regular 65537-gon, too.

23.2 *Composition series, Jordan-Hölder theorem*

Now we should check that the simplicity of A_n really does prevent there being any *other* chain of subgroups with cyclic quotients that might secretly permit a solution in radicals.

A **composition series** for a finite group G is a chain of subgroups

$$\{1\} \subset G_1 \subset \ldots \subset G_m = G$$

where each G_i is normal in G_{i+1} and the quotient G_{i+1}/G_i is either *cyclic of prime order* or *simple*. [1]

23.2.1 Theorem: Let

$$\{1\} = G_0 \subset G_1 \subset \ldots \subset G_m = G$$

$$\{1\} = H_0 \subset H_1 \subset \ldots \subset H_n = G$$

be two composition series for G. Then $m = n$ and the *sets* of quotients $\{G_{i+1}/G_i\}$ and $\{H_{j+1}/G_j\}$ (counting multiplicities) are identical.

Proof: (*Comments*) This theorem is quite non-trivial, and we will not prove it. The key ingredient is the *Jordan-Zassenhaus butterfly lemma*, which itself is technical and non-trivial. The proof of the analogue for modules over a ring is more intuitive, and is a worthwhile result in itself, which we leave to the reader. ///

23.3 *Solving cubics by radicals*

We follow *J.-L. Lagrange* to recover the renaissance Italian formulas of Cardan and Tartaglia in terms of radicals for the zeros of the general cubic

$$x^3 + ax^2 + bx + c$$

with a, b, c in a field k of characteristic neither 3 nor 2. [2] Lagrange's method creates an expression, the *resolvent*, having more accessible symmetries. [3]

Let ω be a primitive cube root of unity. Let α, β, γ be the three zeros of the cubic above. The **Lagrange resolvent** is

$$\lambda = \alpha + \omega \cdot \beta + \omega^2 \gamma$$

The point is that any cyclic permutation of the roots alters λ by a cube root of unity. Thus, λ^3 is *invariant* under cyclic permutations of the roots, so we anticipate that λ^3 lies in a smaller field than do the roots. This is intended to reduce the problem to a simpler one.

---

[1] Again, it is often convenient that the notion of *simple* group makes an exception for cyclic groups of prime order.

[2] In characteristic 3, there are no primitive cube roots of 1, and the whole setup fails. In characteristic 2, unless we are somehow assured that the discriminant is a square in the ground field, the *auxiliary quadratic* which arises does not behave the way we want.

[3] The complication that cube roots of unity are involved was disturbing, historically, since complex number were viewed with suspicion until well into the 19th century.

Compute

$$\lambda^3 = \left(\alpha + \omega\beta + \omega^2\gamma\right)^3$$

$$= \alpha^3 + \beta^3 + \gamma^3 + 3\omega\alpha^2\beta + 3\omega^2\alpha\beta^2 + 3\omega^2\alpha^2\gamma + 3\omega\alpha\gamma^2 + 3\omega\beta^2\gamma + 3\omega^2\beta\gamma^2 + 6\alpha\beta\gamma$$

$$= \alpha^3 + \beta^3 + \gamma^3 + 3\omega(\alpha^2\beta + \beta^2\gamma + \gamma^2\alpha) + 3\omega^2(\alpha\beta^2 + \beta\gamma^2 + \alpha^2\gamma) + 6\alpha\beta\gamma$$

Since $\omega^2 = -1 - \omega$ this is

$$\alpha^3 + \beta^3 + \gamma^3 + 6\alpha\beta\gamma + 3\omega(\alpha^2\beta + \beta^2\gamma + \gamma^2\alpha) - 3\omega(\alpha\beta^2 + \beta\gamma^2 + \alpha^2\gamma) - 3(\alpha\beta^2 + \beta\gamma^2 + \alpha^2\gamma)$$

In terms of the *elementary symmetric polynomials*

$$s_1 = \alpha + \beta, \qquad s_2 = \alpha\beta + \beta\gamma + \gamma\alpha \qquad s_3 = \alpha\beta\gamma$$

we have

$$\alpha^3 + \beta^3 + \gamma^3 = s_1^3 - 3s_1 s_2 + 3s_3$$

Thus,

$$\lambda^3 = s_1^3 - 3s_1 s_2 + 9s_3 + 3\omega(\alpha^2\beta + \beta^2\gamma + \gamma^2\alpha) - 3\omega(\alpha\beta^2 + \beta\gamma^2 + \alpha^2\gamma) - 3(\alpha\beta^2 + \beta\gamma^2 + \alpha^2\gamma)$$

Neither of the two trinomials

$$A = \alpha^2\beta + \beta^2\gamma + \gamma^2\alpha \qquad B = \alpha\beta^2 + \beta\gamma^2 + \gamma\alpha^2$$

is invariant under *all* permutations of α, β, γ, but only under the subgroup generated by *3-cycles*, so we cannot use symmetric polynomial algorithm to express these two trinomials *polynomially* in terms of elementary symmetric polynomials. [4]

But all is not lost, since $A + B$ and AB *are* invariant under *all* permutations of the roots, since any 2-cycle permutes A and B. So both $A + B$ and AB are expressible in terms of elementary symmetric polynomials, and then the two trinomials are the roots of

$$x^2 - (A + B)x + AB = 0$$

which is solvable by radicals in characteristic not 2.

We obtain the expression for $A+B$ in terms of elementary symmetric polynomials. Without even embarking upon the algorithm, a reasonable guess finishes the problem:

$$s_1 s_2 - 3s_3 = (\alpha+\beta+\gamma)(\alpha\beta+\beta\gamma+\gamma\alpha) - 3\alpha\beta\gamma = \alpha^2\beta + \beta^2\gamma + \gamma^2\alpha + \alpha\beta^2 + \beta\gamma^2 + \gamma\alpha^2 = A+B$$

Determining the expression for AB is more work, but not so bad.

$$AB = (\alpha^2\beta + \beta^2\gamma + \gamma^2\alpha) \cdot (\alpha\beta^2 + \beta\gamma^2 + \alpha^2\gamma) = \alpha^3\beta^3 + \beta^3\gamma^3 + \gamma^3\alpha^3 + \alpha^4\beta\gamma + \alpha\beta^4\gamma + \alpha\beta\gamma^4 + 3s_3^2$$

We can observe that already (using an earlier calculation)

$$\alpha^4\beta\gamma + \alpha\beta^4\gamma + \alpha\beta\gamma^4 = s_3 \cdot (\alpha^3 + \beta^3 + \gamma^3) = s_3(s_1^3 - 3s_1 s_2 + 3s_3)$$

---

[4] In an earlier computation regarding the special cubic $x^3 + x^2 - 2x - 1$, we could make use of the connection to the 7th root of unity to obtain explicit expressions for $\alpha^2\beta + \beta^2\gamma + \gamma^2\alpha$ and $\alpha\beta^2 + \beta\gamma^2 + \alpha^2\gamma$, but for the general cubic there are no such tricks available.

For $\alpha^3\beta^3 + \beta^3\gamma^3 + \gamma^3\alpha^3$ follow the algorithm: its value at $\gamma = 0$ is $\alpha^3\beta^3 = s_2^3$ (with the s_2 for α, β alone). Thus, we consider

$$\alpha^3\beta^3 + \beta^3\gamma^3 + \gamma^3\alpha^3 - (\alpha\beta + \beta\gamma + \gamma\alpha)^3$$

$$= -6s_3^2 - 3\left(\alpha^2\beta^3\gamma + \alpha^3\beta^2\gamma + \alpha\beta^3\gamma^2 + \alpha\beta^2\gamma^3 + \alpha^2\beta\gamma^3 + \alpha^3\beta\gamma^2\right)$$

$$= -6s_3^2 - 3s_3\left(\alpha\beta^2 + \alpha^2\beta + \beta^2\gamma + \beta\gamma^2 + \alpha\gamma^2 + \alpha^2\gamma\right) = -6s_3^2 - 3s_3(s_1s_2 - 3s_3)$$

by our computation of $A + B$. Together, the three parts of AB give

$$AB = s_3(s_1^3 - 3s_1s_2 + 3s_3) + \left(s_2^3 - 6s_3^2 - 3s_3(s_1s_2 - 3s_3)\right) + 3s_3^2$$

$$= s_1^3 s_3 - 3s_1s_2s_3 + 3s_3^2 + s_2^3 - 6s_3^2 - 3s_1s_2s_3 + 9s_3^2 + 3s_3^2 = s_1^3 s_3 - 6s_1s_2s_3 + 9s_3^2 + s_2^3$$

That is, A and B are the two zeros of the quadratic

$$x^2 - (s_1s_2 - 3s_3)x + (s_1^3 s_3 - 6s_1s_2s_3 + 9s_3^2 + s_2^3) = x^2 - (-ab + 3c)x + (a^3 c - 6abc + 9c^2 + b^3)$$

The **discriminant** of this monic quadratic is [5]

$$\Delta = (\text{linear coef})^2 - 4(\text{constant coef}) = (-ab + 3c)^2 - 4(a^3 c - 6abc + 9c^2 + b^3)$$

$$= a^2 b^2 - 6abc + 9c^2 - 4a^3 c + 24abc - 36c^2 - 4b^3 = a^2 b^2 - 27c^2 - 4a^3 c + 18abc - 4b^3$$

In particular, the quadratic formula [6] gives

$$A, B = \frac{(ab - 3c) \pm \sqrt{\Delta}}{2}$$

Then

$$\lambda^3 = s_1^3 - 3s_1s_2 + 9s_3 + 3\omega(\alpha^2\beta + \beta^2\gamma + \gamma^2\alpha) - 3\omega(\alpha\beta^2 + \beta\gamma^2 + \alpha^2\gamma) - 3(\alpha\beta^2 + \beta\gamma^2 + \alpha^2\gamma)$$

$$= -a^3 + 3bc - 9c + 3(\omega - 1)A - 3\omega B$$

$$= -a^3 + 3bc - 9c + 3(\omega - 1) \cdot \frac{(ab - 3c) + \sqrt{\Delta}}{2} - 3\omega \cdot \frac{(ab - 3c) - \sqrt{\Delta}}{2}$$

$$= -a^3 + 3bc - 9c - \frac{3}{2}(ab - 3c) + (3\omega - \frac{1}{2})\sqrt{\Delta}$$

That is, now we can solve for λ by taking a cube root of the mess on the right-hand side:

$$\lambda = \sqrt[3]{(\text{right-hand side})}$$

The same computation works for the analogue λ' of λ with ω replaced by the *other* [7] primitive cube root of unity

$$\lambda' = \alpha + \omega^2 \cdot \beta + \omega \cdot \gamma$$

---

[5] When the x^2 coefficient a vanishes, we will recover the better-known special case that the discriminant is $-27c^2 - 4b^3$.

[6] Which is an instance of this general approach, but for quadratics rather than cubics.

[7] In fact, there is no way to distinguish the two primitive cube roots of unity, so neither has primacy over the other. And, still, either is the square of the other.

The analogous computation is much easier when ω is replaced by 1, since

$$\alpha + 1 \cdot \beta + 1^2 \cdot \gamma = s_1 = -a$$

Thus, we have a linear system

The linear system

$$\begin{cases} \alpha + \beta + \gamma & = & -a \\ \alpha + \omega\beta + \omega^2\gamma & = & \lambda \\ \alpha + \omega^2\beta + \omega\gamma & = & \lambda' \end{cases}$$

has coefficients that readily allow solution, since for a primitive n^{th} root of unity ζ the matrix

$$M = \begin{pmatrix} 1 & 1 & 1 & \cdots & 1 \\ 1 & \zeta & \zeta^2 & \cdots & \zeta^{n-1} \\ 1 & \zeta^2 & (\zeta^2)^2 & \cdots & (\zeta^2)^{n-1} \\ 1 & \zeta^3 & (\zeta^3)^2 & \cdots & (\zeta^3)^{n-1} \\ & & \vdots & & \\ 1 & \zeta^{n-1} & (\zeta^{n-1})^2 & \cdots & (\zeta^{n-1})^{n-1} \end{pmatrix}$$

has inverse

$$M^{-1} = \begin{pmatrix} 1 & 1 & 1 & \cdots & 1 \\ 1 & \zeta^{-1} & (\zeta^{-1})^2 & \cdots & (\zeta^{-1})^{n-1} \\ 1 & \zeta^{-2} & (\zeta^{-2})^2 & \cdots & (\zeta^{-2})^{n-1} \\ 1 & \zeta^{-3} & (\zeta^{-3})^2 & \cdots & (\zeta^{-3})^{n-1} \\ & & \vdots & & \\ 1 & \zeta^{-n+1} & (\zeta^{-n+1})^2 & \cdots & (\zeta^{-n+1})^{n-1} \end{pmatrix}$$

In the present simple case this gives the three roots[8] of the cubic as

$$\begin{cases} \alpha & = & \frac{-a+\lambda+\lambda'}{3} \\ \beta & = & \frac{-a+\omega^2\lambda+\omega\lambda'}{3} \\ \gamma & = & \frac{-a+\omega\lambda+\omega^2\lambda'}{3} \end{cases}$$

23.4 *Worked examples*

23.4.1 Example: Let k be a field of characteristic 0. Let f be an irreducible polynomial in $k[x]$. Prove that f has no repeated factors, even over an algebraic closure of k.

If f has a factor P^2 where P is irreducible in $k[x]$, then P divides $\gcd(f, f') \in k[x]$. Since f was monic, and since the characteristic is 0, the derivative of the highest-degree term is of the form nx^{n-1}, and the coefficient is non-zero. Since f' is not 0, the degree of $\gcd(f, f')$ is at most $\deg f'$, which is strictly less than $\deg f$. Since f is irreducible, this *gcd* in $k[x]$ must

---

[8] Again, the seeming asymmetry among the roots is illusory. For example, since λ is a cube root of something, we really cannot distinguish among λ, $\omega\lambda$, and $\omega^2\lambda$. And, again, we cannot distinguish between ω and ω^2.

be 1. Thus, there are polynomials a, b such that $af + bf' = 1$. The latter identity certainly persists in $K[x]$ for any field extension K of k. ///

23.4.2 Example: Let K be a finite extension of a field k of characteristic 0. Prove that K is separable over k.

Since K is finite over k, there is a finite list of elements $\alpha_1, \ldots, \alpha_n$ in K such that $K = k(\alpha_1, \ldots, \alpha_n)$. From the previous example, the minimal polynomial f of α_1 over k has no repeated roots in an algebraic closure $\overline{k}$ of k, so $k(\alpha_1)$ is separable over k.

We recall [9] the fact that we can map $k(\alpha_1) \longrightarrow \overline{k}$ by sending α_1 to any of the $[k(\alpha_1) : k] = \deg f$ distinct roots of $f(x) = 0$ in $\overline{k}$. Thus, there are $[k(\alpha_1) : k] = \deg f$ distinct distinct imbeddings of $k(\alpha_1)$ into $\overline{k}$, so $k(\alpha_1)$ is separable over k.

Next, observe that for any imbedding $\sigma : k(\alpha_1) \longrightarrow \overline{k}$ of $k(\alpha_1)$ into an algebraic closure $\overline{k}$ of k, by proven properties of $\overline{k}$ we know that $\overline{k}$ is an algebraic closure of $\sigma(k(\alpha_1))$. Further, if $g(x) \in k(\alpha_1)[x]$ is the minimal polynomial of α_2 over $k(\alpha_1)$, then $\sigma(g)(x)$ (applying σ to the coefficients) is the minimal polynomial of α_2 over $\sigma(k(\alpha_1))$. Thus, by the same argument as in the previous paragraph we have $[k(\alpha_1, \alpha_2) : k(\alpha_1)]$ distinct imbeddings of $k(\alpha_1, \alpha_2)$ into $\overline{k}$ for a given imbedding of $k(\alpha_1)$. Then use induction. ///

23.4.3 Example: Let k be a field of characteristic $p > 0$. Suppose that k is **perfect**, meaning that for any $a \in k$ there exists $b \in k$ such that $b^p = a$. Let $f(x) = \sum_i c_i x^i$ in $k[x]$ be a polynomial such that its (algebraic) derivative

$$f'(x) = \sum_i c_i\, i\, x^{i-1}$$

is the zero polynomial. Show that there is a unique polynomial $g \in k[x]$ such that $f(x) = g(x)^p$.

For the derivative to be the 0 polynomial it must be that the characteristic p divides the exponent of every term (with non-zero coefficient). That is, we can rewrite

$$f(x) = \sum_i c_{ip}\, x^{ip}$$

Let $b_i \in k$ such that $b_i^p = c_{ip}$, using the perfectness. Since p divides all the inner binomial coefficients $p!/i!(p-i)!$,

$$\left(\sum_i b_i\, x^i\right)^p = \sum_i c_{ip}\, x^{ip}$$

as desired. ///

23.4.4 Example: Let k be a perfect field of characteristic $p > 0$, and f an irreducible polynomial in $k[x]$. Show that f has no repeated factors (even over an algebraic closure of k).

---

[9] Recall the proof: Let β be a root of $f(x) = 0$ in $\overline{k}$. Let $\varphi : k[x] \longrightarrow k[\beta]$ by $x \longrightarrow \beta$. The kernel of φ is the principal ideal generated by $f(x)$ in $k[x]$. Thus, the map φ factors through $k[x]/\langle f \rangle \approx k[\alpha_1]$.

If f has a factor P^2 where P is irreducible in $k[x]$, then P divides $\gcd(f, f') \in k[x]$. If $\deg \gcd(f, f') < \deg f$ then the irreducibility of f in $k[x]$ implies that the *gcd* is 1, so no such P exists. If $\deg \gcd(f, f') = \deg f$, then $f' = 0$, and (from above) there is a polynomial $g(x) \in k[x]$ such that $f(x) = g(x)^p$, contradicting the irreducibility in $k[x]$.   ///

23.4.5 Example: Show that all finite fields $\mathbb{F}_{p^n}$ with p prime and $1 \leq n \in \mathbb{Z}$ are perfect.

Again because the inner binomial coefficients $p!/i!(p - i)!$ are 0 in characteristic p, the (Frobenius) map $\alpha \longrightarrow \alpha^p$ is not only (obviously) multiplicative, but also additive, so is a ring homomorphism of $\mathbb{F}_{p^n}$ to itself. Since $\mathbb{F}_{p^n}^\times$ is cyclic (of order p^n), for any $\alpha \in \mathbb{F}_{p^n}$ (including 0)

$$\alpha^{(p^n)} = \alpha$$

Thus, since the map $\alpha \longrightarrow \alpha^p$ has the (two-sided) inverse $\alpha \longrightarrow \alpha^{p^{n-1}}$, it is a bijection. That is, everything has a pth root.   ///

23.4.6 Example: Let K be a finite extension of a finite field k. Prove that K is separable over k.

That is, we want to prove that the number of distinct imbeddings σ of K into a fixed algebraic closure $\overline{k}$ is $[K : k]$. Let $\alpha \in K$ be a generator for the cyclic group $K^\times$. Then $K = k(\alpha) = k[\alpha]$, since powers of α already give every element but 0 in K. Thus, from basic field theory, the degree of the minimal polynomial $f(x)$ of α over k is $[K : k]$. The previous example shows that k is perfect, and the example before that showed that irreducible polynomials over a perfect field have no repeated factors. Thus, $f(x)$ has no repeated factors in any field extension of k.

We have also already seen that for algebraic α over k, we can map $k(\alpha)$ to $\overline{k}$ to send α to *any* root β of $f(x) = 0$ in $\overline{k}$. Since $f(x)$ has not repeated factors, there are $[K : k]$ distinct roots β, so $[K : k]$ distinct imbeddings.   ////

23.4.7 Example: Find all fields intermediate between $\mathbb{Q}$ and $\mathbb{Q}(\zeta)$ where ζ is a primitive 17th root of unity.

Since 17 is prime, $\mathrm{Gal}(\mathbb{Q}(\zeta)/\mathbb{Q}) \approx (\mathbb{Z}/17)^\times$ is cyclic (of order 16), and we know that a cyclic group has a unique subgroup of each order dividing the order of the whole. Thus, there are intermediate fields corresponding to the proper divisors $2, 4, 8$ of 16. Let σ_a be the automorphism $\sigma_a \zeta = \zeta^a$.

By a little trial and error, 3 is a generator for the cyclic group $(\mathbb{Z}/17)^\times$, so σ_3 is a generator for the automorphism group. Thus, one reasonably considers

$$\begin{aligned}
\alpha_8 &= \zeta + \zeta^{3^2} + \zeta^{3^4} + \zeta^{3^6} + \zeta^{3^8} + \zeta^{3^{10}} + \zeta^{3^{12}} + \zeta^{3^{14}} \\
\alpha_4 &= \zeta + \zeta^{3^4} + \zeta^{3^8} + \zeta^{3^{12}} \\
\alpha_2 &= \zeta + \zeta^{3^8} = \zeta + \zeta^{-1}
\end{aligned}$$

The α_n is visibly invariant under the subgroup of $(\mathbb{Z}/17)^\times$ of order n. The linear independence of $\zeta, \zeta^2, \zeta^3, \ldots, \zeta^{16}$ shows α_n is *not* by accident invariant under any larger subgroup of the Galois group. Thus, $\mathbb{Q}(\alpha_n)$ is (by Galois theory) the unique intermediate field of degree $16/n$ over $\mathbb{Q}$.

We can also give other characterizations of some of these intermediate fields. First, we have already seen (in discussion of Gauss sums) that

$$\sum_{a \bmod 17} \left(\frac{a}{17}\right)_2 \cdot \zeta^a = \sqrt{17}$$

where $\left(\frac{a}{17}\right)_2$ is the quadratic symbol. Thus,

$$\alpha_8 - \sigma_3\alpha_8 = \sqrt{17}$$
$$\alpha_8 + \sigma_3\alpha_8 = 0$$

so α_8 and $\sigma_3\alpha_8$ are $\pm\sqrt{17}/2$. Further computation can likewise express all the intermediate fields as being obtained by adjoining square roots to the next smaller one. ///

23.4.8 Example: Let f, g be *relatively prime* polynomials in n indeterminates $t_1, \ldots, t_n$, with g not 0. Suppose that the ratio $f(t_1, \ldots, t_n)/g(t_1, \ldots, t_n)$ is invariant under all permutations of the t_i. Show that both f and g are polynomials in the elementary symmetric functions in the t_i.

Let s_i be the ith elementary symmetric function in the t_j's. Earlier we showed that $k(t_1, \ldots, t_n)$ has Galois group S_n (the symmetric group on n letters) over $k(s_1, \ldots, s_n)$. Thus, the given ratio lies in $k(s_1, \ldots, s_n)$. Thus, it is *expressible* as a ratio

$$\frac{f(t_1, \ldots, t_n)}{g(t_1, \ldots, t_n)} = \frac{F(s_1, \ldots, s_n)}{G(s_1, \ldots, s_n)}$$

of polynomials F, G in the s_i.

To prove the stronger result that the original f and g were themselves literally polynomials in the t_i, we seem to need the characteristic of k to be not 2, and we certainly must use the unique factorization in $k[t_1, \ldots, t_n]$.

Write

$$f(t_1, \ldots, t_n) = p_1^{e_1} \ldots p_m^{e_m}$$

where the e_i are positive integers and the p_i are irreducibles. Similarly, write

$$g(t_1, \ldots, t_n) = q_1^{f_1} \ldots q_m^{f_n}$$

where the f_i are positive integers and the q_i are irreducibles. The relative primeness says that none of the q_i are *associate* to any of the p_i. The invariance gives, for any permutation π of

$$\pi\left(\frac{p_1^{e_1} \ldots p_m^{e_m}}{q_1^{f_1} \ldots q_m^{f_n}}\right) = \frac{p_1^{e_1} \ldots p_m^{e_m}}{q_1^{f_1} \ldots q_m^{f_n}}$$

Multiplying out,

$$\prod_i \pi(p_i^{e_i}) \cdot \prod_i q_i^{f_i} = \prod_i p_i^{e_i} \cdot \prod_i \pi(q_i^{f_i})$$

By the relative prime-ness, each p_i divides some one of the $\pi(p_j)$. These ring automorphisms preserve irreducibility, and $\gcd(a, b) = 1$ implies $\gcd(\pi a, \pi b) = 1$, so, symmetrically, the $\pi(p_j)$'s divide the p_i's. And similarly for the q_i's. That is, permuting the t_i's must permute the irreducible factors of f (up to units $k^\times$ in $k[t_1, \ldots, t_n]$) among themselves, and likewise for the irreducible factors of g.

If all permutations *literally* permuted the irreducible factors of f (and of g), rather than merely up to *units*, then f and g would be symmetric. However, at this point we can only be confident that they are permuted *up to constants*.

What we have, then, is that for a permutation π

$$\pi(f) = \alpha_\pi \cdot f$$

for some $\alpha \in k^\times$. For another permutation τ, certainly $\tau(\pi(f)) = (\tau\pi)f$. And $\tau(\alpha_\pi f) = \alpha_\pi \cdot \tau(f)$, since permutations of the indeterminates have no effect on elements of k. Thus, we have

$$\alpha_{\tau\pi} = \alpha_\tau \cdot \alpha_\pi$$

That is, $\pi \longrightarrow \alpha_\pi$ is a group homomorphism $S_n \longrightarrow k^\times$.

It is very useful to know that the alternating group A_n is the *commutator subgroup* of S_n. Thus, if f is not actually invariant under S_n, in any case the group homomorphism $S_n \longrightarrow k^\times$ factors through the quotient S_n/A_n, so is the *sign function* $\pi \longrightarrow \sigma(\pi)$ that is $+1$ for $\pi \in A_n$ and -1 otherwise. That is, f is **equivariant** under S_n by the sign function, in the sense that $\pi f = \sigma(\pi) \cdot f$.

Now we claim that if $\pi f = \sigma(\pi) \cdot f$ then the square root

$$\delta = \sqrt{\Delta} = \prod_{i<j} (t_i - t_j)$$

of the discriminant Δ divides f. To see this, let s_{ij} be the 2-cycle which interchanges t_i and t_j, for $i \neq j$. Then

$$s_{ij} f = -f$$

Under any homomorphism which sends $t_i - t_j$ to 0, since the characteristic is not 2, f is sent to 0. That is, $t_i - t_j$ divides f in $k[t_1, \ldots, t_n]$. By unique factorization, since no two of the monomials $t_i - t_j$ are associate (for distinct pairs $i < j$), we see that the square root δ of the discriminant must divide f.

That is, for f with $\pi f = \sigma(\pi) \cdot f$ we know that $\delta | f$. For f/g to be invariant under S_n, it must be that also $\pi g = \sigma(\pi) \cdot g$. But then $\delta | g$ also, contradicting the assumed relative primeness. Thus, in fact, it must have been that both f and g were *invariant* under S_n, not merely equivariant by the sign function. ///

Exercises

23.1 Let k be a field. Let $\alpha_1, \ldots, \alpha_n$ be distinct elements of $k^\times$. Suppose that $c_1, \ldots, c_n$ in k are such that for all positive integers ℓ

$$\sum_i c_i \alpha_i^\ell = 0$$

Show that all the c_i are 0.

23.2 Solve the cubic $x^3 + ax + b = 0$ in terms of radicals.

23.3 Express a primitive 11th root of unity in terms of radicals.

23.4 Solve $x^4 + ax + b = 0$ in terms of radicals.

24. Eigenvectors, spectral theorems

24.1 *Eigenvectors, eigenvalues*

Let k be a field, not necessarily algebraically closed.

Let T be a k-linear endomorphism of a k-vectorspace V to itself, meaning, as usual, that

$$T(v + w) = Tv + TW \quad \text{and} \quad T(cv) = c \cdot Tv$$

for $v, w \in V$ and $c \in k$. The collection of all such T is denoted $\mathrm{End}_k(V)$, and is a vector space over k with the natural operations

$$(S + T)(v) = Sv + Tv \qquad (cT)(v) = c \cdot Tv$$

A vector $v \in V$ is an **eigenvector** for T with **eigenvalue** $c \in k$ if

$$T(v) = c \cdot v$$

or, equivalently, if
$$(T - c \cdot \mathrm{id}_V)\, v = 0$$

A vector v is a **generalized eigenvector** of T with **eigenvalue** $c \in k$ if, for some integer $\ell \geq 1$
$$(T - c \cdot \mathrm{id}_V)^\ell\, v = 0$$

We will often suppress the id_V notation for the identity map on V, and just write c for the scalar operator $c \cdot \mathrm{id}_V$. The collection of all λ-eigenvectors for T is the λ-**eigenspace** for T on V, and the collection of all generalized λ-eigenvectors for T is the **generalized** λ-**eigenspace** for T on V.

24.1.1 Proposition: Let $T \in \mathrm{End}_k(V)$. For fixed $\lambda \in k$ the λ-eigenspace is a vector subspace of V. The generalized λ-eigenspace is also a vector subspace of V. And both the λ-eigenspace and the generalized one are *stable* under the action of T.

Proof: This is just the linearity of T, hence, of $T - \lambda$. Indeed, for v, w λ-eigenvectors, and for $c \in k$,
$$T(v + w) = Tv + TW = \lambda v + \lambda w = \lambda(v + w) \quad \text{and} \quad T(cv) = c \cdot Tv = c \cdot \lambda v = lam \cdot cv$$

If $(T - \lambda)^m v = 0$ and $(T - \lambda)nw = 0$, let $N = \max(m, n)$. Then
$$(T - \lambda)^N(v + w) = (T - \lambda)^N v + (T - \lambda)^N w = (T - \lambda)^{N-m}(T - \lambda)^m v + (T - \lambda)^{N-n}(T - \lambda)^n w$$
$$= (T - \lambda)^{N-m} 0 + (T - \lambda)^{N-n} 0 = 0$$

Similarly, generalized eigenspaces are stable under scalar multiplication.

Since the operator T commutes with any polynomial in T, we can compute, for $(T - \lambda)^n v = 0$,
$$(T - \lambda)^n (Tv) = T \cdot (T - \lambda)^n (v) = T(0) = 0$$

which proves the stability. ///

24.1.2 Proposition: Let $T \in \mathrm{End}_k(V)$ and let $v_1, \ldots, v_m$ be eigenvectors for T, with *distinct* respective eigenvalues $\lambda_1, \ldots, \lambda_m$ in k. Then for scalars c_i
$$c_1 v_1 + \ldots + c_m v_m = 0 \quad \Longrightarrow \quad \text{all } c_i = 0$$

That is, eigenvectors for distinct eigenvalues are linearly independent.

Proof: Suppose that the given relation is the shortest such with all $c_i \neq 0$. Then apply $T - \lambda_1$ to the relation, to obtain
$$0 + (\lambda_2 - \lambda_1)c_2 v_2 \ldots + (\lambda_m - \lambda_1)c_m v_m = 0 \quad \Longrightarrow \quad \text{all } c_i = 0$$

For $i > 1$ the scalars $\lambda_i - \lambda_1$ are not 0, and $(\lambda_i - \lambda_1)v_i$ is again a non-zero λ_i-eigenvector for T. This contradicts the assumption that the relation was the shortest. ///

So far no use was made of finite-dimensionality, and, indeed, all the above arguments are correct without assuming finite-dimensionality. Now, however, we need to assume finite-dimensionality. In particular,

24.1.3 Proposition: Let V be a finite-dimensional vector space over k. Then
$$\dim_k \mathrm{End}_k(V) = (\dim_k V)^2$$

In particular, $\mathrm{End}_k(V)$ is finite-dimensional.

Proof: An endomorphism T is completely determined by where it sends all the elements of a basis $v_1, \ldots, v_n$ of V, and each v_i can be sent to any vector in V. In particular, let E_{ij} be the endomorphism sending v_i to v_j and sending v_ℓ to 0 for $\ell \neq i$. We claim that these endomorphisms are a k-basis for $\mathrm{End}_k(V)$. First, they span, since any endomorphism T is expressible as

$$T = \sum_{ij} c_{ij} E_{ij}$$

where the $c_{ij} \in k$ are determined by the images of the given basis

$$T(v_i) = \sum_j c_{ij} v_j$$

On the other hand, suppose for some coefficients c_{ij}

$$\sum_{ij} c_{ij} E_{ij} = 0 \in \mathrm{End}_k(V)$$

Applying this endomorphism to v_i gives

$$\sum_j c_{ij} v_j = 0 \in V$$

Since the v_j are linearly independent, this implies that all c_{ij} are 0. Thus, the E_{ij} are a basis for the space of endomorphisms, and we have the dimension count.   ///

For V finite-dimensional, the homomorphism

$$k[x] \longrightarrow k[T] \subset \mathrm{End}_k(V) \qquad \text{by} \quad x \longrightarrow T$$

from the polynomial ring $k[x]$ to the ring $k[T]$ of polynomials in T must have a non-trivial kernel, since $k[x]$ is infinite-dimensional and $k[T]$ is finite-dimensional. The **minimal polynomial** $f(x) \in k[x]$ of T is the (unique) monic generator of that kernel.

24.1.4 Proposition: The eigenvalues of a k-linear endomorphism T are exactly the zeros of its minimal polynomial. [1]

Proof: Let $f(x)$ be the minimal polynomial. First, suppose that $x - \lambda$ divides $f(x)$ for some $\lambda \in k$, and put $g(x) = f(x)/(x - \lambda)$. Since $g(x)$ is not divisible by the minimal polynomial, there is $v \in V$ such that $g(T)v \neq 0$. Then

$$(T - \lambda) \cdot g(T)v = f(T) \cdot v = 0$$

so $g(T)v$ is a (non-zero) λ-eigenvector of T. On the other hand, suppose that λ is an eigenvalue, and let v be a non-zero λ-eigenvector for T. If $x - \lambda$ failed to divide $f(x)$, then the *gcd* of $x - \lambda$ and $f(x)$ is 1, and there are polynomials $a(x)$ and $b(x)$ such that

$$1 = a \cdot (x - \lambda) + b \cdot f$$

Mapping $x \longrightarrow T$ gives

$$\mathrm{id}_V = a(T)(T - \lambda) + 0$$

[1] This does not presume that k is algebraically closed.

Applying this to v gives

$$v = a(T)(T - \lambda)(v) = a(T) \cdot 0 = 0$$

which contradicts $v \neq 0$.   ///

24.1.5 Corollary: Let k be algebraically closed, and V a finite-dimensional vector space over k. Then there is at least one eigenvalue and (non-zero) eigenvector for any $T \in \mathrm{End}_k(V)$.

Proof: The minimal polynomial has at least one linear factor over an algebraically closed field, so by the previous proposition has at least one eigenvector.   ///

24.1.6 Remark: The Cayley-Hamilton theorem [2] is often invoked to deduce the existence of at least one eigenvector, but the last corollary shows that this is not necessary.

24.2 *Diagonalizability, semi-simplicity*

A linear operator $T \in \mathrm{End}_k(V)$ on a finite-dimensional vector space V over a field k is **diagonalizable** [3] if V has a basis consisting of eigenvectors of T. Equivalently, T may be said to be **semi-simple**, or sometimes V itself, as a $k[T]$ or $k[x]$ module, is said to be semi-simple.

Diagonalizable operators are good, because their effect on arbitrary vectors can be very clearly described as a superposition of scalar multiplications in an obvious manner, namely, letting $v_1, \ldots, v_n$ be eigenvectors with eigenvalues $\lambda_1, \ldots, \lambda_n$, if we manage to express a given vector v as a linear combination [4]

$$v = c_1 v_1 + \ldots + c_n v_n$$

of the eigenvectors v_i, with $c_i \in k$, then we can completely describe the effect of T, or even iterates T^ℓ, on v, by

$$T^\ell v = \lambda_1^\ell \cdot c_1 v_1 + \ldots + \lambda_n^\ell \cdot c_n v_n$$

24.2.1 Remark: Even over an algebraically closed field k, an endomorphism T of a finite-dimensional vector space may fail to be diagonalizable by having non-trivial Jordan blocks, meaning that some one of its *elementary divisors* has a *repeated factor*. When k is

---

[2] The Cayley-Hamilton theorem, which we will prove later, asserts that the minimal polynomial of an endomorphism T divides the **characteristic polynomial** $\det(T - x \cdot \mathrm{id}_V)$ of T, where det is *determinant*. But this invocation is unnecessary and misleading. Further, it is easy to give false proofs of this result. Indeed, it seems that Cayley and Hamilton only proved the two-dimensional and perhaps three-dimensional cases.

[3] Of course, in coordinates, diagonalizability means that a matrix M giving the endomorphism T can be literally diagonalized by conjugating it by some invertible A, giving diagonal AMA^{-1}. This conjugation amounts to changing coordinates.

[4] The computational problem of expressing a given vector as a linear combination of eigenvectors is not trivial, but is reasonably addressed via *Gaussian elimination*.

not necessarily algebraically closed, T may fail to be diagonalizable by having one (hence, at least two) of the zeros of its minimal polynomial lie in a proper field extension of k. For not finite-dimensional V, there are further ways that an endomorphism may fail to be diagonalizable. For example, on the space V of two-sided sequences $a = (\ldots, a_{-1}, a_0, a_1, \ldots)$ with entries in k, the operator T given by

$$i^{th} \text{ component } (Ta)_i \text{ of } Ta = (i-1)^{th} \text{ component } a_i \text{ of } a$$

24.2.2 Proposition: An operator $T \in \text{End}_k(V)$ with V finite-dimensional over the field k is diagonalizable if and only if the minimal polynomial $f(x)$ of T factors into linear factors in $k[x]$ and has no repeated factors. Further, letting V_λ be the λ-eigenspace, diagonalizability is equivalent to

$$V = \sum_{\text{eigenvalues } \lambda} V_\lambda$$

Proof: Suppose that f factors into linear factors

$$f(x) = (x - \lambda_1)(x - \lambda_2) \ldots (x - \lambda_n)$$

in $k[x]$ and no factor is repeated. We already saw, above, that the zeros of the minimal polynomial are exactly the eigenvalues, whether or not the polynomial factors into linear factors. What remains is to show that there is a basis of eigenvectors if $f(x)$ factors completely into linear factors, and conversely.

First, suppose that there is a basis $v_1, \ldots, v_n$ of eigenvectors, with eigenvalues $\lambda_1, \ldots, \lambda_n$. Let Λ be the *set*[5] of eigenvalues, specifically *not* attempting to count repeated eigenvalues more than once. Again, we already know that all these eigenvalues do occur among the zeros of the minimal polynomial (*not* counting multiplicities!), and that all zeros of the minimal polynomial are eigenvalues. Let

$$g(x) = \prod_{\lambda \in \Lambda} (x - \lambda)$$

Since every eigenvalue is a zero of $f(x)$, $g(x)$ divides $f(x)$. And $g(T)$ annihilates every eigenvector, and since the eigenvectors span V the endomorphism $g(T)$ is 0. Thus, by definition of the minimal polynomial, $f(x)$ divides $g(x)$. They are both monic, so are equal.

Conversely, suppose that the minimal polynomial $f(x)$ factors as

$$f(x) = (x - \lambda_1) \ldots (x - \lambda_n)$$

with *distinct* λ_i. Again, we have already shown that each λ_i is an eigenvalue. Let V_λ be the λ-eigenspace. Let $\{v_{\lambda,1}, \ldots, v_{lam,d_\lambda}\}$ be a basis for V_λ. We claim that the union

$$\{v_{\lambda,i} : \lambda \text{ an eigenvalue }, 1 \le i \le d_\lambda\}$$

---

[5] Strictly speaking, a set cannot possibly keep track of repeat occurrences, since $\{a, a, b\} = \{a, b\}$, and so on. However, in practice, the notion of set often is corrupted to mean to keep track of repeats. More correctly, a notion of *set* enhanced to keep track of number of repeats is a *multi-set*. Precisely, a **mult-set** M is a set S with a non-negative integer-valued function m on S, where the intent is that $m(s)$ (for $s \in S$) is the number of times s occurs in M, and is called the **multiplicity** of s in M. The question of whether or not the multiplicity can be 0 is a matter of convention and/or taste.

of bases for all the (non-trivial) eigenspaces V_λ is a basis for V. We have seen that eigenvectors for *distinct* eigenvalues are linearly independent, so we need only prove

$$\sum_\lambda V_\lambda = V$$

where the sum is over (distinct) eigenvalues. Let $f_\lambda(x) = f(x)/(x - \lambda)$. Since each linear factor occurs only once in f, the *gcd* of the collection of $f_\lambda(x)$ in $k[x]$ is 1. Therefore, there are polynomials $a_\lambda(x)$ such that

$$1 = \gcd(\{f_\lambda : \lambda \ \text{an eigenvector}\}) = \sum_\lambda a_\lambda(x) \cdot f_\lambda(x)$$

Then for any $v \in V$

$$v = \mathrm{id}_V(v) = \sum_\lambda a_\lambda(T) \cdot f_\lambda(T)(v)$$

Since

$$(T - \lambda) \cdot f_\lambda(T) = f(T) = 0 \in \mathrm{End}_k(V)$$

for each eigenvalue λ

$$f_\lambda(T)(V) \subset V_\lambda$$

Thus, in the expression

$$v = \mathrm{id}_V(v) = \sum_\lambda a_\lambda(T) \cdot f_\lambda(T)(v)$$

each $f_\lambda(T)(v)$ is in V_λ. Further, since T and any polynomial in T stabilizes each eigenspace, $a_\lambda(T)f_\lambda(T)(v)$ is in V_λ. Thus, this sum exhibits an arbitrary v as a sum of elements of the eigenspaces, so these eigenspaces do span the whole space.

Finally, suppose that

$$V = \sum_{\text{eigenvalues } \lambda} V_\lambda$$

Then $\prod_\lambda(T - \lambda)$ (product over distinct λ) annihilates the whole space V, so the minimal polynomial of T factors into distinct linear factors.   ///

An endomorphism P is a **projector** or **projection** if it is *idempotent*, that is, if

$$P^2 = P$$

The **complementary** or **dual** idempotent is

$$1 - P = \mathrm{id}_V - P$$

Note that

$$(1 - P)P = P(1 - P) = P - P^2 = 0 \in \mathrm{End}_l(V)$$

Two idempotents P, Q are **orthogonal** if

$$PQ = QP = 0 \in \mathrm{End}_k(V)$$

If we have in mind an endomorphism T, we will usually care only about projectors P *commuting* with T, that is, with $PT = TP$.

24.2.3 Proposition: Let T be a k-linear operator on a finite-dimensional k-vectorspace V. Let λ be an eigenvalue of T, with eigenspace V_λ, and suppose that the factor $x-\lambda$ occurs with multiplicity one in the minimal polynomial $f(x)$ of T. Then there is a polynomial $a(x)$ such that $a(T)$ is a *projector commuting with T*, and is the identity map on the λ-eigenspace.

Proof: Let $g(x) = f(x)/(x - \lambda)$. The multiplicity assumption assures us that $x - \lambda$ and $g(x)$ are relatively prime, so there are $a(x)$ and $b(x)$ such that

$$1 = a(x)g(x) + b(x)(x - \lambda)$$

or

$$1 - b(x)(x - \lambda) = a(x)g(x)$$

As in the previous proof, $(x - \lambda)g(x) = f(x)$, so $(T - \lambda)g(T) = 0$, and $g(T)(V) \subset V_\lambda$. And, further, because T and polynomials in T stabilize eigenspaces, $a(T)g(T)(V) \subset V_\lambda$. And

$$[a(T)g(T)]^2 = a(T)g(T) \cdot [1 - b(T)(T - \lambda)] = a(T)g(T) - 0 = a(T)g(T)$$

since $g(T)(T - \lambda) = f(T) = 0$. That is,

$$P = a(T)g(T).$$

is the desired projector to the λ-eigenspace. ///

24.2.4 Remark: The condition that the projector commute with T is non-trivial, and without it there are many projectors that will not be what we want.

24.3 *Commuting endomorphisms $ST = TS$*

Two endomorphisms $S, T \in \text{End}_k(V)$ are said to **commute** (with each other) if

$$ST = TS$$

This hypothesis allows us to reach some worthwhile conclusions about eigenvectors of the two separately, and jointly. Operators which do not commute are much more complicated to consider from the viewpoint of eigenvectors. [6]

24.3.1 Proposition: Let S, T be commuting endomorphisms of V. Then S stabilizes every eigenspace of T.

Proof: Let v be a λ-eigenvector of T. Then

$$T(Sv) = (TS)v = (ST)v = S(Tv) = S(\lambda v) = \lambda \cdot Sv$$

as desired. ///

---

[6] Indeed, to study non-commutative collections of operators the notion of *eigenvector* becomes much less relevant. Instead, a more complicated (and/but more interesting) notion of *irreducible subspace* is the proper generalization.

24.3.2 Proposition: Commuting *diagonalizable* endomorphisms S and T on V are *simultaneously diagonalizable*, in the sense that there is a basis consisting of vectors which are simultaneously eigenvectors for *both* S and T.

Proof: Since T is diagonalizable, from above V decomposes as

$$V = \sum_{\text{eigenvalues } \lambda} V_\lambda$$

where V_λ is the λ-eigenspace of T on V. From the previous proposition, S stabilizes each V_λ.

Let's (re) prove that for S diagonalizable on a vector space V, that S is diagonalizable on any S-stable subspace W. Let $g(x)$ be the minimal polynomial of S on V. Since W is S-stable, it makes sense to speak of the minimal polynomial $h(x)$ of S on W. Since $g(S)$ annihilates V, it certainly annihilates W. Thus, $g(x)$ is a polynomial multiple of $h(x)$, since the latter is the unique monic generator for the ideal of polynomials $P(x)$ such that $P(S)(W) = 0$. We proved in the previous section that the diagonalizability of S on V implies that $g(x)$ factors into linear factors in $k[x]$ and no factor is repeated. Since $h(x)$ divides $g(x)$, the same is true of $h(x)$. We saw in the last section that this implies that S on W is diagonalizable.

In particular, V_λ has a basis of eigenvectors for S. These are all λ-eigenvectors for T, so are indeed *simultaneous* eigenvectors for the two endomorphisms.   ///

24.4 *Inner product spaces*

Now take the field k to be either $\mathbb{R}$ or $\mathbb{C}$. We use the **positivity property** of $\mathbb{R}$ that for $r_1, \ldots, r_n \in \mathbb{R}$

$$r_1^2 + \ldots + r_n^2 = 0 \quad \Longrightarrow \quad \text{all } r_i = 0$$

The **norm-squared** of a complex number $\alpha = a + bi$ (with $a, b \in \mathbb{R}$) is

$$|\alpha|^2 = \alpha \cdot \overline{\alpha} = a^2 + b^2$$

where $\overline{a + bi} = a - bi$ is the usual **complex conjugative**. The positivity property in $\mathbb{R}$ thus implies an analogous one for $\alpha_1, \ldots, \alpha_n$, namely

$$|\alpha_1|^2 + \ldots + |\alpha_n|^2 = 0 \quad \Longrightarrow \quad \text{all } \alpha_i = 0$$

24.4.1 Remark: In the following, for scalars $k = \mathbb{C}$ we will need to refer to the complex conjugation on it. But when k is $\mathbb{R}$ the conjugation is trivial. To include both cases at once we will systematically refer to *conjugation*, with the reasonable convention that for $k = \mathbb{R}$ this is the do-nothing operation.

Given a k-vectorspace V, an **inner product** or **scalar product** or **dot product** or **hermitian product** (the latter especially if the set k of scalars is $\mathbb{C}$) is a k-valued function

$$\langle,\rangle : V \times V \longrightarrow k$$

written

$$v \times w \longrightarrow \langle v, w \rangle$$

which meets several conditions. First, a mild condition that $\langle , \rangle$ be k-linear in the first argument and k-conjugate-linear in the second, meaning that $\langle , \rangle$ is *additive* in both arguments:

$$\langle v + v', w + w' \rangle = \langle v, w \rangle + \langle v', w \rangle + \langle v, w' \rangle + \langle v', w' \rangle$$

and scalars behave as

$$\langle \alpha v, \beta w \rangle = \alpha \overline{\beta} \langle v, w \rangle$$

The inner product is **hermitian** in the sense that

$$\langle v, w \rangle = \overline{\langle w, v \rangle}$$

Thus, for ground field k either $\mathbb{R}$ or $\mathbb{C}$,

$$\langle v, v \rangle = \overline{\langle v, v \rangle}$$

so $\langle v, v \rangle \in \mathbb{R}$.

The most serious condition on $\langle , \rangle$ is **positive-definiteness**, which is that

$$\langle v, v \rangle \geq 0 \quad \text{with equality only for } v = 0$$

Two vectors v, w are **orthogonal** or **perpendicular** if

$$\langle v, w \rangle = 0$$

We may write $v \perp w$ for the latter condition. There is an associated **norm**

$$|v| = \langle v, v \rangle^{1/2}$$

and **metric**

$$d(v, w) = |v - w|$$

A vector space basis $e_1, e_2, \ldots, e_n$ of V is an **orthonormal basis** for V if

$$\langle e_i, e_j \rangle = \begin{cases} 1 & (\text{for } i = j) \\ 1 & (\text{for } i = j) \\ 0 & (\text{for } i \neq j) \end{cases}$$

24.4.2 Proposition: *(Gram-Schmidt process)* Given a basis $v_1, v_2, \ldots, v_n$ of a finite-dimensional inner product space V, let

$$
\begin{aligned}
& & & & e_1 &= \frac{v_1}{|v_1|} \\
v_2' &= & v_2 - \langle v_2, e_1 \rangle e_1 & \quad \text{and} \quad & e_2 &= \frac{v_2'}{|v_2'|} \\
v_3' &= & v_3 - \langle v_3, e_1 \rangle e_1 - \langle v_3, e_2 \rangle e_2 & \quad \text{and} \quad & e_3 &= \frac{v_3'}{|v_3'|} \\
& & \ldots & & & \\
v_i' &= & v_i - \sum_{j<i} \langle v_i, e_j \rangle e_j & \quad \text{and} \quad & e_i &= \frac{v_i'}{|v_i'|} \\
& & \ldots & & &
\end{aligned}
$$

Then $e_1, \ldots, e_n$ is an orthonormal basis for V.

24.4.3 Remark: One could also give a more existential proof that orthonormal bases exist, but the conversion of arbitrary basis to an orthonormal one is of additional interest.

Proof: Use induction. Note that for any vector e of length 1

$$\langle v - \langle v, e \rangle e, e \rangle = \langle v, e \rangle - \langle v, e \rangle \langle e, e \rangle = \langle v, e \rangle - \langle v, e \rangle \cdot 1 = 0$$

Thus, for $\ell < i$,

$$\langle v_i', e_\ell \rangle = \langle v_i - \sum_{j<i} \langle v_i, e_j \rangle e_j, e_\ell \rangle = \langle v_i, e_\ell \rangle - \langle \langle v_i, e_\ell \rangle e_\ell, e_\ell \rangle - \sum_{j<i,\, j \neq \ell} \langle \langle v_i, e_j \rangle e_j, e_\ell \rangle$$

$$= \langle v_i, e_\ell \rangle - \langle v_i, e_\ell \rangle \langle e_\ell, e_\ell \rangle - \sum_{j<i,\, j \neq \ell} \langle v_i, e_j \rangle \langle e_j, e_\ell \rangle = \langle v_i, e_\ell \rangle - \langle v_i, e_\ell \rangle - 0 = 0$$

since the e_j's are (by induction) mutually orthogonal and have length 1. One reasonable worry is that v_i' is 0. But by induction $e_1, e_2, \ldots, e_{i-1}$ is a basis for the subspace of V for which $v_1, \ldots, v_{i-1}$ is a basis. Thus, since v_i is linearly independent of $v_1, \ldots, v_{i-1}$ it is also independent of $e_1, \ldots, e_{i-1}$, so the expression

$$v_i' = v_i + (\text{linear combination of } e_1, \ldots, e_{i-1})$$

cannot give 0. Further, that expression gives the induction step proving that the span of $e_1, \ldots, e_i$ is the same as that of $v_1, \ldots, v_i$.   ///

Let W be a subspace of a finite-dimensional k-vectorspace (k is $\mathbb{R}$ or $\mathbb{C}$) with a (positive-definite) inner product $\langle, \rangle$. The **orthogonal complement** $W^\top$ is

$$W^\top = \{v \in V : \langle v, w \rangle = 0 \text{ for all } w \in W\}$$

It is easy to check that the orthogonal complement is a vector subspace.

24.4.4 Theorem: In finite-dimensional vector spaces V, for subspaces W [7]

$$W^{\perp\perp} = W$$

In particular, for any W

$$\dim_k W + \dim_k W^\perp = \dim_k V$$

Indeed,

$$V = W \oplus W^\perp$$

There is a unique projector P which is an **orthogonal projector** to W in the sense that on P is the identity on W and is 0 on $W^\perp$.

Proof: First, we verify some relatively easy parts. For $v \in W \cap W^\perp$ we have $0 = \langle v, v \rangle$, so $v = 0$ by the positive-definiteness. Next, for $w \in W$ and $v \in W^\perp$,

$$0 = \langle w, v \rangle = \overline{\langle v, w \rangle} = \overline{0}$$

---

[7] In infinite-dimensional inner-product spaces, the orthogonal complement of the orthogonal complement is the topological *closure* of the original subspace.

which proves this inclusion $W \subset W^{\perp\perp}$.

Next, suppose that for a given $v \in V$ there were two expressions

$$v = w + w' = u + u'$$

with $w, u \in W$ and $w', u' \in W^\perp$. Then

$$W \ni w - u = u' - w' \in W^\perp$$

Since $W \cap W^\perp = 0$, it must be that $w = u$ and $w' = u'$, which gives the uniqueness of such an expression (assuming existence).

Let $e_1, \ldots, e_m$ be an orthogonal basis for W. Given $v \in V$, let

$$x = \sum_{1 \le i \le m} \langle v, e_i \rangle \, e_i$$

and

$$y = v - x$$

Since it is a linear combination of the e_i, certainly $x \in W$. By design, $y \in W^\perp$, since for any e_ℓ

$$\langle y, w \rangle = \langle v - \sum_{1 \le i \le m} \langle v, e_i \rangle \, e_i, \ e_\ell \rangle = \langle v, e_\ell \rangle - \sum_{1 \le i \le m} \langle v, e_i \rangle \, \langle e_i, e_\ell \rangle = \langle v, e_\ell \rangle - \langle v, e_\ell \rangle$$

since the e_i are an orthonormal basis for W. This expresses

$$v = x + y$$

as a linear combination of elements of W and $W^\perp$.

Since the map $v \longrightarrow x$ is expressible in terms of the inner product, as just above, this is the desired projector to W. By the uniqueness of the decomposition into W and $W^\perp$ components, the projector is orthogonal, as desired. ///

24.4.5 Corollary: [8] Suppose that a finite-dimensional vector space V has an inner product $\langle, \rangle$. To every k-linear map $L : V \longrightarrow k$ is attached a unique $w \in V$ such that for all $v \in V$

$$Lv = \langle v, w \rangle$$

24.4.6 Remark: The k-linear maps of a k-vectorspace V to k itself are called **linear functionals** on V.

Proof: If L is the 0 map, just take $w = 0$. Otherwise, since

$$\dim_k \ker L = \dim_k V - \dim_k \operatorname{Im} L = \dim_k V - \dim_k k = \dim_k V - 1$$

---

[8] This is a very simple case of the Riesz-Fischer theorem, which asserts the analogue for *Hilbert* spaces, which are the proper infinite-dimensional version of inner-product spaces. In particular, Hilbert spaces are required, in addition to the properties mentioned here, to be *complete* with respect to the metric $d(x, y) = |x - y|$ coming from the inner product. This completeness is automatic for finite-dimensional inner product spaces.

Take a vector e of length 1 in the orthogonal complement[9] $(\ker L)^\perp$. For arbitrary $v \in V$

$$v - \langle v, e\rangle e \in \ker L$$

Thus,

$$L(v) = L(v - \langle v, e\rangle\, e) + L(\langle v, e\rangle\, e) = 0 + \langle v, e\rangle L(e) = \langle v, \overline{L(e)}e\rangle$$

That is, $w = \overline{L(e)}e$ is the desired element of V.   ///

The **adjoint** T^* of $T \in \mathrm{End}_k(V)$ with respect to an inner product $\langle,\rangle$ is another linear operator in $\mathrm{End}_k(V)$ such that, for all $v, w \in V$,

$$\langle Tv, w\rangle = \langle v, T^* w\rangle$$

24.4.7 Proposition: Adjoint operators (on finite-dimensional inner product spaces) exist and are unique.

Proof: Let T be a linear endomorphism of V. Given $x \in V$, the map $v \longrightarrow \langle Tv, x\rangle$ is a linear map to k. Thus, by the previous corollary, there is a unique $y \in V$ such that for all $v \in V$

$$\langle Tv, x\rangle = \langle v, y\rangle$$

We want to define $T^* x = y$. This is well-defined as a function, but we need to prove linearity, which, happily, is not difficult. Indeed, let $x, x' \in V$ and let y, y' be attached to them as just above. Then

$$\langle Tv, x + x'\rangle = \langle Tv, x\rangle + \langle Tv, x\rangle = \langle v, y\rangle + \langle v, y'\rangle = \langle v, y + y'\rangle$$

proving the additivity $T^*(x + x') = T * x + T^* x'$. Similarly, for $c \in k$,

$$\langle Tv, cx\rangle = \bar{c}\langle Tv, x\rangle = \bar{c}\langle v, y\rangle = \langle v, cy\rangle$$

proving the linearity of T^*.   ///

Note that the direct computation

$$\langle T * v, w\rangle = \overline{\langle w, T * v\rangle} = \overline{\langle Tw, v\rangle} = \langle v, Tw, v\rangle$$

shows that, unsurprisingly,

$$(T^*)^* = T$$

A linear operator T on an inner product space V is **normal**[10]   if it commutes with its adjoint, that is, if

$$TT^* = T^* T$$

An operator T is **self-adjoint** or **hermitian** if it is equal to its adjoint, that is, if

$$T = T^*$$

---

[9] Knowing that the orthogonal complement exists is a crucial point, and that fact contains more information than is immediately apparent.

[10] Yet another oh-so-standard but unhelpful use of this adjective.

An operator T on an inner product space V is **unitary** if[11]

$$T^*T = \mathrm{id}_V$$

Since we are discussing finite-dimensional V, this implies that the kernel of T is trivial, and thus T is invertible, since (as we saw much earlier)

$$\dim \ker T + \dim \operatorname{Im} T = \dim V$$

24.4.8 Proposition: Eigenvalues of self-adjoint operators T on an inner product space V are *real*.

Proof: Let v be a (non-zero) eigenvector for T, with eigenvalue λ. Then

$$\lambda \langle v, v \rangle = \langle \lambda v, v \rangle = \langle Tv, v \rangle = \langle v, T^*v \rangle = \langle v, Tv \rangle = \langle v, \lambda v \rangle = \overline{\lambda} \langle v, v \rangle$$

Since $\langle v, v \rangle \neq 0$, this implies that $\overline{\lambda} = \lambda$.   ///

24.5 *Projections without coordinates*

There is another construction of orthogonal projections and orthogonal complements which is less coordinate-dependent, and which applies to infinite-dimensional [12] inner-product spaces as well. Specifically, using the metric

$$d(x, y) = |x - y| = \langle x - y, x - y \rangle^{1/2}$$

the **orthogonal projection** of a vector x to the subspace W is the vector in W closest to x.

To prove this, first observe the **polarization identity**

$$|x + y|^2 + |x - y|^2 = |x|^2 + \langle x, y \rangle + \langle y, x \rangle + |y|^2 + |x|^2 - \langle x, y \rangle - \langle y, x \rangle + |y|^2 = 2|x|^2 + 2|y|^2$$

Fix x not in W, and let u, v be in W such that $|x - u|^2$ and $|x - v|^2$ are within $\varepsilon > 0$ of the infimum μ of all values $|x - w|^2$ for $w \in W$. Then an application of the previous identity gives

$$|(x - u) + (x - v)|^2 + |(x - u) - (x - v)|^2 = 2|x - u|^2 + 2|x - v|^2$$

so

$$|u - v|^2 = 2|x - u|^2 + 2|x - v|^2 - |(x - u) + (x - v)|^2$$

The further small trick is to notice that

$$(x - u) + (x - v) = 2 \cdot \left(x - \frac{u + v}{2} \right)$$

---

[11] For infinite-dimensional spaces this definition of *unitary* is insufficient. Invertibility must be explicitly required, one way or another.

[12] Precisely, this argument applies to arbitrary inner product spaces that are *complete* in the metric sense, namely, that Cauchy sequences converge in the metric naturally attached to the inner product, namely $d(x, y) = |x - y| = \langle x - y, x - y \rangle^{1/2}$.

which is again of the form $x - w'$ for $w' \in W$. Thus,

$$|u - v|^2 = 2|x - u|^2 + 2|x - v|^2 - 4|x - \frac{u+v}{2}|^2 < 2(\mu + \varepsilon) + 2(\mu + \varepsilon) - 4\mu = 4\varepsilon$$

That is, we can make a Cauchy sequence from the u, v.

Granting that Cauchy sequences converge, this proves *existence* of a closest point of W to x, as well as the *uniqueness* of the closest point.   ///

From this viewpoint, the **orthogonal complement** $W^\perp$ to W can be defined to be the collection of vectors x in V such that the orthogonal projection of x to W is 0.

24.6 *Unitary operators*

It is worthwhile to look at different ways of characterizing and constructing *unitary* operators on a finite-dimensional complex vector space V with a hermitian inner product $\langle,\rangle$. These equivalent conditions are easy to verify once stated, but it would be unfortunate to overlook them, so we make them explicit. Again, the *definition* of the unitariness of $T : V \longrightarrow V$ for finite-dimensional[13] V is that $T^*T = \mathrm{id}_V$.

24.6.1 Proposition: For V finite-dimensional[14] $T \in \mathrm{End}_{\mathbb{C}}(V)$ is unitary if and only if $TT^* = \mathrm{id}_V$. Unitary operators on finite-dimensional spaces are necessarily invertible.

Proof: The condition $T^*T = \mathrm{id}_V$ implies that T is injective (since it has a left inverse), and since V is finite-dimensional T is also surjective, so is an isomorphism. Thus, its left inverse T^* is also its right inverse, by uniqueness of inverses.   ///

24.6.2 Proposition: For V finite-dimensional with hermitian inner product $\langle,\rangle$ an operator $T \in \mathrm{End}_{\mathbb{C}}(V)$ is unitary if and only if

$$\langle Tu, Tv \rangle = \langle u, v \rangle$$

for all $u, v \in V$.

Proof: If $T^*T = \mathrm{id}_V$, then by definition of adjoint

$$\langle Tu, Tv \rangle = \langle T^*Tu, v \rangle = \langle \mathrm{id}_V u, v \rangle = \langle u, v \rangle$$

On the other hand, if

$$\langle Tu, Tv \rangle = \langle u, v \rangle$$

then

$$0 = \langle T^*Tu, v \rangle - \langle u, v \rangle = \langle (T^*T - \mathrm{id}_V)u, v \rangle$$

Take $v = (T^*T - \mathrm{id}_V)u$ and invoke the positivity of $\langle,\rangle$ to conclude that $(T^*T - \mathrm{id}_V)u = 0$ for all u. Thus, as an endomorphism, $T^*T - \mathrm{id}_V = 0$, and T is unitary.   ///

---

[13] For infinite-dimensional V one must also explicitly require that T be invertible to have the best version of unitariness. In the finite-dimensional case the first proposition incidentally shows that invertibility is automatic.

[14] Without finite-dimensionality this assertion is generally false.

24.6.3 Proposition: For a unitary operator $T \in \mathrm{End}_{\mathbb{C}}(V)$ on a finite-dimensional V with hermitian inner product $\langle , \rangle$, and for given orthonormal basis $\{f_i\}$ for V, the set $\{Tf_i\}$ is also an orthonormal basis. Conversely, given two ordered orthonormal bases $e_1, \ldots, e_n$ and $f_1, \ldots, f_n$ for V, the uniquely determined endomorphism T such that $Te_i = f_i$ is unitary.

Proof: The first part is immediate. For an orthonormal basis $\{e_i\}$ and unitary T,

$$\langle Te_i, Te_j \rangle = \langle e_i, e_j \rangle$$

so the images Te_i make up an orthonormal basis.

The other part is still easy, but requires a small computation whose idea is important. First, since e_i form a basis, there is a unique linear endomorphism T sending e_i to any particular chosen ordered list of targets. To prove the unitariness of this T we use the criterion of the previous proposition. Let $u = \sum_i a_i e_i$ and $v = \sum_j b_j e_j$ with a_i and b_j in $\mathbb{C}$. Then, on one hand,

$$\langle Tu, Tv \rangle = \sum_{ij} a_i \bar{b}_j \langle Te_i, Te_j \rangle = \sum_i a_i \bar{b}_i$$

by the hermitian-ness of $\langle , \rangle$ and by the linearity of T. On the other hand, a very similar computation gives

$$\langle u, v \rangle = \sum_{ij} a_i \bar{b}_j \langle Te_i, Te_j \rangle = \sum_i a_i \bar{b}_i$$

Thus, T preserves inner products, so is unitary. ///

24.7 *Spectral theorems*

The spectral theorem [15] for *normal* operators subsumes the spectral theorem for *self-adjoint* operators, but the proof in the self-adjoint case is so easy to understand that we give this proof separately. Further, many of the applications to matrices use only the self-adjoint case, so understanding this is sufficient for many purposes.

24.7.1 Theorem: Let T be a self-adjoint operator on a finite-dimensional complex vector space V with a (hermitian) inner product $\langle , \rangle$. Then there is an orthonormal basis $\{e_i\}$ for V consisting of eigenvectors for T.

Proof: To prove the theorem, we need

24.7.2 Proposition: Let W be a T-stable subspace of V, with $T = T^*$. Then the orthogonal complement $W^\perp$ is also T-stable.

Proof: (of proposition) Let $v \in W^\perp$, and $w \in W$. Then

$$\langle Tv, w \rangle = \langle v, T^*w \rangle = \langle v, Tw \rangle = 0$$

since $Tw \in W$. ///

---

[15] The use of the word *spectrum* is a reference to wave phenomena, and the idea that a complicated wave is a superposition of simpler ones.

To prove the theorem, we do an induction on the dimension of V. Let $v \neq 0$ be any vector of length 1 which is an eigenvector for T. We know that T has eigenvectors simply because $\mathbb{C}$ is algebraically closed (so the minimal polynomial of T factors into linear factors) and V is finite-dimensional. Thus, $\mathbb{C} \cdot v$ is T-stable, and, by the proposition just proved, the orthogonal complement $(\mathbb{C} \cdot v)^\perp$ is also T-stable. With the restriction of the inner product to $(\mathbb{C} \cdot v)^\perp$ the restriction of T is *still* self-adjoint, so by induction on dimension we're done.
                                                                        ///

Now we give the more general, and somewhat more complicated, argument for normal operators. This does include the previous case, as well as the case of unitary operators.

24.7.3 Theorem: Let T be a normal operator on a finite-dimensional complex vector space V with a (hermitian) inner product $\langle , \rangle$. Then there is an orthonormal basis $\{e_i\}$ for V consisting of eigenvectors for T.

Proof: First prove

24.7.4 Proposition: Let T be an operator on V, and W a T-stable subspace. Then the orthogonal complement $W^\perp$ of W is T^*-stable. [16]

Proof: (of proposition) Let $v \in W^\perp$, and $w \in W$. Then

$$\langle T^*v, w \rangle = \langle v, Tw \rangle = 0$$

since $Tw \in W$.   ///

The proof of the theorem is by induction on the dimension of V. Let λ be an eigenvalue of T, and V_λ the λ-eigenspace of T on V. The assumption of normality is that T and T^* commute, so, from the general discussion of commuting operators, T^* stabilizes V_λ. Then, by the proposition just proved, $T = T^{**}$ stabilizes $V_\lambda^\perp$. By induction on dimension, we're done.   ///

24.8 *Corollaries of the spectral theorem*

These corollaries do not mention the spectral theorem, so do not hint that it plays a role.

24.8.1 Corollary: Let T be a self-adjoint operator on a finite-dimensional complex vector space V with inner product $\langle , \rangle$. Let $\{e_i\}$ be an orthonormal basis for V. Then there is a unitary operator k on V (that is, $\langle kv, kw \rangle = \langle v, w \rangle$ for all $v, w \in V$) such that

$$\{ke_i\} \quad \text{is an orthonormal basis of } T\text{-eigenvectors}$$

Proof: Let $\{f_i\}$ be an orthonormal basis of T-eigenvectors, whose existence is assured by the spectral theorem. Let k be a linear endomorphism mapping $e_i \longrightarrow f_i$ for all indices i. We claim that k is unitary. Indeed, letting $v = \sum_i a_i e_i$ and $w = \sum_j b_j e_j$,

---

[16] Indeed, this is the natural extension of the analogous proposition in the theorem for hermitian operators.

$$\langle kv, kw \rangle = \sum_{ij} a_i \bar{b}_j \, \langle ke_i, ke_j \rangle = \sum_{ij} a_i \bar{b}_j \, \langle f_i, f_j \rangle = \sum_{ij} a_i \bar{b}_j \langle e_i, e_j \rangle = \langle v, w \rangle$$

This is the unitariness. ///

A self-adjoint operator T on a finite-dimensional complex vector space V with hermitian inner product is **positive definite** if

$$\langle Tv, v \rangle \geq 0 \quad \text{with equality only for } v = 0$$

The operator T is **positive semi-definite** if

$$\langle Tv, v \rangle \geq 0$$

(that is, equality may occur for non-zero vectors v).

24.8.2 Proposition: The eigenvalues of a positive definite operator T are positive real numbers. When T is merely positive semi-definite, the eigenvalues are non-negative.

Proof: We already showed that the eigenvalues of a self-adjoint operator are real. Let v be a non-zero λ-eigenvector for T. Then

$$\lambda \langle v, v \rangle = \langle Tv, v \rangle > 0$$

by the positive definiteness. Since $\langle v, v \rangle > 0$, necessarily $\lambda > 0$. When T is merely semi-definite, we get only $\lambda \geq 0$ by this argument. ///

24.8.3 Corollary: Let $T = T^*$ be positive semi-definite. Then T has a positive semi-definite square root S, that is, S is self-adjoint, positive semi-definite, and

$$S^2 = T$$

If T is positive definite, then S is positive definite.

Proof: Invoking the spectral theorem, there is an orthonormal basis $\{e_i\}$ for V consisting of eigenvectors, with respective eigenvalues $\lambda_i \geq 0$. Define an operator S by

$$Se_i = \sqrt{\lambda_i} \cdot e_i$$

Clearly S has the same eigenvectors as T, with eigenvalues the non-negative real square roots of those of T, and the square of this operator is T. We check directly that it is self-adjoint: let $v = \sum_i a_i e_i$ and $w = \sum_i b_i e_i$ and compute

$$\langle S^*v, w \rangle = \langle v, Sw \rangle = \sum_{ij} a_i \bar{b}_j \langle e_i, e_j \rangle = \sum_{ij} a_i \bar{b}_j \sqrt{\lambda_j} \langle e_i, e_j \rangle = \sum_i a_i \bar{b}_i \sqrt{\lambda_i} \langle e_i, e_i \rangle$$

by orthonormality and the real-ness of $\sqrt{\lambda_i}$. Going backwards, this is

$$\sum_{ij} a_i \bar{b}_j \langle \sqrt{\lambda_i} e_i, e_j \rangle = \langle Sv, w \rangle$$

Since the adjoint is unique, $S = S^*$. ///

The **standard (hermitian) inner product** on $\mathbb{C}^n$ is

$$\langle (v_1, \ldots, v_n), (w_1, \ldots, w_n) \rangle = \sum_{i=1}^{n} v_i \overline{w_j}$$

In this situation, certainly n-by-n complex matrices give $\mathbb{C}$ linear endomorphisms by left multiplication of column vectors. With this inner product, the adjoint of an endomorphism T is

$$T^* = T - \text{conjugate-transpose}$$

as usual. Indeed, we often write the superscript-star to indicate conjugate-transpose of a matrix, if no other meaning is apparent from context, and say that the matrix T is **hermitian**. Similarly, an n-by-n matrix k is **unitary** if

$$kk^* = 1_n$$

where 1_n is the n-by-n identity matrix. This is readily verified to be equivalent to unitariness with respect to the standard hermitian inner product.

24.8.4 Corollary: Let T be a hermitian matrix. Then there is a unitary matrix k such that

$$k^* T k = \text{ diagonal, with diagonal entries the eigenvalues of } T$$

Proof: Let $\{e_i\}$ be the standard basis for $\mathbb{C}^n$. It is orthonormal with respect to the standard inner product. Let $\{f_i\}$ be an orthonormal basis consisting of T-eigenvectors. From the first corollary of this section, let k be the unitary operator mapping e_i to f_i. Then $k^* T k$ is diagonal, with diagonal entries the eigenvalues.   ///

24.8.5 Corollary: Let T be a positive semi-definite hermitian matrix. Then there is a positive semi-definite hermitian matrix S such that

$$S^2 = T$$

Proof: With respect to the standard inner product T is positive semi-definite self-adjoint, so has such a square root, from above.   ///

24.9 *Worked examples*

24.9.1 Example: Let p be the smallest prime dividing the order of a finite group G. Show that a subgroup H of G of index p is necessarily *normal*.

Let G act on cosets gH of H by left multiplication. This gives a homomorphism f of G to the group of permutations of $[G : H] = p$ things. The kernel $\ker f$ certainly lies inside H, since $gH = H$ only for $g \in H$. Thus, $p | [G : \ker f]$. On the other hand,

$$|f(G)| = [G : \ker f] = |G|/|\ker f|$$

and $|f(G)|$ divides the order $p!$ of the symmetric group on p things, by Lagrange. But p is the smallest prime dividing $|G|$, so $f(G)$ can only have order 1 or p. Since p divides the order of $f(G)$ and $|f(G)|$ divides p, we have equality. That is, H is the kernel of f. Every kernel is normal, so H is normal. ///

24.9.2 Example: Let $T \in \mathrm{Hom}_k(V)$ for a finite-dimensional k-vectorspace V, with k a field. Let W be a T-stable subspace. Prove that the minimal polynomial of T on W is a divisor of the minimal polynomial of T on V. Define a natural action of T on the quotient V/W, and prove that the minimal polynomial of T on V/W is a divisor of the minimal polynomial of T on V.

Let $f(x)$ be the minimal polynomial of T on V, and $g(x)$ the minimal polynomial of T on W. (We need the T-stability of W for this to make sense at all.) Since $f(T) = 0$ on V, and since the restriction map

$$\mathrm{End}_k(V) \longrightarrow \mathrm{End}_k(W)$$

is a ring homomorphism,

$$(\text{restriction of})f(t) = f(\text{restriction of } T)$$

Thus, $f(T) = 0$ on W. That is, by definition of $g(x)$ and the PID-ness of $k[x]$, $f(x)$ is a multiple of $g(x)$, as desired.

Define $\overline{T}(v+W) = Tv+W$. Since $TW \subset W$, this is well-defined. Note that we cannot assert, and do not need, an *equality* $TW = W$, but only containment. Let $h(x)$ be the minimal polynomial of $\overline{T}$ (on V/W). Any polynomial $p(T)$ stabilizes W, so gives a well-defined map $\overline{p(T)}$ on V/W. Further, since the natural map

$$\mathrm{End}_k(V) \longrightarrow \mathrm{End}_k(V/W)$$

is a ring homomorphism, we have

$$\overline{p(T)}(v + W) = p(T)(v) + W = p(T)(v + W) + W = p(\overline{T})(v + W)$$

Since $f(T) = 0$ on V, $f(\overline{T}) = 0$. By definition of minimal polynomial, $h(x)|f(x)$. ///

24.9.3 Example: Let $T \in \mathrm{Hom}_k(V)$ for a finite-dimensional k-vectorspace V, with k a field. Suppose that T is *diagonalizable* on V. Let W be a T-stable subspace of V. Show that T is diagonalizable on W.

Since T is diagonalizable, its minimal polynomial $f(x)$ on V factors into linear factors in $k[x]$ (with zeros exactly the eigenvalues), and no factor is repeated. By the previous example, the minimal polynomial $g(x)$ of T on W divides $f(x)$, so (by unique factorization in $k[x]$) factors into linear factors without repeats. And this implies that T is diagonalizable when restricted to W. ///

24.9.4 Example: Let $T \in \mathrm{Hom}_k(V)$ for a finite-dimensional k-vectorspace V, with k a field. Suppose that T is *diagonalizable* on V, with *distinct eigenvalues*. Let $S \in \mathrm{Hom}_k(V)$ commute with T, in the natural sense that $ST = TS$. Show that S is diagonalizable on V.

The hypothesis of *distinct eigenvalues* means that each eigenspace is *one-dimensional*. We have seen that commuting operators stabilize each other's eigenspaces. Thus, S stabilizes each one-dimensional λ-eigenspaces V_λ for T. By the one-dimensionality of V_λ, S is a scalar μ_λ on V_λ. That is, the basis of eigenvectors for T is unavoidably a basis of eigenvectors for S, too, so S is diagonalizable. ///

24.9.5 Example: Let $T \in \operatorname{Hom}_k(V)$ for a finite-dimensional k-vectorspace V, with k a field. Suppose that T is *diagonalizable* on V. Show that $k[T]$ contains the projectors to the eigenspaces of T.

Though it is only implicit, we only want projectors P which *commute* with T.

Since T is diagonalizable, its minimal polynomial $f(x)$ factors into linear factors and has no repeated factors. For each eigenvalue λ, let $f_\lambda(x) = f(x)/(x - \lambda)$. The hypothesis that no factor is repeated implies that the *gcd* of all these $f_\lambda(x)$ is 1, so there are polynomials $a_\lambda(x)$ in $k[x]$ such that

$$1 = \sum_\lambda a_\lambda(x)\, f_\lambda(x)$$

For $\mu \neq \lambda$, the product $f_\lambda(x) f_\mu(x)$ picks up all the linear factors in $f(x)$, so

$$f_\lambda(T) f_\mu(T) = 0$$

Then for each eigenvalue μ

$$(a_\mu(T)\, f_\mu(T))^2 = (a_\mu(T)\, f_\mu(T))\, (1 - \sum_{\lambda \neq \mu} a_\lambda(T)\, f_\lambda(T)) = (a_\mu(T)\, f_\mu(T))$$

Thus, $P_\mu = a_\mu(T)\, f_\mu(T)$ has $P_\mu^2 = P_\mu$. Since $f_\lambda(T) f_\mu(T) = 0$ for $\lambda \neq \mu$, we have $P_\mu P_\lambda = 0$ for $\lambda \neq \mu$. Thus, these are projectors to the eigenspaces of T, and, being polynomials in T, commute with T.

For uniqueness, observe that the diagonalizability of T implies that V is the sum of the λ-eigenspaces V_λ of T. We know that any endomorphism (such as a projector) commuting with T stabilizes the eigenspaces of T. Thus, given an eigenvalue λ of T, an endomorphism P commuting with T and such that $P(V) = V_\lambda$ must be 0 on T-eigenspaces V_μ with $\mu \neq \lambda$, since

$$P(V_\mu) \subset V_\mu \cap V_\lambda = 0$$

And when restricted to V_λ the operator P is required to be the identity. Since V is the sum of the eigenspaces and P is determined completely on each one, there is only one such P (for each λ). ///

24.9.6 Example: Let V be a complex vector space with a (positive definite) inner product. Show that $T \in \operatorname{Hom}_k(V)$ cannot be a normal operator if it has any non-trivial Jordan block.

The spectral theorem for normal operators asserts, among other things, that normal operators are diagonalizable, in the sense that there is a basis of eigenvectors. We know that this implies that the minimal polynomial has no repeated factors. Presence of a non-trivial Jordan block exactly means that the minimal polynomial *does* have a repeated factor, so this cannot happen for normal operators. ///

24.9.7 Example: Show that a positive-definite hermitian n-by-n matrix A has a unique positive-definite square root B (that is, $B^2 = A$).

Even though the question explicitly mentions matrices, it is just as easy to discuss endomorphisms of the vector space $V = \mathbb{C}^n$.

By the spectral theorem, A is diagonalizable, so $V = \mathbb{C}^n$ is the sum of the eigenspaces V_λ of A. By hermitian-ness these eigenspaces are mutually orthogonal. By positive-definiteness

A has *positive* real eigenvalues λ, which therefore have real square roots. Define B on each orthogonal summand V_λ to be the scalar $\sqrt{\lambda}$. Since these eigenspaces are mutually orthogonal, the operator B so defined really is hermitian, as we now verify. Let $v = \sum_\lambda v_\lambda$ and $w = \sum_\mu w_\mu$ be *orthogonal* decompositions of two vectors into eigenvectors v_λ with eigenvalues λ and w_μ with eigenvalues μ. Then, using the orthogonality of eigenvectors with distinct eigenvalues,

$$\langle Bv, w\rangle = \langle B\sum_\lambda v_\lambda, \sum_\mu w_\mu\rangle = \langle \sum_\lambda \lambda v_\lambda, \sum_\mu w_\mu\rangle = \sum_\lambda \lambda \langle v_\lambda, w_\lambda\rangle$$

$$= \sum_\lambda \langle v_\lambda, \lambda w_\lambda\rangle = \langle \sum_\mu v_\mu, \sum_\lambda \lambda w_\lambda\rangle = \langle v, Bw\rangle$$

Uniqueness is slightly subtler. Since we do not know *a priori* that two positive-definite square roots B and C of A *commute*, we *cannot* immediately say that $B^2 = C^2$ gives $(B+C)(B-C) = 0$, etc. If we *could* do that, then since B and C are both positive-definite, we could say

$$\langle (B+C)v, v\rangle = \langle Bv, v\rangle + \langle Cv, v\rangle > 0$$

so $B+C$ is positive-definite and, hence invertible. Thus, $B-C = 0$. But we cannot directly do this. We must be more circumspect.

Let B be a positive-definite square root of A. Then B commutes with A. Thus, B stabilizes each eigenspace of A. Since B is diagonalizable on V, it is diagonalizable on each eigenspace of A (from an earlier example). Thus, since all eigenvalues of B are *positive*, and $B^2 = \lambda$ on the λ-eigenspace V_λ of A, it must be that B is the scalar $\sqrt{\lambda}$ on V_λ. That is, B is uniquely determined. ///

24.9.8 Example: Given a square n-by-n complex matrix M, show that there are unitary matrices A and B such that AMB is *diagonal*.

We prove this for not-necessarily square M, with the unitary matrices of appropriate sizes.

This asserted expression
$$M = \text{unitary} \cdot \text{diagonal} \cdot \text{unitary}$$
is called a **Cartan decomposition** of M.

First, if M is *(square) invertible*, then $T = MM^*$ is self-adjoint and invertible. From an earlier example, the spectral theorem implies that there is a self-adjoint (necessarily invertible) square root S of T. Then

$$1 = S^{-1}TS^{-1} = (S^{-1}M)(^{-1}SM)^*$$

so $k_1 = S^{-1}M$ is unitary. Let k_2 be unitary such that $D = k_2 S k_2^*$ is diagonal, by the spectral theorem. Then

$$M = Sk_1 = (k_2 D k_2^*)k_1 = k_2 \cdot D \cdot (k_2^* k_1)$$

expresses M as
$$M = \text{unitary} \cdot \text{diagonal} \cdot \text{unitary}$$
as desired.

358 Eigenvectors, spectral theorems

In the case of m-by-n (not necessarily invertible) M, we want to reduce to the invertible case by showing that there are m-by-m unitary A_1 and n-by-n unitary B_1 such that

$$A_1 M B_1 = \begin{pmatrix} M' & 0 \\ 0 & 0 \end{pmatrix}$$

where M' is *square* and invertible. That is, we can (in effect) do column and row reduction with *unitary* matrices.

Nearly half of the issue is showing that by left (or right) multiplication by a suitable unitary matrix A an arbitrary matrix M may be put in the form

$$AM = \begin{pmatrix} M_{11} & M_{12} \\ 0 & 0 \end{pmatrix}$$

with 0's below the rth row, where the column space of M has dimension r. To this end, let $f_1, \ldots, f_r$ be an orthonormal basis for the *column space* of M, and extend it to an orthonormal basis $f_1, \ldots, f_m$ for the whole $\mathbb{C}^m$. Let $e_1, \ldots, e_m$ be the standard orthonormal basis for $\mathbb{C}^m$. Let A be the linear endomorphism of $\mathbb{C}^m$ defined by $Af_i = e_i$ for all indices i. We claim that this A is unitary, and has the desired effect on M. That it has the desired effect on M is by design, since any column of the original M will be mapped by A to the span of $e_1, \ldots, e_r$, so will have all 0's below the rth row. A linear endomorphism is determined exactly by where it sends a basis, so all that needs to be checked is the unitariness, which will result from the orthonormality of the bases, as follows. For $v = \sum_i a_i f_i$ and $w = \sum_i b_i f_i$,

$$\langle Av, Aw \rangle = \langle \sum_i a_i\, Af_i, \sum_j b_j\, Af_j \rangle = \langle \sum_i a_i\, e_i, \sum_j b_j\, e_j \rangle = \sum_i a_i \overline{b_i}$$

by orthonormality. And, similarly,

$$\sum_i a_i \overline{b_i} = \langle \sum_i a_i\, f_i, \sum_j b_j\, f_j \rangle = \langle v, w \rangle$$

Thus, $\langle Av, Aw \rangle = \langle v, w \rangle$. To be completely scrupulous, we want to see that the latter condition implies that $A^*A = 1$. We have $\langle A^*Av, w \rangle = \langle v, w \rangle$ for all v and w. If $A^*A \neq 1$, then for some v we would have $A^*Av \neq v$, and for that v take $w = (A^*A - 1)v$, so

$$\langle (A^*A - 1)v, w \rangle = \langle (A^*A - 1)v, (A^*A - 1)v \rangle > 0$$

contradiction. That is, A is certainly unitary.

If we had had the foresight to prove that row rank is always equal to column rank, then we would know that a combination of the previous left multiplication by unitary and a corresponding right multiplication by unitary would leave us with

$$\begin{pmatrix} M' & 0 \\ 0 & 0 \end{pmatrix}$$

with M' *square* and invertible, as desired. ///

24.9.9 Example: Given a square n-by-n complex matrix M, show that there is a unitary matrix A such that AM is *upper triangular*.

Let $\{e_i\}$ be the standard basis for $\mathbb{C}^n$. To say that a matrix is upper triangular is to assert that (with left multiplication of column vectors) each of the maximal family of nested subspaces (called a **maximal flag**)

$$V_0 = 0 \subset V_1 = \mathbb{C}e_1 \subset \mathbb{C}e_1 + \mathbb{C}e_2 \subset \ldots \subset \mathbb{C}e_1 + \ldots + \mathbb{C}e_{n-1} \subset V_n = \mathbb{C}^n$$

is stabilized by the matrix. Of course

$$MV_0 \subset MV_1 \subset MV_2 \subset \ldots \subset MV_{n-1} \subset V_n$$

is another maximal flag. Let f_{i+1} be a unit-length vector in the orthogonal complement to MV_i inside MV_{i+1} Thus, these f_i are an orthonormal basis for V, and, in fact, $f_1, \ldots, f_t$ is an orthonormal basis for MV_t. Then let A be the unitary endomorphism such that $Af_i = e_i$. (In an earlier example and in class we checked that, indeed, a linear map which sends one orthonormal basis to another is unitary.) Then

$$AMV_i = V_i$$

so AM is upper-triangular. ///

24.9.10 Example: Let Z be an m-by-n complex matrix. Let Z^* be its conjugate-transpose. Show that

$$\det(1_m - ZZ^*) = \det(1_n - Z^*Z)$$

Write Z in the (rectangular) Cartan decomposition

$$Z = ADB$$

with A and B unitary and D is m-by-n of the form

$$D = \begin{pmatrix} d_1 & & & & & \\ & d_2 & & & & \\ & & \ddots & & & \\ & & & d_r & & \\ & & & & 0 & \\ & & & & & \ddots \end{pmatrix}$$

where the diagonal d_i are the only non-zero entries. We grant ourselves that $\det(xy) = \det(x) \cdot \det(y)$ for square matrices x, y of the same size. Then

$$\det(1_m - ZZ^*) = \det(1_m - ADBB^*D^*A^*) = \det(1_m - ADD^*A^*) = \det(A \cdot (1_m - DD^*) \cdot A^*)$$

$$= \det(AA^*) \cdot \det(1_m - DD^*) = \det(1_m - DD^*) = \prod_i (1 - d_i \overline{d_i})$$

Similarly,

$$\det(1_n - Z^*Z) = \det(1_n - B^*D^*A^*ADB) = \det(1_n - B^*D^*DB) = \det(B^* \cdot (1_n - D^*D) \cdot B)$$

$$= \det(B^*B) \cdot \det(1_n - D^*D) = \det(1_n - D^*D) = \prod_i (1 - d_i \overline{d_i})$$

which is the same as the first computation. ///

Exercises

24.1 Let B be a bilinear form on a vector space V over a field k. Suppose that for $x, y \in V$ if $B(x, y) = 0$ then $B(y, x) = 0$. Show that B is either *symmetric* or *alternating*, that is, either $B(x, y) = B(y, x)$ for all $x, y \in V$ or $B(x, y) = -B(y, x)$ for all $x, y \in V$.

24.2 Let R be a commutative ring of endomorphisms of a finite-dimensional vector space V over $\mathbb{C}$ with a hermitian inner product $\langle, \rangle$. Suppose that R is closed under taking adjoints with respect to $\langle, \rangle$. Suppose that the only R-stable subspaces of V are $\{0\}$ and V itself. Prove that V is one-dimensional.

24.3 Let T be a self-adjoint operator on a complex vector space V with hermitian inner product $\bar{,}\rangle$. Let W be a T-stable subspace of V. Show that the restriction of T to W is self-adjoint.

24.4 Let T be a diagonalizable k-linear endomorphism of a k-vectorspace V. Let W be a T-stable subspace of V. Show that T is diagonalizable on W.

24.5 Let V be a finite-dimensional vector space over an algebraically closed field k. Let T be a k-linear endomorphism of V. Show that T can be written uniquely as $T = D + N$ where D is diagonalizable, N is nilpotent, and $DN = ND$.

24.6 Let S, T be commuting k-linear endomorphisms of a finite-dimensional vector space V over an algebraically closed field k. Show that S, T have a common non-zero eigenvector.

25. Duals, naturality, bilinear forms

25.1 *Dual vector spaces*

A **(linear) functional** $\lambda : V \longrightarrow k$ on a vector space V over k is a linear map from V to the field k itself, viewed as a one-dimensional vector space over k. The collection V^* of all such linear functionals is the **dual space** of V.

25.1.1 Proposition: The collection V^* of linear functionals on a vector space V over k is itself a vector space over k, with the addition

$$(\lambda + \mu)(v) = \lambda(v) + \mu(v)$$

and scalar multiplication

$$(\alpha \cdot \lambda)(v) = \alpha \cdot \lambda(v)$$

Proof: The 0-vector in V^* is the linear functional which sends every vector to 0. The additive inverse $-\lambda$ is defined by

$$(-\lambda)(v) = -\lambda(v)$$

The distributivity properties are readily verified:

$$(\alpha(\lambda + \mu))(v) = \alpha(\lambda + \mu)(v) = \alpha(\lambda(v) + \mu(v)) = \alpha\lambda(v) + \alpha\mu(v) = (\alpha\lambda)(v) + (\alpha\mu)(v)$$

and

$$((\alpha + \beta) \cdot \lambda)(v) = (\alpha + \beta)\lambda(v) = \alpha\lambda(v) + \beta\lambda(v) = (\alpha\lambda)(v) + (\beta\lambda)(v)$$

as desired.   ///

Let V be a finite-dimensional[1] vector space, with a basis $e_1, \ldots, e_n$ for V. A **dual basis** $\lambda_1, \ldots, \lambda_n$ for V^* (and $\{e_i\}$) is a basis for V^* with the property that

$$\lambda_j(e_i) = \begin{cases} 1 & \text{(for } i = j) \\ 0 & \text{(for } i \neq j) \end{cases}$$

From the definition alone it is not at all clear that a dual basis exists, but the following proposition proves that it does.

25.1.2 Proposition: The dual space V^* to an n-dimensional vector space V (with n a positive integer) is also n-dimensional. Given a basis $e_1, \ldots, e_n$ for V, there exists a unique corresponding **dual basis** $\lambda_1, \ldots, \lambda_n$ for V^*, namely a basis for V^* with the property that

$$\lambda_j(e_i) = \begin{cases} 1 & \text{(for } i = j) \\ 0 & \text{(for } i \neq j) \end{cases}$$

Proof: Proving the existence of a dual basis corresponding to the given basis will certainly prove the dimension assertion. Using the *uniqueness* of expression of a vector in V as a linear combination of the basis vectors, we can unambiguously define a linear functional λ_j by

$$\lambda_j \left(\sum_i c_i e_i \right) = c_j$$

These functionals certainly have the desired relation to the basis vectors e_i. We must prove that the λ_j are a basis for V^*. If

$$\sum_j b_j \lambda_j = 0$$

then apply this functional to e_i to obtain

$$b_i = \left(\sum_j b_j \lambda_j \right)(e_i) = 0(e_i) = 0$$

This holds for every index i, so all coefficients are 0, proving the linear independence of the λ_j. To prove the spanning property, let λ be an arbitrary linear functional on V. We claim that

$$\lambda = \sum_j \lambda(e_j) \cdot \lambda_j$$

---

[1] Some of the definitions and discussion here make sense for infinite-dimensional vector spaces V, but many of the conclusions are either false or require substantial modification to be correct. For example, by contrast to the proposition here, for infinite-dimensional V the (infinite) dimension of V^* is strictly larger than the (infinite) dimension of V. Thus, for example, the natural inclusion of V into its second dual V^{**} would fail to be an isomorphism.

Indeed, evaluating the left-hand side on $\sum_i a_i e_i$ gives $\sum_i a_i \lambda(e_i)$, and evaluating the right-hand side on $\sum_i a_i e_i$ gives

$$\sum_j \sum_i a_i \, \lambda(e_j) \, \lambda_j(e_i) = \sum_i a_i \lambda(e_i)$$

since $\lambda_j(e_i) = 0$ for $i \neq j$. This proves that any linear functional is a linear combination of the λ_j.   ///

Let W be a subspace of a vector space V over k. The **orthogonal complement** $W^\perp$ of W in V^* is

$$W^\perp = \{\lambda \in V^* : \lambda(w) = 0, \text{ for all } w \in W\}$$

• The orthogonal complement $W^\perp$ of a subspace W of a vector space V is a vector subspace of V^*.

Proof: Certainly $W^\perp$ contains 0. If $\lambda(w) = 0$ and $\mu(w) = 0$ for all $w \in W$, then certainly $(\lambda + \mu)(w) = 0$. Likewise, $(-\lambda)(w) = \lambda(-w)$, so $W^\perp$ is a subspace.   ///

25.1.3 Corollary: Let W be a subspace of a finite-dimensional vector space V over k.

$$\dim W + \dim W^\perp = \dim V$$

Proof: Let $e_1, \ldots, e_m$ be a basis of W, and extend it to a basis $e_1, \ldots, e_m, f_{m+1}, \ldots, f_n$ of V. Let $\lambda_1, \ldots, \lambda_m, \mu_{m+1}, \ldots, \mu_n$ be the corresponding dual basis of V^*. To prove the corollary it would suffice to prove that $\mu_{m+1}, \ldots, \mu_n$ form a basis for $W^\perp$. First, these functionals do lie in $W^\perp$, since they are all 0 on the basis vectors for W. To see that they span $W^\perp$, let

$$\lambda = \sum_{1 \leq i \leq m} a_i \lambda_i + \sum_{m+1 \leq j \leq n} b_j \mu_j$$

be a functional in $W^\perp$. Evaluating both sides on $e_\ell \in W$ gives

$$0 = \lambda(e_\ell) = \sum_{1 \leq i \leq m} a_i \lambda_i(e_\ell) + \sum_{m+1 \leq j \leq n} b_j \mu_j(e_\ell) = a_\ell$$

by the defining property of the dual basis. That is, every functional in $W^\perp$ is a linear combination of the μ_j, and thus the latter form a basis for $W^\perp$. Then

$$\dim W + \dim W^\perp = m + (n - m) = n = \dim V$$

as claimed.   ///

The **second dual** V^{**} of a vector space V is the dual of its dual. There is a natural vector space homomorphism $\varphi : V \longrightarrow V^{**}$ of a vector space V to its second V^{**} by [2]

$$\varphi(v)(\lambda) = \lambda(v) \qquad \text{(for } v \in V, \lambda \in V^*\text{)}$$

---

[2] The austerity or starkness of this map is very different from formulas written in terms of matrices and column or row vectors. Indeed, this is a different sort of assertion. Further, the sense of *naturality* here might informally be construed exactly as that the formula does *not* use a basis, matrices, or any other manifestation of choices. Unsurprisingly, but unfortunately, very elementary mathematics does not systematically present us with good examples of naturality, since the emphasis is more often on computation. Indeed, we often take for granted the idea that two different sorts of computations will ineluctably yield the same result. Luckily, this is often the case, but becomes increasingly less obvious in more complicated situations.

25.1.4 Corollary: Let V be a finite-dimensional vector space. Then the natural map of V to V^{**} is an isomorphism.

Proof: If v is in the kernel of the linear map $v \longrightarrow \varphi(v)$, then $\varphi(v)(\lambda) = 0$ for all λ, so $\lambda(v) = 0$ for all λ. But if v is non-zero then v can be part of a basis for V, which has a dual basis, among which is a functional λ such that $\lambda(v) = 1$. Thus, for $\varphi(v)(\lambda)$ to be 0 for all λ it must be that $v = 0$. Thus, the kernel of φ is $\{0\}$, so (from above) φ is an injection. From the formula

$$\dim \ker \varphi + \dim \operatorname{Im} \varphi = \dim V$$

it follows that $\dim \operatorname{Im} \varphi = \dim V$. We showed above that the dimension of V^* is the same as that of V, since V is finite-dimensional. Likewise, the dimension of $V^{**} = (V^*)^*$ is the same as that of V^*, hence the same as that of V. Since the dimension of the image of φ in V^{**} is equal to the dimension of V, which is the same as the dimension of V^{**}, the image must be all of V^{**}. Thus, $\varphi : V \longrightarrow V^{**}$ is an isomorphism.   ///

25.1.5 Corollary: Let W be a subspace of a finite-dimensional vector space V over k. Let $\varphi : V \longrightarrow V^{**}$ be the isomorphism of the previous corollary. Then

$$(W^\perp)^\perp = \varphi(W)$$

Proof: First, show that

$$\varphi(W) \subset (W^\perp)^\perp$$

Indeed, for $\lambda \in W^\perp$,

$$\varphi(w)(\lambda) = \lambda(w) = 0$$

On the other hand,

$$\dim W + \dim W^\perp = \dim V$$

and likewise

$$\dim W^\perp + \dim(W^\perp)^\perp = \dim V^* = \dim V$$

Thus, $\varphi(W) \subset (W^\perp)^\perp$ and

$$\dim(W^\perp)^\perp = \dim \varphi(W)$$

since φ is an isomorphism. Therefore, $\varphi(W) = (W^\perp)^\perp$.   ///

As an illustration of the efficacy of the present viewpoint, we can prove a useful result about matrices.

25.1.6 Corollary: Let M be an m-by-n matrix with entries in a field k. Let R be the subspace of k^n spanned by the rows of M. Let C be the subspace of k^m spanned by the columns of M. Let

$$\begin{aligned} \text{column rank of } M &= \dim C \\ \text{row rank of } M &= \dim R \end{aligned}$$

Then

$$\text{column rank of } M = \text{row rank of } M$$

Proof: The matrix M gives a linear transformation $T : k^n \longrightarrow k^m$ by $T(v) = Mv$ where v is a column vector of length n. It is easy to see that the column space of M is the image of T. It is a little subtler that the row space is $(\ker T)^\perp$. From above,

$$\dim \ker T + \dim \operatorname{Im} T = \dim V$$

and also

$$\dim \ker T + \dim(\ker T)^\perp = \dim V$$

Thus,

$$\text{column rank } M = \dim \operatorname{Im} T = \dim(\ker T)^\perp = \text{ row rank } M$$

as claimed. ///

25.2 *First example of naturality*

We have in hand the material to illustrate a simple case of a **natural isomorphism** versus not-natural isomorphisms. This example could be given in the context of *category theory*, and in fact could be a first example, but it is possible to describe the phenomenon without the larger context. [3]

Fix a field k, and consider the map [4]

$$D : \{k\text{-vectorspaces}\} \longrightarrow \{k\text{-vectorspaces}\}$$

from the class of k-vectorspaces to itself given by **duality**, namely [5]

$$DV = V^* = \operatorname{Hom}_k(V, k)$$

Further, for a k-vectorspace homomorphism $f : V \longrightarrow W$ we have an associated map [6] f^* of the duals spaces

$$f^* : W^* \longrightarrow V^* \quad \text{by} \quad f^*(\mu)(v) = \mu(fv) \quad \text{for} \quad \mu \in W^*, \ v \in V$$

Note that f^* **reverses** direction, going from W^* to V^*, while the original f goes from V to W.

The map [7] $F = D \circ D$ associating to vector spaces V their *double* duals V^{**} also gives maps

$$f^{**} : V^{**} \longrightarrow W^{**}$$

---

[3] Indeed, probably a collection of such examples should precede a development of general category theory, else there is certainly insufficient motivation to take the care necessary to develop things in great generality.

[4] In category-theory language a map on objects *and* on the maps among them is a **functor**. We will not emphasize this language just now.

[5] Certainly this class is not a *set*, since it is far too large. This potentially worrying foundational point is another feature of nascent category theory, as opposed to development of mathematics based as purely as possible on set theory.

[6] We might write $Df : DW \longrightarrow DV$ in other circumstances, in order to emphasize the fact that D maps both objects and the homomorphisms among them, but at present this is not the main point.

[7] Functor.

for any k-vectorspace map $f : V \longrightarrow W$. (The direction of the arrows has been reversed twice, so is back to the original direction.)

And for each k-vectorspace V we have a k-vectorspace map [8]

$$\eta_V : V \longrightarrow V^{**} = (V^*)^*$$

given by

$$\eta_V(v)(\lambda) = \lambda(v)$$

The aggregate η of all the maps $\eta_V : V \longrightarrow V^{**}$ is a **natural transformation**[9] meaning that for all k-vectorspace maps

$$f : V \longrightarrow W$$

the diagram

$$
\begin{array}{ccc}
V & \overset{\eta_V}{\longrightarrow} & V^{**} \\
f \downarrow & & \downarrow f^{**} \\
W & \overset{\eta_W}{\longrightarrow} & W^{**}
\end{array}
$$

commutes. The commutativity of the diagram involving a particular M and N is called **functoriality** in M and in N. That the diagram commutes is verified very simply, as follows. Let $v \in V$, $\mu \in W^*$. Then

$$
\begin{aligned}
((f^{**} \circ \eta_V)(v))(\mu) &= (f^{**}(\eta_V v))(\mu) && \text{(definition of composition)} \\
&= (\eta_V v)(f^* \mu) && \text{(definition of } f^{**}) \\
&= (f^* \mu)(v) && \text{(definition of } \eta_V) \\
&= \mu(fv) && \text{(definition of } f^*) \\
&= (\eta_W(fv))(\mu) && \text{(definition of } \eta_W) \\
&= ((\eta_W \circ f)(v))(\mu) && \text{(definition of composition)}
\end{aligned}
$$

Since equality of elements of W^{**} is implied by equality of values on elements of W^*, this proves that the diagram commutes.

Further, for V finite-dimensional, we have

$$\dim_k V = \dim_k V^* = \dim_k V^{**}$$

which implies that each η_V must be an *isomorphism*. Thus, the aggregate η of the isomorphisms $\eta_V : V \longrightarrow V^{**}$ is called a **natural equivalence**. [10]

---

[8] The austere or stark nature of this map certainly should be viewed as being in extreme contrast to the coordinate-based linear maps encountered in introductory linear algebra. The very austerity itself, while being superficially simple, may cause some vertigo or cognitive dissonance for those completely unacquainted with the possibility of writing such things. Rest assured that this discomfort will pass.

[9] We should really speak of a natural transformation η *from a functor to another functor*. Here, η is from the identity functor on k-vectorspaces (which associates each V to itself), *to* the functor that associates to V its second dual V^{**}.

[10] More precisely, on the category of finite-dimensional k-vectorspaces, η is a natural equivalence of the identity functor with the second-dual functor.

25.3 *Bilinear forms*

Abstracting the notion of **inner product** or **scalar product** or **dot product** on a vector space V over k is that of **bilinear form** or **bilinear pairing**. For purpose of *this* section, a bilinear form on V is a k-valued function of two V-variables, written $v \cdot w$ or $\langle v, w \rangle$, with the following properties for $u, v, w \in V$ and $\alpha \in k$

• (Linearity in both arguments) $\langle \alpha u + v, w \rangle = \alpha \langle u, w \rangle + \langle v, w \rangle$ and $\langle \alpha u, \beta v + v' \rangle = \beta \langle u, v \rangle + \langle u, v' \rangle$

• (Non-degeneracy) For all $v \neq 0$ in V there is $w \in V$ such that $\langle v, w \rangle \neq 0$. Likewise, for all $w \neq 0$ in V there is $v \in V$ such that $\langle v, w \rangle \neq 0$.

The two linearity conditions together are **bilinearity**.

In some situations, we may also have

• (Symmetry) $\langle u, v \rangle = \langle v, u \rangle$ However, the symmetry condition is not necessarily critical in many applications.

25.3.1 Remark: When the scalars are the complex numbers $\mathbb{C}$, sometimes a variant of the symmetry condition is useful, namely a *hermitian* condition that $\langle u, v \rangle = \overline{\langle v, u \rangle}$ where the bar denotes complex conjugation.

25.3.2 Remark: When the scalars are real or complex, sometimes, but not always, the non-degeneracy and symmetry are usefully replaced by a *positive-definiteness* condition, namely that $\langle v, v \rangle \geq 0$ and is 0 only for $v = 0$.

When a vector space V has a non-degenerate bilinear form $\langle, \rangle$, there are two natural linear maps $v \longrightarrow \lambda_v$ and $v \longrightarrow \mu_v$ from V to its dual V^*, given by

$$\lambda_v(w) = \langle v, w \rangle$$

$$\mu_v(w) = \langle w, v \rangle$$

That λ_v and μ_v are linear functionals on V is an immediate consequence of the linearity of $\langle, \rangle$ in its arguments, and the linearity of the map $v \longrightarrow \lambda_v$ itself is an immediate consequence of the linearity of $\langle, \rangle$ in its arguments.

25.3.3 Remark: All the following assertions for $L : v \longrightarrow \lambda_v$ have completely analogous assertions for $v \longrightarrow \mu_v$, and we leave them to the reader.

25.3.4 Corollary: Let V be a finite-dimensional vector space with a non-degenerate bilinear form $\langle, \rangle$. The linear map $L : v \longrightarrow \lambda_v$ above is an isomorphism $V \longrightarrow V^*$.

Proof: The non-degeneracy means that for $v \neq 0$ the linear functional λ_v is not 0, since there is $w \in V$ such that $\lambda_v(w) \neq 0$. Thus, the linear map $v \longrightarrow \lambda_v$ has kernel $\{0\}$, so $v \longrightarrow \lambda_v$ is *injective*. Since V is finite-dimensional, from above we know that it and its dual have the same dimension. Let $L(v) = \lambda_v$. Since

$$\dim \operatorname{Im} L + \dim \ker L = \dim V$$

the image of V under $v \longrightarrow \lambda_v$ in V is that of V. Since proper subspaces have strictly smaller dimension it must be that $L(V) = V^*$. ///

Let V be a finite-dimensional vector space with non-degenerate form $\langle, \rangle$, and W a subspace. Define the **orthogonal complement**

$$W^\perp = \{\lambda \in V^* : \lambda(w) = 0, \text{ for all } w \in W\}$$

25.3.5 Corollary: Let V be a finite-dimensional vector space with a non-degenerate form $\langle,\rangle$, and W a subspace. Under the isomorphism $L : v \longrightarrow \lambda_v$ of V to its dual,

$$L(\{v \in V : \langle v, w\rangle = 0 \text{ for all } w \in W\}) = W^\perp$$

Proof: Suppose that $L(v) \in W^\perp$. Thus, $\lambda_v(w) = 0$ for all $w \in W$. That is, $\langle v, w\rangle = 0$ for all $w \in W$. On the other hand, suppose that $\langle v, w\rangle = 0$ for all $w \in W$. Then $\lambda_v(w) = 0$ for all $w \in W$, so $\lambda_v \in W^\perp$. ///

25.3.6 Corollary: Now suppose that $\langle,\rangle$ is *symmetric*, meaning that $\langle v, w\rangle = \langle w, v\rangle$ for all $v, w \in V$. Redefine

$$W^\perp = \{v \in V : \langle v, w\rangle = 0 \text{ for all } w \in W\}$$

Then
$$\dim W + \dim W^\perp = \dim V$$

and
$$W^{\perp\perp} = W$$

Proof: With our original definition of $W_{\text{orig}}^\perp$ as

$$W_{\text{orig}}^\perp = \{\lambda \in V^* : \lambda(w) = 0 \text{ for all } w \in W\}$$

we had proven
$$\dim W + \dim W_{\text{orig}}^\perp = \dim V$$

We just showed that $L(W^\perp) = W_{\text{orig}}^\perp$, and since the map $L : V \longrightarrow V^*$ by $v \longrightarrow \lambda_v$ is an isomorphism
$$\dim W^\perp = \dim W_{\text{orig}}^\perp$$

Thus,
$$\dim W + \dim W^\perp = \dim V$$

as claimed.

Next, we claim that $W \subset W^{\perp\perp}$. Indeed, for $w \in W$ it is certainly true that for $v \in W^\perp$

$$\langle v, w\rangle = \langle v, w\rangle = 0$$

That is, we see easily that $W \subset W^{\perp\perp}$. On the other hand, from

$$\dim W + \dim W^\perp = \dim V$$

and
$$\dim W^\perp + \dim W^{\perp\perp} = \dim V$$

we see that $\dim W^{\perp\perp} = \dim W$. Since W is a subspace of $W^{\perp\perp}$ with the same dimension, the two must be equal (from our earlier discussion). ///

25.3.7 Remark: When a non-degenerate bilinear form on V is *not* symmetric, there are two different versions of $W^\perp$, depending upon which argument in $\langle,\rangle$ is used:

$$W^{\perp, \mathrm{rt}} = \{v \in V : \langle v, w \rangle = 0, \ \text{for all } w \in W\}$$

$$W^{\perp, \mathrm{lft}} = \{v \in V : \langle w, v \rangle = 0, \ \text{for all } w \in W\}$$

And then there are two correct statements about $W^{\perp\perp}$, namely

$$\left(W^{\perp, \mathrm{rt}}\right)^{\perp, \mathrm{lft}} = W$$

$$\left(W^{\perp, \mathrm{lft}}\right)^{\perp, \mathrm{rt}} = W$$

These are proven in the same way as the last corollary, but with more attention to the lack of symmetry in the bilinear form. In fact, to more scrupulously consider possible asymmetry of the form, we proceed as follows.

For many purposes we can consider bilinear maps[11] (that is, k-valued maps linear in each argument)

$$\langle , \rangle : V \times W \longrightarrow k$$

where V and W are vectorspaces over the field k. [12]

The most common instance of such a pairing is that of a vector space and its dual

$$\langle , \rangle : V \times V^* \longrightarrow k$$

by

$$\langle v, \lambda \rangle = \lambda(v)$$

This notation and viewpoint helps to emphasize the near-symmetry[13] of the relationship between V and V^*.

Rather than simply *assume* non-degeneracy conditions, let us give ourselves a language to talk about such issues. Much as earlier, define

$$W^{\perp} = \{v \in V : \langle v, w \rangle = 0 \ \text{for all } w \in W\}$$

$$V^{\perp} = \{w \in W : \langle v, w \rangle = 0 \ \text{for all } v \in V\}$$

Then we have

25.3.8 Proposition: A bilinear form $\langle , \rangle : V \times W \longrightarrow k$ induces a bilinear form, still denoted $\langle , \rangle$,

$$\langle , \rangle : V/W^{\perp} \times W/V^{\perp} \longrightarrow k$$

defined in the natural manner by

$$\langle v + W^{\perp}, w + V^{\perp} \rangle = \langle v, w \rangle$$

for any representatives v, w for the cosets. This form is *non-degenerate* in the sense that, on the quotient, given $x \in V/W^{\perp}$, there is $y \in W/V^{\perp}$ such that $\langle x, y \rangle \neq 0$, and symmetrically.

---

[11] Also called bilinear *forms*, or bilinear *pairings*, or simply *pairings*.

[12] Note that now the situation is unsymmetrical, insofar as the first and second arguments to $\langle , \rangle$ are from different spaces, so that there is no obvious sense to any property of *symmetry*.

[13] The second dual V^{**} is naturally isomorphic to V if and only if $\dim V < \infty$.

Proof: The first point is that the bilinear form on the quotients is *well-defined*, which is immediate from the definition of $W^\perp$ and $V^\perp$. Likewise, the non-degeneracy follows from the definition: given $x = v + W^\perp$ in $V/W^\perp$, take $w \in W$ such that $\langle v, w \rangle \neq 0$, and let $y = w + V^\perp$. ///

25.3.9 Remark: The pairing of a vector space V and its dual is non-degenerate, even if the vector space is infinite-dimensional.

In fact, the pairing of (finite-dimensional) V and V^* is the universal example of a non-degenerate pairing:

25.3.10 Proposition: For finite-dimensional V and W, a non-degenerate pairing

$$\langle , \rangle : V \times W \longrightarrow k$$

gives natural isomorphisms

$$V \overset{\approx}{\longrightarrow} W^*$$
$$W \overset{\approx}{\longrightarrow} V^*$$

via

$$v \longrightarrow \lambda_v \quad \text{where} \quad \lambda_v(w) = \langle v, w \rangle$$
$$w \longrightarrow \lambda_w \quad \text{where} \quad \lambda_w(v) = \langle v, w \rangle$$

Proof: The indicated maps are easily seen to be linear, with trivial kernels because the pairing is non-degenerate. The dimensions match, so these maps are isomorphisms. ///

25.4 *Worked examples*

25.4.1 Example: Let k be a field, and V a finite-dimensional k-vectorspace. Let Λ be a subset of the dual space V^*, with $|\Lambda| < \dim V$. Show that the **homogeneous system of equations**

$$\lambda(v) = 0 \quad \text{(for all } \lambda \in \Lambda)$$

has a non-trivial (that is, non-zero) solution $v \in V$ (meeting all these conditions).

The dimension of the span W of Λ is strictly less than $\dim V^*$, which we've proven is $\dim V^* = \dim V$. We may also identify $V \approx V^{**}$ via the natural isomorphism. With that identification, we may say that the set of solutions is $W^\perp$, and

$$\dim(W^\perp) + \dim W = \dim V^* = \dim V$$

Thus, $\dim W^\perp > 0$, so there are non-zero solutions. ///

25.4.2 Example: Let k be a field, and V a finite-dimensional k-vectorspace. Let Λ be a *linearly independent* subset of the dual space V^*. Let $\lambda \longrightarrow a_\lambda$ be a set map $\Lambda \longrightarrow k$. Show that an **inhomogeneous system of equations**

$$\lambda(v) = a_\lambda \quad \text{(for all } \lambda \in \Lambda)$$

has a solution $v \in V$ (meeting all these conditions).

Let $m = |\Lambda|$, $\Lambda = \{\lambda_1, \ldots, \lambda_m\}$. One way to use the linear independence of the functionals in Λ is to extend Λ to a basis $\lambda_1, \ldots, \lambda_n$ for V^*, and let $e_1, \ldots, e_n \in V^{**}$ be the corresponding dual basis for V^{**}. Then let $v_1, \ldots, v_n$ be the images of the e_i in V under the natural isomorphism $V^{**} \approx V$. (This achieves the effect of making the λ_i be a dual basis to the v_i. We had only literally proven that one can go from a basis of a vector space to a dual basis of its dual, and not the reverse.) Then

$$v = \sum_{1 \leq i \leq m} a_{\lambda_i} \cdot v_i$$

is a solution to the indicated set of equations, since

$$\lambda_j(v) = \sum_{1 \leq i \leq m} a_{\lambda_i} \cdot \lambda_j(v_i) = a_{\lambda_j}$$

for all indices $j \leq m$. ///

25.4.3 Example: Let T be a k-linear endomorphism of a finite-dimensional k-vectorspace V. For an eigenvalue λ of T, let V_λ be the generalized λ-eigenspace

$$V_\lambda = \{v \in V : (T - \lambda)^n v = 0 \text{ for some } 1 \leq n \in \mathbb{Z}\}$$

Show that the projector P of V to V_λ (commuting with T) lies inside $k[T]$.

First we do this assuming that the minimal polynomial of T factors into linear factors in $k[x]$.

Let $f(x)$ be the minimal polynomial of T, and let $f_\lambda(x) = f(x)/(x - \lambda)^e$ where $(x - \lambda)^e$ is the precise power of $(x - \lambda)$ dividing $f(x)$. Then the collection of all $f_\lambda(x)$'s has *gcd* 1, so there are $a_\lambda(x) \in k[x]$ such that

$$1 = \sum_\lambda a_\lambda(x) f_\lambda(x)$$

We claim that $E_\lambda = a_\lambda(T)f_\lambda(T)$ is a projector to the generalized λ-eigenspace V_λ. Indeed, for $v \in V_\lambda$,

$$v = 1_V \cdot v = \sum_\mu a_\mu(T)f_\mu(T) \cdot v = \sum_\mu a_\mu(T)f_\mu(T) \cdot v = a_\lambda(T)f_\lambda(T) \cdot v$$

since $(x - \lambda)^e$ divides $f_\mu(x)$ for $\mu \neq \lambda$, and $(T - \lambda)^e v = 0$. That is, it acts as the identity on V_λ. And

$$(T - \lambda)^e \circ E_\lambda = a_\lambda(T)\, f(T) = 0 \in \text{End}_k(V)$$

so the image of E_λ is inside V_λ. Since E_λ is the identity on V_λ, it must be that the image of E_λ is *exactly* V_λ. For $\mu \neq \lambda$, since $f(x)|f_\mu(x)f_\lambda(x)$, $E_\mu E_\lambda = 0$, so these idempotents are *mutually orthogonal*. Then

$$(a_\lambda(T)f_\lambda(T))^2 = (a_\lambda(T)f_\lambda(T)) \cdot (1 - \sum_{\mu \neq \lambda} a_\mu(T)f_\mu(T)) = a_\lambda(T)f_\lambda(T) - 0$$

That is, $E_\lambda^2 = E_\lambda$, so E_λ is *a projector to V_λ.*

The mutual orthogonality of the idempotents will yield the fact that V is the direct sum of all the generalized eigenspaces of T. Indeed, for any $v \in V$,

$$v = 1 \cdot v = \left(\sum_\lambda E_\lambda \right) v = \sum_\lambda (E_\lambda v)$$

and $E_\lambda v \in V_\lambda$. Thus,

$$\sum_\lambda V_\lambda = V$$

To check that the sum is (unsurprisingly) direct, let $v_\lambda \in V_\lambda$, and suppose

$$\sum_\lambda v_\lambda = 0$$

Then $v_\lambda = E_\lambda v_\lambda$, for all λ. Then apply E_μ and invoke the orthogonality of the idempotents to obtain

$$v_\mu = 0$$

This proves the linear independence, and that the sum is direct.

To prove *uniqueness* of a projector E to V_λ commuting with T, note that any operator S commuting with T necessarily stabilizes all the generalized eigenspaces of T, since for $v \in V_\mu$

$$(T - \lambda)^e \, Sv = S \, (T - \lambda)^e v = S \cdot 0 = 0$$

Thus, E stabilizes all the V_μs. Since V is the direct sum of the V_μ and E maps V to V_λ, it must be that E is 0 on V_μ for $\mu \neq \lambda$. Thus,

$$E = 1 \cdot E_\lambda + \sum_{\mu \neq \lambda} 0 \cdot E_\mu = E_\lambda$$

That is, there is just one projector to V_λ that also commutes with T. *This finishes things under the assumption that $f(x)$ factors into linear factors in $k[x]$.*

The more general situation is similar. More generally, for a monic irreducible $P(x)$ in $k[x]$ dividing $f(x)$, with $P(x)^e$ the precise power of $P(x)$ dividing $f(x)$, let

$$f_P(x) = f(x)/P(x)^e$$

Then these f_P have *gcd* 1, so there are $a_P(x)$ in $k[x]$ such that

$$1 = \sum_P a_P(x) \cdot f_P(x)$$

Let $E_P = a_P(T) f_P(T)$. Since $f(x)$ divides $f_P(x) \cdot f_Q(x)$ for distinct irreducibles P, Q, we have $E_P \circ E_Q = 0$ for $P \neq Q$. And

$$E_P^2 = E_P (1 - \sum_{Q \neq P} E_Q) = E_P$$

so (as in the simpler version) the E_P's are mutually orthogonal idempotents. And, similarly, V is the direct sum of the subspaces

$$V_P = E_P \cdot V$$

We can also characterize V_P as the kernel of $P^e(T)$ on V, where $P^e(x)$ is the power of $P(x)$ dividing $f(x)$. If $P(x) = (x - \lambda)$, then V_P is the generalized λ-eigenspace, and E_P is the projector to it.

If E were another projector to V_λ commuting with T, then E stabilizes V_P for all irreducibles P dividing the minimal polynomial f of T, and E is 0 on V_Q for $Q \neq (x - \lambda)$, and E is 1 on V_λ. That is,

$$E = 1 \cdot E_{x-\lambda} + \sum_{Q \neq x-\lambda} 0 \cdot E_Q = E_P$$

This proves the uniqueness even in general. ///

25.4.4 Example: Let T be a matrix in Jordan normal form with entries in a field k. Let T_{ss} be the matrix obtained by converting all the off-diagonal 1's to 0's, making T diagonal. Show that T_{ss} is in $k[T]$.

This implicitly demands that the minimal polynomial of T factors into linear factors in $k[x]$.

Continuing as in the previous example, let $E_\lambda \in k[T]$ be the projector to the generalized λ-eigenspace V_λ, and keep in mind that we have shown that V is the direct sum of the generalized eigenspaces, equivalent, that $\sum_\lambda E_\lambda = 1$. By definition, the operator T_{ss} is the scalar operator λ on V_λ. Then

$$T_{ss} = \sum_\lambda \lambda \cdot E_\lambda \in k[T]$$

since (from the previous example) each E_λ is in $k[T]$. ///

25.4.5 Example: Let $M = \begin{pmatrix} A & B \\ 0 & D \end{pmatrix}$ be a matrix in a block decomposition, where A is m-by-m and D is n-by-n. Show that

$$\det M = \det A \cdot \det D$$

One way to prove this is to use the formula for the determinant of an N-by-N matrix

$$\det C = \sum_{\pi \in S_N} \sigma(\pi) \, a_{\pi(1),1} \dots a_{\pi(N),N}$$

where c_{ij} is the $(i,j)^{th}$ entry of C, π is summed over the symmetric group S_N, and σ is the sign homomorphism. Applying this to the matrix M,

$$\det M = \sum_{\pi \in S_{m+n}} \sigma(\pi) \, M_{\pi(1),1} \dots M_{\pi(m+n),m+n}$$

where M_{ij} is the $(i,j)^{th}$ entry. Since the entries M_{ij} with $1 \leq j \leq m$ and $m < i \leq m + n$ are all 0, we should only sum over π with the property that

$$\pi(j) \leq m \quad \text{for} \quad 1 \leq j \leq m$$

That is, π stabilizes the subset $\{1, \dots, m\}$ of the indexing set. Since π is a bijection of the index set, necessarily such π stabilizes $\{m + 1, m + 2, \dots, m + n\}$, also. Conversely, each pair (π_1, π_2) of permutation π_1 of the first m indices and π_2 of the last n indices gives a permutation of the whole set of indices.

Let X be the set of the permutations $\pi \in S_{m+n}$ that stabilize $\{1, \ldots, m\}$. For each $\pi \in X$, let π_1 be the restriction of π to $\{1, \ldots, m\}$, and let π_2 be the restriction to $\{m+1, \ldots, m+n\}$. And, in fact, if we plan to index the entries of the block D in the usual way, we'd better be able to think of π_2 as a permutation of $\{1, \ldots, n\}$, also. Note that $\sigma(\pi) = \sigma(\pi_1)\sigma(\pi_2)$. Then

$$\det M = \sum_{\pi \in X} \sigma(\pi) \, M_{\pi(1),1} \cdots M_{\pi(m+n),m+n}$$

$$= \sum_{\pi \in X} \sigma(\pi) \, (M_{\pi(1),1} \cdots M_{\pi(m),m}) \cdot (M_{\pi(m+1),m+1} \cdots M_{\pi(m+n),m+n})$$

$$= \left(\sum_{\pi_1 \in S_m} \sigma(\pi_1) \, M_{\pi_1(1),1} \cdots M_{\pi_1(m),m} \right) \cdot \left(\sum_{\pi_2 \in S_n} \sigma(\pi_2)(M_{\pi_2(m+1),m+1} \cdots M_{\pi_2(m+n),m+n}) \right)$$

$$= \left(\sum_{\pi_1 \in S_m} \sigma(\pi_1) \, A_{\pi_1(1),1} \cdots A_{\pi_1(m),m} \right) \cdot \left(\sum_{\pi_2 \in S_n} \sigma(\pi_2) D_{\pi_2(1),1} \cdots D_{\pi_2(n),n} \right) = \det A \cdot \det D$$

where in the last part we have mapped $\{m+1, \ldots, m+n\}$ bijectively by $\ell \longrightarrow \ell - m$.   ///

25.4.6 Example: The so-called *Kronecker product*[14] of an m-by-m matrix A and an n-by-n matrix B is

$$A \otimes B = \begin{pmatrix} A_{11} \cdot B & A_{12} \cdot B & \cdots & A_{1m} \cdot B \\ A_{21} \cdot B & A_{22} \cdot B & \cdots & A_{2m} \cdot B \\ & & \vdots & \\ A_{m1} \cdot B & A_{m2} \cdot B & \cdots & A_{mm} \cdot B \end{pmatrix}$$

where, as it may appear, the matrix B is inserted as n-by-n blocks, multiplied by the respective entries A_{ij} of A. Prove that

$$\det(A \otimes B) = (\det A)^n \cdot (\det B)^m$$

at least for $m = n = 2$.

If no entry of the first row of A is non-zero, then both sides of the desired equality are 0, and we're done. So suppose some entry A_{1i} of the first row of A is non-zero. If $i \neq 1$, then for $\ell = 1, \ldots, n$ interchange the ℓth and $(i-1)n + \ell^{th}$ columns of $A \otimes B$, thus multiplying the determinant by $(-1)^n$. This is compatible with the formula, so we'll assume that $A_{11} \neq 0$ to do an induction on m.

We will manipulate n-by-n *blocks* of scalar multiples of B rather than actual scalars.

Thus, assuming that $A_{11} \neq 0$, we want to subtract multiples of the left column of n-by-n blocks from the blocks further to the right, to make the top n-by-n blocks all 0 (apart from the leftmost block, $A_{11}B$). In terms of manipulations of columns, for $\ell = 1, \ldots, n$ and $j = 2, 3, \ldots, m$ subtract A_{1j}/A_{11} times the ℓth column of $A \otimes B$ from the $((j-1)n + \ell)^{th}$.

---

[14]   As we will see shortly, this is really a **tensor product**, and we will treat this question more sensibly.

Since for $1 \leq \ell \leq n$ the ℓth column of $A \otimes B$ is A_{11} times the ℓth column of B, and the $((j-1)n + \ell)^{th}$ column of $A \otimes B$ is A_{1j} times the ℓth column of B, this has the desired effect of killing off the n-by-n blocks along the top of $A \otimes B$ except for the leftmost block. And the $(i,j)^{th}$ n-by-n block of $A \otimes B$ has become $(A_{ij} - A_{1j}A_{i1}/A_{11}) \cdot B$. Let

$$A'_{ij} = A_{ij} - A_{1j}A_{i1}/A_{11}$$

and let D be the $(m-1)$-by-$(m-1)$ matrix with $(i,j)^{th}$ entry $D_{ij} = A'_{(i-1),(j-1)}$. Thus, the manipulation so far gives

$$\det(A \otimes B) = \det \begin{pmatrix} A_{11}B & 0 \\ * & D \otimes B \end{pmatrix}$$

By the previous example (or its tranpose)

$$\det \begin{pmatrix} A_{11}B & 0 \\ * & D \otimes B \end{pmatrix} = \det(A_{11}B) \cdot \det(D \otimes B) = A_{11}^n \det B \cdot \det(D \otimes B)$$

by the multilinearity of det.

And, at the same time subtracting A_{1j}/A_{11} times the first column of A from the jth column of A for $2 \leq j \leq m$ does not change the determinant, and the new matrix is

$$\begin{pmatrix} A_{11} & 0 \\ * & D \end{pmatrix}$$

Also by the previous example,

$$\det A = \det \begin{pmatrix} A_{11} & 0 \\ * & D \end{pmatrix} = A_{11} \cdot \det D$$

Thus, putting the two computations together,

$$\det(A \otimes B) = A_{11}^n \det B \cdot \det(D \otimes B) = A_{11}^n \det B \cdot (\det D)^n (\det B)^{m-1}$$

$$= (A_{11} \det D)^n \det B \cdot (\det B)^{m-1} = (\det A)^n (\det B)^m$$

as claimed.

Another approach to this is to observe that, in these terms, $A \otimes B$ is

$$\left(\begin{array}{ccccccccc} A_{11} & 0 & \cdots & 0 & & A_{1m} & 0 & \cdots & 0 \\ 0 & A_{11} & & & & 0 & A_{1m} & & \\ \vdots & & \ddots & & \cdots & \vdots & & \ddots & \\ 0 & & & A_{11} & & 0 & & & A_{1m} \\ & \vdots & & & & & \vdots & & \\ A_{m1} & 0 & \cdots & 0 & & A_{mm} & 0 & \cdots & 0 \\ 0 & A_{m1} & & & & 0 & A_{mm} & & \\ \vdots & & \ddots & & \cdots & \vdots & & \ddots & \\ 0 & & & A_{m1} & & 0 & & & A_{mm} \end{array} \right) \left(\begin{array}{cccc} B & 0 & \cdots & 0 \\ 0 & B & & \\ \vdots & & \ddots & \\ 0 & & & B \end{array} \right)$$

where there are m copies of B on the diagonal. By suitable permutations of rows and columns (with an interchange of rows for each interchange of columns, thus giving no net change of sign), the matrix containing the A_{ij}s becomes

$$\begin{pmatrix} A & 0 & \cdots & 0 \\ 0 & A & & \\ \vdots & & \ddots & \\ 0 & & & A \end{pmatrix}$$

with n copies of A on the diagonal. Thus,

$$\det(A \otimes B) = \det \begin{pmatrix} A & 0 & \cdots & 0 \\ 0 & A & & \\ \vdots & & \ddots & \\ 0 & & & A \end{pmatrix} \cdot \det \begin{pmatrix} B & 0 & \cdots & 0 \\ 0 & B & & \\ \vdots & & \ddots & \\ 0 & & & B \end{pmatrix} = (\det A)^n \cdot (\det B)^m$$

This might be more attractive than the first argument, depending on one's tastes.   ///

Exercises

25.1 Let T be a hermitian operator on a finite-dimensional complex vector space V with a positive-definite inner product $\langle,\rangle$. Let P be an orthogonal projector to the λ-eigenspace V_λ of T. (This means that P is the identity on V_λ and is 0 on the orthogonal complement $V_\lambda^\perp$ of V_λ.) Show that $P \in \mathbb{C}[T]$.

25.2 Let T be a diagonalizable operator on a finite-dimensional vector space V over a field k. Show that there is a *unique* projector P to the λ-eigenspace V_λ of T such that $TP = PT$.

25.3 Let k be a field, and V, W finite-dimensional vector spaces over k. Let S be a k-linear endomorphism of V, and T a k-linear endomorphism of W. Let $S \oplus T$ be the k-linear endomorphism of $V \oplus W$ defined by

$$(S \oplus T)(v \oplus w) = S(v) \oplus T(w) \qquad \text{(for } v \in V \text{ and } w \in W)$$

Show that the minimal polynomial of $S \oplus T$ is the least common multiple of the minimal polynomials of S and T.

25.4 Let T be an n-by-n matrix with entries in a commutative ring R, with non-zero entries only *above* the diagonal. Show that $T^n = 0$.

25.5 Let T be an endomorphism of a finite-dimensional vector space V over a field k. Suppose that T is *nilpotent*, that is, that $T^n = 0$ for some positive integer n. Show that $\operatorname{tr} T = 0$.

25.6 Let k be a field of characteristic 0, and T a k-linear endomorphism of an n-dimensional vector space V over k. Show that T is nilpotent if and only if $\operatorname{trace}(T^i) = 0$ for $1 \le i \le n$.

25.7 Fix a field k of characteristic not 2, and let $K = k(\sqrt{D})$ where D is a non-square in k. Let σ be the non-trivial automorphism of K over k. Let $\Delta \in k^\times$. Let A be the

k-subalgebra of 2-by-2 matrices over K generated by

$$\begin{pmatrix} 0 & 1 \\ \Delta & 0 \end{pmatrix} \quad \begin{pmatrix} \alpha & 0 \\ 0 & \alpha^\sigma \end{pmatrix}$$

where α ranges over K. Find a condition relating D and Δ necessary and sufficient for A to be a division algebra.

25.8 A *Lie algebra* (named after the mathematician Sophus Lie) over a field k of characteristic 0 is a k-vectorspace with a k-bilinear map $[,]$ (the *Lie bracket*) such that $[x, y] = -[y, x]$, and satisfying the *Jacobi identity*

$$[[x, y], z] = [x, [y, z]] - [y, [x, z]]$$

Let A be an (associative) k-algebra. Show that A can be made into a Lie algebra by defining $[x, y] = xy - yx$.

25.9 Let $\mathfrak{g}$ be a Lie algebra over a field k. Let A be the associative algebra of k-vectorspace endomorphisms of $\mathfrak{g}$. The *adjoint* action of $\mathfrak{g}$ on itself is defined by

$$(\mathrm{ad}x)(y) = [x, y]$$

Show that the map $\mathfrak{g} \longrightarrow \mathrm{Aut}_k G$ defined by $x \longrightarrow \mathrm{ad}x$ is a *Lie homomorphism*, meaning that

$$[\mathrm{ad}x, \mathrm{ad}y] = \mathrm{ad}[x, y]$$

(The latter property is the *Jacobi identity*.)

26. Determinants I

Both as a careful review of a more pedestrian viewpoint, and as a transition to a coordinate-independent approach, we roughly follow Emil Artin's rigorization of determinants *of matrices* with entries in a *field*. Standard properties are derived, in particular *uniqueness*, from simple assumptions. We also prove *existence*. Soon, however, we will want to develop corresponding intrinsic versions of ideas about *endomorphisms*. This is *multilinear algebra*. Further, for example to treat the Cayley-Hamilton theorem in a forthright manner, we will want to consider modules over commutative rings, not merely vector spaces over fields.

26.1 *Prehistory*

Determinants arose many years ago in *formulas* for solving linear equations. This is **Cramer's Rule**, described as follows. [1]   Consider a system of n linear equations in n unknowns $x_1, \ldots, x_n$

$$
\begin{array}{ccccccccc}
a_{11}x_1 & + & a_{12}x_2 & + & \ldots & + & a_{1n}x_n & = & c_1 \\
a_{21}x_1 & + & a_{22}x_2 & + & \ldots & + & a_{2n}x_n & = & c_2 \\
a_{31}x_1 & + & a_{32}x_2 & + & \ldots & + & a_{3n}x_n & = & c_3 \\
\vdots & & & \ddots & & & \vdots & & \vdots \\
a_{n1}x_1 & + & a_{n2}x_2 & + & \ldots & + & a_{nn}x_n & = & c_n
\end{array}
$$

---

[1] We will prove Cramer's Rule just a little later. In fact, quite contrary to a naive intuition, the proof is very easy from an only slightly more sophisticated viewpoint.

Let A be the matrix with $(i,j)^{th}$ entry a_{ij}. Let $A^{(\ell)}$ be the matrix A with its ℓth column replaced by the c_is, that is, the $(i,\ell)^{th}$ entry of $A^{(\ell)}$ is c_ℓ. Then Cramer's Rule asserts that

$$x_\ell = \frac{\det A^{(\ell)}}{\det A}$$

where det is determinant, at least for $\det A \neq 0$. It is implicit that the *coefficients* a_{ij} and the *constants* c_ℓ are in a *field*. As a practical method for solving linear systems Cramer's Rule is far from optimal. Gaussian elimination is much more efficient, but is less *interesting*.

Ironically, in the context of very elementary mathematics it seems difficult to give an intelligible definition or formula for determinants of arbitrary sizes, so typical discussions are limited to very small matrices. For example, in the 2-by-2 case there is the palatable formula

$$\det \begin{pmatrix} a & b \\ c & d \end{pmatrix} = ad - bc$$

Thus, for the linear system

$$ax + by = c_1$$
$$cx + dy = c_2$$

by Cramer's Rule

$$x = \frac{\det \begin{pmatrix} c_1 & b \\ c_2 & d \end{pmatrix}}{\det \begin{pmatrix} a & b \\ c & d \end{pmatrix}} \qquad y = \frac{\det \begin{pmatrix} a & c_1 \\ c & c_2 \end{pmatrix}}{\det \begin{pmatrix} a & b \\ c & d \end{pmatrix}}$$

In the 3-by-3 case there is the still-barely-tractable formula (reachable by a variety of elementary mnemonics)

$$\det \begin{pmatrix} a_{11} & a_{12} & a_{13} \\ a_{21} & a_{22} & a_{23} \\ a_{31} & a_{32} & a_{33} \end{pmatrix}$$

$$= (a_{11}a_{22}a_{33} + a_{12}a_{23}a_{31} + a_{13}a_{21}a_{32}) - (a_{31}a_{22}a_{13} + a_{32}a_{23}a_{11} + a_{31}a_{21}a_{12})$$

Larger determinants are defined ambiguously by induction as *expansions by minors*. [2]

Inverses of matrices are expressible, inefficiently, in terms of determinants. The **cofactor matrix** or **adjugate matrix** A^{adjg} of an n-by-n matrix A has $(i,j)^{th}$ entry

$$A_{ij}^{\text{adjg}} = (-1)^{i+j} \det A^{(ji)}$$

where A^{ji} is the matrix A with j^{th} row and i^{th} column removed. [3]   Then

$$A \cdot A^{\text{adjg}} = (\det A) \cdot 1_n$$

where 1_n is the n-by-n identity matrix. That is, if A is invertible,

$$A^{-1} = \frac{1}{\det A} \cdot A^{\text{adjg}}$$

---

[2] We describe expansion by minors just a little later, and prove that it is in fact unambiguous and correct.

[3] Yes, there is a reversal of indices: the $(ij)^{th}$ entry of A^{adjg} is, up to sign, the determinant of A with j^{th} row and i^{th} column removed. Later discussion of *exterior algebra* will clarify this construction/formula.

In the 2-by-2 case this formula is useful:

$$\begin{pmatrix} a & b \\ c & d \end{pmatrix}^{-1} = \frac{1}{ad - bc} \cdot \begin{pmatrix} d & -b \\ -c & a \end{pmatrix}$$

Similarly, a matrix (with entries in a field) is invertible if and only if its determinant is non-zero. [4]

The **Cayley-Hamilton theorem** is a widely misunderstood result, often given with seriously flawed proofs. [5] The **characteristic polynomial** $P_T(x)$ of an n-by-n matrix T is defined to be

$$P_T(x) = \det(x \cdot 1_n - T)$$

The assertion is that

$$P_T(T) = 0_n$$

where 0_n is the n-by-n zero matrix. The main use of this is that the **eigenvalues** of T are the roots of $P_T(x) = 0$. However, except for very small matrices, this is a suboptimal computational approach, *and* the **minimal polynomial** is far more useful for demonstrating qualitative facts about endomorphisms. Nevertheless, because there is a *formula* for the characteristic polynomial, it has a substantial popularity.

The easiest false proof of the Cayley-Hamilton Theorem is to apparently compute

$$P_T(T) = \det(T \cdot 1_n - T) = \det(T - T) = \det(0_n) = 0$$

The problem is that the substitution $x \cdot 1_n \longrightarrow T \cdot 1_n$ is not legitimate. The operation cannot be any kind of *scalar multiplication* after T is substituted for x, nor can it be composition of endomorphisms (nor multiplication of matrices). Further, there are interesting fallacious explanations of this incorrectness. For example, to say that we cannot substitute the *non-scalar* T for the *scalar variable* x fails to recognize that this is exactly what happens in the assertion of the theorem, *and* fails to see that the real problem is in the notion of the scalar multiplication of 1_n by x. That is, the *correct* objection is that $x \cdot 1_n$ is no longer a matrix with entries in the original field k (whatever that was), but in the polynomial ring $k[x]$, or in its field of fractions $k(x)$. But then it is much less clear what it might mean to substitute T for x, if x has become a kind of scalar.

Indeed, Cayley and Hamilton only proved the result in the 2-by-2 and 3-by-3 cases, by direct computation.

Often a correct argument is given that invokes the (existence part of the) structure theorem for finitely-generated modules over PIDs. A little later, our discussion of *exterior algebra* will allow a more direct argument, using the adjugate matrix. More importantly, the exterior algebra will make possible the long-postponed *uniqueness* part of the proof of the structure theorem for finitely-generated modules over PIDs.

---

[4] We prove this later in a much broader context.

[5] We give two different correct proofs later.

26.2 *Definitions*

For the present discussion, a **determinant** is a function D of *square matrices* with entries in a field k, taking values in that field, satisfying the following properties.

- **Linearity** as a function of each column: letting $C_1, \ldots, C_n$ in k^n be the columns of an n-by-n matrix C, for each $1 \le i \le n$ the function

$$C_i \longrightarrow D(C)$$

is a k-linear map $k^n \longrightarrow k$. [6]   That is, for scalar b and for two columns C_i and C_i'

$$D(\ldots, bC_i, \ldots) = b \cdot D(\ldots, C_i, \ldots)$$

$$D(\ldots, C_i + C_i', \ldots) = D(\ldots, C_i, \ldots) + D(\ldots, C_i', \ldots)$$

- **Alternating property:** [7]   If two adjacent columns of a matrix are equal, the determinant is 0.

- **Normalization:** The determinant of an identity matrix is 1:

$$D \begin{pmatrix} 1 & 0 & 0 & \cdots & 0 \\ 0 & 1 & 0 & \cdots & 0 \\ 0 & 0 & 1 & \cdots & 0 \\ \vdots & \vdots & & \ddots & \vdots \\ 0 & 0 & 0 & \cdots & 1 \end{pmatrix} = 1$$

That is, as a function of the columns, if the columns are the standard basis vectors in k^n then the value of the determinant is 1.

26.3 *Uniqueness and other properties*

- If two columns of a matrix are interchanged the value of the determinant is multiplied by -1. That is, writing the determinant as a function of the columns

$$D(C) = D(C_1, \ldots, C_n)$$

we have

$$D(C_1, \ldots, C_{i-1}, C_i, C_{i+1}, C_{i+2}, \ldots, C_n) = -D(C_1, \ldots, C_{i-1}, C_{i+1}, C_i, C_{i+2}, \ldots, C_n)$$

---

[6] Linearity as a function of several vector arguments is called **multilinearity**.

[7] The etymology of *alternating* is somewhat obscure, but does have a broader related usage, referring to rings that are *anti-commutative*, that is, in which $x \cdot y = -y \cdot x$. We will see how this is related to the present situation when we talk about *exterior algebras*. Another important family of alternating rings is *Lie algebras*, named after Sophus Lie, but in these the product is written $[x, y]$ rather than $x \cdot y$, both by convention and for functional reasons.

Proof: There is a little trick here. Consider the matrix with $C_i + C_j$ at both the ith and j^{th} columns. Using the linearity in both i^{th} and j^{th} columns, we have

$$0 = D(\ldots, C_i + C_j, \ldots, C_i + C_j, \ldots)$$

$$= D(\ldots, C_i, \ldots, C_i, \ldots) + D(\ldots, C_i, \ldots, C_j, \ldots)$$

$$+ D(\ldots, C_j, \ldots, C_i, \ldots) + D(\ldots, C_j, \ldots, C_j, \ldots)$$

The first and last determinants on the right are also 0, since the matrices have two identical columns. Thus,

$$0 = D(\ldots, C_i, \ldots, C_j, \ldots) + D(\ldots, C_j, \ldots, C_i, \ldots)$$

as claimed. ///

26.3.1 Remark: If the characteristic of the underlying field k is not 2, then we can replace the requirement that equality of two columns forces a determinant to be 0 by the requirement that interchange of two columns multiplies the determinant by -1. But this latter is a strictly weaker condition when the characteristic is 2.

• For any permutation π of $\{1, 2, 3, \ldots, n\}$ we have

$$D(C_{\pi(1)}, \ldots, C_{\pi(n)}) = \sigma(\pi) \cdot D(C_1, \ldots, C_n)$$

where C_i are the columns of a square matrix and σ is the *sign* function on S_n.

Proof: This argument is completely natural. The *adjacent transpositions* generate the permutation group S_n, and the sign function $\sigma(\pi)$ evaluated on a permutation π is $(-1)^t$ where t is the number of adjacent transpositions used to express π in terms of adjacent permutations. ///

• The value of a determinant is *unchanged* if a multiple of one column is added to another. That is, for indices $i < j$, with columns C_i considered as vectors in k^n, and for $b \in k$,

$$D(\ldots, C_i, \ldots, C_j, \ldots) = D(\ldots, C_i, \ldots, C_j + bC_i, \ldots)$$

$$D(\ldots, C_i, \ldots, C_j, \ldots) = D(\ldots, C_i + bC_j, \ldots, C_j, \ldots)$$

Proof: Using the linearity in the jth column,

$$D(\ldots, C_i, \ldots, C_j + bC_i, \ldots) = D(\ldots, C_i, \ldots, C_j, \ldots) + b \cdot D(\ldots, C_i, \ldots, C_i, \ldots)$$

$$= D(\ldots, C_i, \ldots, C_j, \ldots) + b \cdot 0 = D(\ldots, C_i, \ldots, C_j, \ldots)$$

since a determinant is 0 if two columns are equal. ///

• Let

$$C_j = \sum_i b_{ij} A_i$$

where b_{ij} are in k and $A_i \in k^n$. Let C be the matrix with i^{th} column C_i, and let A the the matrix with i^{th} column A_i. Then

$$D(C) = \left(\sum_{\pi \in S_n} \sigma(\pi) \, b_{\pi(1),1} \cdots, b_{\pi(n),n} \right) \cdot D(A)$$

and also

$$D(C) = \left(\sum_{\pi \in S_n} \sigma(\pi) \, b_{1,\pi(1),1} \cdots, b_{n,\pi(n)} \right) \cdot D(A)$$

Proof: First, expanding using (multi-) linearity, we have

$$D(\ldots, C_j, \ldots) = D(\ldots, \sum_i b_{ij} A_i, \ldots) = \sum_{i_1, \ldots, i_n} b_{i_1,1} \ldots b_{i_n,n} \, D(A_{i_1}, \ldots, A_{i_n})$$

where the ordered n-tuple $i_1, \ldots, i_n$ is summed over all choices of ordered n-tupes with entries from $\{1, \ldots, n\}$. If any two of i_p and i_q with $p \neq q$ are equal, then the matrix formed from the A_i will have two identical columns, and will be 0. Thus, we may as well sum over *permutations* of the ordered n-tuple $1, 2, 3, \ldots, n$. Letting π be the permutation which takes ℓ to i_ℓ, we have

$$D(A_{i_1}, \ldots, A_{i_n}) = \sigma(\pi) \cdot D(A_1, \ldots, A_n)$$

Thus,

$$D(C) = D(\ldots, C_j, \ldots) = \left(\sum_{\pi \in S_n} \sigma(\pi) \, b_{\pi(1),1} \cdots, b_{n,\pi(n)} \right) \cdot D(A)$$

as claimed.   For the second, complementary, formula, since multiplication in k is commutative,

$$b_{\pi(1),1} \ldots b_{\pi(n),n} = b_{1,\pi^{-1}(1)} \ldots b_{n,\pi^{-1}(n)}$$

Also,

$$1 = \sigma(1) = \sigma(\pi \circ \pi^{-1}) = \sigma(\pi) \cdot \sigma(\pi^{-1})$$

And the map $\pi \longrightarrow \pi^{-1}$ is a bijecton of S_n to itself, so

$$\sum_{\pi \in S_n} \sigma(\pi) \, b_{\pi(1),1} \cdots, b_{n,\pi(n)} = \sum_{\pi \in S_n} \sigma(\pi) \, b_{1,\pi(1)} \cdots, b_{\pi(n),n}$$

which yields the second formula.   ///

26.3.2 Remark: So far we have *not* used the normalization that the determinant of the identity matrix is 1. Now we will use this.

• Let c_{ij} be the $(i,j)^{th}$ entry of an n-by-n matrix C. Then

$$D(C) = \sum_{\pi \in S_n} \sigma(\pi) \, c_{\pi(1),1} \cdots, c_{n,\pi(n)}$$

Proof: In the previous result, take A to be the identity matrix.   ///

• (**Uniqueness**) There is at most one one determinant function on n-by-n matrices.

Proof: The previous formula is valid once we prove that determinants exist. ///

• The transpose $C^\top$ of C has the same determinant as does C

$$D(C^\top) = D(C)$$

Proof: Let c_{ij} be the $(i,j)^{th}$ entry of C. The $(i,j)^{th}$ entry $c_{ij}^\top$ of $C^\top$ is c_{ji}, and we have shown that

$$D(C^\top) = \sum_{\pi \in S_n} \sigma(\pi)\, c_{\pi(1),1}^\top \cdots c_{\pi(n),n}^\top$$

Thus,

$$D(C^\top) = \sum_{\pi \in S_n} \sigma(\pi)\, c_{\pi(1),1} \cdots c_{n,\pi(n)}$$

which is also $D(C)$, as just shown. ///

• (**Multiplicativity**) For two square matrices A, B with entries a_{ij} and b_{ij} and product $C = AB$ with entries c_{ij}, we have

$$D(AB) = D(A) \cdot D(B)$$

Proof: The jth column C_j of the product C is the linear combination

$$A_1 \cdot b_{1,j} + \ldots + A_n \cdot b_{n,j}$$

of the columns $A_1, \ldots, A_n$ of A. Thus, from above,

$$D(AB) = D(C) = \left(\sum_\pi \sigma(\pi)\, b_{\pi(1),1} \cdots b_{\pi(n),1} \right) \cdot D(A)$$

And we know that the sum is $D(B)$. ///

• If two rows of a matrix are identical, then its determinant is 0.

Proof: Taking transpose leaves the determinant alone, and a matrix with two identical columns has determinant 0. ///

• **Cramer's Rule** Let A be an n-by-n matrix with j^{th} column A_j. Let b be a column vector with i^{th} entry b_i. Let x be a column vector with i^{th} entry x_i. Let $A^{(\ell)}$ be the matrix obtained from A by replacing the jth column A_j by b. Then a solution x to an equation

$$Ax = b$$

is given by

$$x_\ell = \frac{D(A^{(\ell)})}{D(A)}$$

if $D(A) \neq 0$.

Proof: This follows directly from the alternating multilinear nature of determinants. First, the equation $Ax = b$ can be rewritten as an expression of b as a linear combination of the columns of A, namely

$$b = x_1 A_1 + x_2 A_2 + \ldots + x_n A_n$$

Then

$$D(A^{(\ell)}) = D(\ldots, A_{\ell-1}, \sum_j x_j A_j, A_{\ell+1}, \ldots) = \sum_j x_j \cdot D(\ldots, A_{\ell-1}, A_j, A_{\ell+1}, \ldots)$$

$$= x_\ell \cdot D(\ldots, A_{\ell-1}, A_\ell, A_{\ell+1}, \ldots) = x_\ell \cdot D(A)$$

since the determinant is 0 whenever two columns are identical, that is, unless $\ell = j$.
///

26.3.3 Remark: In fact, this proof of Cramer's Rule does a little more than verify the formula. First, even if $D(A) = 0$, still

$$D(A^{(\ell)}) = x_\ell \cdot D(A)$$

Second, for $D(A) \neq 0$, the computation actually shows that the solution x is *unique* (since any solutions x_ℓs satisfy the indicated relation).

• An n-by-n matrix is invertible if and only if its determinant is non-zero.

Proof: If A has an inverse A^{-1}, then from $A \cdot A^{-1} = 1_n$ and the multiplicativity of determinants,

$$D(A) \cdot D(A^{-1}) = D(1_n) = 1$$

so $D(A) \neq 0$. On the other hand, suppose $D(A) \neq 0$. Let e_i be the ith standard basis element of k^n, as a column vector. For each $j = 1, \ldots, n$ Cramer's Rule gives us a solution b_j to the equation

$$A b_j = e_j$$

Let B be the matrix whose j^{th} column is b_j. Then

$$AB = 1_n$$

To prove that also $BA = 1_n$ we proceed a little indirectly. Let T_M be the endomorphism of k^n given by a matrix M. Then

$$T_A \circ T_B = T_{AB} = T_{1_n} = \mathrm{id}_{k^n}$$

Thus, T_A is surjective. Since

$$\dim \mathrm{Im}\, T_A + \dim \ker T_A = n$$

necessarily T_A is also injective, so is an isomorphism of k^n. In particular, a right inverse is a left inverse, so also

$$T_{BA} = T_B \circ T_A = \mathrm{id}_{k^n}$$

The only matrix that gives the identity map on k^n is 1_n, so $BA = 1_n$. Thus, A is invertible.

$/\!/\!/$

26.3.4 Remark: All the above discussion assumes *existence* of determinants.

26.4 Existence

The standard *ad hoc* argument for existence is ugly, and we won't write it out. If one must a way to proceed is to check directly by induction on size that an *expansion by minors* along any row or column meets the requirements for a determinant function. Then invoke uniqueness.

This argument might be considered acceptable, but, in fact, it is much less illuminating than the use above of the key idea of *multilinearity* to prove properties of determinants before we're sure they exist. With hindsight, the capacity to talk about a determinant function $D(A)$ which is *linear* as a function of each column (and is alternating) is very effective in proving properties of determinants.

That is, without the notion of *linearity* a derivation of properties of determinants is much clumsier. This is why high-school treatments (and 200-year-old treatments) are awkward.

By contrast, we need a more sophisticated viewpoint than basic linear algebra in order to give a conceptual reason for the *existence* of determinants. Rather than muddle through expansion by minors, we will wait until we have developed the exterior algebra that makes this straightforward.

Exercises

26.1 Prove the *expansion by minors* formula for determinants, namely, for an n-by-n matrix A with entries a_{ij}, letting A^{ij} be the matrix obtained by deleting the ith row and j^{th} column, for any fixed row index i,

$$\det A = (-1)^i \sum_{j=1}^{n} (-1)^j a_{ij} \det A^{ij}$$

and symmetrically for expansion along a column. (*Hint:* Prove that this formula is linear in each row/column, and invoke the uniqueness of determinants.)

26.2 From just the most basic properties of determinants of matrices, show that the determinant of an upper-triangular matrix is the product of its diagonal entries. That is, show that

$$\det \begin{pmatrix} a_{11} & a_{12} & a_{13} & \cdots & a_{1n} \\ 0 & a_{22} & a_{23} & \cdots & a_{2n} \\ 0 & 0 & a_{33} & & \\ \vdots & & & \ddots & \vdots \\ 0 & \cdots & & 0 & a_{nn} \end{pmatrix} = a_{11}a_{22}a_{33}\ldots a_{nn}$$

26.3 Show that determinants *respect block decompositions*, at least to the extent that

$$\det \begin{pmatrix} A & B \\ 0 & D \end{pmatrix} = \det A \cdot \det D$$

where A is an m-by-n matrix, B is m-by-n, and D is n-by-n.

26.4 By an example, show that it is *not* always the case that

$$\det \begin{pmatrix} A & B \\ C & D \end{pmatrix} = \det A \cdot \det D - \det B \cdot \det C$$

for blocks A, B, C, D.

26.5 Let $x_1, \ldots, x_n$ and $y_1, \ldots, y_n$ be two orthonormal bases in a real inner-product space. Let M be the matrix whose ij^{th} entry is

$$M_{ij} = \langle x_i, y_j \rangle$$

Show that $\det M = 1$.

26.6 For real numbers a, b, c, d, prove that

$$\left| \det \begin{pmatrix} a & b \\ c & d \end{pmatrix} \right| = (\text{area of parallelogram spanned by } (a,b) \text{ and } (c,d))$$

26.7 For real vectors $v_i = (x_i, y_i, z_i)$ with $i = 1, 2, 3$, show that

$$\left| \det \begin{pmatrix} x_1 & y_1 & z_1 \\ x_2 & y_2 & z_2 \\ x_3 & y_3 & z_3 \end{pmatrix} \right| = (\text{volume of parallelogram spanned by } v_1, v_2, v_3)$$

27. Tensor products

In this first pass at tensor products, we will only consider tensor products of modules over commutative rings with identity. This is not at all a critical restriction, but does offer many simplifications, while still illuminating many important features of tensor products and their applications.

27.1 *Desiderata*

It is time to take stock of what we are missing in our development of linear algebra and related matters.

Most recently, we are missing the proof of existence of determinants, although linear algebra is sufficient to give palatable proofs of the properties of determinants.

We want to be able to give a direct and natural proof of the Cayley-Hamilton theorem (without using the structure theorem for finitely-generated modules over PIDs). This example suggests that linear algebra over *fields* is insufficient.

We want a sufficient conceptual situation to be able to finish the *uniqueness* part of the structure theorem for finitely-generated modules over PIDs. Again, linear or multi-linear algebra over fields is surely insufficient for this.

We might want an antidote to the antique styles of discussion of *vectors v_i* [sic], *covectors v^i* [sic], *mixed tensors T^{ij}_k*, and other vague entities whose nature was supposedly specified by the number and pattern of upper and lower subscripts. These often-ill-defined notions came into existence in the mid-to-late 19th century in the development of geometry. Perhaps the impressive point is that, even without adequate algebraic grounding, people managed to envision and roughly formulate geometric notions.

In a related vein, at the beginning of calculus of several variables, one finds ill-defined notions and ill-made distinctions between

$$dx \, dy$$

and

$$dx \wedge dy$$

with the nature of the so-called *differentials dx* and *dy* even less clear. For a usually unspecified reason,

$$dx \wedge dy = -dy \wedge dx$$

though perhaps

$$dx \, dy = dy \, dx$$

In other contexts, one may find confusion between the integration of *differential forms* versus integration with respect to a *measure*. We will not resolve *all* these confusions here, only the question of what $a \wedge b$ might mean.

Even in fairly concrete linear algebra, the question of **extension of scalars** to convert a real vector space to a complex vector space is possibly mysterious. On one hand, if we are content to say that vectors are *column* vectors or *row* vectors, then we might be equally content in allowing complex entries. For that matter, once a basis for a real vector space is chosen, to write apparent linear combinations with complex coefficients (rather than merely real coefficients) is easy, as symbol manipulation. However, it is quite unclear what meaning can be attached to such expressions. Further, it is unclear what effect a different choice of basis might have on this process. Finally, without a choice of basis, these *ad hoc* notions of extension of scalars are stymied. Instead, the construction below of the tensor product

$$V \otimes_R \mathbb{C} = \text{complexification of } V$$

of a real vectorspace V with $\mathbb{C}$ over $\mathbb{R}$ is exactly right, as will be discussed later.

The notion of *extension of scalars* has important senses in situations which are qualitatively different than complexification of real vector spaces. For example, there are several reasons to want to convert *abelian groups A* ($\mathbb{Z}$-modules) into $\mathbb{Q}$-vectorspaces in some reasonable, natural manner. After explicating a minimalist notion of reasonability, we will see that a tensor product

$$A \otimes_{\mathbb{Z}} \mathbb{Q}$$

is just right.

There are many examples of application of the construction and universal properties of tensor products.

27.2 *Definitions, uniqueness, existence*

Let R be a commutative ring with 1. We will only consider R-modules M with the property [1] that $1 \cdot m = m$ for all $m \in M$. Let M, N, and X be R-modules. A map

$$B : M \times N \longrightarrow X$$

is R-**bilinear** if it is R-linear separately in each argument, that is, if

$$
\begin{aligned}
B(m + m', n) &= B(m, n) + B(m', n) \\
B(rm, n) &= r \cdot B(m, n) \\
B(m, n + n') &= B(m, n) + B(m, n') \\
B(m, rn) &= r \cdot B(m, n)
\end{aligned}
$$

for all $m, m' \in M$, $n, n' \in N$, and $r \in R$.

As in earlier discussion of free modules, and in discussion of polynomial rings as free algebras, we will define tensor products by *mapping properties*. This will allow us an easy proof that tensor products (if they exist) are *unique* up to *unique isomorphism*. Thus, whatever construction we contrive must inevitably yield the same (or, better, *equivalent*) object. Then we give a modern construction.

A **tensor product** of R-modules M, N is an R-module denoted $M \otimes_R N$ together with an R-bilinear map

$$\tau : M \times N \longrightarrow M \otimes_R N$$

such that, for every R-bilinear map

$$\varphi : M \times N \longrightarrow X$$

there is a unique *linear* map

$$\Phi : M \otimes_{\mathbb{R}} N \longrightarrow X$$

such that the diagram

commutes, that is, $\varphi = \Phi \circ \tau$.

The usual notation does not involve any symbol such as τ, but, rather, denotes the image $\tau(m \times n)$ of $m \times n$ in the tensor product by

$$m \otimes n = \text{image of } m \times n \text{ in } M \otimes_R N$$

In practice, the implied R-bilinear map

$$M \times N \longrightarrow M \otimes_R N$$

---

[1] Sometimes such a module M is said to be **unital**, but this terminology is not universal, and, thus, somewhat unreliable. Certainly the term is readily confused with other usages.

is often left anonymous. This seldom causes serious problems, but we will be temporarily more careful about this while setting things up and proving basic properties.

The following proposition is typical of uniqueness proofs for objects defined by mapping property requirements. Note that internal details of the objects involved play no role. Rather, the argument proceeds by manipulation of arrows.

27.2.1 Proposition: Tensor products $M \otimes_R N$ are unique up to unique isomorphism. That is, given two tensor products

$$\tau_1 : M \times N \longrightarrow T_1$$

$$\tau_2 : M \times N \longrightarrow T_2$$

there is a *unique isomorphism* $i : T_1 \longrightarrow T_2$ such that the diagram

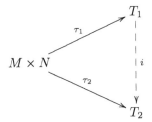

commutes, that is, $\tau_2 = i \circ \tau_1$.

Proof: First, we show that for a tensor product $\tau : M \times N \longrightarrow T$, the only map $f : T \longrightarrow T$ compatible with τ is the identity. That is the identity map is the only map f such that

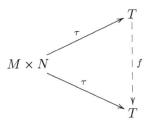

commutes. Indeed, the definition of a tensor product demands that, given the bilinear map

$$\tau : M \times N \longrightarrow T$$

(with T in the place of the earlier X) there is a unique linear map $\Phi : T \longrightarrow T$ such that the diagram

$$
\begin{array}{ccc}
T & & \\
\tau \uparrow & \diagdown \raisebox{0pt}{Φ} & \\
M \times N & \xrightarrow{\ \ \tau\ \ } & T
\end{array}
$$

commutes. The identity map on T certainly has this property, so is the *only* map $T \longrightarrow T$ with this property.

Looking at two tensor products, first take $\tau_2 : M \times N \longrightarrow T_2$ in place of the $\varphi : M \times N \longrightarrow X$. That is, there is a unique linear $\Phi_1 : T_1 \longrightarrow T_2$ such that

$$
\begin{array}{ccc}
T_1 & & \\
\tau_1 \uparrow & \overset{\Phi_1}{\dashrightarrow} & \\
M \times N & \xrightarrow{\ \tau_2\ } & T_2
\end{array}
$$

commutes. Similarly, reversing the roles, there is a unique linear $\Phi_2 : T_2 \longrightarrow T_1$ such that

$$
\begin{array}{ccc}
T_2 & & \\
\tau_2 \uparrow & \overset{\Phi_2}{\dashrightarrow} & \\
M \times N & \xrightarrow{\ \tau_1\ } & T_1
\end{array}
$$

commutes. Then $\Phi_2 \circ \Phi_1 : T_1 \longrightarrow T_1$ is compatible with τ_1, so is the identity, from the first part of the proof. And, symmetrically, $\Phi_1 \circ \Phi_2 : T_2 \longrightarrow T_2$ is compatible with τ_2, so is the identity. Thus, the maps Φ_i are mutual inverses, so are isomorphisms. ///

For existence, we will give an argument in what might be viewed as an *extravagant* modern style. Its extravagance is similar to that in E. Artin's proof of the existence of algebraic closures of fields, in which we create an indeterminate for each irreducible polynomial, and look at the polynomial ring in these myriad indeterminates. In a similar spirit, the tensor product $M \otimes_R N$ will be created as a quotient of a truly huge module by an only slightly less-huge module.

27.2.2 Proposition: Tensor products $M \otimes_R N$ exist.

Proof: Let $i : M \times N \longrightarrow F$ be the free R-module on the *set* $M \times N$. Let Y be the R-submodule generated by all elements

$$
\begin{aligned}
& i(m + m', n) - i(m, n) - i(m', n) \\
& i(rm, n) - r \cdot i(m, n) \\
& i(m, n + n') - i(m, n) - i(m, n') \\
& i(m, rn) - r \cdot i(m, n)
\end{aligned}
$$

for all $r \in R$, $m, m' \in M$, and $n, n' \in N$. Let

$$
q : F \longrightarrow F/Y
$$

be the quotient map. We claim that $\tau = q \circ i : M \times N \longrightarrow F/Y$ is a tensor product.

Given a bilinear map $\varphi : M \times N \longrightarrow X$, by properties of free modules there is a unique $\Psi : F \longrightarrow X$ such that the diagram

$$
\begin{array}{ccc}
F & & \\
i \uparrow & \overset{\Psi}{\dashrightarrow} & \\
M \times N & \xrightarrow{\ \varphi\ } & X
\end{array}
$$

commutes. We claim that Ψ factors through F/Y, that is, that there is $\Phi : F/Y \longrightarrow X$ such that

$$
\Psi = \Phi \circ q : F \longrightarrow X
$$

Indeed, since $\varphi : M \times N \longrightarrow X$ is bilinear, we conclude that, for example,

$$\varphi(m + m', n) = \varphi(m, n) + \varphi(m', n)$$

Thus,

$$(\Psi \circ i)(m + m', n) = (\Psi \circ i)(m, n) + (\Psi \circ i)(m', n)$$

Thus, since Ψ is linear,

$$\Psi(\, i(m + m', n) - i(m, n) - i(m', n) \,) = 0$$

A similar argument applies to all the generators of the submodule Y of F, so Ψ does factor through F/Y. Let Φ be the map such that $\Psi = \Phi \circ q$.

A similar argument on the generators for Y shows that the composite

$$\tau = q \circ i : M \times N \longrightarrow F/Y$$

is bilinear, even though i was only a set map.

The uniqueness of Ψ yields the uniqueness of Φ, since q is a surjection, as follows. For two maps Φ_1 and Φ_2 with

$$\Phi_1 \circ q = \Psi = \Phi_2 \circ q$$

given $x \in F/Y$ let $y \in F$ be such that $q(y) = x$. Then

$$\Phi_1(x) = (\Phi_1 \circ q)(y) = \Psi(y) = (\Phi_2 \circ q)(y) = \Phi_2(x)$$

Thus, $\Phi_1 = \Phi_2$.   ///

27.2.3 Remark: It is worthwhile to contemplate the many things we did *not* do to prove the uniqueness and the existence.

Lest anyone think that tensor products $M \otimes_R N$ contain anything not implicitly determined by the behavior of the **monomial tensors** [2] $m \otimes n$, we prove

27.2.4 Proposition: The monomial tensors $m \otimes n$ (for $m \in M$ and $n \in N$) generate $M \otimes_R N$ as an R-module.

Proof: Let X be the submodule of $M \otimes_R N$ generated by the monomial tensors, $Q = M \otimes_R N)/X$ the quotient, and $q : M \otimes_R N \longrightarrow Q$ the quotient map. Let

$$B : M \times N \longrightarrow Q$$

be the 0-map. A defining property of the tensor product is that there is a unique R-linear

$$\beta : M \otimes_R N \longrightarrow Q$$

making the usual diagram commute, that is, such that $B = \beta \circ \tau$, where $\tau : M \times N \longrightarrow M \otimes_R N$. Both the quotient map q and the 0-map $M \otimes_R N \longrightarrow Q$ allow the 0-map $M \times N \longrightarrow Q$ to factor through, so by the uniqueness the quotient map is the 0-map. That is, Q is the 0-module, so $X = M \otimes_R N$.   ///

---

[2] Again, $m \otimes n$ is the image of $m \times n \in M \times N$ in $M \otimes_R N$ under the map $\tau : M \times N \longrightarrow M \otimes_R N$.

27.2.5 Remark: Similarly, define the tensor product

$$\tau : M_1 \times \ldots \times M_n \longrightarrow M_1 \otimes_R \ldots \otimes_R M_n$$

of an arbitrary finite number of R-modules as an R-module and multilinear map τ such that, for any R-multilinear map

$$\varphi : M_1 \times M_2 \times \ldots \times M_n \longrightarrow X$$

there is a unique R-linear map

$$\Phi : M_1 \otimes_R M_2 \otimes_R \ldots \otimes_R M_n \longrightarrow X$$

such that $\varphi = \Phi \circ \tau$. That is, the diagram

$$M_1 \otimes_R M_2 \otimes_R \ldots \otimes_R M_n$$

$$M_1 \times M_2 \times \ldots \times M_n \xrightarrow{\quad \varphi \quad} X$$

commutes. There is the subordinate issue of proving **associativity**, namely, that there are natural isomorphisms

$$(M_1 \otimes_R \ldots \otimes_R M_{n-1}) \otimes_R M_n \approx M_1 \otimes_R (M_2 \otimes_R \ldots \otimes_R M_n)$$

to be sure that we need not worry about parentheses.

27.3 *First examples*

We want to illustrate the possibility of computing [3] tensor products without needing to make any use of any suppositions about the *internal* structure of tensor products.

First, we emphasize that to show that a tensor product $M \otimes_R N$ of two R-modules (where R is a commutative ring with identity) is 0, it suffices to show that all monomial tensors are 0, since these generate the tensor product (as R-module). [4]

Second, we emphasize [5] that in $M \otimes_R N$, with $r \in R$, $m \in M$, and $n \in N$, we can always rearrange

$$(rm) \otimes n = r(m \otimes n) = m \otimes (rn)$$

Also, for $r, s \in R$,

$$(r + s)(m \otimes n) = rm \otimes n + sm \otimes n$$

---

[3] Of course, it is unclear in what *sense* we are computing. In the simpler examples the tensor product is the 0 module, which needs no further explanation. However, in other cases, we will see that a certain tensor product is the *right answer* to a natural question, without necessarily determining what the tensor product is in some *other* sense.

[4] This was proven just above.

[5] These are merely translations into this notation of part of the definition of the tensor product, but deserve emphasis.

27.3.1 Example: Let's experiment[6]   first with something like

$$\mathbb{Z}/5 \otimes_{\mathbb{Z}} \mathbb{Z}/7$$

Even a novice may anticipate that the fact that 5 annihilates the left factor, while 7 annihilates the right factor, creates an interesting dramatic tension. What will come of this? For any $m \in \mathbb{Z}/5$ and $n \in \mathbb{Z}/7$, we can do things like

$$0 = 0 \cdot (m \otimes n) = (0 \cdot m) \otimes n = (5 \cdot m) \otimes n = m \otimes 5n$$

and

$$0 = 0 \cdot (m \otimes n) = m \otimes (0 \cdot n) = m \otimes (7 \cdot n) = 7m \otimes n = 2m \otimes n$$

Then

$$(5m \otimes n) - 2 \cdot (2m \otimes n) = (5 - 2 \cdot 2)m \otimes n = m \otimes n$$

but also

$$(5m \otimes n) - 2 \cdot (2m \otimes n) = 0 - 2 \cdot 0 = 0$$

That is, every monomial tensor in $\mathbb{Z}/5 \otimes_{\mathbb{Z}} \mathbb{Z}/7$ is 0, so the whole tensor product is 0.

27.3.2 Example: More systematically, given relatively prime integers[7] a, b, we claim that

$$\mathbb{Z}/a \otimes_{\mathbb{Z}} \mathbb{Z}/b = 0$$

Indeed, using the Euclidean-ness of $\mathbb{Z}$, let $r, s \in \mathbb{Z}$ such that

$$1 = ra + sb$$

Then

$$m \otimes n = 1 \cdot (m \otimes n) = (ra + sb) \cdot (m \otimes n) = ra(m \otimes n) + s$$

$$= b(m \otimes n) = a(rm \otimes n) + b(m \otimes sn) = a \cdot 0 + b \cdot 0 = 0$$

Thus, every monomial tensor is 0, so the whole tensor product is 0.

27.3.3 Remark: Yes, it somehow not visible that these should be 0, since we probably think of tensors are complicated objects, not likely to be 0. But this vanishing is an assertion that there are *no* non-zero $\mathbb{Z}$-bilinear maps from $\mathbb{Z}/5 \times \mathbb{Z}/7$, which is a plausible more-structural assertion.

27.3.4 Example: Refining the previous example: let a, b be arbitrary non-zero integers. We claim that

$$\mathbb{Z}/a \otimes_{\mathbb{Z}} \mathbb{Z}/b \approx \mathbb{Z}/\gcd(a, b)$$

---

[6] Or pretend, disingenuously, that we don't know what will happen? Still, some tangible numerical examples are worthwhile, much as a picture may be worth many words.

[7] The same argument obviously works as stated in Euclidean rings R, rather than just $\mathbb{Z}$. Further, a restated form works for arbitrary commutative rings R with identity: given two ring elements a, b such that the ideal $Ra + Rb$ generated by *both* is the whole ring, we have $R/a \otimes_R R/b = 0$. The point is that this adjusted hypothesis again gives us $r, s \in R$ such that $1 = ra + sb$, and then the same argument works.

First, take $r, s \in \mathbb{Z}$ such that

$$\gcd(a, b) = ra + sb$$

Then the same argument as above shows that this *gcd* annihilates every monomial

$$(ra + sb)(m \otimes n) = r(am \otimes n) + s(m \otimes bn) = r \cdot 0 + s \cdot 0 = 0$$

Unlike the previous example, we are not entirely done, since we didn't simply prove that the tensor product is 0. We need something like

27.3.5 Proposition: Let $\{m_\alpha : \alpha \in A\}$ be a set of generators for an R-module M, and $\{n_\beta : \beta \in B\}$ a set of generators for an R-module N. Then

$$\{m_\alpha \otimes n_\beta : \alpha \in A, \ \beta \in B\}$$

is a set of generators [8] for $M \otimes_R N$.

Proof: Since monomial tensors generate the tensor product, it suffices to show that every monomial tensor is expressible in terms of the $m_\alpha \otimes n_\beta$. Unsurprisingly, taking r_α and s_β in R (0 for all but finitely-many indices), by multilinearity

$$\left(\sum_\alpha r_\alpha m_\alpha\right) \otimes \left(\sum_\beta s_\beta n_\beta\right) = \sum_{\alpha,\beta} r_\alpha s_\beta \, m_\alpha \otimes n_\beta$$

This proves that the special monomials $m_\alpha \otimes n_\beta$ generate the tensor product. ///

Returning to the example, since $1 + a\mathbb{Z}$ generates $\mathbb{Z}/a$ and $1 + b\mathbb{Z}$ generates $\mathbb{Z}/b$, the proposition assures us that $1 \otimes 1$ generates the tensor product. We already know that

$$\gcd(a, b) \cdot 1 \otimes 1 = 0$$

Thus, we know that $\mathbb{Z}/a \otimes \mathbb{Z}/b$ is isomorphic to *some quotient of* $\mathbb{Z}/\gcd(a, b)$.

But this does not preclude the possibility that something else is 0 for a reason we didn't anticipate. One more ingredient is needed to prove the claim, namely exhibition of a sufficiently non-trivial bilinear map to eliminate the possibility of any further collapsing. One might naturally contrive a $\mathbb{Z}$-blinear map with formulaic expression

$$B(x, y) = xy \dots$$

but there may be some difficulty in intuiting where that xy resides. To understand this, we must be scrupulous about cosets, namely

$$(x + a\mathbb{Z}) \cdot (y + b\mathbb{Z}) = xy + ay\mathbb{Z} + bx\mathbb{Z} + ab\mathbb{Z} \subset xy + a\mathbb{Z} + b\mathbb{Z} = xy + \gcd(a,b)\mathbb{Z}$$

That is, the bilinear map is

$$B : \mathbb{Z}/a \times \mathbb{Z}/b \longrightarrow \mathbb{Z}/\gcd(a, b)$$

---

[8] It would be unwise, and generally very difficult, to try to give generators *and relations* for tensor products.

By construction,
$$B(1,1) = 1 \in \mathbb{Z}/\gcd(a,b)$$
so
$$\beta(1 \otimes 1) = B(1,1) = 1 \in \mathbb{Z}/\gcd(a,b)$$

In particular, the map is a surjection. Thus, knowing that the tensor product is generated by $1 \otimes 1$, and that this element has order *dividing* $\gcd(a,b)$, we find that it has order *exactly* $\gcd(a,b)$, so is *isomorphic* to $\mathbb{Z}/\gcd(a,b)$, by the map

$$x \otimes y \longrightarrow xy$$

27.4 *Tensor products $f \otimes g$ of maps*

Still R is a commutative ring with 1.

An important type of map on a tensor product arises from pairs of R-linear maps on the modules in the tensor product. That is, let

$$f : M \longrightarrow M' \quad g : N \longrightarrow N'$$

be R-module maps, and attempt to define

$$f \otimes g : M \otimes_R N \longrightarrow M' \otimes_R N'$$

by

$$(f \otimes g)(m \otimes n) = f(m) \otimes g(n)$$

Justifiably interested in being sure that this formula makes sense, we proceed as follows.

If the map is *well-defined* then it is defined completely by its values on the monomial tensors, since these generate the tensor product. To prove well-definedness, we invoke the defining property of the tensor product, by first considering a bilinear map

$$B : M \times N \longrightarrow M' \otimes_R N'$$

given by

$$B(m \times n) = f(m) \otimes g(n)$$

To see that *this* bilinear map is well-defined, let

$$\tau' : M' \times N' \longrightarrow M' \otimes_R N'$$

For fixed $n \in N$, the composite

$$m \longrightarrow f(m) \longrightarrow \tau'(f(m), g(n)) = f(m) \otimes g(n)$$

is certainly an R-linear map in m. Similarly, for fixed $m \in M$,

$$n \longrightarrow g(n) \longrightarrow \tau'(f(m), g(n)) = f(m) \otimes g(n)$$

is an R-linear map in n. Thus, B is an R-bilinear map, and the formula for $f \otimes g$ expresses the induced linear map on the tensor product.

Similarly, for an n-tuple of R-linear maps

$$f_i : M_i \longrightarrow N_i$$

there is an associated linear

$$f_1 \otimes \ldots \otimes f_n : M_1 \otimes \ldots \otimes M_n \longrightarrow N_1 \otimes \ldots \otimes N_n$$

27.5 *Extension of scalars, functoriality, naturality*

How to turn an R-module M into an S-module? [9] We assume that R and S are commutative rings with unit, and that there is a ring homomorphism $\alpha : R \longrightarrow S$ such that $\alpha(1_R) = 1_S$. For example the situation that $R \subset S$ with $1_R = 1_S$ is included. But also we want to allow not-injective maps, such as quotient maps $\mathbb{Z} \longrightarrow \mathbb{Z}/n$. This makes S an R-algebra, by

$$r \cdot s = \alpha(r)s$$

Before describing the *internal details* of this conversion, we should tell what criteria it should meet. Let

$$F : \{R - \text{modules}\} \longrightarrow \{S - \text{modules}\}$$

be this conversion. [10] Our main requirement is that for R-modules M and S-modules N, there should be a natural[11] isomorphism[12]

$$\text{Hom}_S(FM, N) \approx \text{Hom}_R(M, N)$$

where on the right side we *forget* that N is an S-module, remembering only the action of R on it. If we want to make more explicit this *forgetting*, we can write

$$\text{Res}^S_R N = R\text{-module obtained by forgetting } S\text{-module structure on } N$$

---

[9] As an alert reader can guess, the anticipated answer involves tensor products. However, we can lend some dignity to the proceedings by explaining *requirements* that should be met, rather than merely contriving from an R-module a thing that happens to be an S-module.

[10] This F would be an example of a **functor** from the **category** of R-modules and R-module maps to the **category** of S-modules and S-module maps. To be a genuine functor, we should also tell how F converts R-module *homomorphisms* to S-module homomorphisms. We do not need to develop the formalities of category theory just now, so will not do so. In fact, direct development of a variety of such examples surely provides the only sensible and genuine motivation for a later formal development of category theory.

[11] This sense of *natural* will be made precise shortly. It is the same sort of naturality as discussed earlier in the simplest example of second duals of finite-dimensional vector spaces over fields.

[12] It suffices to consider the map as an isomorphism of abelian groups, but, in fact, the isomorphism potentially makes sense as an S-module map, if we give both sides S-module structures. For $\Phi \in \text{Hom}_S(FM, N)$, there is an unambiguous and unsurprising S-module structure, namely $(s\Phi)(m') = s \cdot \Phi(m') = s \cdot \Phi(m')$ for $m' \in FM$ and $s \in S$. For $\varphi \in \text{Hom}_R(M, N)$, since N does have the additional structure of S-module, we have $(s \cdot \varphi)(m) = s \cdot \varphi(m)$.

and then, more carefully, write what we want for extension of scalars as

$$\mathrm{Hom}_S(FM, N) \approx \mathrm{Hom}_R(M, \mathrm{Res}_R^S N)$$

Though we'll not use it much in the immediate sequel, this extra notation does have the virtue that it makes clear that *something happened* to the module N.

This association of an S-module FM to an R-module M is *not* itself a module map. Instead, it is a **functor** from R-modules to S-modules, meaning that for an R-module map $f : M \longrightarrow M'$ there should be a naturally associated S-module map $Ff : FM \longrightarrow FM'$. Further, F should respect the composition of module homomorphisms, namely, for R-module homomorphisms

$$M \xrightarrow{f} M' \xrightarrow{g} M''$$

it should be that

$$F(g \circ f) = Fg \circ Ff : FM \longrightarrow FM''$$

This already makes clear that we shouldn't be completely cavalier in converting R-modules to S-modules.

Now we are able to describe the **naturality** we require of the desired isomorphism

$$\mathrm{Hom}_S(FM, N) \xrightarrow{i_{M,N}} \mathrm{Hom}_R(M, N)$$

One part of the naturality is **functoriality in** N, which requires that for every R-module map $g : N \longrightarrow N'$ the diagram

$$
\begin{array}{ccc}
\mathrm{Hom}_S(FM, N) & \xrightarrow{\ i_{M,N}\ } & \mathrm{Hom}_R(M, N) \\
\downarrow{\scriptstyle g \circ -} & & \downarrow{\scriptstyle g \circ -} \\
\mathrm{Hom}_S(FM, N') & \xrightarrow{\ i_{M,N'}\ } & \mathrm{Hom}_R(M, N')
\end{array}
$$

commutes, where the map $g \circ -$ is (post-) composition with g, by

$$g \circ - \ : \varphi \longrightarrow g \circ \varphi$$

Obviously one oughtn't imagine that it is easy to haphazardly guess a functor F possessing such virtues. [13]   There is also the requirement of **functoriality in** M, which requires for every $f : M \longrightarrow M'$ that the diagram

$$
\begin{array}{ccc}
\mathrm{Hom}_S(FM, N) & \xrightarrow{\ i_{M,N}\ } & \mathrm{Hom}_R(M, N) \\
\uparrow{\scriptstyle - \circ Ff} & & \uparrow{\scriptstyle - \circ f} \\
\mathrm{Hom}_S(FM', N) & \xrightarrow{\ i_{M',N}\ } & \mathrm{Hom}_R(M', N)
\end{array}
$$

commutes, where the map $- \circ Ff$ is (pre-) composition with Ff, by

$$- \circ Ff \ : \varphi \longrightarrow \varphi \circ Ff$$

---

[13] Further, the same attitude might demand that we worry about the *uniqueness* of such F. Indeed, there is such a uniqueness statement that can be made, but more preparation would be required than we can afford just now. The assertion would be about uniqueness of *adjoint functors*.

After all these demands, it is a relief to have

27.5.1 Theorem: The extension-of-scalars (from R to S) module FM attached to an R-module M is

$$\text{extension-of-scalars-}R\text{-to-}S \text{ of } M = M \otimes_R S$$

That is, for every R-module M and S-module N there is a *natural* isomorphism

$$\text{Hom}_S(M \otimes_R S, N) \xrightarrow{i_{M,N}} \text{Hom}_R(M, \text{Res}_R^S N)$$

given by

$$i_{M,N}(\Phi)(m) = \Phi(m \otimes 1)$$

for $\Phi \in \text{Hom}_S(M \otimes_R S, N)$, with inverse

$$j_{M,N}(\varphi)(m \otimes s) = s \cdot \varphi(m)$$

for $s \in S$, $m \in M$.

Proof: First, we verify that the map $i_{M,N}$ given in the statement is an isomorphism, and then prove the functoriality in N, and functoriality in M.

For the moment, write simply i for $i_{M,N}$ and j for $j_{M,N}$. Then

$$((j \circ i)\Phi)(m \otimes s) = (j(i\Phi))(m \otimes s) = s \cdot (i\Phi)(m) = s \cdot \Phi(m \otimes 1) = \Phi(m \otimes s)$$

and

$$((i \circ j)\varphi)(m) = (i(j\varphi))(m) = (j\varphi)(m \otimes 1) = 1 \cdot \varphi(m) = \varphi(m)$$

This proves that the maps are isomorphisms.

For functoriality in N, we must prove that for every R-module map $g : N \longrightarrow N'$ the diagram

$$
\begin{array}{ccc}
\text{Hom}_S(M \otimes_R S, N) & \xrightarrow{i_{M,N}} & \text{Hom}_R(M, N) \\
\downarrow{\scriptstyle g \circ -} & & \downarrow{\scriptstyle g \circ -} \\
\text{Hom}_S(M \otimes_R S, N') & \xrightarrow{i_{M,N'}} & \text{Hom}_R(M, N')
\end{array}
$$

commutes. For brevity, let $i = i_{M,N}$ and $i' = i_{M,N'}$. Directly computing, using the definitions,

$$((i' \circ (g \circ -))\Phi)(m) = (i' \circ (g \circ \Phi))(m) = (g \circ \Phi)(m \otimes 1)$$

$$= g(\Phi(m \otimes 1)) = g(i\Phi(m)) = ((g \circ -) \circ i)\Phi)(m)$$

For functoriality in M, for each R-module homomorphism $f : M \longrightarrow M'$ we must prove that the diagram

$$
\begin{array}{ccc}
\text{Hom}_S(M \otimes_R S, N) & \xrightarrow{i_{M,N}} & \text{Hom}_R(M, N) \\
\uparrow{\scriptstyle - \circ Ff} & & \uparrow{\scriptstyle - \circ f} \\
\text{Hom}_S(M' \otimes_R S, N) & \xrightarrow{i_{M',N}} & \text{Hom}_R(M', N)
\end{array}
$$

commutes, where $f \otimes 1$ is the map of $M \otimes_R S$ to itself determined by

$$(f \otimes 1)(m \otimes s) = f(m) \otimes s$$

and $- \circ (f \otimes 1)$ is (pre-) composition with this function. Again, let $i = i_{M,N}$ and $i' = i_{M',N}$, and compute directly

$$(((- \circ f) \circ i')\Psi)(m) = ((- \circ f)(i'\Psi)(m) = (i'\Psi \circ f)(m) = (i'\Psi)(fm)$$

$$= \Psi(fm \otimes 1) = (\Psi \circ (f \otimes 1))(m \otimes 1) = (i(\Psi \circ (f \otimes 1)))(m) = ((i \circ (- \circ (f \otimes 1)))\Psi)(m)$$

Despite the thicket of parentheses, this does prove what we want, namely, that

$$(- \circ f) \circ i' = i \circ (- \circ (f \otimes 1))$$

proving the functoriality of the isomorphism in M.   ///

27.6 *Worked examples*

27.6.1 Example: For distinct primes p, q, compute

$$\mathbb{Z}/p \otimes_{\mathbb{Z}/pq} \mathbb{Z}/q$$

where for a divisor d of an integer n the abelian group $\mathbb{Z}/d$ is given the $\mathbb{Z}/n$-module structure by

$$(r + n\mathbb{Z}) \cdot (x + d\mathbb{Z}) = rx + d\mathbb{Z}$$

We claim that this tensor product is 0. To prove this, it suffices to prove that every $m \otimes n$ (the image of $m \times n$ in the tensor product) is 0, since we have shown that these *monomial* tensors always generate the tensor product.

Since p and q are relatively prime, there exist integers a, b such that $1 = ap + bq$. Then for all $m \in \mathbb{Z}/p$ and $n \in \mathbb{Z}/q$,

$$m \otimes n = 1 \cdot (m \otimes n) = (ap + bq)(m \otimes n) = a(pm \otimes n) + b(m \otimes qn) = a \cdot 0 + b \cdot 0 = 0$$

An auxiliary point is to recognize that, indeed, $\mathbb{Z}/p$ and $\mathbb{Z}/q$ really are $\mathbb{Z}/pq$-modules, and that the equation $1 = ap + bq$ still does make sense inside $\mathbb{Z}/pq$.   ///

27.6.2 Example: Compute $\mathbb{Z}/n \otimes_{\mathbb{Z}} \mathbb{Q}$ with $0 < n \in \mathbb{Z}$.

We claim that the tensor product is 0. It suffices to show that every $m \otimes n$ is 0, since these monomials generate the tensor product. For any $x \in \mathbb{Z}/n$ and $y \in \mathbb{Q}$,

$$x \otimes y = x \otimes (n \cdot \frac{y}{n}) = (nx) \otimes \frac{y}{n} = 0 \otimes \frac{y}{n} = 0$$

as claimed.   ///

27.6.3 Example: Compute $\mathbb{Z}/n \otimes_{\mathbb{Z}} \mathbb{Q}/\mathbb{Z}$ with $0 < n \in \mathbb{Z}$.

We claim that the tensor product is 0. It suffices to show that every $m \otimes n$ is 0, since these monomials generate the tensor product. For any $x \in \mathbb{Z}/n$ and $y \in \mathbb{Q}/\mathbb{Z}$,

$$x \otimes y = x \otimes (n \cdot \frac{y}{n}) = (nx) \otimes \frac{y}{n} = 0 \otimes \frac{y}{n} = 0$$

as claimed. ///

27.6.4 Example: Compute $\operatorname{Hom}_{\mathbb{Z}}(\mathbb{Z}/n, \mathbb{Q}/\mathbb{Z})$ for $0 < n \in \mathbb{Z}$.

Let $q : \mathbb{Z} \longrightarrow \mathbb{Z}/n$ be the natural quotient map. Given $\varphi \in \operatorname{Hom}_{\mathbb{Z}}(\mathbb{Z}/n, \mathbb{Q}/\mathbb{Z})$, the composite $\varphi \circ q$ is a $\mathbb{Z}$-homomorphism from the free $\mathbb{Z}$-module $\mathbb{Z}$ (on one generator 1) to $\mathbb{Q}/\mathbb{Z}$. A homomorphism $\Phi \in \operatorname{Hom}_{\mathbb{Z}}(\mathbb{Z}, \mathbb{Q}/\mathbb{Z})$ is completely determined by the image of 1 (since $\Phi(\ell) = \Phi(\ell \cdot 1) = \ell \cdot \Phi(1)$), and since $\mathbb{Z}$ is *free* this image can be *anything* in the target $\mathbb{Q}/\mathbb{Z}$.

Such a homomorphism $\Phi \in \operatorname{Hom}_{\mathbb{Z}}(\mathbb{Z}, \mathbb{Q}/\mathbb{Z})$ factors through $\mathbb{Z}/n$ if and only if $\Phi(n) = 0$, that is, $n \cdot \Phi(1) = 0$. A complete list of representatives for equivalence classes in $\mathbb{Q}/\mathbb{Z}$ annihilated by n is $0, \frac{1}{n}, \frac{2}{n}, \frac{3}{n}, \ldots, \frac{n-1}{n}$. Thus, $\operatorname{Hom}_{\mathbb{Z}}(\mathbb{Z}/n, \mathbb{Q}/\mathbb{Z})$ is in bijection with this set, by

$$\varphi_{i/n}(x + n\mathbb{Z}) = ix/n + \mathbb{Z}$$

In fact, we see that $\operatorname{Hom}_{\mathbb{Z}}(\mathbb{Z}/n, \mathbb{Q}/\mathbb{Z})$ is an abelian group isomorphic to $\mathbb{Z}/n$, with

$$\varphi_{1/n}(x + n\mathbb{Z}) = x/n + \mathbb{Z}$$

as a generator. ///

27.6.5 Example: Compute $\mathbb{Q} \otimes_{\mathbb{Z}} \mathbb{Q}$.

We claim that this tensor product is isomorphic to $\mathbb{Q}$, via the $\mathbb{Z}$-linear map β induced from the $\mathbb{Z}$-bilinar map $B : \mathbb{Q} \times \mathbb{Q} \longrightarrow \mathbb{Q}$ given by

$$B : x \times y \longrightarrow xy$$

First, observe that the monomials $x \otimes 1$ generate the tensor product. Indeed, given $a/b \in \mathbb{Q}$ (with a, b integers, $b \neq 0$) we have

$$x \otimes \frac{a}{b} = (\frac{x}{b} \cdot b) \otimes \frac{a}{b} = \frac{x}{b} \otimes (b \cdot \frac{a}{b}) = \frac{x}{b} \otimes a = \frac{x}{b} \otimes a \cdot 1 = (a \cdot \frac{x}{b}) \otimes 1 = \frac{ax}{b} \otimes 1$$

proving the claim. Further, any finite $\mathbb{Z}$-linear combination of such elements can be rewritten as a single one: letting $n_i \in \mathbb{Z}$ and $x_i \in \mathbb{Q}$, we have

$$\sum_i n_i \cdot (x_i \otimes 1) = (\sum_i n_i x_i) \otimes 1$$

This gives an outer bound for the size of the tensor product. Now we need an inner bound, to know that there is no *further* collapsing in the tensor product.

From the defining property of the tensor product there *exists* a (unique) $\mathbb{Z}$-linear map from the tensor product to $\mathbb{Q}$, through which B factors. We have $B(x, 1) = x$, so the induced $\mathbb{Z}$-linear map β is a bijection on $\{x \otimes 1 : x \in \mathbb{Q}\}$, so it is an isomorphism. ///

27.6.6 Example: Compute $(\mathbb{Q}/\mathbb{Z}) \otimes_{\mathbb{Z}} \mathbb{Q}$.

We claim that the tensor product is 0. It suffices to show that every $m \otimes n$ is 0, since these monomials generate the tensor product. Given $x \in \mathbb{Q}/\mathbb{Z}$, let $0 < n \in \mathbb{Z}$ such that $nx = 0$. For any $y \in \mathbb{Q}$,

$$x \otimes y = x \otimes (n \cdot \frac{y}{n}) = (nx) \otimes \frac{y}{n} = 0 \otimes \frac{y}{n} = 0$$

as claimed.   ///

27.6.7 Example: Compute $(\mathbb{Q}/\mathbb{Z}) \otimes_\mathbb{Z} (\mathbb{Q}/\mathbb{Z})$.

We claim that the tensor product is 0. It suffices to show that every $m \otimes n$ is 0, since these monomials generate the tensor product. Given $x \in \mathbb{Q}/\mathbb{Z}$, let $0 < n \in \mathbb{Z}$ such that $nx = 0$. For any $y \subset \mathbb{Q}/\mathbb{Z}$,

$$x \otimes y = x \otimes (n \cdot \frac{y}{n}) = (nx) \otimes \frac{y}{n} = 0 \otimes \frac{y}{n} = 0$$

as claimed. Note that we do *not* claim that $\mathbb{Q}/Z$ is a $\mathbb{Q}$-module (which it is not), but only that for given $y \in \mathbb{Q}/\mathbb{Z}$ there is another element $z \in \mathbb{Q}/\mathbb{Z}$ such that $nz = y$. That is, $\mathbb{Q}/Z$ is a **divisible** $\mathbb{Z}$-module.   ///

27.6.8 Example: Prove that for a subring R of a commutative ring S, with $1_R = 1_S$, polynomial rings $R[x]$ behave well with respect to tensor products, namely that (as rings)

$$R[x] \otimes_R S \approx S[x]$$

Given an R-algebra homomorphism $\varphi : R \longrightarrow A$ and $a \in A$, let $\Phi : R[x] \longrightarrow A$ be the unique R-algebra homomorphism $R[x] \longrightarrow A$ which is φ on R and such that $\varphi(x) = a$. In particular, this works for A an S-algebra and φ the restriction to R of an S-algebra homomorphism $\varphi : S \longrightarrow A$. By the defining property of the tensor product, the bilinear map $B : R[x] \times S \longrightarrow A$ given by

$$B(P(x) \times s) = s \cdot \Phi(P(x))$$

gives a unique R-module map $\beta : R[x] \otimes_R S \longrightarrow A$. Thus, the tensor product has most of the properties necessary for it to be the free S-algebra on one generator $x \otimes 1$.

27.6.9 Remark: However, we might be concerned about verification that each such β is an S-algebra map, rather than just an R-module map. We can certainly write an expression that appears to describe the multiplication, by

$$(P(x) \otimes s) \cdot (Q(x) \otimes t) = P(x)Q(x) \otimes st$$

for polynomials P, Q and $s, t \in S$. *If* it is *well-defined*, then it is visibly associative, distributive, etc., as required.

27.6.10 Remark: The S-module structure itself is more straightforward: for any R-module M the tensor product $M \otimes_R S$ has a natural S-module structure given by

$$s \cdot (m \otimes t) = m \otimes st$$

for $s, t \in S$ and $m \in M$. But one could object that this structure is chosen at random. To argue that this is a *good* way to convert M into an S-module, we claim that for any other S-module N we have a natural isomorphism of abelian groups

$$\text{Hom}_S(M \otimes_R S, N) \approx \text{Hom}_R(M, N)$$

(where on the right-hand side we simply *forget* that N had more structure than that of R-module). The map is given by

$$\Phi \longrightarrow \varphi_\Phi \quad \text{where} \quad \varphi_\Phi(m) = \Phi(m \otimes 1)$$

and has inverse

$$\Phi_\varphi \longleftarrow \varphi \quad \text{where} \quad \Phi_\varphi(m \otimes s) = s \cdot \varphi(m)$$

One might further carefully verify that these two maps are inverses.

27.6.11 Remark: The definition of the tensor product does give an $\mathbb{R}$-linear map

$$\beta : R[x] \otimes_R S \longrightarrow S[x]$$

associated to the R-bilinear $B : R[x] \times S \longrightarrow S[x]$ by

$$B(P(x) \otimes s) = s \cdot P(x)$$

for $P(x) \in R[x]$ and $s \in S$. But it does not seem trivial to prove that this gives an isomorphism. Instead, it may be better to use the universal mapping property of a free algebra. In any case, there would still remain the issue of proving that the induced maps are S-algebra maps.

27.6.12 Example: Let K be a field extension of a field k. Let $f(x) \in k[x]$. Show that

$$k[x]/f \otimes_k K \approx K[x]/f$$

where the indicated quotients are by the ideals generated by f in $k[x]$ and $K[x]$, respectively.

Upon reflection, one should realize that we want to prove isomorphism as $K[x]$-modules. Thus, we implicitly use the facts that $k[x]/f$ is a $k[x]$-module, that $k[x] \otimes_k K \approx K[x]$ as K-algebras, and that $M \otimes_k K$ gives a $k[x]$-module M a $K[x]$-module structure by

$$\left(\sum_i s_i x^i\right) \cdot (m \otimes 1) = \sum_i (x^i \cdot m) \otimes s_i$$

The map

$$k[x] \otimes_k K \approx_{\text{ring}} K[x] \longrightarrow K[x]/f$$

has kernel (in $K[x]$) exactly of multiples $Q(x) \cdot f(x)$ of $f(x)$ by polynomials $Q(x) = \sum_i s_i x^i$ in $K[x]$. The inverse image of such a polynomial via the isomorphism is

$$\sum_i x^i f(x) \otimes s_i$$

Let I be the ideal generated in $k[x]$ by f, and $\tilde{I}$ the ideal generated by f in $K[x]$. The k-bilinear map

$$k[x]/f \times K \longrightarrow K[x]/f$$

by

$$B : (P(x) + I) \times s \longrightarrow s \cdot P(x) + \tilde{I}$$

gives a map $\beta : k[x]/f \otimes_k K \longrightarrow K[x]/f$. The map β is *surjective*, since

$$\beta\left(\sum_i (x^i + I) \otimes s_i\right) = \sum_i s_i x^i + \tilde{I}$$

hits every polynomial $\sum_i s_i x^i$ mod $\tilde{I}$. On the other hand, if

$$\beta(\sum_i (x^i + I) \otimes s_i) \in \tilde{I}$$

then $\sum_i s_i x^i = F(x) \cdot f(x)$ for some $F(x) \in K[x]$. Let $F(x) = \sum_j t_j x^j$. With $f(x) = \sum_\ell c_\ell x^\ell$, we have

$$s_i = \sum_{j+\ell=i} t_j c_\ell$$

Then, using k-linearity,

$$\sum_i (x^i + I) \otimes s_i = \sum_i \left(x^i + I \otimes (\sum_{j+\ell=i} t_j c_\ell) \right) = \sum_{j,\ell} (x^{j+\ell} + I \otimes t_j c_\ell)$$

$$= \sum_{j,\ell} (c_\ell x^{j+\ell} + I \otimes t_j) = \sum_j (\sum_\ell c_\ell x^{j+\ell} + I) \otimes t_j = \sum_j (f(x) x^j + I) \otimes t_j = \sum_j 0 = 0$$

So the map is a bijection, so is an isomorphism.   ///

27.6.13 Example: Let K be a field extension of a field k. Let V be a finite-dimensional k-vectorspace. Show that $V \otimes_k K$ is a good definition of the **extension of scalars** of V from k to K, in the sense that for any K-vectorspace W

$$\mathrm{Hom}_K(V \otimes_k K, W) \approx \mathrm{Hom}_k(V, W)$$

where in $\mathrm{Hom}_k(V, W)$ we *forget* that W was a K-vectorspace, and only think of it as a k-vectorspace.

This is a special case of a general phenomenon regarding *extension of scalars*. For any k-vectorspace V the tensor product $V \otimes_k K$ has a natural K-module structure given by

$$s \cdot (v \otimes t) = v \otimes st$$

for $s, t \in K$ and $v \in V$. To argue that this is a *good* way to convert k-vectorspaces V into K-vectorspaces, claim that for any other K-module W have a natural isomorphism of abelian groups

$$\mathrm{Hom}_K(V \otimes_k K, W) \approx \mathrm{Hom}_k(V, W)$$

On the right-hand side we *forget* that W had more structure than that of k-vectorspace. The map is

$$\Phi \longrightarrow \varphi_\Phi \quad \text{where} \quad \varphi_\Phi(v) = \Phi(v \otimes 1)$$

and has inverse

$$\Phi_\varphi \longleftarrow \varphi \quad \text{where} \quad \Phi_\varphi(v \otimes s) = s \cdot \varphi(v)$$

To verify that these are mutual inverses, compute

$$\varphi_{\Phi_\varphi}(v) = \Phi_\varphi(v \otimes 1) = 1 \cdot \varphi(v) = \varphi(v)$$

and

$$\Phi_{\varphi_\Phi}(v \otimes 1) = 1 \cdot \varphi_\Phi(v) = \Phi(v \otimes 1)$$

which proves that the maps are inverses.   ///

27.6.14 Remark: In fact, the two spaces of homomorphisms in the isomorphism can be given natural structures of K-vectorspaces, and the isomorphism just constructed can be verified to respect this additional structure. The K-vectorspace structure on the left is clear, namely

$$(s \cdot \Phi)(m \otimes t) = \Phi(m \otimes st) = s \cdot \Phi(m \otimes t)$$

The structure on the right is

$$(s \cdot \varphi)(m) = s \cdot \varphi(m)$$

The latter has only the one presentation, since only W is a K-vectorspace.

27.6.15 Example: Let M and N be free R-modules, where R is a commutative ring with identity. Prove that $M \otimes_R N$ is free and

$$\operatorname{rank} M \otimes_R N = \operatorname{rank} M \cdot \operatorname{rank} N$$

Let M and N be free on generators $i : X \longrightarrow M$ and $j : Y \longrightarrow N$. We claim that $M \otimes_R N$ is free on a set map

$$\ell : X \times Y \longrightarrow M \otimes_R N$$

To verify this, let $\varphi : X \times Y \longrightarrow Z$ be a set map. For each fixed $y \in Y$, the map $x \longrightarrow \varphi(x, y)$ factors through a unique R-module map $B_y : M \longrightarrow Z$. For each $m \in M$, the map $y \longrightarrow B_y(m)$ gives rise to a unique R-linear map $n \longrightarrow B(m, n)$ such that

$$B(m, j(y)) = B_y(m)$$

The linearity in the second argument assures that we still have the linearity in the first, since for $n = \sum_t r_t\, j(y_t)$ we have

$$B(m, n) = B(m, \sum_t r_t j(y_t)) = \sum_t r_t B_{y_t}(m)$$

which is a linear combination of linear functions. Thus, there is a unique map to Z induced on the tensor product, showing that the tensor product with set map $i \times j : X \times Y \longrightarrow M \otimes_R N$ is free. ///

27.6.16 Example: Let M be a free R-module of rank r, where R is a commutative ring with identity. Let S be a commutative ring with identity containing R, such that $1_R = 1_S$. Prove that as an S module $M \otimes_R S$ is free of rank r.

We prove a bit more. First, instead of simply an *inclusion* $R \subset S$, we can consider any ring homomorphism $\psi : R \longrightarrow S$ such that $\psi(1_R) = 1_S$.

Also, we can consider arbitrary sets of generators, and give more details. Let M be free on generators $i : X \longrightarrow M$, where X is a set. Let $\tau : M \times S \longrightarrow M \otimes_R S$ be the canonical map. We claim that $M \otimes_R S$ is free on $j : X \longrightarrow M \otimes_R S$ defined by

$$j(x) = \tau(i(x) \times 1_S)$$

Given an S-module N, we can be a little forgetful and consider N as an R-module via ψ, by $r \cdot n = \psi(r)n$. Then, given a set map $\varphi : X \longrightarrow N$, since M is free, there is a unique R-module map $\Phi : M \longrightarrow N$ such that $\varphi = \Phi \circ i$. That is, the diagram

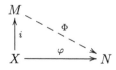

commutes. Then the map
$$\psi : M \times S \longrightarrow N$$
by
$$\psi(m \times s) = s \cdot \Phi(m)$$
induces (by the defining property of $M \otimes_R S$) a unique $\Psi : M \otimes_R S \longrightarrow N$ making a commutative diagram

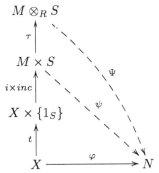

where *inc* is the inclusion map $\{1_S\} \longrightarrow S$, and where $t : X \longrightarrow X \times \{1_S\}$ by $x \longrightarrow x \times 1_S$. Thus, $M \otimes_R S$ is free on the composite $j : X \longrightarrow M \otimes_R S$ defined to be the composite of the vertical maps in that last diagram. This argument does not depend upon finiteness of the generating set.   ///

27.6.17 Example: For finite-dimensional vectorspaces V, W over a field k, prove that there is a natural isomorphism

$$(V \otimes_k W)^* \approx V^* \otimes W^*$$

where $X^* = \mathrm{Hom}_k(X, k)$ for a k-vectorspace X.

For finite-dimensional V and W, since $V \otimes_k W$ is free on the cartesian product of the generators for V and W, the dimensions of the two sides match. We make an isomorphism from right to left. Create a bilinear map

$$V^* \times W^* \longrightarrow (V \otimes_k W)^*$$

as follows. Given $\lambda \in V^*$ and $\mu \in W^*$, as usual make $\Lambda_{\lambda,\mu} \in (V \otimes_k W)^*$ from the bilinear map
$$B_{\lambda,\mu} : V \times W \longrightarrow k$$
defined by
$$B_{\lambda,\mu}(v, w) = \lambda(v) \cdot \mu(w)$$
This induces a unique functional $\Lambda_{\lambda,\mu}$ on the tensor product. This induces a unique linear map
$$V^* \otimes W^* \longrightarrow (V \otimes_k W)^*$$
as desired.

Since everything is finite-dimensional, bijectivity will follow from injectivity. Let $e_1, \ldots, e_m$ be a basis for V, $f_1, \ldots, f_n$ a basis for W, and $\lambda_1, \ldots, \lambda_m$ and $\mu_1, \ldots, \mu_n$ corresponding dual bases. We have shown that a basis of a tensor product of free modules is free on the cartesian product of the generators. Suppose that $\sum_{ij} c_{ij} \lambda_i \otimes \mu_j$ gives the 0 functional on

$V \otimes W$, for some scalars c_{ij}. Then, for every pair of indices s, t, the function is 0 on $e_s \otimes f_t$. That is,

$$0 = \sum_{ij} c_{ij} \lambda_i(e_s)\, \lambda_j(f_t) = c_{st}$$

Thus, all constants c_{ij} are 0, proving that the map is injective. Then a dimension count proves the isomorphism. ///

27.6.18 Example: For a finite-dimensional k-vectorspace V, prove that the bilinear map

$$B : V \times V^* \longrightarrow \mathrm{End}_k(V)$$

by

$$B(v \times \lambda)(x) = \lambda(x) \cdot v$$

gives an isomorphism $V \otimes_k V^* \longrightarrow \mathrm{End}_k(V)$. Further, show that the composition of endormorphisms is the same as the map induced from the map on

$$(V \otimes V^*) \times (V \otimes V^*) \longrightarrow V \otimes V^*$$

given by

$$(v \otimes \lambda) \times (w \otimes \mu) \longrightarrow \lambda(w)v \otimes \mu$$

The bilinear map $v \times \lambda \longrightarrow T_{v,\lambda}$ given by

$$T_{v,\lambda}(w) = \lambda(w) \cdot v$$

induces a *unique* linear map $j : V \otimes V^* \longrightarrow \mathrm{End}_k(V)$.

To prove that j is injective, we may use the fact that a basis of a tensor product of free modules is free on the cartesian product of the generators. Thus, let $e_1, \ldots, e_n$ be a basis for V, and $\lambda_1, \ldots, \lambda_n$ a dual basis for V^*. Suppose that

$$\sum_{i,j=1}^{n} c_{ij}\, e_i \otimes \lambda_j \longrightarrow 0\mathrm{End}_k(V)$$

That is, for every e_ℓ,

$$\sum_{ij} c_{ij}\lambda_j(e_\ell)e_i = 0 \in V$$

This is

$$\sum_i c_{ij}e_i = 0 \quad \text{(for all } j)$$

Since the e_is are linearly independent, all the c_{ij}s are 0. Thus, the map j is injective. Then counting k-dimensions shows that this j is a k-linear isomorphism.

Composition of endomorphisms is a bilinear map

$$\mathrm{End}_k(V) \times \mathrm{End}_k(V) \overset{\circ}{\longrightarrow} \mathrm{End}_k(V)$$

by

$$S \times T \longrightarrow S \circ T$$

Denote by
$$c : (v \otimes \lambda) \times (w \otimes \mu) \longrightarrow \lambda(w)v \otimes \mu$$
the allegedly corresonding map on the tensor products. The induced map on $(V \otimes V^*) \otimes (V \otimes V^*)$ is an example of a **contraction map** on tensors. We want to show that the diagram

$$
\begin{array}{ccc}
\mathrm{End}_k(V) \times \mathrm{End}_k(V) & \xrightarrow{\;\;\circ\;\;} & \mathrm{End}_k(V) \\
{\scriptstyle j \times j} \uparrow & & \uparrow {\scriptstyle j} \\
(V \otimes_k V^*) \times (V \otimes_k V^*) & \xrightarrow{\;\;c\;\;} & V \otimes_k V^*
\end{array}
$$

commutes. It suffices to check this starting with $(v \otimes \lambda) \times (w \otimes \mu)$ in the lower left corner. Let $x \in V$. Going up, then to the right, we obtain the endomorphism which maps x to

$$j(v \otimes \lambda) \circ j(w \otimes \mu)\,(x) = j(v \otimes \lambda)(j(w \otimes \mu)(x)) = j(v \otimes \lambda)(\mu(x)\,w)$$

$$= \mu(x)\,j(v \otimes \lambda)(w) = \mu(x)\,\lambda(w)\,v$$

Going the other way around, to the right then up, we obtain the endomorphism which maps x to
$$j\big(\,c((v \otimes \lambda) \times (w \otimes \mu))\,\big)\,(x) = j\big(\,\lambda(w)(v \otimes \mu)\,\big)\,(x) = \lambda(w)\,\mu(x)\,v$$

These two outcomes are the same.   ///

27.6.19 Example: Under the isomorphism of the previous problem, show that the linear map
$$\mathrm{tr} : \mathrm{End}_k(V) \longrightarrow k$$
is the linear map
$$V \otimes V^* \longrightarrow k$$
induced by the bilinear map $v \times \lambda \longrightarrow \lambda(v)$.

Note that the induced map

$$V \otimes_k V^* \longrightarrow k \quad \text{by} \quad v \otimes \lambda \longrightarrow \lambda(v)$$

is another **contraction map** on tensors. Part of the issue is to compare the coordinate-bound trace with the induced (contraction) map $t(v \otimes \lambda) = \lambda(v)$ determined uniquely from the bilinear map $v \times \lambda \longrightarrow \lambda(v)$. To this end, let $e_1, \ldots, e_n$ be a basis for V, with dual basis $\lambda_1, \ldots, \lambda_n$. The corresponding matrix coefficients $T_{ij} \in k$ of a k-linear endomorphism T of V are
$$T_{ij} = \lambda_i(Te_j)$$
(Always there is the worry about interchange of the indices.) Thus, in these coordinates,

$$\mathrm{tr}\,T = \sum_i \lambda_i(Te_i)$$

Let $T = j(e_s \otimes \lambda_t)$. Then, since $\lambda_t(e_i) = 0$ unless $i = t$,

$$\mathrm{tr}\,T = \sum_i \lambda_i(Te_i) = \sum_i \lambda_i(j(e_s \otimes \lambda_t)e_i) = \sum_i \lambda_i(\lambda_t(e_i) \cdot e_s) = \lambda_t(\lambda_t(e_t) \cdot e_s) = \begin{cases} 1 & (s = t) \\ 0 & (s \neq t) \end{cases}$$

On the other hand,

$$t(e_s \otimes \lambda_t) = \lambda_t(e_s) = \begin{cases} 1 & (s = t) \\ 0 & (s \neq t) \end{cases}$$

Thus, these two k-linear functionals agree on the monomials, which span, they are equal.

///

27.6.20 Example: Prove that $\operatorname{tr}(AB) = \operatorname{tr}(BA)$ for two endomorphisms of a finite-dimensional vector space V over a field k, with trace defined as just above.

Since the maps

$$\operatorname{End}_k(V) \times \operatorname{End}_k(V) \longrightarrow k$$

by

$$A \times B \longrightarrow \operatorname{tr}(AB) \quad \text{and/or} \quad A \times B \longrightarrow \operatorname{tr}(BA)$$

are bilinear, it suffices to prove the equality on (images of) monomials $v \otimes \lambda$, since these span the endomophisms over k. Previous examples have converted the issue to one concerning $V_k^{\otimes} V^*$. (We have already shown that the isomorphism $V \otimes_k V^* \approx \operatorname{End}_k(V)$ is converts a *contraction* map on tensors to composition of endomorphisms, and that the trace on tensors defined as another contraction corresponds to the trace of matrices.) Let tr now denote the contraction-map trace on tensors, and (temporarily) write

$$(v \otimes \lambda) \circ (w \otimes \mu) = \lambda(w)\, v \otimes \mu$$

for the contraction-map composition of endomorphisms. Thus, we must show that

$$\operatorname{tr}\,(v \otimes \lambda) \circ (w \otimes \mu) = \operatorname{tr}\,(w \otimes \mu) \circ (v \otimes \lambda)$$

The left-hand side is

$$\operatorname{tr}\,(v \otimes \lambda) \circ (w \otimes \mu) = \operatorname{tr}\,(\lambda(w)\, v \otimes \mu) = \lambda(w)\operatorname{tr}\,(v \otimes \mu) = \lambda(w)\,\mu(v)$$

The right-hand side is

$$\operatorname{tr}\,(w \otimes \mu) \circ (v \otimes \lambda) = \operatorname{tr}\,(\mu(v)\, w \otimes \lambda) = \mu(v)\operatorname{tr}\,(w \otimes \lambda) = \mu(v)\,\lambda(w)$$

These elements of k are the same.

///

27.6.21 Example: Prove that tensor products are *associative*, in the sense that, for R-modules A, B, C, we have a *natural isomorphism*

$$A \otimes_R (B \otimes_R C) \approx (A \otimes_R B) \otimes_R C$$

In particular, *do* prove the *naturality*, at least the one-third part of it which asserts that, for every R-module homomorphism $f : A \longrightarrow A'$, the diagram

$$\begin{array}{ccc} A \otimes_R (B \otimes_R C) & \xrightarrow{\approx} & (A \otimes_R B) \otimes_R C \\ \downarrow{\scriptstyle f \otimes (1_B \otimes 1_C)} & & \downarrow{\scriptstyle (f \otimes 1_B) \otimes 1_C} \\ A' \otimes_R (B \otimes_R C) & \xrightarrow{\approx} & (A' \otimes_R B) \otimes_R C \end{array}$$

commutes, where the two horizontal isomorphisms are those determined in the first part of the problem. (One might also consider maps $g : B \longrightarrow B'$ and $h : C \longrightarrow C'$, but these behave similarly, so there's no real compulsion to worry about them, apart from awareness of the issue.)

Since all tensor products are over R, we drop the subscript, to lighten the notation. As usual, to make a (linear) map *from* a tensor product $M \otimes N$, we induce uniquely from a bilinear map on $M \times N$. We have done this enough times that we will suppress this part now.

The thing that is slightly less trivial is construction of maps *to* tensor products $M \otimes N$. These are always obtained by composition with the canonical bilinear map

$$M \times N \longrightarrow M \otimes N$$

Important at present is that we can create n-fold tensor products, as well. Thus, we prove the indicated isomorphism by proving that both the indicated iterated tensor products are (naturally) isomorphic to the un-parenthesis'd tensor product $A \otimes B \otimes C$, with canonical map $\tau : A \times B \times C \longrightarrow A \otimes B \otimes C$, such that for every trilinear map $\varphi : A \times B \times C \longrightarrow X$ there is a unique linear $\Phi : A \otimes B \otimes C \longrightarrow X$ such that

The set map

$$A \times B \times C \approx (A \times B) \times C \longrightarrow (A \otimes B) \otimes C$$

by

$$a \times b \times c \longrightarrow (a \times b) \times c \longrightarrow (a \otimes b) \otimes c$$

is linear in each single argument (for fixed values of the others). Thus, we are assured that there is a unique induced linear map

$$A \otimes B \otimes C \longrightarrow (A \otimes B) \otimes C$$

such that

commutes.

Similarly, from the set map

$$(A \times B) \times C \approx A \times B \times C \longrightarrow A \otimes B \otimes C$$

by

$$(a \times b) \times c \longrightarrow a \times b \times c \longrightarrow a \otimes b \otimes c$$

is linear in each single argument (for fixed values of the others). Thus, we are assured that there is a unique induced linear map

$$(A \otimes B) \otimes C \longrightarrow A \otimes B \otimes C$$

such that

$$(A \otimes B) \otimes C$$

$$(A \times B) \times C \xrightarrow{\qquad} A \otimes B \otimes C$$

with dashed map j

commutes.

Then $j \circ i$ is a map of $A \otimes B \otimes C$ to itself compatible with the canonical map $A \times B \times C \longrightarrow A \otimes B \otimes C$. By uniqueness, $j \circ i$ is the identity on $A \otimes B \otimes C$. Similarly (just very slightly more complicatedly), $i \circ j$ must be the identity on the iterated tensor product. Thus, these two maps are mutual inverses.

To prove naturality in one of the arguments A, B, C, consider $f : C \longrightarrow C'$. Let j_{ABC} be the isomorphism for a fixed triple A, B, C, as above. The diagram of maps of cartesian products (of sets, at least)

$$(A \times B) \times C \xrightarrow{j_{ABC}} A \times B \times C$$
$$\downarrow{(1_A \times 1_B) \times f} \qquad \downarrow{1_A \times 1_B \times f}$$
$$(A \times B) \times C \xrightarrow{j} A \times B \times C$$

does commute: going down, then right, is

$$j_{ABC'} \left((1_A \times 1_B) \times f \right)((a \times b) \times c)) = j_{ABC'} \left((a \times b) \times f(c) \right) = a \times b \times f(c)$$

Going right, then down, gives

$$(1_A \times 1_B \times f) \left(j_{ABC}((a \times b) \times c) \right) = (1_A \times 1_B \times f) \left(a \times b \times c \right) = a \times b \times f(c)$$

These are the same. ///

Exercises

27.1 Let I and J be two ideals in a PID R. Determine

$$R/I \otimes_R R/J$$

27.2 For an R-module M and an ideal I in R, show that

$$M/I \cdot M \approx M \otimes_R R/I$$

27.3 Let R be a commutative ring with unit, and S a commutative R algebra. Given an R-bilinear map $B : V \times W \longrightarrow R$, give a natural S-blinear extension of B to the S-linear extensions $S \otimes_R V$ and $S \otimes_R W$.

27.4 A **multiplicative subset** S of a commutative ring R with unit is a subset of R closed under multiplication. The **localization** $S^{-1}R$ of R at S is the collection of ordered pairs (r, s) with $r \in R$ and $s \in S$, modulo the equivalence relation that $(r, s) \sim (r', s')$ if and only if there is $s'' \in S$ such that

$$s'' \cdot (rs' - r's) = 0$$

Let P be a prime ideal in R. Show that $S^{-1}P$ is a prime ideal in $S^{-1}R$.

27.5 In the situation of the previous exercise, show that the field of fractions of $(S^{-1}R)/(S^{-1}P)$ is naturally isomorphic to the field of fractions of R/P.

27.6 In the situation of the previous two exercises, for an R-module M, define a reasonable notion of $S^{-1}M$.

27.7 In the situation of the previous three exercises, for an R-module M, show that

$$S^{-1}M \approx M \otimes_R S^{-1}R$$

27.8 Identify the commutative $\mathbb{Q}$-algebra $\mathbb{Q}(\sqrt{2}) \otimes_{\mathbb{Q}} \mathbb{Q}(\sqrt{2})$ as a sum of fields.

27.9 Identify the commutative $\mathbb{Q}$-algebra $\mathbb{Q}(\sqrt[3]{2}) \otimes_{\mathbb{Q}} \mathbb{Q}(\sqrt[3]{2})$ as a sum of fields.

27.10 Let ζ be a primitie 5th root of unity. Identify the commutative $\mathbb{Q}$-algebra $\mathbb{Q}(\sqrt{5}) \otimes_{\mathbb{Q}} \mathbb{Q}(\zeta)$ as a sum of fields.

27.11 Let $\mathfrak{H}$ be the Hamiltonian quaternions. Identify $\mathfrak{H} \otimes_{\mathbb{R}} \mathbb{C}$ in familiar terms.

27.12 Let $\mathfrak{H}$ be the Hamiltonian quaternions. Identify $\mathfrak{H} \otimes_{\mathbb{R}} \mathfrak{H}$ in familiar terms.

28. Exterior powers

While many of the arguments here have analogues for tensor products, it is worthwhile to repeat these arguments with the relevant variations, both for practice, and to be sensitive to the differences.

28.1 *Desiderata*

Again, we review missing items in our development of linear algebra.

We are missing a development of determinants of matrices whose entries may be in commutative rings, rather than fields. We would like an *intrinsic* definition of determinants of endomorphisms, rather than one that depends upon a choice of coordinates, even if we eventually prove that the determinant is independent of the coordinates. We anticipate that Artin's axiomatization of determinants of matrices should be mirrored in much of what we do here.

We want a direct and natural proof of the Cayley-Hamilton theorem. Linear algebra over *fields* is insufficient, since the introduction of the indeterminate x in the definition of the characteristic polynomial takes us outside the class of vector spaces over fields.

We want to give a conceptual proof for the *uniqueness* part of the structure theorem for finitely-generated modules over principal ideal domains. Multi-linear algebra over fields is surely insufficient for this.

28.2 *Definitions, uniqueness, existence*

Let R be a commutative ring with 1. We only consider R-modules M with the property that $1 \cdot m = m$ for all $m \in M$. Let M and X be R-modules. An R-multilinear map

$$B : \underbrace{M \times \ldots \times M}_{n} \longrightarrow X$$

is **alternating** if $B(m_1, \ldots, m_n) = 0$ whenever $m_i = m_j$ for two indices $i \neq j$.

As in earlier discussion of free modules, and in discussion of polynomial rings as free algebras, we will define exterior powers by *mapping properties*. As usual, this allows an easy proof that exterior powers (if they exist) are *unique* up to *unique isomorphism*. Then we give a modern construction.

An exterior n^{th} power $\bigwedge_R^n M$ over R of an R-module M is an R-module $\bigwedge_R^n M$ with an alternating R-multilinear map (called the **canonical map**) [1]

$$\alpha : \underbrace{M \times \ldots \times M}_{n} \longrightarrow \bigwedge_R^n M$$

such that, for every alternating R-multilinear map

$$\varphi : \underbrace{M \times \ldots \times M}_{n} \longrightarrow X$$

there is a *unique* R-linear map

$$\Phi : \bigwedge_R^n M \longrightarrow X$$

such that $\varphi = \Phi \circ \alpha$, that is, such that the diagram

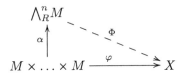

commutes.

28.2.1 Remark: If there is no ambiguity, we may drop the subscript R on the exterior power $\bigwedge_R^n M$, writing simply $\bigwedge^n M$.

---

[1] There are many different *canonical maps* in different situations, but context should always make clear what the properties are that are expected. Among other things, this potentially ambiguous phrase allows us to avoid trying to give a permanent symbolic name to the maps in question.

The usual notation does not involve any symbol such as α, but in our development it is handy to have a name for this map. The standard notation denotes the image $\alpha(m \times n)$ of $m \times n$ in the exterior product by

$$\text{image of } m_1 \times \ldots \times m_n \text{ in } \bigwedge\nolimits^n M = m_1 \wedge \ldots \wedge m_n$$

In practice, the implied R-multilinear alternating map

$$M \times \ldots \times M \longrightarrow \bigwedge\nolimits^n M$$

called α here is often left anonymous.

The following proposition is typical of uniqueness proofs for objects defined by mapping property requirements. It is essentially identical to the analogous argument for tensor products. Note that internal details of the objects involved play no role. Rather, the argument proceeds by manipulation of arrows.

28.2.2 Proposition: Exterior powers $\alpha : M \times \ldots \times M \longrightarrow \bigwedge^n M$ are unique up to unique isomorphism. That is, given two exterior n^{th} powers

$$\alpha_1 : M \times \ldots \times M \longrightarrow E_1$$

$$\alpha_2 : M \times \ldots \times M \longrightarrow E_2$$

there is a *unique R-linear isomorphism* $i : E_1 \longrightarrow E_2$ such that the diagram

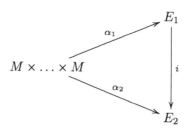

commutes, that is, $\alpha_2 = i \circ \alpha_1$.

Proof: First, we show that for a nth exterior power $\alpha : M \times \ldots \times M \longrightarrow T$, the only map $f : E \longrightarrow E$ compatible with α is the identity. That is, the identity map is the only map f such that

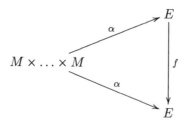

commutes. Indeed, the definition of a nth exterior power demands that, given the alternating multilinear map

$$\alpha : M \times \ldots \times M \longrightarrow E$$

(with E in the place of the earlier X) there is a unique linear map $\Phi : E \longrightarrow E$ such that the diagram

$$
\begin{array}{ccc}
E & & \\
\alpha \uparrow & \diagdown \ \ {}^{\Phi} & \\
M \times \ldots \times M & \xrightarrow{\ \ \alpha \ \ } & E
\end{array}
$$

commutes. The identity map on E certainly has this property, so is the *only* map $E \longrightarrow E$ with this property.

Looking at two n^{th} exterior powers, first take $\alpha_2 : M \times \ldots \times M \longrightarrow E_2$ in place of the $\varphi : M \times \ldots \times M \longrightarrow X$. That is, there is a unique linear $\Phi_1 : E_1 \longrightarrow E_2$ such that the diagram

$$
\begin{array}{ccc}
E_1 & & \\
\alpha_1 \uparrow & \diagdown \ \ {}^{\Phi_1} & \\
M \times \ldots \times M & \xrightarrow{\ \ \alpha_2 \ \ } & E_2
\end{array}
$$

commutes. Similarly, reversing the roles, there is a unique linear $\Phi_2 : E_2 \longrightarrow E_1$ such that

$$
\begin{array}{ccc}
E_2 & & \\
\alpha_2 \uparrow & \diagdown \ \ {}^{\Phi_2} & \\
M \times \ldots \times M & \xrightarrow{\ \ \alpha_1 \ \ } & E_1
\end{array}
$$

commutes. Then $\Phi_2 \circ \Phi_1 : E_1 \longrightarrow E_1$ is compatible with α_1, so is the identity, from the first part of the proof. And, symmetrically, $\Phi_1 \circ \Phi_2 : E_2 \longrightarrow E_2$ is compatible with α_2, so is the identity. Thus, the maps Φ_i are mutual inverses, so are isomorphisms.   ///

For existence, we express the nth exterior power $\bigwedge^n M$ as a quotient of the tensor power

$$
\overset{n}{\bigotimes} M = \underbrace{M \otimes \ldots \otimes M}_{n}
$$

28.2.3 Proposition: n^{th} exterior powers $\bigwedge^n M$ exist. In particular, let I be the submodule of $\bigotimes^n M$ generated by all tensors

$$
m_1 \otimes \ldots \otimes m_n
$$

where $m_i = m_j$ for some $i \neq j$. Then

$$
\bigwedge^n M = \overset{n}{\bigotimes} M/I
$$

The alternating map

$$
\alpha : M \times \ldots \times M \longrightarrow \bigwedge^n M
$$

is the composite of the quotient map $\bigotimes^n \longrightarrow \bigwedge^n M$ with the canonical multilinear map $M \times \ldots \times M \longrightarrow \bigotimes^n M$.

Proof: Let $\varphi : M \times \ldots \times M \longrightarrow X$ be an alternating R-multilinear map. Let $\tau : M \times \ldots \times M \longrightarrow \bigotimes^n M$ be the tensor product. By properties of the tensor product there is a unique R-linear $\Psi : \bigotimes^n M \longrightarrow X$ through which φ factors, namely $\varphi = \Psi \circ \tau$.

Let $q : \bigotimes^n \longrightarrow \bigwedge^n M$ be the quotient map. We claim that Ψ factors through q, as $\Psi = \Phi \circ q$, for a linear map $\Phi : \bigwedge^n M \longrightarrow X$. That is, we claim that there is a commutative diagram

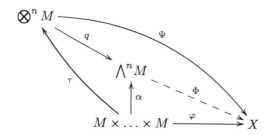

Specifically, we claim that $\Psi(I) = 0$, where I is the submodule generated by $m_1 \otimes \ldots \otimes m_n$ with $m_i = m_j$ for some $i \neq j$. Indeed, using the fact that φ is alternating,

$$\Psi(m_1 \otimes \ldots \otimes m) = \Psi(\tau(m_1 \times \ldots \times m_n)) = \varphi(m_1 \times \ldots \times m_n) = 0$$

That is, $\ker \Psi \supset I$, so Ψ factors through the quotient $\bigwedge^n M$.

Last, we must check that the map $\alpha = q \circ \tau$ is alternating. Indeed, with $m_i = m_j$ (and $i \neq j$),

$$\alpha(m_1 \times \ldots \times m_n) = (q \circ \tau)(m_1 \times \ldots \times m_n) = q(m_1 \otimes \ldots \otimes m_n)$$

Since $m_i = m_j$, that monomial tensor is in the submodule I, which is the kernel of the quotient map q. Thus, α is alternating.   ///

28.3 *Some elementary facts*

Again,[2] the naive notion of *alternating* would entail that, for example, in $\bigwedge^2 M$

$$x \wedge y = -y \wedge x$$

More generally, in $\bigwedge^n M$,

$$\ldots \wedge m_i \wedge \ldots \wedge m_j \wedge \ldots = -\ldots \wedge m_j \wedge \ldots \wedge m_i \wedge \ldots$$

(interchanging the ith and j^{th} elements) for $i \neq j$. However, this isn't the definition. Again, the *definition* is that

$$\ldots \wedge m_i \wedge \ldots \wedge m_j \wedge \ldots = 0 \quad \text{if } m_i = m_j \text{ for any } i \neq j$$

This latter condition is strictly stronger than the change-of-sign requirement if 2 is a 0-divisor in the underlying ring R. As in Artin's development of determinants from the alternating property, we do recover the change-of-sign property, since

$$0 = (x + y) \wedge (x + y) = x \wedge x + x \wedge y + y \wedge x + y \wedge y = 0 + x \wedge y + y \wedge x + 0$$

---

[2] We already saw this refinement in the classical context of determinants of matrices, as axiomatized in the style of Emil Artin.

which gives
$$x \wedge y = -y \wedge x$$

The natural induction on the number of 2-cycles in a permutation π proves

28.3.1 Proposition: For $m_1, \ldots, m_n$ in M, and for a permutation π of n things,

$$m_{\pi(1)} \wedge \ldots \wedge m_{\pi(n)} = \sigma(\pi) \cdot m_1 \wedge \ldots \wedge m_n$$

Proof: Let $\pi = s\tau$, where s is a 2-cycle and τ is a permutation expressible as a product of fewer 2-cycles than π. Then

$$m_{\pi(1)} \wedge \ldots \wedge m_{\pi(n)} = m_{s\tau(1)} \wedge \ldots \wedge m_{s\tau(n)} = -m_{\tau(1)} \wedge \ldots \wedge m_{\tau(n)}$$

$$= -\sigma(\tau) \cdot m_1 \wedge \ldots \wedge m_n = \sigma(\pi) \cdot m_1 \wedge \ldots \wedge m_n$$

as asserted.   ///

28.3.2 Proposition: The *monomial* exterior products $m_1 \wedge \ldots \wedge m_n$ generate $\bigwedge^n M$ as an R-module, as the m_i run over all elements of M.

Proof: Let X be the submodule of $\bigwedge^n M$ generated by the monomial tensors, $Q = (\bigwedge^n M)/X$ the quotient, and $q : \bigwedge^n M \longrightarrow X$ the quotient map. Let

$$B : M \times \ldots \times M \longrightarrow Q$$

be the 0-map. A defining property of the nth exterior power is that there is a unique R-linear
$$\beta : \bigwedge^n M \longrightarrow Q$$

making the usual diagram commute, that is, such that $B = \beta \circ \alpha$, where $\alpha : M \times \ldots \times M \longrightarrow \bigwedge^n M$. Both the quotient map q and the 0-map $\bigwedge^n M \longrightarrow Q$ allow the 0-map $M \times \ldots \times M \longrightarrow Q$ to factor through, so by the uniqueness the quotient map is the 0-map. That is, Q is the 0-module, so $X = \bigwedge^n M$.   ///

28.3.3 Proposition: Let $\{m_\beta : \beta \in B\}$ be a set of generators for an R-module M, where the index set B is *ordered*. Then the monomials

$$m_{\beta_1} \wedge \ldots \wedge m_{\beta_n} \quad \text{with} \quad \beta_1 < \beta_2 < \ldots < \beta_n$$

generate $\bigwedge^n M$.

Proof: First, claim that the monomials

$$m_{\beta_1} \wedge \ldots \wedge m_{\beta_n} \quad \text{(no condition on } \beta_i\text{s)}$$

generate the exterior power. Let I be the submodule generated by them. If I is proper, let $X = (\bigwedge^n M)/I$ and let $q : \bigwedge^n M \longrightarrow X$ be the quotient map. The composite

$$q \circ \alpha : \underbrace{M \times \ldots \times M}_{n} \longrightarrow \bigwedge^n M \longrightarrow X$$

is an alternating map, and is 0 on any $m_{\beta_1} \times \ldots \times m_{\beta_n}$. In each variable, separately, the map is linear, and vanishes on generators for M, so is 0. Thus, $q \circ \alpha = 0$. This map certainly factors through the 0-map $\bigwedge^n M \longrightarrow X$. But, using the defining property of the exterior power, the uniqueness of a map $\bigwedge^n M \longrightarrow X$ through which $q \circ \alpha$ factors implies that $q = 0$, and $X = 0$. Thus, these monomials generate the whole.

Now we will see that we can reorder monomials to put the indices in ascending order. First, since

$$m_{\beta_1} \wedge \ldots \wedge m_{\beta_n} = \alpha(m_{\beta_1} \times \ldots \times m_{\beta_n})$$

and α is alternating, the monomial is 0 if $m_{\beta_i} = m_{\beta_j}$ for $\beta_i \neq \beta_j$. And for a permutation π of n things, as observed just above,

$$m_{\beta_{\pi(1)}} \wedge \ldots \wedge m_{\beta_{\pi(n)}} = \sigma(\pi) \cdot m_{\beta_1} \wedge \ldots \wedge m_{\beta_n}$$

where σ is the parity function on permutations. Thus, to express elements of $\bigwedge^n M$ it suffices to use only monomials with indices in ascending order. ///

28.4 Exterior powers $\bigwedge^n f$ of maps

Still R is a commutative ring with 1.

An important type of map on an exterior power $\bigwedge^n M$ arises from R-linear maps on the module M. That is, let

$$f : M \longrightarrow N$$

be an R-module map, and attempt to define

$$\bigwedge^n f : \bigwedge^n M \longrightarrow \bigwedge^n N$$

by

$$(\textstyle\bigwedge^n f)(m_1 \wedge \ldots \wedge m_n) = f(m_1) \wedge \ldots \wedge f(m_n)$$

Justifiably interested in being sure that this formula makes sense, we proceed as follows.

If the map is *well-defined* then it is defined completely by its values on the monomial exterior products, since these generate the exterior power. To prove well-definedness, we invoke the defining property of the nth exterior power. Let $\alpha' : N \times \ldots \times N \longrightarrow \bigwedge^n N$ be the canonical map. Consider

$$B : \underbrace{M \times \ldots \times M}_{n} \overset{f \times \ldots \times f}{\longrightarrow} \underbrace{N \times \ldots \times N}_{n} \overset{\alpha'}{\longrightarrow} \bigwedge^n N$$

given by

$$B(m_1 \times \ldots \times m_n) = f(m_1) \wedge \ldots \wedge f(m_n)$$

For fixed index i, and for fixed $m_j \in M$ for $j \neq i$, the composite

$$m \longrightarrow \alpha'(\ldots \times f(m_{i-1}) \times f(m) \times f(m_{i+1}) \wedge \ldots)$$

is certainly an R-linear map in m. Thus, B is R-multilinear. As a function of each single argument in $M \times \ldots \times M$, the map B is linear, so B is multilinear. Since α' is alternating, B is alternating. Then (by the defining property of the exterior power) there is a unique R-linear map Φ giving a commutative diagram

$$
\begin{array}{ccccc}
\bigwedge^n M & \dashrightarrow & & \Phi = \bigwedge^n f & \\
\alpha \uparrow & & & & \searrow \\
M \times \ldots \times M & \xrightarrow{\ f \times \ldots \times f\ } & N \times \ldots \times N & \xrightarrow{\ \alpha'\ } & \bigwedge^n N
\end{array}
$$

the formula for $\bigwedge^n f$ is the induced linear map Φ on the nth exterior power. Since the map arises as the unique induced map via the defining property of $\bigwedge^n M$, it is certainly well-defined.

28.5 *Exterior powers of free modules*

The main point here is that free modules over commutative rings with identity behave much like vector spaces over fields, with respect to multilinear algebra operations. In particular, we prove **non-vanishing** of the nth exterior power of a free module of rank n, which (as we will see) proves the existence of determinants.

At the end, we discuss the natural bilinear map

$$\bigwedge^s M \times \bigwedge^t M \longrightarrow \bigwedge^{s+t} M$$

by

$$(m_1 \wedge \ldots \wedge m_s) \times (m_{s+1} \wedge \ldots \wedge m_{s+t}) \longrightarrow m_1 \wedge \ldots \wedge m_s \wedge m_{s+1} \wedge \ldots \wedge m_{s+t}$$

which does not require free-ness of M.

28.5.1 Theorem: Let F be a free module of rank n over a commutative ring R with identity. Then $\bigwedge^\ell F$ is free of rank $\binom{n}{\ell}$. In particular, if $m_1, \ldots, m_n$ form an R-basis for F, then the monomials

$$m_{i_1} \wedge \ldots \wedge m_{i_\ell} \quad \text{with } i_1 < \ldots < i_\ell$$

are an R basis for $\bigwedge^\ell F$.

Proof: The elementary discussion just above shows that the monomials involving the basis and with strictly ascending indices *generate* $\bigwedge^\ell F$. The remaining issue is to prove linear independence.

First, we prove that $\bigwedge^n F$ is free of rank 1. We know that it is generated by

$$m_1 \wedge \ldots \wedge m_n$$

But for all we know it might be that

$$r \cdot m_1 \wedge \ldots \wedge m_n = 0$$

for some $r \neq 0$ in R. We must prove that this does not happen. To do so, we make a non-trivial alternating (multilinear) map

$$\varphi : \underbrace{F \times \ldots \times F}_{n} \longrightarrow R$$

To make this, let $\lambda_1, \ldots, \lambda_n$ be a *dual basis*[3] for $\text{Hom}_R(F, R)$, namely,

$$\lambda_i(m_j) = \begin{cases} 1 & i = j \\ 0 & \text{(else)} \end{cases}$$

---

[3] These exist, since (by definition of free-ness of F) given a set of desired images $\varphi(m_i) \in R$ of the basis m_i, there is a unique map $\Phi : F \longrightarrow R$ such that $\Phi(m_i) = \varphi(m_i)$.

For arbitrary $x_1, \ldots, x_n$ in F, let [4]

$$\varphi(x_1 \times \ldots \times x_n) = \sum_{\pi \in S_n} \sigma(\pi) \lambda_1(x_{\pi(1)}) \ldots \lambda_n(x_{\pi(n)})$$

where S_n is the group of permutations of n things. Suppose that for some $i \neq j$ we have $x_i = x_j$. Let i' and j' be indices such that $\pi(i') = i$ and $\pi(j') = j$. Let s still be the 2-cycle that interchanges i and j. Then the $n!$ summands can be seen to cancel in pairs, by

$$\sigma(\pi) \lambda_1(x_{\pi(1)}) \ldots \lambda_n(x_{\pi(n)}) + \sigma(s\pi) \lambda_1(x_{s\pi(1)}) \ldots \lambda_n(x_{s\pi(n)})$$

$$= \sigma(\pi) \left(\prod_{\ell \neq i', j'} \lambda_\ell(x_{\pi(\ell)}) \right) \cdot \left(\lambda_i(x_{\pi(i')} \lambda_i(x_{\pi(j')}) - \lambda_i(x_{s\pi(i')}) \lambda_i(x_{s\pi(j')}) \right)$$

Since s just interchanges $i = \pi(i')$ and $j = \pi(j')$, the rightmost sum is 0. This proves the alternating property of φ.

To see that φ is not trivial, note that when the arguments to φ are the basis elements $m_1, \ldots, m_n$, in the expression

$$\varphi(m_1 \times \ldots \times m_n) = \sum_{\pi \in S_n} \sigma(\pi) \lambda_1(m_{\pi(1)}) \ldots \lambda_n(m_{\pi(n)})$$

$\lambda_i(m_{\pi(i)}) = 0$ unless $\pi(i) = i$. That is, the only non-zero summand is with $\pi = 1$, and we have

$$\varphi(m_1 \times \ldots \times m_n) = \lambda_1(m_1) \ldots \lambda_n(m_n) = 1 \in R$$

Then φ induces a map $\Phi : \bigwedge^n F \longrightarrow R$ such that

$$\Phi(m_1 \wedge \ldots \wedge m_n) = 1$$

For $r \in R$ such that $r \cdot (m_1 \wedge \ldots \wedge m_n) = 0$, apply Φ to obtain

$$0 = \Phi(0) = \Phi(r \cdot m_1 \wedge \ldots \wedge m_n) = r \cdot \Phi(m_1 \wedge \ldots \wedge m_n) = r \cdot 1 = r$$

This proves that $\bigwedge^n F$ is free of rank 1.

The case of $\bigwedge^\ell F$ with $\ell < n$ reduces to the case $\ell = n$, as follows. We already know that monomials $m_{i_1} \wedge \ldots \wedge m_{i_\ell}$ with $i_1 < \ldots < i_\ell$ *span* $\bigwedge^\ell F$. Suppose that

$$\sum_{i_1 < \ldots < i_\ell} r_{i_1 \ldots i_\ell} \cdot m_{i_1} \wedge \ldots \wedge m_{i_\ell} = 0$$

The trick is to consider, for a fixed ℓ-tuple $j_1 < \ldots < j_\ell$ of indices, the R-linear map

$$f : \bigwedge^\ell F \longrightarrow \bigwedge^n F$$

given by

$$f(x) = x \wedge (m_1 \wedge m_2 \wedge \ldots \wedge \widehat{m_{j_1}} \wedge \ldots \wedge \widehat{m_{j_\ell}} \wedge \ldots \wedge m_n)$$

---

[4] This formula is suggested by the earlier discussion of determinants of *matrices* following Artin.

where
$$m_1 \wedge m_2 \wedge \ldots \wedge \widehat{m_{j_1}} \wedge \ldots \wedge \widehat{m_{j_\ell}} \wedge \ldots \wedge m_n)$$

is the monomial with exactly the m_{j_t}s *missing*. Granting that this map is well-defined,

$$0 = f(0) = f\left(\sum_{i_1 < \ldots < i_\ell} r_{i_1 \ldots i_\ell} \cdot m_{i_1} \wedge \ldots \wedge m_{i_\ell}\right) = \pm r_{j_1 \ldots j_\ell} m_1 \wedge \ldots \wedge m_n$$

since all the other monomials have some repeated m_t, so are 0. That is, any such relation must have all coefficients 0. This proves the linear independence of the indicated monomials.

To be sure that these maps f are well-defined, [5] we prove a more systematic result, which will finish the proof of the theorem.

28.5.2 Proposition: Let M be an R-module. [6] Let s, t be positive integers. The canonical alternating multilinear map

$$\alpha : M \times \ldots \times M \longrightarrow \bigwedge^{s+t} M$$

induces a natural bilinear map

$$B : (\bigwedge^s M) \times (\bigwedge^t M) \longrightarrow \bigwedge^{s+t} M$$

by

$$(m_1 \wedge \ldots \wedge m_s) \times (m_{s+1} \wedge \ldots \wedge m_{s+t}) \longrightarrow m_1 \wedge \ldots \wedge m_s \wedge m_{s+1} \wedge \ldots \wedge m_{s+t}$$

Proof: For fixed choice of the last t arguments, the map α on the first s factors is certainly alternating multilinear. Thus, from the defining property of $\bigwedge^s M$, α factors uniquely through the map

$$\bigwedge^s M \times \underbrace{M \times \ldots \times M}_{t} \longrightarrow \bigwedge^{s+t} M$$

defined (by linearity) by

$$(m_1 \wedge \ldots \wedge m_s) \times m_{s+1} \times \ldots \times m_{s+t} = m_1 \wedge \ldots \wedge m_s \wedge m_{s+1} \wedge \ldots \wedge m_{s+t}$$

Indeed, by the defining property of the exterior power, for each fixed choice of last t arguments the map is linear on $\bigwedge^s M$. Further, for fixed choice of first arguments α on the last t arguments is alternating multilinear, so α factors through the expected map

$$(\bigwedge^s M) \times (\bigwedge^t M) \longrightarrow \bigwedge^{s+t} M$$

linear in the $\bigwedge^t M$ argument for each choice of the first. That is, this map is bilinear.   ///

---

[5] The importance of verifying that symbolically reasonable expressions make sense is often underestimated. Seemingly well-defined things can easily be ill-defined. For example, $f : \mathbb{Z}/3 \longrightarrow \mathbb{Z}/5$ defined [sic] by $f(x) = x$, or, seemingly more clearly, by $f(x + 3\mathbb{Z}) = x + 5\mathbb{Z}$. This is not well-defined, since $0 = f(0) = f(3) = 3 \neq 0$.

[6] In particular, M need not be free, and need not be finitely-generated.

28.6 *Determinants revisited*

The fundamental idea is that for an endomorphism T of a free R-module M of rank n (with R commutative with unit), $\det T \in R$ is determined as

$$Tm_1 \wedge \ldots \wedge Tm_n = (\det T) \cdot (m_1 \wedge \ldots \wedge m_n)$$

Since $\bigwedge^n M$ is free of rank 1, all R-linear endomorphisms are given by scalars: indeed, for an endomorphism A of a rank-1 R-module with generator e,

$$A(re) = r \cdot Ae = r \cdot (s \cdot e)$$

for all $r \in$, for some $s \in R$, since $Ae \in R \cdot e$.

This gives a *scalar* $\det T$, intrinsically defined, assuming that we verify that this does what we want.

And certainly this would give a pleasant proof of the *multiplicativity* of determinants, since

$$(\det ST) \cdot (m_1 \wedge \ldots \wedge m_n) = (ST)m_1 \wedge \ldots \wedge (ST)m_n = S(Tm_1) \wedge \ldots \wedge S(Tm_n)$$

$$= (\det S)(Tm_1 \wedge \ldots \wedge Tm_n) = (\det S)(\det T)(m_1 \wedge \ldots \wedge m_n)$$

Note that we use the fact that

$$(\det T) \cdot (m_1 \wedge \ldots \wedge m_n) = Tm_1 \wedge \ldots \wedge Tm_n$$

for *all* n-tuples of elements m_i in F.

Let $e_1, \ldots, e_n$ be the standard basis of k^n. Let $v_1, \ldots, v_n$ be the columns of an n-by-n matrix. Let T be the endomorphism (of column vectors) given by (left multiplication by) that matrix. That is, $Te_i = v_i$. Then

$$v_1 \wedge \ldots \wedge v_n = Te_1 \wedge \ldots \wedge Te_n = (\det T) \cdot (e_1 \wedge \ldots \wedge e_n)$$

The leftmost expression in the latter line is an alternating multilinear $\bigwedge^n(k^n)$-valued function. (Not k-valued.) But since we know that $\bigwedge^n(k^n)$ is one-dimensional, and is spanned by $e_1 \wedge \ldots \wedge e_n$, (once again) we know that there is a unique scalar $\det T$ such that the right-hand equality holds. That is, the map

$$v_1 \times \ldots \times v_n \longrightarrow \det T$$

where T is the endomorphism given by the matrix with columns v_i, is an alternating k-valued map. And it is 1 for $v_i = e_i$.

This translation back to matrices verifies that our intrinsic determinant meets our earlier axiomatized requirements for a determinant. ///

Finally we note that the basic formula for determinants of matrices that followed from Artin's axiomatic characterization, at least in the case of entires in *fields*, is valid for matrices with entries in commutative rings (with units). That is, for an n-by-n matrix A with entries A_{ij} in a commutative ring R with unit,

$$\det A = \sum_{\pi \in S_n} \sigma(\pi) A_{\pi(1),1} \ldots A_{\pi(n),n}$$

where S_n is the symmetric group on n things and $\sigma(\pi)$ is the sign function on permutations. Indeed, let $v_1, \ldots, v_n$ be the rows of A, let $e_1, \ldots, e_n$ be the standard basis (row) vectors for R^n, and consider A as an endomorphism of R^n. As in the previous argument, $A \cdot e_j = e_j A = v_j$ (where A acts by right matrix multiplication). And $v_i = \sum_j A_{ij} e_j$. Then

$$(\det A)\, e_1 \wedge \ldots \wedge e_n = (A \cdot e_1) \wedge \ldots \wedge (A \cdot e_n) = v_1 \wedge \ldots \wedge v_n = \sum_{i_1, \ldots, i_n} (A_{1i_1} e_{i_1}) \wedge \ldots \wedge (A_{ni_n} e_{i_n})$$

$$= \sum_{\pi \in S_n} \left(A_{1\pi(1)} e_{\pi(1)} \right) \wedge \ldots \wedge \left(A_{n\pi(n)} e_{\pi(n)} \right) = \sum_{\pi \in S_n} \left(A_{1\pi(1)} \ldots A_{n\pi(n)} \right) e_{\pi(1)} \wedge \ldots \wedge e_{\pi(n)}$$

$$= \sum_{\pi \in S_n} (A_{\pi^{-1}(1),1} \ldots A_{\pi^{-1}(n),n})\, \sigma(\pi) e_1 \wedge \ldots \wedge e_n$$

by reordering the e_is, using the alternating multilinear nature of $\bigwedge^n(R^n)$. Of course $\sigma(\pi) = \sigma(\pi^{-1})$. Replacing π by π^{-1} (thus replacing π^{-1} by π) gives the desired

$$(\det A)\, e_1 \wedge \ldots \wedge e_n = \sum_{\pi \in S_n} (A_{\pi(1),1} \ldots A_{\pi(n),n})\, \sigma(\pi) e_1 \wedge \ldots \wedge e_n$$

Since $e_1 \wedge \ldots \wedge e_n$ is an R-basis for the free rank-one R-module $\bigwedge^n(R^n)$, this proves that $\det A$ is given by the asserted formula. ///

28.6.1 Remark: Indeed, the point that $e_1 \wedge \ldots \wedge e_n$ is an R-basis for the free rank-one R-module $\bigwedge^n(R^n)$, as opposed to being 0 or being annihilated by some non-zero elements of R, is exactly what is needed to make the earlier seemingly field-oriented arguments work more generally.

28.7 *Minors of matrices*

At first, one might be surprised at the following phenomenon.

Let

$$M = \begin{pmatrix} a & b & c \\ x & y & z \end{pmatrix}$$

with entries in some commutative ring R with unit. Viewing each of the two rows as a vector in R^3, inside $\bigwedge^2 R^3$ we compute (letting e_1, e_2, e_3 be the standard basis)

$$(ae_1 + be_2 + ce_3) \wedge (xe_1 + ye_2 + ze_3)$$

$$= \begin{cases} & axe_1 \wedge e_1 & + & aye_1 \wedge e_2 & + & aze_1 \wedge e_3 \\ + & bxe_2 \wedge e_1 & + & bye_2 \wedge e_2 & + & bze_2 \wedge e_3 \\ + & cxe_3 \wedge e_1 & + & cye_3 \wedge e_2 & + & cze_3 \wedge e_3 \end{cases}$$

$$= \begin{cases} & 0 & + & aye_1 \wedge e_2 & + & aze_1 \wedge e_3 \\ -bxe_1 \wedge e_2 & + & & 0 & + & bze_2 \wedge e_3 \\ -cxe_1 \wedge e_3 & + & -cye_2 \wedge e_3 & + & & 0 \end{cases}$$

$$= (ay - bx)\, e_1 \wedge e_2 + (az - cx)\, e_1 \wedge e_3 + (bz - cy)\, e_2 \wedge e_3$$

$$= \begin{vmatrix} a & b \\ x & y \end{vmatrix} e_1 \wedge e_2 + \begin{vmatrix} a & c \\ x & z \end{vmatrix} e_1 \wedge e_3 + \begin{vmatrix} b & c \\ y & z \end{vmatrix} e_2 \wedge e_3$$

where, to fit it on a line, we have written

$$\begin{vmatrix} a & b \\ x & y \end{vmatrix} = \det \begin{pmatrix} a & b \\ x & y \end{pmatrix}$$

That is, *the coefficients in the second exterior power are the determinants of the two-by-two minors.*

At some point it becomes unsurprising to have

28.7.1 Proposition: Let M be an m-by-n matrix with $m < n$, entries in a commutative ring R with identity. Viewing the rows $M_1, \ldots, M_m$ of M as elements of R^n, and letting $e_1, \ldots, e_n$ be the standard basis of R^n, in $\bigwedge^m R^n$

$$M_1 \wedge \ldots \wedge M_n = \sum_{i_1 < \ldots < i_m} \det(M^{i_1 \cdots i_m}) \cdot e_{i_1} \wedge \ldots \wedge e_{i_m}$$

where $M^{i_1 \cdots i_m}$ is the m-by-m matrix consisting of the i_1^{th}, i_2^{th}, ..., i_m^{th} columns of M.

Proof: Write

$$M_i = \sum_j r_{ij} e_j$$

Then

$$M_1 \wedge \ldots \wedge M_m = \sum_{i_1, \ldots, i_m} (M_{1i_1} e_{i_1}) \wedge (M_{2i_2} e_{i_2}) \wedge \ldots \wedge (M_{mi_m} e_{i_n})$$

$$= \sum_{i_1, \ldots, i_m} M_{1i_1} \ldots M_{mi_m} \, e_{i_1} \wedge e_{i_2} \wedge \ldots \wedge e_{i_m}$$

$$= \sum_{i_1 < \ldots < i_m} \sum_{\pi \in S_m} \sigma(\pi) \, M_{1, i_{\pi(1)}} \ldots M_{m, i_{\pi(i)}} \, e_{i_1} \wedge \ldots \wedge e_{i_m}$$

$$= \sum_{i_1 < \ldots < i_m} \det M^{i_1 \cdots i_m} \, e_{i_1} \wedge \ldots \wedge e_{i_m}$$

where we reorder the e_{i_j}s via π in the permutations group S_m of $\{1, 2, \ldots, m\}$ and $\sigma(\pi)$ is the sign function on permutation. This uses the general formula for the determinant of an n-by-n matrix, from above. ///

28.8 *Uniqueness in the structure theorem*

Exterior powers give a decisive trick to give an *elegant* proof of the uniqueness part of the structure theorem for finitely-generated modules over principal ideal domains. This will be the immediate application of

28.8.1 Proposition: Let R be a commutative ring with identity. Let M be a free R-module with R-basis $m_1, \ldots, m_n$. Let $d_1, \ldots, d_n$ be elements of R, and let

$$N = R \cdot d_1 m_1 \oplus \ldots \oplus R \cdot d_n m_n \subset M$$

Then, for any $1 < \ell \in \mathbb{Z}$, we have

$$\bigwedge^\ell N = \bigoplus_{j_1 < \ldots < j_\ell} R \cdot (d_{j_1} \ldots d_{j_\ell}) \cdot (m_{j_1} \wedge \ldots \wedge m_{j_\ell}) \subset \bigwedge^\ell M$$

28.8.2 Remark: We do not need to assume that R is a PID, nor that $d_1|\ldots|d_n$, in this proposition.

Proof: Without loss of generality, by re-indexing, suppose that $d_1,\ldots,d_t$ are non-zero and $d_{t+1} = d_{t+2} = \ldots = d_n = 0$. We have already shown that the ordered monomials $m_{j_1} \wedge \ldots \wedge m_{j_\ell}$ are a basis for the free R-module $\bigwedge^\ell M$, whether or not R is a PID. Similarly, the basis $d_1 m_1, \ldots, d_t m_t$ for N yields a basis of the ℓ-fold monomials for $\bigwedge^\ell N$, namely

$$d_{j_1} m_{j_1} \wedge \ldots \wedge d_{j_\ell} m_{j_\ell} \quad \text{with} \quad j_1 < \ldots < j_\ell \leq t$$

By the multilinearity,

$$d_{j_1} m_{j_1} \wedge \ldots \wedge d_{j_\ell} m_{j_\ell} = (d_{j_1} d_{j_2} \ldots d_{j_\ell}) \cdot (m_{j_1} \wedge \ldots \wedge m_{j_\ell})$$

This is all that is asserted.   ///

At last, we prove the uniqueness of elementary divisors.

28.8.3 Corollary: Let R be a principal ideal domain. Let M be a finitely-generated free R-module, and N a submodule of M. Then there is a basis $m_1, \ldots, m_n$ of M and *elementary divisors* $d_1|\ldots|d_n$ in R such that

$$N = Rd_1 m_1 \oplus \ldots \oplus Rd_n m_n$$

The ideals Rd_i are uniquely determined by M, N.

Proof: The *existence* was proven much earlier. Note that the *highest* elementary divisor d_n, or, really, the ideal Rd_n, is determined *intrinsically* by the property

$$Rd_n = \{r \in R : r \cdot (M/N) = 0\}$$

since d_n is a least common multiple of all the d_is. That is, Rd_n is the **annihilator** of M/N.

Suppose that t is the last index so that $d_t \neq 0$, so $d_1, \ldots, d_t$ are non-zero and $d_{t+1} = d_{t+2} = \ldots = d_n = 0$. Using the proposition, the annihilator of $\bigwedge^2 M/\bigwedge^2 N$ is $R \cdot d_{t-1} d_t$, since d_{t-1} and d_t are the two largest non-zero elementary divisors. Since Rd_t is uniquely determined, Rd_{t-1} is uniquely determined.

Similarly, the annihilator of $\bigwedge^i M/\bigwedge^i N$ is $Rd_{t-i+1} \ldots d_{t-1} d_t$, which is uniquely determined. By induction, d_t, d_{t-1}, $\ldots$, d_{t-i+2} are uniquely determined. Thus, d_{t-i+1} is uniquely determined.   ///

28.9 *Cartan's lemma*

To further illustrate computations in exterior algebra, we prove a result that arises in differential geometry, often accidentally disguised as something more than the simple exterior algebra it is.

28.9.1 Proposition: *(Cartan)* Let V be a vector space over a field k. Let $v_1, \ldots, v_n$ be linearly independent vectors in V. Let $w_1, \ldots, w_n$ be any vectors in V. Then

$$v_1 \wedge w_1 + \ldots + v_n \wedge w_n = 0$$

if and only if there is a *symmetric* matrix with entries $A_{ij} \in k$ such that

$$w_i = \sum_i A_{ij} \, v_j$$

Proof: First, prove that if the identity holds, then the w_j's lie in the span of the v_i's. Suppose not. Then, by renumbering for convenience, we can suppose that $w_1, v_1, \ldots, v_n$ are linearly independent. Let $\eta = v_2 \wedge \ldots \wedge v_n$. Then

$$\left(v_1 \wedge w_1 + \ldots + v_n \wedge w_n \right) \wedge \eta = 0 \wedge \eta = 0 \in \textstyle\bigwedge^{n+1} V$$

On the other hand, the exterior products of η with all summands but the first are 0, since some v_i with $i \geq 2$ is repeated. Thus,

$$\left(v_1 \wedge w_1 + \ldots + v_n \wedge w_n \right) \wedge \eta = v_1 \wedge w_1 \wedge \eta = v_1 \wedge w_1 \wedge v_2 \wedge \ldots \wedge v_n \neq 0$$

This contradiction proves that the w_j's do all lie in the span of the v_i's if the identity is satisfied. Let A_{ij} be elements of k expressing the w_j's as linear combinations

$$w_i = \sum_i A_{ij} \, v_j$$

We need to prove that $A_{ij} = A_{ji}$.

Let

$$\omega = v_1 \wedge \ldots \wedge v_n \in \textstyle\bigwedge^n V$$

By our general discussion of exterior powers, by the linear independence of the v_i this is non-zero. For $1 \leq i \leq n$, let

$$\omega_i = v_1 \wedge \ldots \wedge \widehat{v_i} \wedge \ldots \wedge v_n \in \textstyle\bigwedge^{n-1} V$$

where the hat indicates omission. In any linear combination $v = \sum_j c_j v_j$ we can pick out the ith coefficient by exterior product with ω_i, namely

$$v \wedge \omega_i = \left(\sum_j c_j v_j \right) \wedge \omega_i = \sum_j c_j v_j \wedge \omega_i = c_i v_i \wedge \omega_i = c_i \, \omega$$

For $i < j$, let

$$\omega_{ij} = v_1 \wedge \ldots \wedge \widehat{v_i} \wedge \ldots \wedge \widehat{v_j} \wedge \ldots \wedge v_n \in \textstyle\bigwedge^{n-2} V$$

Then

$$\left(v_1 \wedge w_1 + \ldots + v_n \wedge w_n \right) \wedge \omega_{ij} = v_1 \wedge w_1 \wedge \omega_{ij} + \ldots + v_n \wedge w_n \wedge \omega_{ij}$$

$$= v_i \wedge w_i \wedge \omega_{ij} + v_j \wedge w_j \wedge \omega_{ij}$$

since all the other monomials vanish, having repeated factors. Thus, moving things around slightly,

$$w_i \wedge v_i \wedge \omega_{ij} = - w_j \wedge v_j \wedge \omega_{ij}$$

By moving the v_i and v_j across, flipping signs as we go, with $i < j$, we have

$$v_i \wedge \omega_{ij} = (-1)^{i-1}w_j \qquad\qquad v_j \wedge \omega_{ij} = (-1)^{j-2}w_i$$

Thus, using ω_ℓ to pick off the ℓth coefficient,

$$(-1)^{i-1}A_{ij}(-1)^{i-1}w_i \wedge w_j = w_i \wedge v_i \wedge \omega_{ij} = - w_j \wedge v_j \wedge \omega_{ij}$$

$$= -(-1)^{j-2}w_j \wedge w_i = (-1)^{j-1}A_{ji}$$

From this, we get the claimed symmetry of the collection of A_{ij}'s, namely,

$$A_{ij} = A_{ji}$$

Reversing this argument gives the converse. Specifically, suppose that $w_i = \sum_j A_{ij} v_j$ with $A_{ij} = A_{ji}$. Let W be the span of $v_1, \ldots, v_n$ inside W. Then running the previous computation backward directly yields

$$\left(v_1 \wedge w_1 + \ldots + v_n \wedge w_n\right) \wedge \omega_{ij} = 0$$

for all $i < j$. The monomials ω_{ij} span $\bigwedge^{n-2}W$ and we have shown the non-degeneracy of the pairing

$$\bigwedge^{n-2}W \times \bigwedge^2 W \longrightarrow \bigwedge^n W \qquad \text{by} \qquad \alpha \times \beta \longrightarrow \alpha \wedge \beta$$

Thus,

$$v_1 \wedge w_1 + \ldots + v_n \wedge w_n = 0 \in \bigwedge^2 W \subset \bigwedge^2 V$$

as claimed.   ///

28.10 *Cayley-Hamilton theorem*

28.10.1 Theorem: (*Cayley-Hamilton*) Let T be a k-linear endomorphism of a finite-dimensional vector space V over a field k. Let $P_T(x)$ be the characteristic polynomial

$$P_T(x) = \det(x \cdot 1_V - T)$$

Then

$$P_T(T) = 0 \in \text{End}_k(V)$$

28.10.2 Remarks: Cayley and Hamilton proved the cases with $n = 2, 3$ by direct computation. The theorem can be made a corollary of the structure theorem for finitely-generated modules over principal ideal domains, if certain issues are glossed over. For example, how should an indeterminate x act on a vectorspace? It would be premature to say that $x \cdot 1_V$ acts as T on V, even though at the end this is *exactly* what is supposed to happen, because, if $x = T$ at the outset, then $P_T(x)$ is simply 0, and the theorem asserts nothing. Various misconceptions can be turned into false proofs. For example, it is *not* correct to argue that

$$P_T(T) = \det(T - T) = \det 0 = 0 \qquad \text{(incorrect)}$$

However, the argument given just below *is* a *correct* version of this idea. Indeed, in light of these remarks, we must clarify what it means to *substitute* T for x. Incidental to the argument, *intrinsic* versions of *determinant* and *adjugate* (or *cofactor*) endomorphism are described, in terms of multi-linear algebra.

Proof: The module $V \otimes_k k[x]$ is free of rank $\dim_k V$ over $k[x]$, and is the object associated to V on which the indeterminate x reasonably acts. Also, V is a $k[T]$-module by the action $v \longrightarrow Tv$, so $V \otimes_k k[x]$ is a $k[T] \otimes_k k[x]$-module. The **characteristic polynomial** $P_T(x) \in k[x]$ of $T \in \text{End}_k(V)$ is the determinant of $1 \otimes x - T \otimes 1$, defined intrinsically by

$$\textstyle\bigwedge^n_{k[x]} (T \otimes 1 - 1 \otimes x) = P_T(x) \cdot 1 \qquad \text{(where } n = \dim_k V = \text{rk}_{k[x]} V \otimes_k k[x]\text{)}$$

where the first 1 is the identity in $k[x]$, the second 1 is the identity map on V, and the last 1 is the identity map on $\bigwedge^n_{k[x]}(V \otimes_k k[x])$.

To *substitute* T for x is a special case of the following procedure. Let R be a commutative ring with 1, and M an R-module with $1 \cdot m = m$ for all $m \in M$. For an ideal I of R, the quotient $M/I \cdot M$ is the natural associated R/I-module, and every R-endomorphism α of M such that

$$\alpha(I \cdot M) \subset I \cdot M$$

descends to an R/I-endomorphism of $M/I \cdot M$. In the present situation,

$$R = k[T] \otimes_k k[x] \qquad M = V \otimes_k k[x]$$

and I is the ideal generated by $1 \otimes x - T \otimes 1$. Indeed, $1 \otimes x$ is the image of x in this ring, and $T \otimes 1$ is the image of T. Thus, $1 \otimes x - T \otimes 1$ should map to 0.

To prove that $P_T(T) = 0$, we will *factor* $P_T(x) \cdot 1$ so that after substituting T for x the resulting endomorphism $P_T(T) \cdot 1$ has a literal factor of $T - T = 0$. To this end, consider the natural $k[x]$-bilinear map

$$\langle,\rangle : \textstyle\bigwedge^{n-1}_{k[x]} V \otimes_k k[x] \ \times \ V \otimes_k k[x] \ \longrightarrow \ \bigwedge^n_{k[x]} V \otimes_k k[x]$$

of free $k[x]$-modules, identifying $V \otimes_k k[x]$ with its first exterior power. Letting $A = 1 \otimes x - T \otimes 1$, for all $m_1, \ldots, m_n$ in $V \otimes_k k[x]$,

$$\langle \textstyle\bigwedge^{n-1} A(m_1 \wedge \ldots \wedge m_{n-1}), \ A m_n \rangle \ = \ P_T(x) \cdot m_1 \wedge \ldots \wedge m_n$$

By definition, the *adjugate* or *cofactor* endomorphism A^{adg} of A is the adjoint of $\bigwedge^{n-1} A$ with respect to this pairing. Thus,

$$\langle m_1 \wedge \ldots \wedge m_{n-1}, \ (A^{\text{adg}} \circ A) \, m_n \rangle \ = \ P_T(x) \cdot m_1 \wedge \ldots \wedge m_n$$

and, therefore,

$$A^{\text{adg}} \circ A \ = \ P_T(x) \cdot 1 \qquad \text{(on } V \otimes_k k[x]\text{)}$$

Since $\langle,\rangle$ is $k[x]$-bilinear, A^{adg} is a $k[x]$-endomorphism of $V \otimes_k k[x]$. To verify that A^{adg} commutes with $T \otimes 1$, it suffices to verify that A^{adg} commutes with A. To this end, further extend scalars on all the free $k[x]$-modules $\bigwedge_{k[x]}^{\ell} V \otimes_k k[x]$ by tensoring with the field of fractions $k(x)$ of $k[x]$. Then

$$A^{\text{adg}} \cdot A \;=\; P_T(x) \cdot 1 \qquad\qquad (\text{now on } V \otimes_k k(x))$$

Since $P_T(x)$ is monic, it is non-zero, hence, invertible in $k(x)$. Thus, A is invertible on $V \otimes_k k(x)$, and

$$A^{\text{adg}} \;=\; P_T(x) \cdot A^{-1} \qquad\qquad (\text{on } V \otimes_k k(x))$$

In particular, the corresponding version of A^{adg} commutes with A on $V \otimes_k k(x)$, and, thus, A^{adg} commutes with A on $V \otimes_k k[x]$.

Thus, A^{adg} descends to an R/I-linear endomorphism of $M/I \cdot M$, where

$$R = k[T] \otimes_k k[x] \qquad M = V \otimes_k k[x] \qquad I = R \cdot A \qquad\qquad (\text{with } A = 1 \otimes x - T \otimes 1)$$

That is, on the quotient $M/I \cdot M$,

$$(\text{image of })A^{\text{adg}} \cdot (\text{image of })(1 \otimes x - T \otimes 1) \;=\; P_T(T) \cdot 1_{M/IM}$$

The image of $1 \otimes x - T \otimes 1$ here is 0, so

$$(\text{image of })A^{\text{adg}} \cdot 0 \;=\; P_T(T) \cdot 1_{M/IM}$$

This implies that

$$P_T(T) \;=\; 0 \qquad\qquad (\text{on } M/IM)$$

Note that the composition

$$V \longrightarrow V \otimes_k k[x] = M \longrightarrow M/IM$$

is an isomorphism of $k[T]$-modules, and, *a fortiori*, of k-vectorspaces.   ///

28.10.3 Remark: This should not be the first discussion of this result seen by a novice. However, all the issues addressed are genuine!

28.11 *Worked examples*

28.11.1 Example: Consider the injection $\mathbb{Z}/2 \overset{t}{\longrightarrow} \mathbb{Z}/4$ which maps

$$t : x + 2\mathbb{Z} \longrightarrow 2x + 4\mathbb{Z}$$

Show that the induced map

$$t \otimes 1_{\mathbb{Z}/2} : \mathbb{Z}/2 \otimes_{\mathbb{Z}} \mathbb{Z}/2 \longrightarrow \mathbb{Z}/4 \otimes_{\mathbb{Z}} \mathbb{Z}/2$$

is no longer an injection.

We claim that $t \otimes 1$ is the 0 map. Indeed,

$$(t \otimes 1)(m \otimes n) = 2m \otimes n = 2 \cdot (m \otimes n) = m \otimes 2n = m \otimes 0 = 0$$

for all $m \in \mathbb{Z}/2$ and $n \in \mathbb{Z}/2$. ///

28.11.2 Example: Prove that if $s : M \longrightarrow N$ is a *surjection* of $\mathbb{Z}$-modules and X is any other $\mathbb{Z}$ module, then the induced map

$$s \otimes 1_Z : M \otimes_{\mathbb{Z}} X \longrightarrow N \otimes_{\mathbb{Z}} X$$

is still surjective.

Given $\sum_i n_i \otimes x_i$ in $N \otimes_{\mathbb{Z}} X$, let $m_i \in M$ be such that $s(m_i) = n_i$. Then

$$(s \otimes 1)(\sum_i m_i \otimes x_i) = \sum_i s(m_i) \otimes x_i = \sum_i n_i \otimes x_i$$

so the map is surjective. ///

28.11.3 Remark: Note that the only issue here is hidden in the verification that the induced map $s \otimes 1$ exists.

28.11.4 Example: Give an example of a surjection $f : M \longrightarrow N$ of $\mathbb{Z}$-modules, and another $\mathbb{Z}$-module X, such that the induced map

$$f \circ - : \mathrm{Hom}_{\mathbb{Z}}(X, M) \longrightarrow \mathrm{Hom}_{\mathbb{Z}}(X, N)$$

(by post-composing) *fails* to be surjective.

Let $M = \mathbb{Z}$ and $N = \mathbb{Z}/n$ with $n > 0$. Let $X = \mathbb{Z}/n$. Then

$$\mathrm{Hom}_{\mathbb{Z}}(X, M) = \mathrm{Hom}_{\mathbb{Z}}(\mathbb{Z}/n, \mathbb{Z}) = 0$$

since

$$0 = \varphi(0) = \varphi(nx) = n \cdot \varphi(x) \in \mathbb{Z}$$

so (since n is not a 0-divisor in $\mathbb{Z}$) $\varphi(x) = 0$ for all $x \in \mathbb{Z}/n$. On the other hand,

$$\mathrm{Hom}_{\mathbb{Z}}(X, N) = \mathrm{Hom}_{\mathbb{Z}}(\mathbb{Z}/n, \mathbb{Z}/n) \approx \mathbb{Z}/n \neq 0$$

Thus, the map cannot possibly be surjective. ///

28.11.5 Example: Let $G : \{\mathbb{Z} - \text{modules}\} \longrightarrow \{\text{sets}\}$ be the functor that forgets that a module is a module, and just retains the underlying set. Let $F : \{\text{sets}\} \longrightarrow \{\mathbb{Z} - \text{modules}\}$ be the functor which creates the free module FS on the set S (*and* keeps in mind a map $i : S \longrightarrow FS$). Show that for any set S and any $\mathbb{Z}$-module M

$$\mathrm{Hom}_{\mathbb{Z}}(FS, M) \approx \mathrm{Hom}_{\text{sets}}(S, GM)$$

Prove that the isomorphism you describe is *natural* in S. (It is also natural in M, but don't prove this.)

Our definition of *free module* says that $FS = X$ is free on a (set) map $i : S \longrightarrow X$ if for every set map $\varphi : S \longrightarrow M$ with R-module M gives a unique R-module map $\Phi : X \longrightarrow M$ such that the diagram

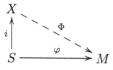

commutes. Of course, given Φ, we obtain $\varphi = \Phi \circ i$ by composition (in effect, restriction). We claim that the required isomorphism is

$$\text{Hom}_{\mathbb{Z}}(FS, M) \xleftarrow{\;\Phi \longleftrightarrow \varphi\;} \text{Hom}_{\text{sets}}(S, GM)$$

Even prior to naturality, we must prove that this is a bijection. Note that the set of maps of a set into an R-module has a natural structure of R-module, by

$$(r \cdot \varphi)(s) = r \cdot \varphi(s)$$

The map in the direction $\varphi \longrightarrow \Phi$ is an *injection*, because two maps φ, ψ mapping $S \longrightarrow M$ that induce the same map Φ on X give $\varphi = \Phi \circ i = \psi$, so $\varphi = \psi$. And the map $\varphi \longrightarrow \Phi$ is *surjective* because a given Φ is induced from $\varphi = \Phi \circ i$.

For naturality, for fixed S and M let the map $\varphi \longrightarrow \Phi$ be named $j_{S,M}$. That is, the isomorphism is

$$\text{Hom}_{\mathbb{Z}}(FS, M) \xleftarrow{\;j_{S,X}\;} \text{Hom}_{\text{sets}}(S, GM)$$

To show naturality in S, let $f : S \longrightarrow S'$ be a set map. Let $i' : S' \longrightarrow X'$ be a free module on S'. That is, $X' = FS'$. We must show that

$$
\begin{array}{ccc}
\text{Hom}_{\mathbb{Z}}(FS, M) & \xleftarrow{\;j_{S,M}\;} & \text{Hom}_{\text{sets}}(S, GM) \\
{\scriptstyle -\circ Ff}\big\uparrow & & \big\uparrow{\scriptstyle -\circ f} \\
\text{Hom}_{\mathbb{Z}}(FS', M) & \xleftarrow{\;j_{S',M}\;} & \text{Hom}_{\text{sets}}(S', GM)
\end{array}
$$

commutes, where $- \circ f$ is pre-composition by f, and $- \circ Ff$ is pre-composition by the induced map $Ff : FS \longrightarrow FS'$ on the free modules $X = FS$ and $X' = FS'$. Let $\varphi \in \text{Hom}_{\text{set}}(S', GM)$, and $x = \sum_s r_s \cdot i(s) \in X = FS$, Go up, then left, in the diagram, computing,

$$(j_{S,M} \circ (- \circ f))\,(\varphi)(x) = j_{S,M}\,(\varphi \circ f)\,(x) = j_{S,M}\,(\varphi \circ f)\left(\sum_s r_s i(s)\right) = \sum_s r_s(\varphi \circ f)(s)$$

On the other hand, going left, then up, gives

$$((- \circ Ff) \circ j_{S',M})\,(\varphi)(x) = (j_{S',M}(\varphi) \circ Ff)\,(x) = (j_{S',M}(\varphi))\,Ff(x)$$

$$= (j_{S',M}(\varphi))\left(\sum_s r_s i'(fs)\right) = \sum_s r_s \varphi(fs)$$

These are the same. ///

28.11.6 Example: Let $M = \begin{pmatrix} m_{21} & m_{22} & m_{23} \\ m_{31} & m_{32} & m_{33} \end{pmatrix}$ be a 2-by-3 integer matrix, such that the *gcd* of the three 2-by-2 minors is 1. Prove that there exist three integers m_{11}, m_{12}, m_{33} such that

$$\det \begin{pmatrix} m_{11} & m_{12} & m_{13} \\ m_{21} & m_{22} & m_{23} \\ m_{31} & m_{32} & m_{33} \end{pmatrix} = 1$$

This is the easiest of this and the following two examples. Namely, let M_i be the 2-by-2 matrix obtained by omitting the ith column of the given matrix. Let a, b, c be integers such that

$$a \det M_1 - b \det M_2 + c \det M_3 = \gcd(\det M_1, \det M_2, \det M_3) = 1$$

Then, expanding by minors,

$$\det \begin{pmatrix} a & b & c \\ m_{21} & m_{22} & m_{23} \\ m_{31} & m_{32} & m_{33} \end{pmatrix} = a \det M_1 - b \det M_2 + c \det M_3 = 1$$

as desired.   ///

28.11.7 Example: Let a, b, c be integers whose *gcd* is 1. Prove (without manipulating matrices) that there is a 3-by-3 integer matrix with top row $(a \; b \; c)$ with determinant 1.

Let $F = \mathbb{Z}^3$, and $E = \mathbb{Z} \cdot (a, b, c)$. We claim that, since $\gcd(a, b, c) = 1$, F/E is torsion-free. Indeed, for $(x, y, z) \in F = \mathbb{Z}^3$, $r \in \mathbb{Z}$, and $r \cdot (x, y, z) \in E$, there must be an integer t such that $ta = rx$, $tb = ry$, and $tc = rz$. Let u, v, w be integers such that

$$ua + vb + wz = \gcd(a, b, c) = 1$$

Then the usual stunt gives

$$t = t \cdot 1 = t \cdot (ua + vb + wz) = u(ta) + v(tb) + w(tc) = u(rx) + v(ry) + w(rz) = r \cdot (ux + vy + wz)$$

This implies that $r|t$. Thus, dividing through by r, $(x, y, z) \in \mathbb{Z} \cdot (a, b, c)$, as claimed.

Invoking the Structure Theorem for finitely-generated $\mathbb{Z}$-modules, there is a basis f_1, f_2, f_3 for F and $0 < d_1 \in \mathbb{Z}$ such that $E = \mathbb{Z} \cdot d_1 f_1$. Since F/E is torsionless, $d_1 = 1$, and $E = \mathbb{Z} \cdot f_1$. Further, since both (a, b, c) and f_1 generate E, and $\mathbb{Z}^\times = \{\pm 1\}$, without loss of generality we can suppose that $f_1 = (a, b, c)$.

Let A be an endomorphism of $F = \mathbb{Z}^3$ such that $A f_i = e_i$. Then, writing A for the matrix giving the endomorphism A,

$$(a, b, c) \cdot A = (1, 0, 0)$$

Since A has an inverse B,

$$1 = \det 1_3 = \det(AB) = \det A \cdot \det B$$

so the determinants of A and B are in $\mathbb{Z}^\times = \{\pm 1\}$. We can adjust A by right-multiplying by

$$\begin{pmatrix} 1 & 0 & 0 \\ 0 & 1 & 0 \\ 0 & 0 & -1 \end{pmatrix}$$

to make $\det A = +1$, and retaining the property $f_1 \cdot A = e_1$. Then

$$A^{-1} = 1_3 \cdot A^{-1} = \begin{pmatrix} e_1 \\ e_2 \\ e_3 \end{pmatrix} \cdot A^{-1} = \begin{pmatrix} a & b & c \\ * & * & * \\ * & * & * \end{pmatrix}$$

That is, the original (a, b, c) is the top row of A^{-1}, which has integer entries and determinant 1.   ///

28.11.8 Example: Let

$$M = \begin{pmatrix} m_{11} & m_{12} & m_{13} & m_{14} & m_{15} \\ m_{21} & m_{22} & m_{23} & m_{24} & m_{25} \\ m_{31} & m_{32} & m_{33} & m_{34} & m_{35} \end{pmatrix}$$

and suppose that the *gcd* of all determinants of 3-by-3 minors is 1. Prove that there exists a 5-by-5 integer matrix $\tilde{M}$ with M as its top 3 rows, such that $\det \tilde{M} = 1$.

Let $F = \mathbb{Z}^5$, and let E be the submodule generated by the rows of the matrix. Since $\mathbb{Z}$ is a PID and F is free, E is free.

Let $e_1, \ldots, e_5$ be the standard basis for $\mathbb{Z}^5$. We have shown that the monomials $e_{i_1} \wedge e_{i_2} \wedge e_{i_3}$ with $i_1 < i_2 < i_3$ are a basis for $\bigwedge^3 F$. Since the *gcd* of the determinants of 3-by-3 minors is 1, some determinant of 3-by-3 minor is non-zero, so the rows of M are linearly independent over $\mathbb{Q}$, so E has rank 3 (rather than something less). The structure theorem tells us that there is a $\mathbb{Z}$-basis $f_1, \ldots, f_5$ for F and divisors $d_1 | d_2 | d_3$ (all non-zero since E is of rank 3) such that

$$E = \mathbb{Z} \cdot d_1 f_1 \oplus \mathbb{Z} \cdot d_2 f_2 \oplus \mathbb{Z} \cdot d_3 f_3$$

Let $i : E \longrightarrow F$ be the inclusion. Consider $\bigwedge^3 : \bigwedge^3 E \longrightarrow \bigwedge^3 F$. We know that $\bigwedge^3 E$ has $\mathbb{Z}$-basis

$$d_1 f_1 \wedge d_2 f_2 \wedge d_3 f_3 = (d_1 d_2 d_3) \cdot (f_1 \wedge f_2 \wedge f_3)$$

On the other hand, we claim that the coefficients of $(d_1 d_2 d_3) \cdot (f_1 \wedge f_2 \wedge f_3)$ in terms of the basis $e_{i_1} \wedge e_{i_2} \wedge e_{i_3}$ for $\bigwedge^3 F$ are exactly (perhaps with a change of sign) the determinants of the 3-by-3 minors of M. Indeed, since both f_1, f_2, f_3 and the three rows of M are bases for the rowspace of M, the f_is are linear combinations of the rows, and *vice versa* (with integer coefficients). Thus, there is a 3-by-3 matrix with determinant ± 1 such that left multiplication of M by it yields a new matrix with rows f_1, f_2, f_3. At the same time, this changes the determinants of 3-by-3 minors by at most $\pm$, by the multiplicativity of determinants.

The hypothesis that the *gcd* of all these coordinates is 1 means exactly that $\bigwedge^3 F / \bigwedge^3 E$ is torsion-free. (If the coordinates had a common factor $d > 1$, then d would annihilate the quotient.) This requires that $d_1 d_2 d_3 = 1$, so $d_1 = d_2 = d_3 = 1$ (since we take these divisors to be positive). That is,

$$E = \mathbb{Z} \cdot f_1 \oplus \mathbb{Z} \cdot f_2 \oplus \mathbb{Z} \cdot f_3$$

Writing f_1, f_2, and f_3 as row vectors, they are $\mathbb{Z}$-linear combinations of the rows of M, which is to say that there is a 3-by-3 integer matrix L such that

$$L \cdot M = \begin{pmatrix} f_1 \\ f_2 \\ f_3 \end{pmatrix}$$

Since the f_i are also a $\mathbb{Z}$-basis for E, there is another 3-by-3 integer matrix K such that

$$M = K \cdot \begin{pmatrix} f_1 \\ f_2 \\ f_3 \end{pmatrix}$$

Then $LK = LK = 1_3$. In particular, taking determinants, both K and L have determinants in $\mathbb{Z}^\times$, namely, ± 1.

Let A be a $\mathbb{Z}$-linear endomorphism of $F = \mathbb{Z}^5$ mapping f_i to e_i. Also let A be the 5-by-5 integer matrix such that right multiplication of a row vector by A gives the effect of the endomorphism A. Then

$$L \cdot M \cdot A = \begin{pmatrix} f_1 \\ f_2 \\ f_3 \end{pmatrix} \cdot A = \begin{pmatrix} e_1 \\ e_2 \\ e_3 \end{pmatrix}$$

Since the endormorphism A is invertible on $F = \mathbb{Z}^5$, it has an inverse endomorphism A^{-1}, whose matrix has integer entries. Then

$$M = L^{-1} \cdot \begin{pmatrix} e_1 \\ e_2 \\ e_3 \end{pmatrix} \cdot A^{-1}$$

Let

$$\Lambda = \begin{pmatrix} L^{-1} & 0 & 0 \\ 0 & 1 & 0 \\ 0 & 0 & \pm 1 \end{pmatrix}$$

where the $\pm 1 = \det A = \det A^{-1}$. Then

$$\Lambda \cdot \begin{pmatrix} e_1 \\ e_2 \\ e_3 \\ e_4 \\ e_5 \end{pmatrix} \cdot A^{-1} = \Lambda \cdot 1_5 \cdot A^{-1} = \Lambda \cdot A^{-1}$$

has integer entries and determinant 1 (since we adjusted the ± 1 in Λ). At the same time, it is

$$\Lambda \cdot A^{-1} = \begin{pmatrix} L^{-1} & 0 & 0 \\ 0 & 1 & 0 \\ 0 & 0 & \pm 1 \end{pmatrix} \cdot \begin{pmatrix} e_1 \\ e_2 \\ e_3 \\ * \\ * \end{pmatrix} \cdot A^{-1} = \begin{pmatrix} M \\ * \\ * \end{pmatrix} = \text{5-by-5}$$

This is the desired integer matrix $\tilde{M}$ with determinant 1 and upper 3 rows equal to the given matrix.   ///

28.11.9 Example: Let R be a commutative ring with unit. For a *finitely-generated* free R-module F, prove that there is a (natural) isomorphism

$$\text{Hom}_R(F, R) \approx F$$

Or is it only

$$\text{Hom}_R(R, F) \approx F$$

instead? (*Hint:* Recall the definition of a free module.)

For *any* R-module M, there is a (natural) isomorphism

$$i : M \longrightarrow \text{Hom}_R(R, M)$$

given by

$$i(m)(r) = r \cdot m$$

This is *injective*, since if $i(m)(r)$ were the 0 homomorphism, then $i(m)(r) = 0$ for all r, which is to say that $r \cdot m = 0$ for all $r \in R$, in particular, for $r = 1$. Thus, $m = 1 \cdot m = 0$, so $m = 0$. (Here we use the standing assumption that $1 \cdot m = m$ for all $m \in M$.) The map is *surjective*, since, given $\varphi \in \operatorname{Hom}_R(R, M)$, we have

$$\varphi(r) = \varphi(r \cdot 1) = r \cdot \varphi(1)$$

That is, $m = \varphi(1)$ determines φ completely. Then $\varphi = i(\varphi(m))$ and $m = i(m)(1)$, so these are mutually inverse maps. This did *not* use finite generation, nor free-ness.   ///

Consider now the other form of the question, namely whether or not

$$\operatorname{Hom}_R(F, R) \approx F$$

is valid for F finitely-generated and free. Let F be free on $i : S \longrightarrow F$, with finite S. Use the natural isomorphism

$$\operatorname{Hom}_R(F, R) \approx \operatorname{Hom}_{\text{sets}}(S, R)$$

discussed earlier. The right-hand side is the collection of R-valued functions on S. Since S is finite, the collection of *all* R-valued functions on S is just the collection of functions which vanish off a finite subset. The latter was our construction of the free R-module on S. So we have the isomorphism.   ///

28.11.10 Remark: Note that if S is not finite, $\operatorname{Hom}_R(F, R)$ is too large to be isomorphic to F. If F is not free, it may be too small. Consider $F = \mathbb{Z}/n$ and $R = \mathbb{Z}$, for example.

28.11.11 Remark: And this discussion needs a *choice* of the generators $i : S \longrightarrow F$. In the language style which speaks of generators as being chosen elements of the module, we have most certainly *chosen a basis*.

28.11.12 Example: Let R be an integral domain. Let M and N be free R-modules of finite ranks r, s, respectively. Suppose that there is an R-bilinear map

$$B : M \times N \longrightarrow R$$

which is *non-degenerate* in the sense that for every $0 \neq m \in M$ there is $n \in N$ such that $B(m, n) \neq 0$, and *vice versa*. Prove that $r = s$.

All tensors and homomorphisms are over R, so we suppress the subscript and other references to R when reasonable to do so. We use the important identity (proven afterward)

$$\operatorname{Hom}(A \otimes B, C) \xrightarrow{\;\;i_{A,B,C}\;\;} \operatorname{Hom}(A, \operatorname{Hom}(B, C))$$

by

$$i_{A,B,C}(\Phi)(a)(b) = \Phi(a \otimes b)$$

We also use the fact (from an example just above) that for F free on $t : S \longrightarrow F$ there is the natural (given $t : S \longrightarrow F$, anyway!) isomorphism

$$j : \operatorname{Hom}(F, R) \approx \operatorname{Hom}_{\text{sets}}(S, R) = F$$

for modules E, given by

$$j(\psi)(s) = \psi(t(s))$$

where we use construction of free modules on sets S that they are R-valued functions on S taking non-zero values at only finitely-many elements.

Thus,

$$\text{Hom}(M \otimes N, R) \xrightarrow{\ i\ } \text{Hom}(M, \text{Hom}(N, R)) \xrightarrow{\ j\ } \text{Hom}(M, N)$$

The bilinear form B induces a linear functional β such that

$$\beta(m \otimes n) = B(m, n)$$

The hypothesis says that for each $m \in M$ there is $n \in N$ such that

$$i(\beta)(m)(n) \neq 0$$

That is, for all $m \in M$, $i(\beta)(m) \in \text{Hom}(N, R) \approx N$ is 0. That is, the map $m \longrightarrow i(\beta)(m)$ is *injective*. So the existence of the non-degenerate bilinear pairing yields an injection of M to N. Symmetrically, there is an injection of N to M.

Using the assumption that R is a PID, we know that a submodule of a free module is free of lesser-or-equal rank. Thus, the two inequalities

$$\text{rank}\, M \leq \text{rank}\, N \qquad \text{rank}\, N \leq \text{rank}\, M$$

from the two inclusions imply equality. ///

28.11.13 Remark: The hypothesis that R is a PID may be too strong, but I don't immediately see a way to work around it.

Now let's prove (again?) that

$$\text{Hom}(A \otimes B, C) \xrightarrow{\ i\ } \text{Hom}(A, \text{Hom}(B, C))$$

by

$$i(\Phi)(a)(b) = \Phi(a \otimes b)$$

is an isomorphism. The map in the other direction is

$$j(\varphi)(a \otimes b) = \varphi(a)(b)$$

First,

$$i(j(\varphi))(a)(b) = j(\varphi)(a \otimes b) = \varphi(a)(b)$$

Second,

$$j(i(\Phi))(a \otimes b) = i(\Phi)(a)(b) = \Phi(a \otimes b)$$

Thus, these maps are mutual inverses, so each is an isomorphism. ///

28.11.14 Example: Write an explicit isomorphism

$$\mathbb{Z}/a \otimes_{\mathbb{Z}} \mathbb{Z}/b \longrightarrow \mathbb{Z}/\gcd(a, b)$$

and verify that it is what is claimed.

First, we know that monomial tensors generate the tensor product, and for any $x, y \in \mathbb{Z}$

$$x \otimes y = (xy) \cdot (1 \otimes 1)$$

so the tensor product is generated by $1 \otimes 1$. Next, we claim that $g = \gcd(a, b)$ annihilates every $x \otimes y$, that is, $g \cdot (x \otimes y) = 0$. Indeed, let r, s be integers such that $ra + sb = g$. Then

$$g \cdot (x \otimes y) = (ra + sb) \cdot (x \otimes y) = r(ax \otimes y) = s(x \otimes by) = r \cdot 0 + s \cdot 0 = 0$$

So the generator $1 \otimes 1$ has order dividing g. To prove that that generator has order *exactly* g, we construct a bilinear map. Let

$$B : \mathbb{Z}/a \times \mathbb{Z}/b \longrightarrow \mathbb{Z}/g$$

by

$$B(x \times y) = xy + g\mathbb{Z}$$

To see that this is well-defined, first compute

$$(x + a\mathbb{Z})(y + b\mathbb{Z}) = xy + xb\mathbb{Z} + ya\mathbb{Z} + ab\mathbb{Z}$$

Since

$$xb\mathbb{Z} + ya\mathbb{Z} \subset b\mathbb{Z} + a\mathbb{Z} = \gcd(a, b)\mathbb{Z}$$

(and $ab\mathbb{Z} \subset g\mathbb{Z}$), we have

$$(x + a\mathbb{Z})(y + b\mathbb{Z}) + g\mathbb{Z} = xy + xb\mathbb{Z} + ya\mathbb{Z} + ab\mathbb{Z} + \mathbb{Z}$$

and well-definedness. By the defining property of the tensor product, this gives a unique linear map β on the tensor product, which on monomials is

$$\beta(x \otimes y) = xy + \gcd(a, b)\mathbb{Z}$$

The generator $1 \otimes 1$ is mapped to 1, so the image of $1 \otimes 1$ has order $\gcd(a, b)$, so $1 \otimes 1$ has order divisible by $\gcd(a, b)$. Thus, having already proven that $1 \otimes 1$ has order at most $\gcd(a, b)$, this must be its order.

In particular, the map β is injective on the cyclic subgroup generated by $1 \otimes 1$. That cyclic subgroup is the whole group, since $1 \otimes 1$. The map is also surjective, since $\cdot 1 \otimes 1$ hits $r \bmod \gcd(a, b)$. Thus, it is an isomorphism.   ///

28.11.15 Example: Let $\varphi : R \longrightarrow S$ be commutative rings with unit, and suppose that $\varphi(1_R) = 1_S$, thus making S an R-algebra. For an R-module N prove that $\mathrm{Hom}_R(S, N)$ is (*yet another*) good definition of *extension of scalars* from R to S, by checking that for every S-module M there is a natural isomorphism

$$\mathrm{Hom}_R(\mathrm{Res}_R^S M, N) \approx \mathrm{Hom}_S(M, \mathrm{Hom}_R(S, N))$$

where $\mathrm{Res}_R^S M$ is the R-module obtained by forgetting S, and letting $r \in R$ act on M by $r \cdot m = \varphi(r)m$. (*Do* prove naturality in M, also.)

Let

$$i : \mathrm{Hom}_R(\mathrm{Res}_R^S M, N) \longrightarrow \mathrm{Hom}_S(M, \mathrm{Hom}_R(S, N))$$

be defined for $\varphi \in \operatorname{Hom}_R(\operatorname{Res}^S_R M, N)$ by

$$i(\varphi)(m)(s) = \varphi(s \cdot m)$$

This makes *some* sense, at least, since M is an S-module. We must verify that $i(\varphi) : M \longrightarrow \operatorname{Hom}_R(S, N)$ is S-linear. Note that the S-module structure on $\operatorname{Hom}_R(S, N)$ is

$$(s \cdot \psi)(t) = \psi(st)$$

where $s, t \in S$, $\psi \in \operatorname{Hom}_R(S, N)$. Then we check:

$$(i(\varphi)(sm))(t) = i(\varphi)(t \cdot sm) = i(\varphi)(stm) = i(\varphi)(m)(st) = (s \cdot i(\varphi)(m))(t)$$

which proves the S-linearity.

The map j in the other direction is described, for $\Phi \in \operatorname{Hom}_S(M, \operatorname{Hom}_R(S, N))$, by

$$j(\Phi)(m) = \Phi(m)(1_S)$$

where 1_S is the identity in S. Verify that these are mutual inverses, by

$$i(j(\Phi))(m)(s) = j(\Phi)(s \cdot m) = \Phi(sm)(1_S) = (s \cdot \Phi(m))(1_S) = \Phi(m)(s \cdot 1_S) = \Phi(m)(s)$$

as hoped. (Again, the equality

$$(s \cdot \Phi(m))(1_S) = \Phi(m)(s \cdot 1_S)$$

is the definition of the S-module structure on $\operatorname{Hom}_R(S, N)$.) In the other direction,

$$j(i(\varphi))(m) = i(\varphi)(m)(1_S) = \varphi(1 \cdot m) = \varphi(m)$$

Thus, i and j are mutual inverses, so are isomorphisms.

For naturality, let $f : M \longrightarrow M'$ be an S-module homomorphism. Add indices to the previous notation, so that

$$i_{M,N} : \operatorname{Hom}_R(\operatorname{Res}^S_R M, N) \longrightarrow \operatorname{Hom}_S(M, \operatorname{Hom}_R(S, N))$$

is the isomorphism discussed just above, and $i_{M',N}$ the analogous isomorphism for M' and N. We must show that the diagram

$$\begin{array}{ccc} \operatorname{Hom}_R(\operatorname{Res}^S_R M, N) & \xrightarrow{i_{M,N}} & \operatorname{Hom}_S(M, \operatorname{Hom}_R(S, N)) \\ {\scriptstyle -\circ f}\uparrow & & \uparrow{\scriptstyle -\circ f} \\ \operatorname{Hom}_R(\operatorname{Res}^S_R M', N)) & \xrightarrow{i_{M',N}} & \operatorname{Hom}_S(M', \operatorname{Hom}_R(S, N)) \end{array}$$

commutes, where $-\circ f$ is pre-composition with f. (We use the same symbol for the map $f : M \longrightarrow M'$ on the modules whose S-structure has been forgotten, leaving only the R-module structure.) Starting in the lower left of the diagram, going up then right, for $\varphi \in \operatorname{Hom}_R(\operatorname{Res}^S_R M', N)$,

$$(i_{M,N} \circ (-\circ f)\,\varphi)(m)(s) = (i_{M,N}(\varphi \circ f))(m)(s) = (\varphi \circ f)(s \cdot m) = \varphi(f(s \cdot m))$$

On the other hand, going right, then up,

$$((-\circ f) \circ i_{M',N} \, \varphi)\,(m)(s) = (i_{M',N} \, \varphi)\,(fm)(s) = \varphi(s \cdot fm) = \varphi(f(s \cdot m))$$

since f is S-linear. That is, the two outcomes are the same, so the diagram commutes, proving functoriality in M, which is a part of the naturality assertion.   ///

28.11.16 Example: Let

$$M = \mathbb{Z} \oplus \mathbb{Z} \oplus \mathbb{Z} \oplus \mathbb{Z} \qquad N = \mathbb{Z} \oplus 4\mathbb{Z} \oplus 24\mathbb{Z} \oplus 144\mathbb{Z}$$

What are the elementary divisors of $\bigwedge^2(M/N)$?

First, note that this is *not* the same as asking about the structure of $(\bigwedge^2 M)/(\bigwedge^2 N)$. Still, we can address that, too, after dealing with the question that *was* asked.

First,

$$M/N = \mathbb{Z}/\mathbb{Z} \oplus \mathbb{Z}/4\mathbb{Z} \oplus \mathbb{Z}/24\mathbb{Z} \oplus \mathbb{Z}/144\mathbb{Z} \approx \mathbb{Z}/4 \oplus \mathbb{Z}/24 \oplus \mathbb{Z}/144$$

where we use the obvious slightly lighter notation. Generators for M/N are

$$m_1 = 1 \oplus 0 \oplus 0 \qquad m_2 = 0 \oplus 1 \oplus 0 \qquad m_3 = 0 \oplus 0 \oplus 1$$

where the 1s are respectively in $\mathbb{Z}/4$, $\mathbb{Z}/24$, and $\mathbb{Z}/144$. We know that $e_i \wedge e_j$ *generate* the exterior square, for the 3 pairs of indices with $i < j$. Much as in the computation of $\mathbb{Z}/a \otimes \mathbb{Z}/b$, for e in a $\mathbb{Z}$-module E with $a \cdot e = 0$ and f in E with $b \cdot f = 0$, let r, s be integers such that

$$ra + sb = \gcd(a, b)$$

Then

$$\gcd(a, b) \cdot e \wedge f = r(ae \wedge f) + s(e \wedge bf) = r \cdot 0 + s \cdot 0 = 0$$

Thus, $4 \cdot e_1 \wedge e_2 = 0$ and $4 \cdot e_1 \wedge e_3 = 0$, while $24 \cdot e_2 \wedge e_3 = 0$. If there are no further relations, then we could have

$$\bigwedge^2(M/N) \approx \mathbb{Z}/4 \oplus \mathbb{Z}/4 \oplus \mathbb{Z}/24$$

(so the elementary divisors would be $4, 4, 24$.)

To prove, in effect, that there are no further relations than those just indicated, we must construct suitable alternating bilinear maps. Suppose for $r, s, t \in \mathbb{Z}$

$$r \cdot e_1 \wedge e_2 + s \cdot e_1 \wedge e_3 + t \cdot e_2 \wedge e_3 = 0$$

Let

$$B_{12} : (\mathbb{Z}e_1 \oplus \mathbb{Z}e_2 \oplus \mathbb{Z}e_3) \times (\mathbb{Z}e_1 \oplus \mathbb{Z}e_2 \oplus \mathbb{Z}e_3) \longrightarrow \mathbb{Z}/4$$

by

$$B_{12}(xe_1 + ye_2 + ze_3, \, \xi e_1 + \eta e_2 + \zeta e_3) = (x\eta - \xi y) + 4\mathbb{Z}$$

(As in earlier examples, since $4|4$ and $4|24$, this is *well-defined*.) By arrangement, this B_{12} is alternating, and induces a unique linear map β_{12} on $\bigwedge^2(M/N)$, with

$$\beta_{12}(e_1 \wedge e_2) = 1 \qquad \beta_{12}(e_1 \wedge e_3) = 0 \qquad \beta_{12}(e_2 \wedge e_3) = 0$$

Applying this to the alleged relation, we find that $r = 0 \bmod 4$. Similar contructions for the other two pairs of indices $i < j$ show that $s = 0 \bmod 4$ and $t = 0 \bmod 24$. This shows that we have all the relations, and

$$\textstyle\bigwedge^2(M/N) \approx \mathbb{Z}/4 \oplus \mathbb{Z}/4 \oplus \mathbb{Z}/24$$

as hoped/claimed.   ///

Now consider the other version of this question. Namely, letting

$$M = \mathbb{Z} \oplus \mathbb{Z} \oplus \mathbb{Z} \oplus \mathbb{Z} \qquad N = \mathbb{Z} \oplus 4\mathbb{Z} \oplus 24\mathbb{Z} \oplus 144\mathbb{Z}$$

compute the elementary divisors of $(\bigwedge^2 M)/(\bigwedge^2 N)$.

Let e_1, e_2, e_3, e_4 be the standard basis for $\mathbb{Z}^4$. Let $i : N \longrightarrow M$ be the inclusion. We have shown that exterior powers of free modules are free with the expected generators, so M is free on

$$e_1 \wedge e_2, \quad e_1 \wedge e_3, \quad e_1 \wedge e_4, \quad e_2 \wedge e_3, \quad e_2 \wedge e_4, \quad e_3 \wedge e_4$$

and N is free on

$$(1 \cdot 4)\, e_1 \wedge e_2, \quad (1 \cdot 24)\, e_1 \wedge e_3, \quad (1 \cdot 144)\, e_1 \wedge e_4, \quad (4 \cdot 24)\, e_2 \wedge e_3, \quad (4 \cdot 144)\, e_2 \wedge e_4, \quad (24 \cdot 144)\, e_3 \wedge e_4$$

The inclusion $i : N \longrightarrow M$ induces a natural map $\bigwedge^2 i : \bigwedge^2 \longrightarrow \bigwedge^2 M$, taking $r \cdot e_i \wedge e_j$ (in N) to $r \cdot e_i \wedge e_j$ (in M). Thus, the quotient of $\bigwedge^2 M$ by (the image of) $\bigwedge^2 N$ is visibly

$$\mathbb{Z}/4 \oplus \mathbb{Z}/24 \oplus \mathbb{Z}/144 \oplus \mathbb{Z}/96 \oplus \mathbb{Z}/576 \oplus \mathbb{Z}/3456$$

The integers $4, 24, 144, 96, 576, 3456$ do not quite have the property $4|24|144|96|576|3456$, so are not elementary divisors. The problem is that neither $144|96$ nor $96|144$. The only primes dividing all these integers are 2 and 3, and, in particular,

$$4 = 2^2, \quad 24 = 2^3 \cdot 3, \quad 144 = 2^4 \cdot 3^2, \quad 96 = 2^5 \cdot 3, \quad 576 = 2^6 \cdot 3^2, \quad 3456 = 2^7 \cdot 3^3,$$

From Sun-Ze's theorem,

$$\mathbb{Z}/(2^a \cdot 3^b) \approx \mathbb{Z}/2^a \oplus \mathbb{Z}/3^b$$

so we can rewrite the summands $\mathbb{Z}/144$ and $\mathbb{Z}/96$ as

$$\mathbb{Z}/144 \oplus \mathbb{Z}/96 \approx (\mathbb{Z}/2^4 \oplus \mathbb{Z}/3^2) \oplus (\mathbb{Z}/2^5 \oplus \mathbb{Z}/3) \approx (\mathbb{Z}/2^4 \oplus \mathbb{Z}/3) \oplus (\mathbb{Z}/2^5 \oplus \mathbb{Z}/3^2) \approx \mathbb{Z}/48 \oplus \mathbb{Z}/288$$

Now we do have $4|24|48|288|576|3456$, and

$$(\textstyle\bigwedge^2 M)/(\bigwedge^2 N) \approx \mathbb{Z}/4 \oplus \mathbb{Z}/24 \oplus \mathbb{Z}/48 \oplus \mathbb{Z}/288 \oplus \mathbb{Z}/576 \oplus \mathbb{Z}/3456$$

is in elementary divisor form.   ///

Exercises

28.1   Show that there is a natural isomorphism

$$f_X : \Pi_s \operatorname{Hom}_R(M_s, X) \approx \operatorname{Hom}_R(\oplus_s M_s, X)$$

where everything is an R-module, and R is a commutative ring.

28.2 For an abelian group A (equivalently, $\mathbb{Z}$-module), the *dual* group ($\mathbb{Z}$-module) is

$$A^* = \operatorname{Hom}(A, \mathbb{Q}/\mathbb{Z})$$

Prove that the dual group of a direct sum is the direct product of the duals. Prove that the dual group of a *finite* abelian group A is isomorphic to A (although not *naturally* isomorphic).

28.3 Let R be a commutative ring with unit. Let M be a finitely-generated free module over R. Let $M^* = \operatorname{Hom}_R(M, R)$ be the dual. Show that, for each integer $\ell \geq 1$, the module $\bigwedge^\ell M$ is dual to $\bigwedge^\ell M^*$, under the bilinear map induced by

$$\langle m_1 \wedge \ldots \wedge m_\ell, \ \mu_1 \wedge \ldots \wedge \mu_\ell \rangle \ = \ \det\{\langle m_i, \mu_j \rangle\}$$

for $m_i \in M$ and $\mu_j \in M^*$.

28.4 Let $v_1, \ldots, v_n$ be linearly independent vectors in a vector space V over a field k. For each pair of indices $i < j$, take another vector $w_{ij} \in V$. Suppose that

$$\sum_{i<j} v_i \wedge v_j \wedge w_{ij} = 0$$

Show that the w_{ij}'s are in the span of the v_k's. Let

$$w_{ij} = \sum_k c_{ij}^k v_k$$

Show that, for $i < j < k$,

$$c_{ij}^k - c_{ik}^j + c_{jk}^i = 0$$

28.5 Show that the *adjugate* (that is, *cofactor*) matrix of a 2-by-2 matrix with entries in a commutative ring R is

$$\begin{pmatrix} a & b \\ c & d \end{pmatrix}^{\mathrm{adg}} = \begin{pmatrix} d & -b \\ -c & a \end{pmatrix}$$

28.6 Let M be an n-by-n matrix with entries in a commutative ring R with unit, viewed as an endomorphism of the free R-module R^n by left matrix multiplication. Determine the matrix entries for the adjugate matrix M^{adg} in terms of those of M.